AA002869

34th Ethylene Producers Conference 2022

Topical Conference at the 2022 AIChE Spring Meeting and
18th Global Congress on Process Safety

San Antonio, Texas, USA
10 – 14 April 2022

ISBN: 978-1-7138-6816-3

Printed from e-media with permission by:

Curran Associates, Inc.
57 Morehouse Lane
Red Hook, NY 12571

Some format issues inherent in the e-media version may also appear in this print version.

Printed with permission by Curran Associates, Inc. (2023)

For permission requests, please contact AIChE
at the address below.

AIChE
120 Wall Street, FL 23
New York, NY 10005-4020

Phone: (800) 242-4363
Fax: (203) 775-5177

www.aiche.org

Additional copies of this publication are available from:

Curran Associates, Inc.
57 Morehouse Lane
Red Hook, NY 12571 USA
Phone: 845-758-0400
Fax: 845-758-2633
Email: curran@proceedings.com
Web: www.proceedings.com

TABLE OF CONTENTS

AIChE PAPER NUMBER: 5a 639383

METHOD OF FAN CONTROLS

Brad Smith
Senior Sales Manager
Twin City Clarage, LLP

Mike Lowe
Senior Sales Application Engineer
Twin City Clarage, LLP

Prepared for Presentation at the 2022 Spring Meeting
San Antonio, Texas, April 14-14, 2022

METHODS OF FAN CONTROLS

Brad Smith, P.Eng.
Senior Sales Manager
Twin City Clarage, LLP
Mike Lowe
Senior Sales Application Engineer
Twin City Clarage, LLP

Abstract:

With the increase in environmental emission regulations, designs of the Ethylene plant auxiliary equipment (such as fired heaters, furnaces, and fired boilers) have placed more strain on the centrifugal fan turn down requirements. The low NOx emission regulations sometimes require low turn down to start up the auxiliary equipment. This paper explains the various methods to control the centrifugal fan, including pre-spin and resistance dampers, speed control, and airflow recirculation, as follows:

- Inlet louver damper (ILD)
- Variable inlet vane damper (VIV)
- Outlet louver damper (OLD)
- Variable frequency drive (VFD)
- By-pass recirculation duct and flue gas recirculation (FGR) dampers

In addition to providing a complete description of each control method, this paper also uses the following combinations of two methods:

- ILD or VIV and VFD
- ILD and by-pass recirculation duct

Finally, we present a case study where a combustion air forced draft (FD) fan on an industrial boiler is designed using an ILD, OLD, and by-pass recirculation duct installed to reach a low-fire startup condition.

Introduction:

In many Ethylene plant applications, centrifugal fans are used to support production equipment such as furnaces, fired heaters, steam generation boilers, and other process or environmental equipment. In many cases the above equipment will require the fan(s) to operate under a wide range of performance conditions for a short or long period of time. The fan is sized to operate at a design point in all applications, and thus, the normal and additional performance points will be below the design point. Consequently, the combination of the extended operating times for the performance points and the tendency to oversize the fan(s) create a need for efficient flow control.

This paper will discuss the following five principal methods of flow control for centrifugal fans:

- Inlet louver damper (ILD)
- Variable inlet vane damper (VIV)
- Outlet louver damper (OLD)
- Variable frequency drive (VFD)
- By-pass recirculation duct and flue gas recirculation (FGR) dampers

Figure 1 illustrates the damper locations on a centrifugal fan.

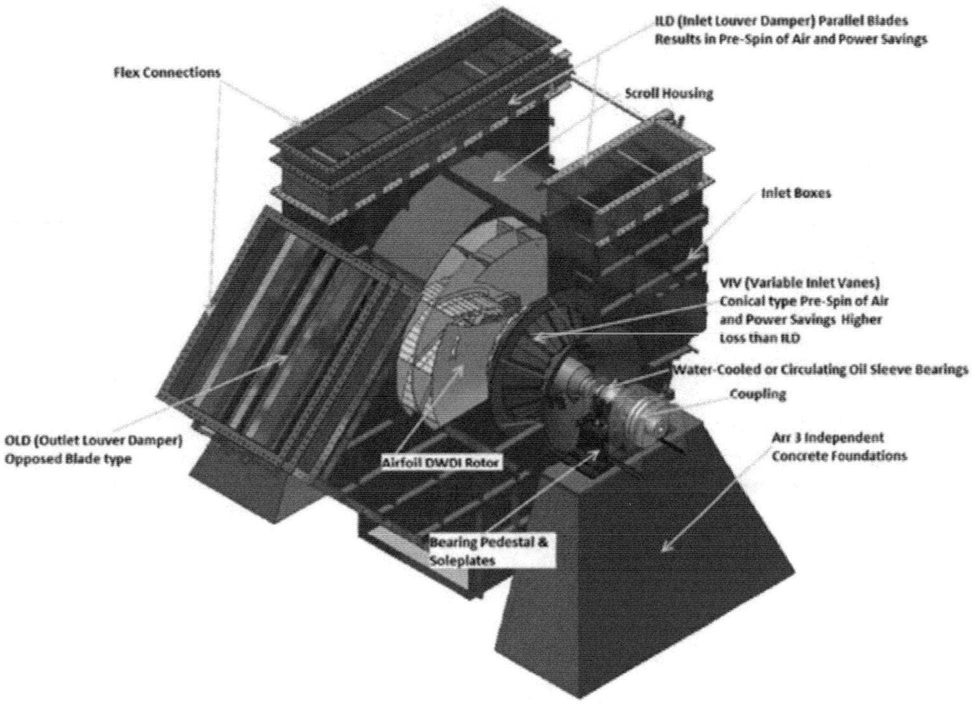

Figure 1: Damper locations on a centrifugal fan.

Inlet Box Louver Damper (ILD):

The ILD is a parallel airfoil blade designed to pre-spin or create a swirl at the inlet fan impeller. It is usually constructed using two-piece parallel blades that run along the narrow axis of the damper. Typically, the standard construction material is mild steel ASTM A-36, but special materials can be used based on the airstream requirements. When airstream temperatures are over 300°F or when the system pressure dictates, the special graphite bearings and stuffing boxes would be used to protect the bearings from harsh airstream conditions. The ILD can be modulated either manually with the leveler or handwheel, or automatically using a pneumatic or electric actuator with a positioner. The actuator or handwheel is mounted on one of the damper blade shafts and linkages are attached to the other damper shafts to allow them to modulate at the same time. Figure 2 shows the installation of an actuator and the linkages.

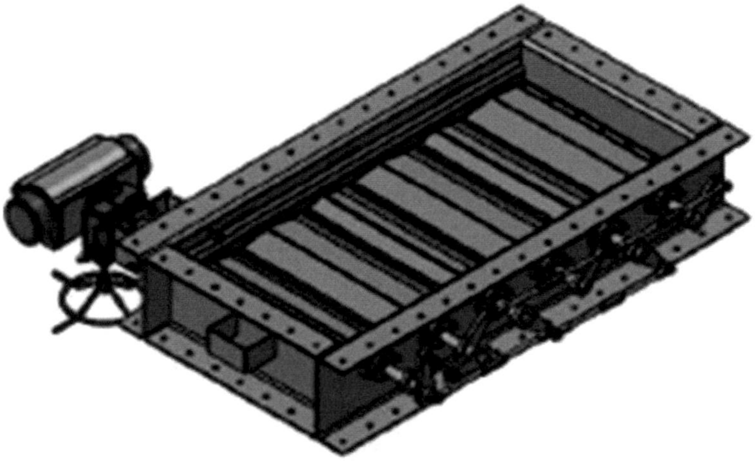

Figure 2: Assembly drawing of an inlet box louver damper

An ILD is a method to control the centrifugal fan volume flow rate in the system. It typically offers a three to one turndown ratio of the fan airflow. As the ILD is modulated, the fan curve characteristics change. The fan efficiency will be maintained, and therefore power consumption is reduced by lowering the fan flow rate. A sample fan performance curve with various operating points that are controlled using an ILD is shown in Appendix A. The typical leakage rate of a standard constructed ILD is 10%-15%, but low leakage designs are available that can reduce the leakage rate to under 5%. The ILD is installed directly upstream of the fan inlet box, as shown in Figure 3.

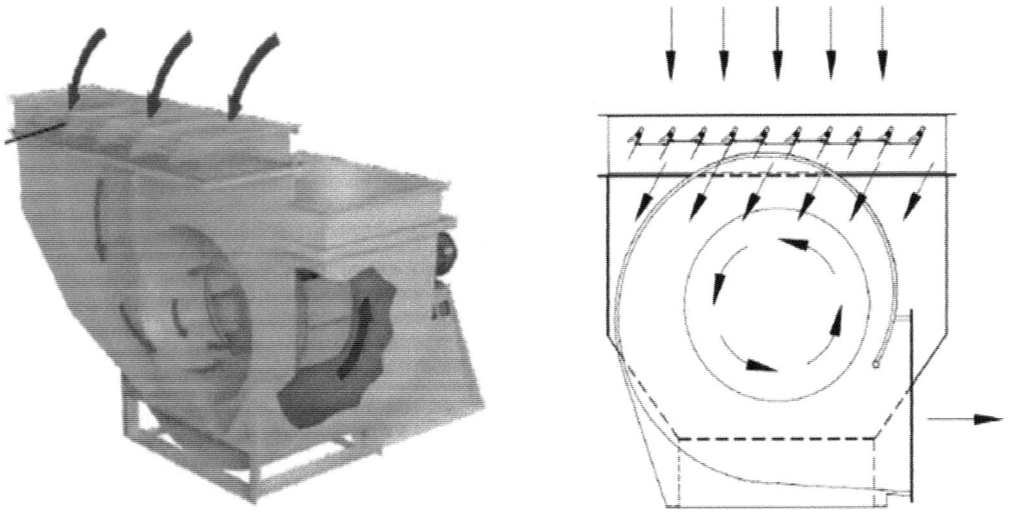

Figure 3: Location of the inlet box louver damper

There are some instances where ILDs are fitted with opposed blades. This is usually used as a fresh air mixing box damper where a more stratified air stream is required instead of a pre-spin. The resistance causes air to draw from the FGR duct as opposed to pulling fresh air.

The advantages of using an ILD are described below:

- It is an efficient method for modulating air flow and is more efficient than an outlet damper.
- It provides operating cost savings compared to an outlet damper.
- There is an improvement in high turndown load stability, especially when used in conjunction with a speed control (VFD).
- The damper can be used as an isolation damper to aid in starting up an induced draft fan.
- Bearing are located outside of the airstream.

The disadvantages of using an ILD are presented below:

- Due to the reduced pre-spin intensity the control efficiency is not as good as VIV.
- The fan sound levels increase as the ILD is closed.

Variable Inlet Damper (VIV):

The VIV can be a nested internal design or an external bolt-on design to pre-spin or create a swirl at the inlet fan impeller. The nested VIV is installed in the inlet cone, as shown in Figure 4B. The nested design should only be used for clean and low temperature air because the center hub and linkages are in the air stream. The external bolt-on design is mounted in between the inlet cone of the fan and the inlet box or ductwork, as shown in Figure 4A. The external design has radial vanes with the linkages located on the outside of the damper. This allows the VIV to be used in higher air temperature applications. Special designed External VIV's can be made with cantilever shaft and bearing arrangement to keep bearings out of the airstream.

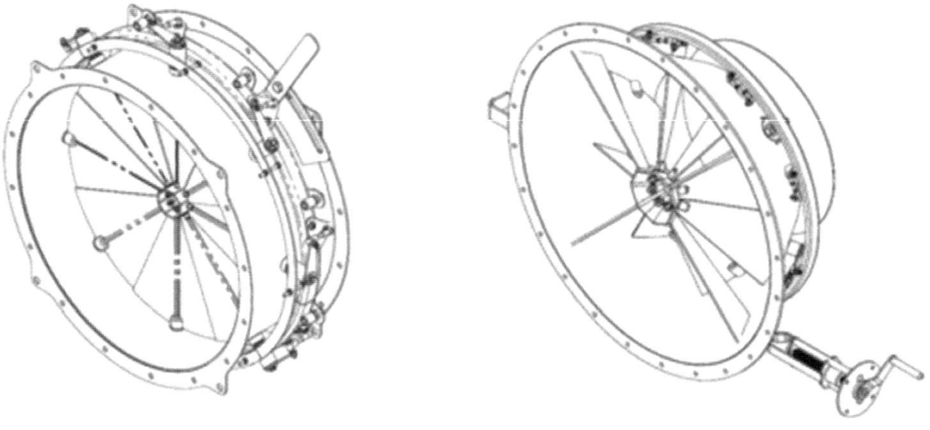

4A: External or bolt **4B: Internal or nested**

Figure 4: Assembly drawing of a variable inlet vane damper

Typically, the VIV offers a better turndown ratio, of four to one, compared to an ILD due to its closer proximity to the fan inlet. Similar to the ILD, the VIV modulates the fan curve characteristic changes, and the fan efficiency will be maintained. Therefore, the power consumption is reduced when the fan flow rate is lowered. A sample fan performance curve with various operating points that can be controlled using a VIV is shown in Appendix B. The typical leakage rate of a standard constructed ILD is 15%-20%, but unlike the ILD it is difficult to design a low leakage VIV damper. Installation of both the bolt-on and nested VIV is shown in Figure 5.

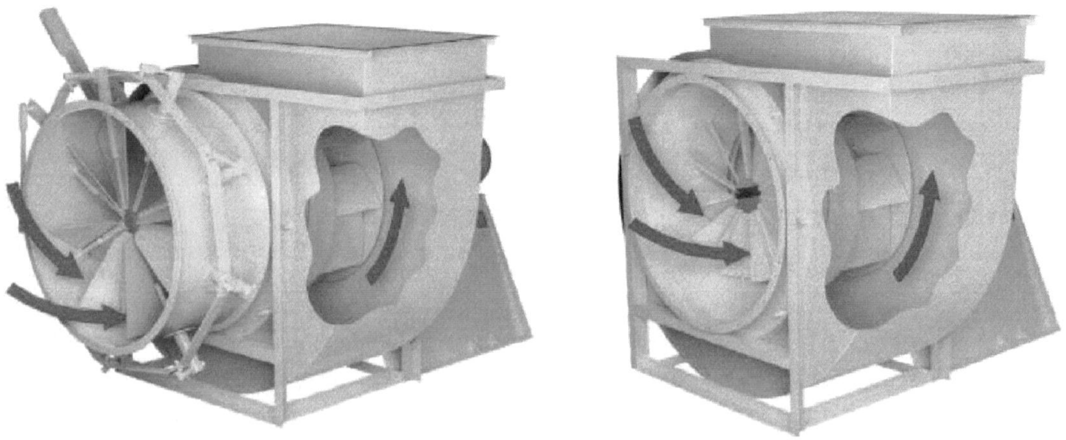

Figure 5: Installation of the bolt-on and nested VIV damper

The VIV damper can be modulated either manually using the leveler or handwheel, or automatically using a pneumatic or electric actuator with a positioner. The actuator or handwheel is mounted on one of damper blade shafts and linkages are attached to the other damper shafts to allow them to modulate at the same time.

The advantages of using a VIV are described below:

- It is the most efficient method of modulating air flow using a damper.
- It provides operating cost savings compared to an inlet louver damper and an outlet damper.
- It improves the high turndown load stability, especially when used in conjunction with speed control (VFD).
- The damper can be used as an isolation damper to aid in starting up an induced draft fan.

The disadvantages of using a VIV are presented below:

- It is a more complicated operating mechanism, especially when vanes are located in the inlet cone.
- It has a higher pressure loss, especially when vanes are located in the inlet cone.
- The fan sound levels increase as the VIV damper is closed.
- It has higher leakage rates compared to the ILD.

Outlet Louver Damper (OLD):

The outlet damper is also known as a system resistance damper, and as the OLD is modulated it changes the system curve characteristics. The OLD can also be used as a fan isolation damper. The most common design for the OLD is an opposed blade design. It is typically a two-piece airfoil blade that runs horizontal to the discharge of the fan duct. Similar to the ILD, the standard construction materials used is mild steel ASTM A-36, but special materials can be used based on the airstream requirements. When airstream temperatures are over 300°F or when the system pressure dictates, the special graphite bearings and stuffing boxes can be used to protect the bearings from the harsh airstream conditions.

The OLD can be modulated either manually using the leveler or handwheel, or automatically using a pneumatic or electric actuator with a positioner. When the OLD is used as an isolation damper an ON-OFF actuator is used. The actuator or handwheel is mounted on one of damper blade shafts and linkages are attached to the other damper shafts to allow them to modulate at the same time. As the name suggests the outlet damper is mounted on the discharge of the fan, but it can be installed further down the ductwork. Typically, it is installed directly to the discharge of the fan to keep the size as small as possible. Figure 6 shows an illustration of an OLD installation on a fan and the orientation of the blades and linkages.

Figure 6: Assembly drawing of an outlet louver damper

The OLD provides flow control by modulating the system resistance pressure in the air flow path. As the damper closes it increases the pressure on the discharge side of the fan thereby reducing the flow rate. By increasing the system resistance pressure, the OLD causes the fan to work against a back pressure and shifts the system curve operating point to the left along its fan performance curve. The effects of the OLD on the system curve are shown in Appendix C. The typical leakage rate of a standard constructed OLD is 10%-15%, but low leakage designs are available that can reduce the leakage rate to under 5%.

The OLD is typically an opposed blade design for fans with discharge ductwork. The opposed blade damper provides the quickest way to stratify the air profile across the ductwork which will reduce the opportunity for system effect. A parallel blade can be used in cases where there is no discharge ductwork or where there is a bend in the duct work close to the fan discharge. The direction of the airflow path with a parallel bladed OLD are difficult to predict so it is not recommended; Figure 7 illustrates the air profile of an opposed blade and parallel blade OLD design.

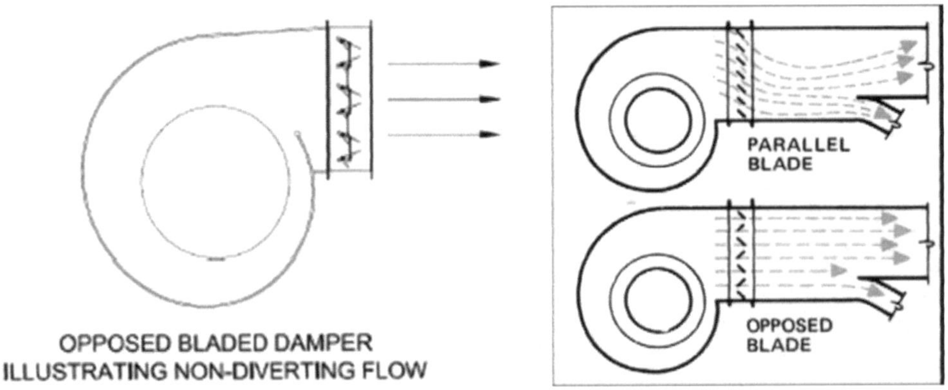

Figure 7: Illustration of the air profile of an outlet louver damper

The advantages of using an OLD are described below:

- The initial cost is lower compared to an ILD or VIV.
- It has a simple operation mechanism.
- It can be used as an isolation damper.

The disadvantages of using an OLD are presented below:

- This is the least efficient method of fan control compared to the other forms of flow control.
- It has a higher operating cost compared to an ILD or VIV.
- Higher Pressure drop due to higher velocity at fan outlet, and "system effect".
- There are possible aerodynamic instability problems if an inappropriate vane configuration is used.

Variable Frequency Drive (VFD):

A VFD is a motor controller that drives an electric motor by varying the frequency and voltage supplied to the electric motor. VFD's are typically installed in the control room or in a controlled environment room. Varying the fan speed effectively creates a new fan performance curve with respect to the flow rate or fan pressure and the fan power absorbed. Appendix D shows an example fan performance curve and how it changes with the varying speed. It typically offers a five to one turndown ratio of the fan airflow. Lower turndowns are available with motors that have external cooling and special design couplings. VFD's provides an efficient method of flow control compared to dampers. Figure 8 presents a graph comparing the VFD, ILD, and VIV.

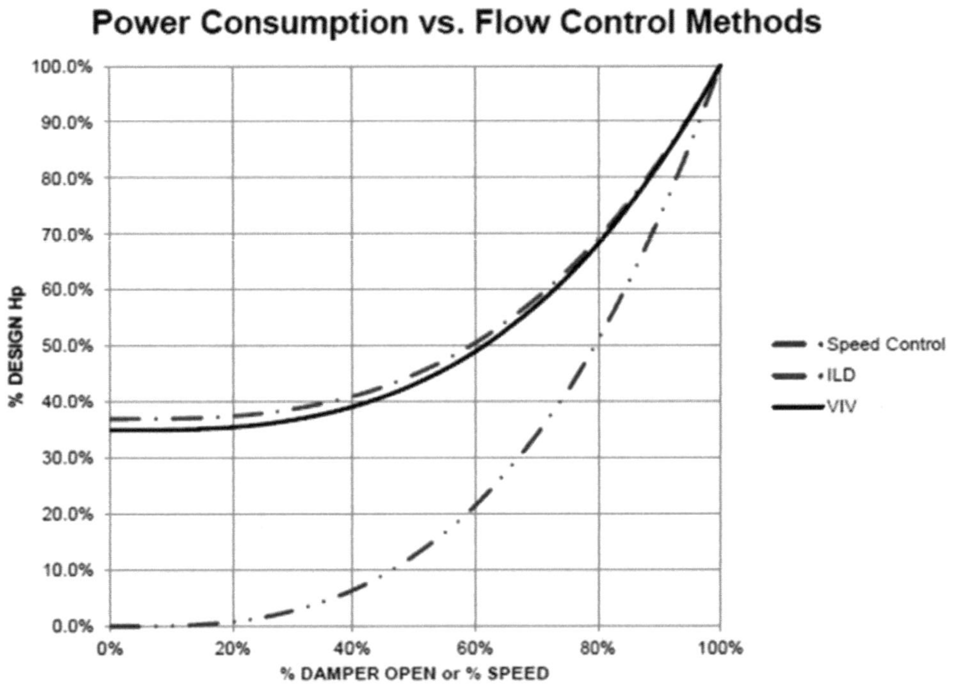

Figure 8: Comparison of VFD, VIV, and ILD in power consumption

The fan laws are the basic formulas that are used to calculate a fan's performance. The fan law fan power consumption with varying speed has the largest impact in fan design and control. The power, H, (HP or kW) is required to drive the change in fan speed, N, (RPM) of the fan impeller varies by the cube, as follows:

$$H_2 = H_1 \times (N_2/N_1)^3$$

where:

H_1 = original fan power (HP or kW)
H_2 = new fan power (HP or kW)
N_1 = original fan speed (rpm)
N_2 = new fan speed (rpm)

The advantages of using a VFD are described as follows:

- The most efficient method of flow control.
- Largest savings on operation cost.
- Programmable operation when used with an inverter; and
- The fan sound levels decrease as the fan speed is lowered.

The disadvantages of using a VFD are presented below:

- There are possible problems with fan over speeding. (i.e., if the speed control is variable, a maximum speed must be determined and built into the operating system).

- A VFD has potential problems due to the excitation of the fan natural frequencies at certain speeds. (It is the responsibility of the fan manufacturer to guarantee that the fan is stable at the maximum speed and any other specified duty speeds when it is operating on the supplied mounting structure. It is possible that the foundation or mounting structure has a resonant frequency within the running range of the fan. It is important that fan is not operated for extended periods under this condition). There may be speeds that must be dialed out of the operating range. Additionally, there are possible aerodynamic instability problems if an inappropriate vane configuration is used.
- It has a higher upfront capital cost, especially medium voltage, compared to dampers.

Combination Method of Flow Control:

Some Ethylene plant systems such as an induced draft fan for a furnace and a forced draft fan for an industrial boiler require the fan to have the ability to achieve a 10 to 1 turndown ratio of the fan airflow. In these cases, a combination of two methods of control is required. The two common combination methods that will be discussed are:

- ILD or VIV damper and VFD; and
- ILD and by-pass recirculation duct.

ILD or VIV and VFD Combination:

Using the combination of an ILD or VIV and a VFD, the primary flow control method would be the VFD because it provides the best efficiencies and power cost savings. The VFD would be used to modulate the fan speed, which in turn modulates the air flow, until the fan motor reaches its minimum operating speed. The VIV or ILD is then used to modulate the fan's air flow. The minimum operating speed of each fan is determined by the fan OEM engineer in conjunction with the motor supplier. Operating the fan and motor below the minimum operating speed could lead to the following issues:

- Coupling issues. Some couplings could chatter at low operating speeds, which can lead to coupling failure or induce vibration in the fan bearings; and
- The motor may not be able to cool adequately at low operating speeds. The increased temperature may lead to motor damage.
- Sleeve oil bearings may not get proper lubrication unless forced oil systems are provided.

The advantages of using the combination of ILD or VIV and VFD are described below:

- It allows the fan system to have turn down capabilities of 10 or 15 to 1.
- It allows the fan to easily operate on fan points that do not follow the design system curve.

The disadvantages of using the combination of ILD or VIV and VFD are presented below:

- If the fan uses a medium voltage motor and VFD, the VFD has a high upfront capital cost.

ILD and By-Pass Recirculation Duct Combination:

For the combination of an ILD and a by-pass recirculation duct (BPRD), the primary method of flow control would be the ILD. The ILD is used to modulate the fan's air flow until it reaches the minimum closed position, with the BPRD in the full closed position. The minimum closed position for the ILD is 20% open. Once the ILD reaches the 20% open position and further turndown of the fan is required, the BPRD begins to open and redirect a portion of air at the discharge of the fan back into the inlet box. Recirculating more air back into the fan's inlet box reduces the volume of air exiting the fan discharge and into the fan system. Figure 9 shows an illustration of a BPRD installation (the BPRD is shown in blue and the inlet damper is shown in green).

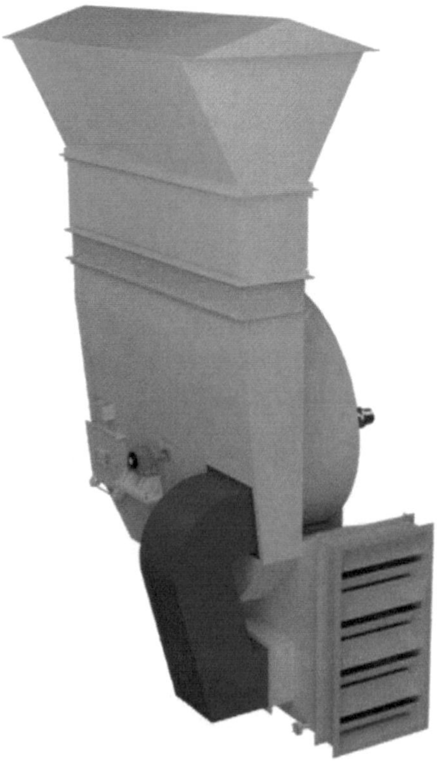

Figure 9: By-pass recirculation duct assembly installation.

The advantages of using the combination of an ILD and BPRD are described below:

- It allows the fan system to have turn down capabilities of 10 or 15 to 1.

- It allows the fan to easily operate on fan points that do not follow the design system curve.
- Lower upfront capital cost versus the cost of a VFD.

The disadvantage of using the combination of an ILD and BPRD are presented below:

- It does not have the same energy savings capabilities as the ILD or VIV and VFD combination.

Case Study:

A chemical company, in the southern United States, is installing a large new steam generator. The steam generator design is a natural gas boiler. The boiler design requires a burner system, and a forced draft. The burner system design requires a large forced draft fan with multiple operating points. See Appendix E for the case study fan performance data and fan curve. The customer's preference is to control the forced draft fan using dampers instead of a VFD. The best method to control the fan is using an ILD, OLD and a BPRD, see Figure 10 for the layout of the dampers. The ILD works best with a BPRD compared to a VIV, because the ILD creates a better draw through recirculation duct. The forced draft fan will use the ILD to operate at the following performance points: design point, P1, P2, P6, P7, and P10. The ILD will then stop closing at 20% open to prevent any vibration due to surge. The BPRD will then begin to open to operate at the performance points: P3, P4, P5, P6, P8, and P9.

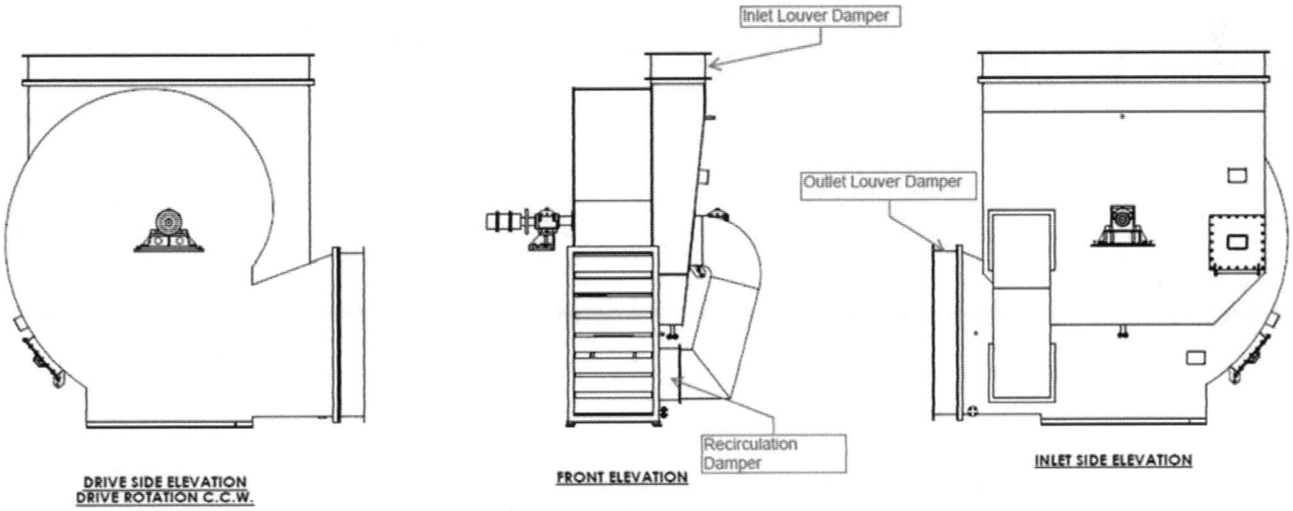

Figure 10: Case study fan control design.

Summary:

In summarize, there are numerous methods and combinations are available to control a centrifugal fan. To determine the best method of control for any given system the following questions should be discussed with the fan OEM, engineering firm, and end user:

- Does the system require the fans to be designed with large turndown capabilities with frequent changes in performance?
  - Then VFD should considered. Energy savings can offset the upfront cost with-in a few years.
- A VFD should be considered if there is particulate in the air stream.
- Does the system require the fans to be design with minimal change and small turndown requirements?
  - Then a damper could be considered.
- If the upfront capital cost or the location does not best suit for speed control and the system requires the fans to be designed with large turndown capabilities with frequent changes in performance?
  - Then a combination of dampers should be considered; ILD and BPRD or ILD and OLD.
- A typical combustion air fan on a boiler system can use:
  - a VFD or VIV or a combination of ILD and BPRD, for the large turndown range.
  - an OLD to increase the turndown range.
  - an ILD to help draw in FGR gas.
- A typical induced draft fan used on an ethylene furnace system can use:
  - a VFD or ILD or a combination of ILD and BPRD, for the large turndown range.
  - an OLD to increase the turndown range.

In any case the fan OEM can design and select the best method of control for any given system in consolation of the engineering firm and the end user.

Appendix A: Fan performance curve controlled with an inlet box louver damper

Performance Ratings				
Ratings	Rating Point	P1 P2222A00 100% Load	P2 P2222A80 80% Load	
Volume:	187393 Acfm	154867 Acfm	123276 Acfm	
Static Pressure:	40.35 in.wg.	30.57 in.wg.	18.77 in.wg.	
Density:	0.0701 lb/ft3	0.073 lb/ft3	0.073 lb/ft3	
Temperature:	102 F	80 F	80 F	
Speed:	1180 rpm	1180 rpm	1180 rpm	
Brake Horsepower:	1461 hp	1201 hp	993 hp	
Static Efficiency:	82 %	62 %	37 %	
Noise Level:	92 dBA	98 dBA	95 dBA	

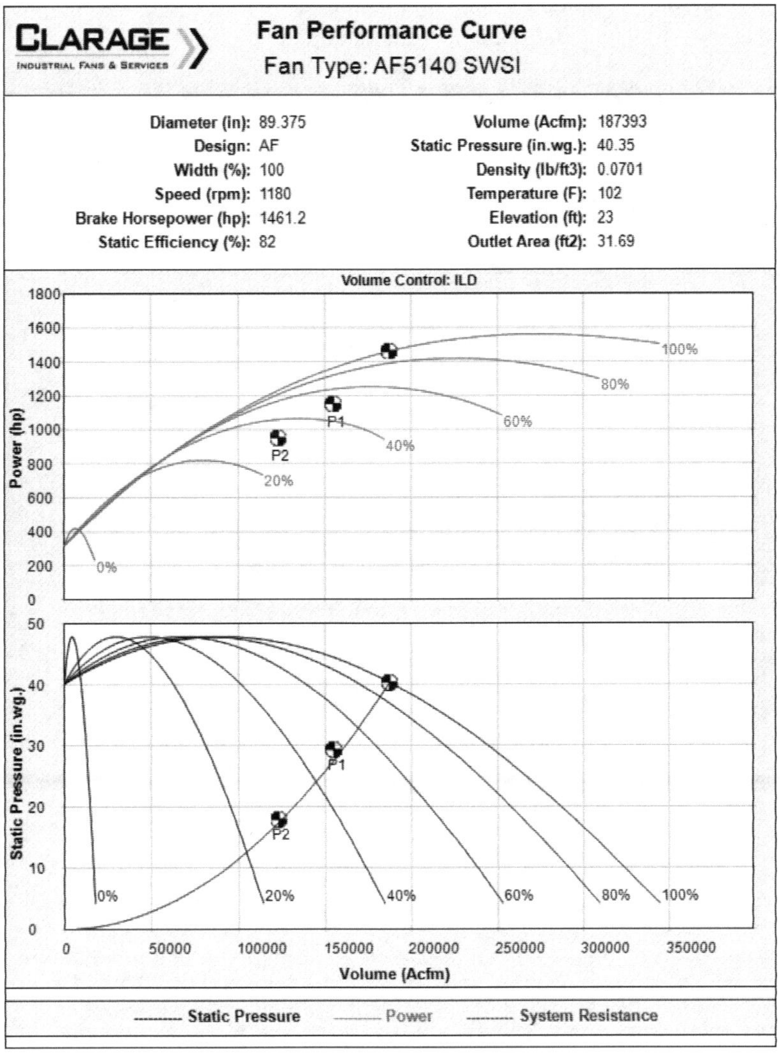

Appendix B: Fan performance curve controlled with a variable inlet vane damper

Performance Ratings				
Ratings	Rating Point	P1 P2222A00 100% Load	P2 P2222A80 80% Load	
Volume:	187393 Acfm	154867 Acfm	123276 Acfm	
Static Pressure:	40.35 in.wg.	30.57 in.wg.	18.77 in.wg.	
Density:	0.0701 lb/ft3	0.073 lb/ft3	0.073 lb/ft3	
Temperature:	102 F	80 F	80 F	
Speed:	1180 rpm	1180 rpm	1180 rpm	
Brake Horsepower:	1460 hp	1130 hp	865 hp	
Static Efficiency:	82 %	66 %	42 %	
Noise Level:	92 dBA	98 dBA	96 dBA	

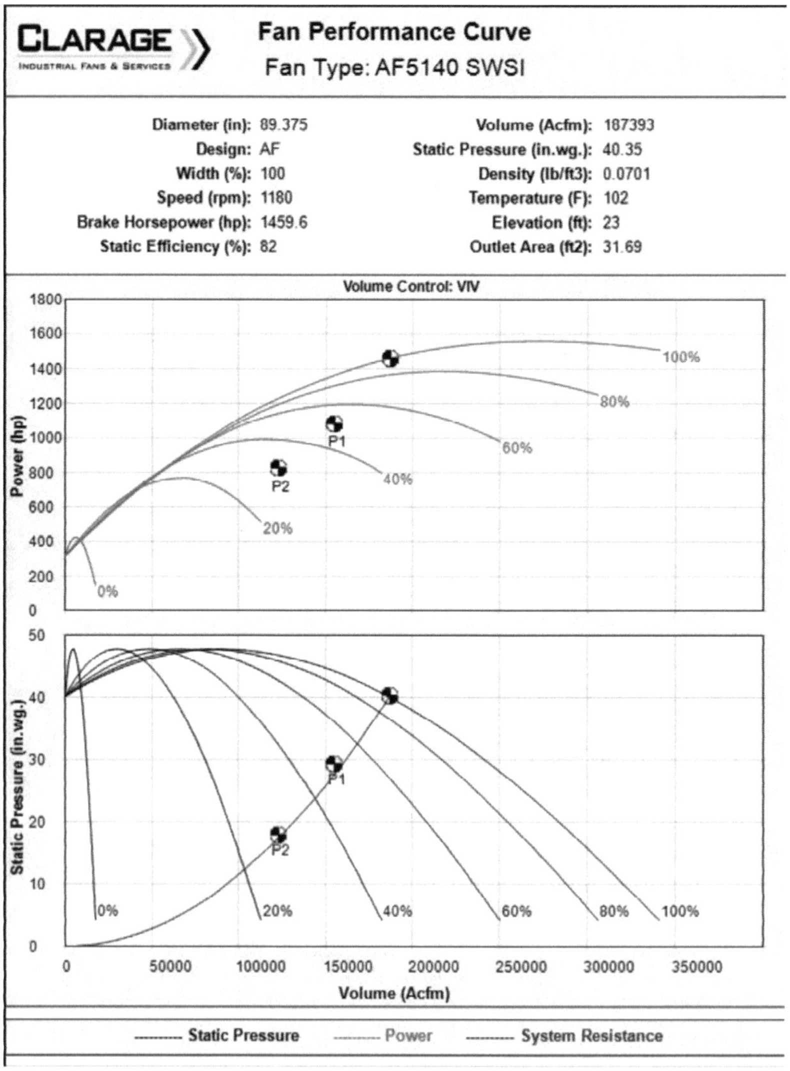

Appendix C: Effects of modulating the outlet damper

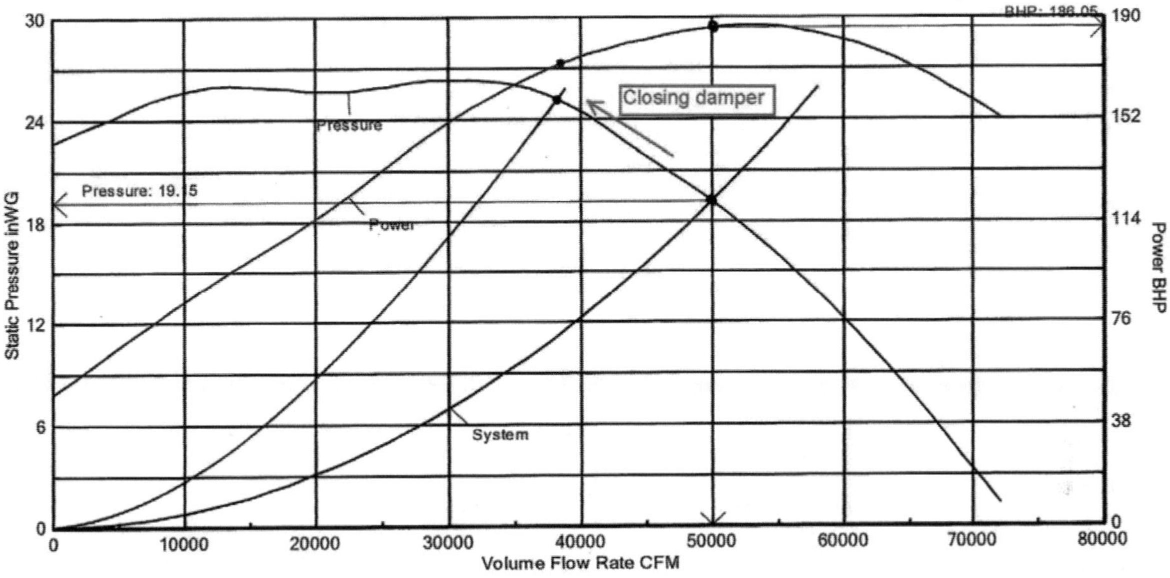

Appendix D: Fan performance curve controlled with a variable frequency drive

Performance Ratings				
Ratings	Rating Point	P1 P2222A00 100% Load	P2 P2222A80 80% Load	
Volume:	187393 Acfm	154867 Acfm	123276 Acfm	
Static Pressure:	40.35 in.wg.	30.57 in.wg.	18.77 in.wg.	
Density:	0.0701 lb/ft3	0.073 lb/ft3	0.073 lb/ft3	
Temperature:	102 F	80 F	80 F	
Speed:	1180 rpm	998 rpm	787 rpm	
Brake Horsepower:	1460 hp	907 hp	443 hp	
Static Efficiency:	82 %	82 %	82 %	
Noise Level:	92 dBA	89 dBA	83 dBA	

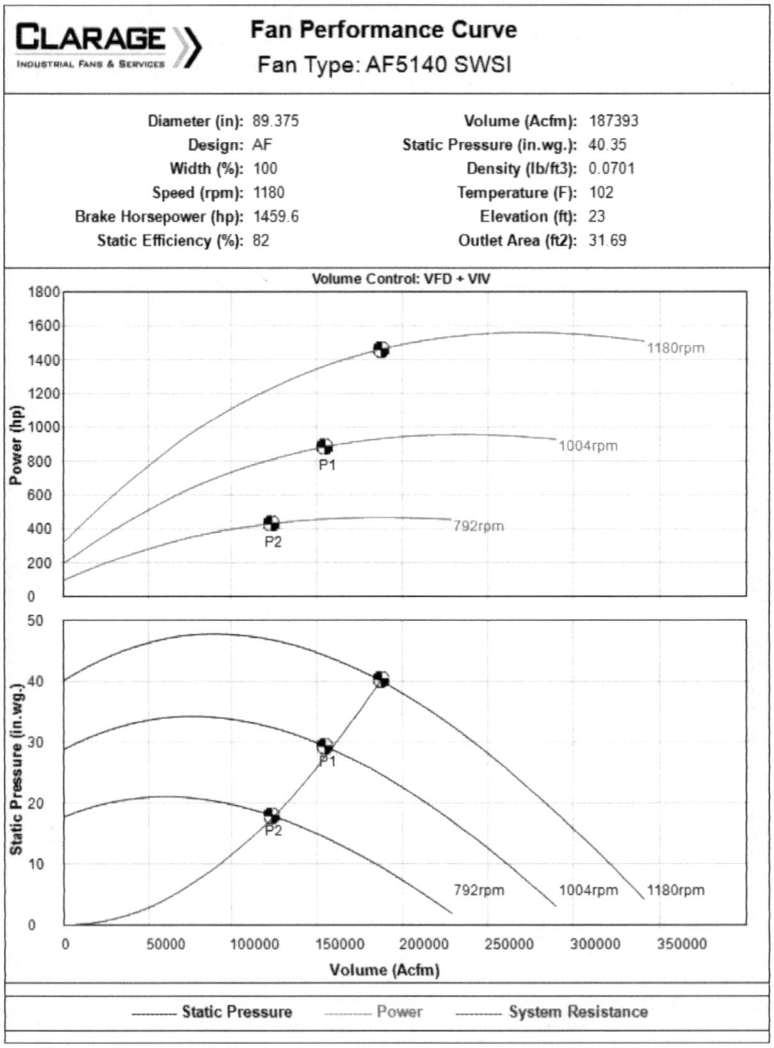

Appendix E: Case study steam generator FD fan performance

Performance Ratings

Ratings	Rating Point	P1 P2222A00 100% Load	P2 P2222A80 80% Load	P3 P2222A60 60% Load
Volume:	187393 Acfm	154867 Acfm	123276 Acfm	92025 Acfm
Static Pressure:	40.35 in.wg.	30.57 in.wg.	18.77 in.wg.	9.24 in.wg.
Density:	0.0701 lb/ft3	0.073 lb/ft3	0.073 lb/ft3	0.073 lb/ft3
Temperature:	102 F	80 F	80 F	80 F
Speed:	1180 rpm	1180 rpm	1180 rpm	1180 rpm
Brake Horsepower:	1461 hp	1201 hp	993 hp	892 hp
Static Efficiency:	82 %	62 %	37 %	15 %
Noise Level:	92 dBA	98 dBA	95 dBA	91 dBA

Ratings	P4 P2222A40 40% Load	P5 P2222A20 20% Load	P6 P2222B00 100% Load	P7 P2222B80 80% Load
Volume:	61139 Acfm	47270 Acfm	146286 Acfm	116464 Acfm
Static Pressure:	4.3 in.wg.	2.51 in.wg.	27.09 in.wg.	16.63 in.wg.
Density:	0.074 lb/ft3	0.074 lb/ft3	0.073 lb/ft3	0.073 lb/ft3
Temperature:	80 F	80 F	80 F	80 F
Speed:	1180 rpm	1180 rpm	1180 rpm	1180 rpm
Brake Horsepower:	845 hp	832 hp	1131 hp	963 hp
Static Efficiency:	5 %	2 %	55 %	32 %
Noise Level:	90 dBA	90 dBA	97 dBA	95 dBA

Ratings	P8 P2222B60 60% Load	P9 P2222B40 40% Load	P10 P2222A0V 100% Load	
Volume:	86987 Acfm	57809 Acfm	168067 Acfm	
Static Pressure:	8.97 in.wg.	3.82 in.wg.	33.64 in.wg.	
Density:	0.073 lb/ft3	0.073 lb/ft3	0.073 lb/ft3	
Temperature:	80 F	80 F	102 F	
Speed:	1180 rpm	1180 rpm	1180 rpm	
Brake Horsepower:	880 hp	840 hp	1292 hp	
Static Efficiency:	14 %	69 %	69 %	
Noise Level:	91 dBA	90 dBA	98 dBA	

CLARAGE
INDUSTRIAL FANS & SERVICES

Fan Performance Curve

Fan Type: AF5140 SWSI

Diameter (in):	89.375	**Volume (Acfm):**	187393
Design:	AF	**Static Pressure (in.wg.):**	40.35
Width (%):	100	**Density (lb/ft3):**	0.0701
Speed (rpm):	1180	**Temperature (F):**	102
Brake Horsepower (hp):	1461.2	**Elevation (ft):**	23
Static Efficiency (%):	82	**Outlet Area (ft2):**	31.69

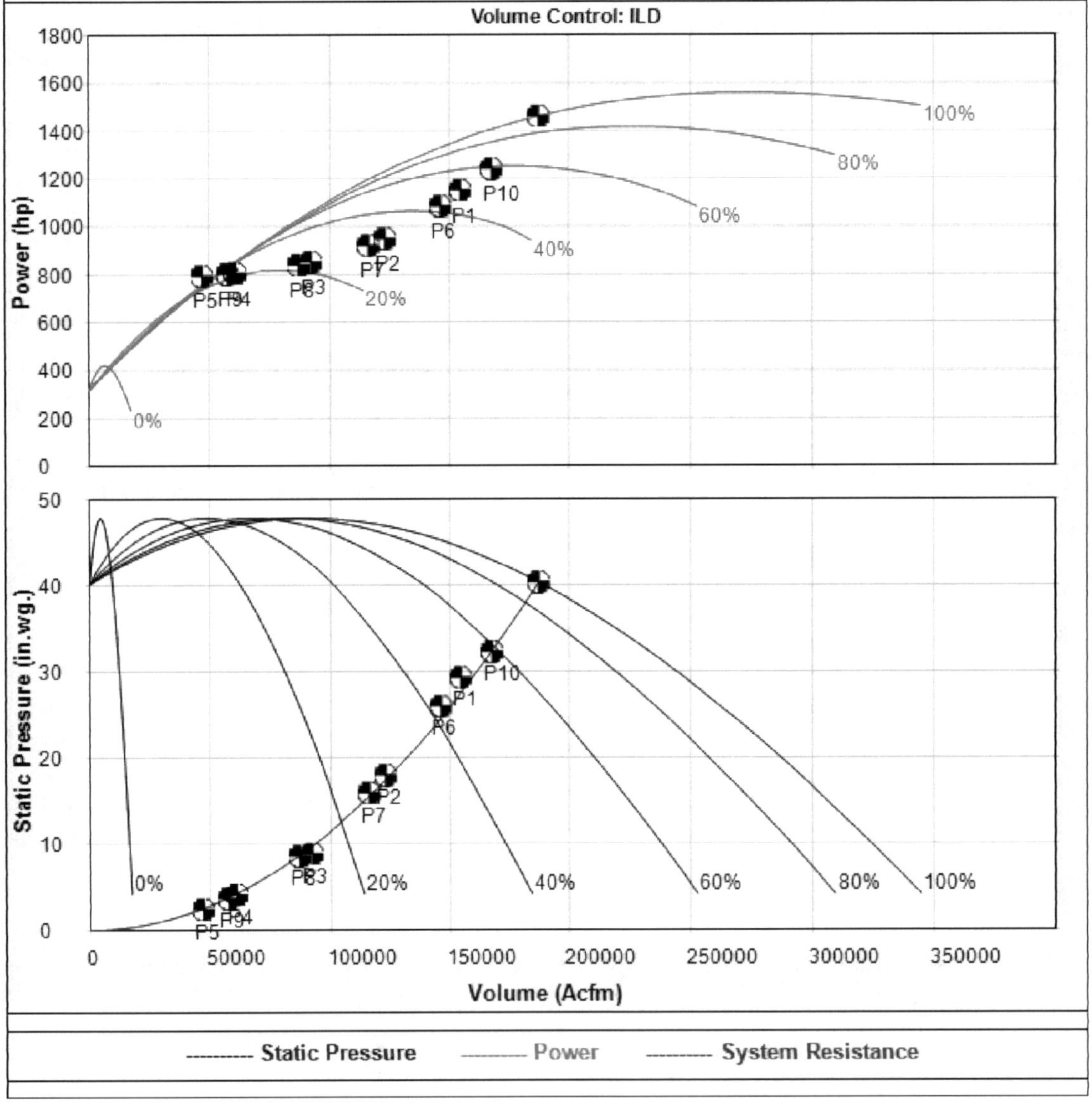

AIChE PAPER NUMBER: 5b

MULTIPOINT GROUND FLARE TURNAROUND PLANNING

Vance Varner
Business Development and Technology Director
John Zink Hamworthy Combustion

Prepared for Presentation at the 2022 Spring National Meeting
San Antonio, TX April 11-14, 2022

ABSTRACT
The maintenance shutdown of an ethylene production unit flare is challenging in several ways. Operating companies and units have continued to extend the time between major outages. Flares are essential equipment, almost always without on-line spares. Thus, this maintenance time extension applies for flares as well. The planning is critical to ensure the time for the flare "out of service" is as short as possible, especially since flares are generally the last equipment taken out of service, and the first back in service during these outages.

The use of multipoint ground flares (MPGF) technology has become more common as the main flare equipment for new ethylene units. Maintenance planning for this equipment is different than for elevated flares. MPGF are more complex when compared with an elevated flare, containing many different types of components. Some of these items can be maintained outside of a turnaround, while other items are only available during a full flare outage.

This paper examines the overall maintenance strategy for a typical MPGF, focusing on the turnaround planning to work toward a successful outcome. A good working knowledge of the flare's operation and maintenance history can aid making good maintenance decisions.

Further details on information presented in this paper can be found in the *John Zink Combustion Handbook*, Volume 2[1].

---

[1] *John Zink Combustion Handbook*, Volume 2, Charles E. Baukal, Jr., editor, Boca Raton, FL, CRC Press, pp. 377-394

MPGF / ELEVATED FLARE EQUIPMENT

Both MPGF and elevated flares utilize combustion to destroy unusable gases sent from the operating unit[2]. Some of the components within each type of flare are similar, but others are very different. These differences make the maintenance strategy different.

Major parts of a MPGF (refer to Figure 1):

- Manifold (main inlet header)
- Manifold drain pot
- Manifold pressure transmitters
- Staging headers / runners
- Automated staging valves
- Bypass devices to handle over-pressure
- Automated post-purge valves
- *Pilots*
- *Fuel gas supply system for pilots (regulators / valves / meters / transmitters / strainers)*
- Flare burners
- Runner shielding – SS sheet metal OR rocks
- *Pilot monitoring – normally retractable thermocouple*
- *Electronic pilot ignition*
- *Flame front generator (FFG) for pilot ignition*
- *Steam control valve (for the low pressure steam assisted burners)*
- *Steam piping system and devices (meters, traps, etc.)*
- *Utility piping – air, nitrogen, fuel gas*
- *Control system (PLC / DCS)*
- Rock covering for ground
- Fence surrounding burner field
- *Cameras*

NOTE: items in *italics* are common with elevated flares

Depending on the design and layout, a few of these items can be maintained / repaired while the flare system is still in service. All the items will be discussed in further detail below.

[2] *John Zink Combustion Handbook*, Volume 3, Chapter 11 "Flares", Charles E. Baukal, Jr., editor, Boca Raton, FL, CRC Press, pp. 251-297

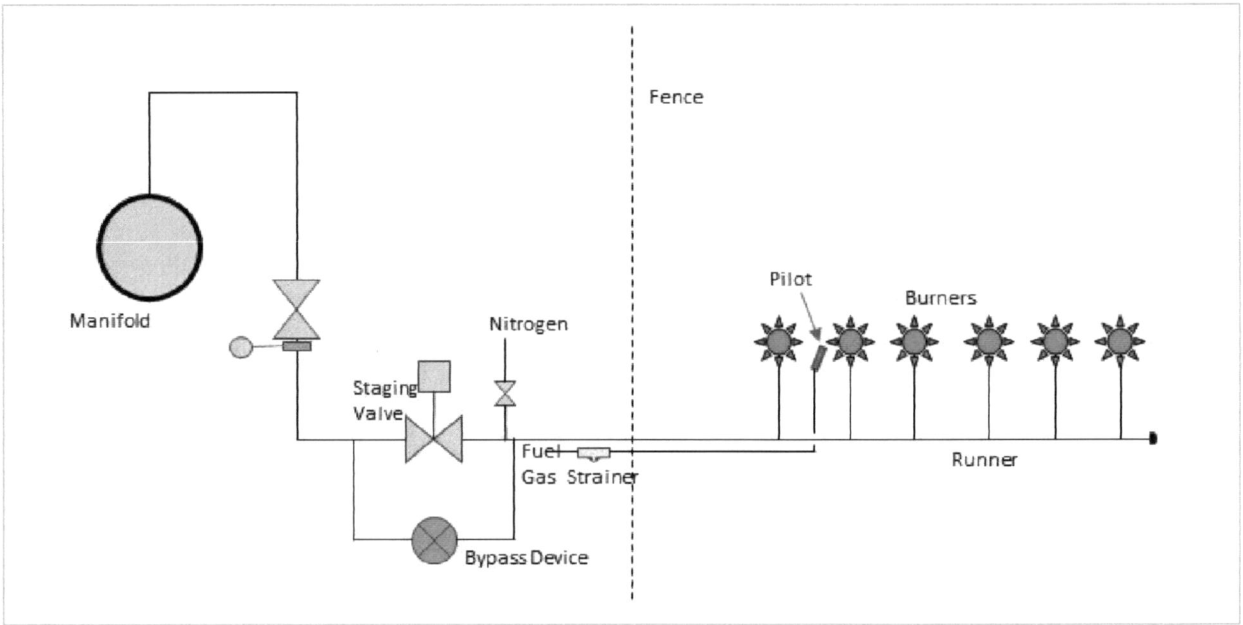

Figure 1 – MPGF single runner / parts

Other items that may be part of the flare system that should be considered in planning:

- Knock-out drums
- Liquid seal
- Header instrumentation (flow meter, analyzers)

There are a few other differences that are important to point out here without delving into the technical reasons which have been covered in other papers / articles / discussions with flare suppliers.

COMPONENT	**MPGF**	**Elevated**
Header pressure	10+ psig	2-3 psig
Burner exit velocity	Sonic (~1000 ft/s)	<400 ft/s
Minimum NHV_{CZ}	800 Btu/SCF	270 Btu/SCF
Smoke	None	Yes at high rates
Gas exit	Near grade	>250 feet for typical ethylene plant

A few of these contribute to important considerations when planning the maintenance.

PLANNING

As with any plant turnaround, one of the key decisions is to review all the equipment that will need work, categorize the work as turnaround or pre-turnaround, and develop the schedule to minimize the time for the equipment to be out of service. Other plans include the concept of discovery work during the outage, which parts to pre-purchase for replacement (or potential replacement), available time for the turnaround work, and work schedule (12 or 24 hour). Examples of long lead time items include valves, soft parts for valves, flare burners, pilots etc. Items such as these should be pre-purchased.

A full planning checklist should be developed which covers the pre-turnaround inspections, any pre-turnaround work to be performed, work during the turnaround, final completion checklist, and commissioning / start-up. It is recommended that the end-user work with the Original Equipment Manufacturer (OEM) for assistance in developing this checklist. The OEM may have a pre-existing checklist to use as a template. Additional information may also be found in the flare operating manual provided by the OEM.

INSPECTIONS – PRE-TURNAROUND

These pre-turnaround inspections are important steps in understanding the scope of work to be performed. However, useful visual inspections inside the flare fence (flare field) are difficult.

One of the challenges is obtaining a good vantage point to perform these inspections. Viewing the flare field both under and over the fence with cameras or binoculars can sometimes be effective; however, the size of the flare field and available viewing angles make this task difficult. In addition, remote inspections, either by drone or other unmanned entry vehicle, have become more practical and increasingly more affordable over the years.

As noted, any of these methods have limitations. Even with these limitations, plant staff should be prepared to employ a combination of inspection methods to help ensure the most effective maintenance during the outage.

OUTSIDE THE FENCE CONSIDERATIONS

The design of the MPGF allows personnel to be close to the flare system, even when the flare is in operation. As such, inspection of the flare items outside the fence should be done ahead of any flare outage. Many of the MPGF flare designs employ the use of a spare stage / runner. Having this available would allow operations to isolate one of the runners at a time with no loss of capacity. Isolation of a runner allows the user to maintain equipment like the automated valves, bypass devices, and other items. Once a stage is isolated, the valves / devices can be fully removed for maintenance / rework / replacement if needed.

The typical design for pilot flame detection is the use of thermocouples. Other options include ionization detection, sound, or IR camera, but these are not as common. For the MPGF, the majority of the systems utilize a "retractable" system, where the thermocouple is inserted into a SS tube which extends from the fence edge to the pilot. The thermocouple can be replaced at any time by removing / pulling out the old one and inserting / pushing the new one into place.

The FFG system is generally the same as one would see for an elevated flare. These are typically the back-up ignition device. As such, those FFG components can be isolated and worked on at any time. Periodic testing a minimum of once every 6 months of the FFG unit is highly recommended. This procedure involves proceeding through the steps, including sending a fireball to one pilot.

Other instruments that can normally be maintained with the flare in service are any cameras and the flare manifold pressure transmitters. Cameras are usually in a location accessible but protected from the flare burners. For the manifold pressure measurement, there are usually three transmitters. This setup allows them to be removed from service one at a time with no loss of flare functionality, with the control system utilizing the other two transmitters.

The automated staging valves are open / close valves. These are usually metal seated, tight shut-off butterfly valves. Normal valve maintenance procedures apply to these valves. Preventative maintenance includes the operation (stroking) of each valve at least once per year. This helps ensure the ability for the valve to move (no interference from debris), checking of the connected instrumentation (valve position indicators), and confirming that the actuator will operate within the designed time limit. These valves are robust; however, pipe strain can cause the valves to either stick closed, or not fully seat when attempting to close.

The automated post-purge valves are ball valves operating as open / close valves. Normal valve maintenance procedures apply to these valves. Some designs utilize a bypass orifice around these valves, allowing a continuous flow. Maintenance would include confirming the cleanliness of the orifice.

The safety bypass valves / devices for each stage have a history in the industry of issues with leakage as well as seat damage. These devices are even more susceptible to pipe strain than the staging valves, causing seat mis-alignment or damage. The most common type is a pin-type, which utilizes a pin that is under compression to keep the valve closed during operation. Once the pressure rises to the set point, the pin will deform (buckle) or shear, allowing the valve to open. Preventative maintenance includes the operation (stroking) of each device at least once per year. The pin is removed for this operation and re-inserted when the testing is complete. This action helps ensure the ability for the valve to move (no interference from debris) and will check the connected instrumentation (valve position indicator). A relatively new device for this service is made by Total Valve, model TRV-6820. This is a pressure actuated butterfly valve opens at the set point which is independent of power or air, and has no items which are damaged.

It is recommended to rework these devices during the flare outage to ensure the proper condition.

Figure 2 – Bypass Valve types: buckling pin, shear pin, TRV

In some cases, the bypass device is a rupture disk. If used, there is no testing to be performed. Once burst, the disk must be replaced. The user may consider replacement of all the disks as a preventative measure during the flare outage. This will help avoid early bursting. The consideration is that during the normal operation, these disks experience typical pressure cycles and could experience potential fatigue.

Maintenance of the manual valves isolation valves to flare runners are only accessible during the flare outage. It should be noted if any of these have been difficult to operate. The flare outage provides the opportunity for maintenance of these valves.

There are other components outside the fence that should fall under the normal unit maintenance program – instrumentation, piping components, paint, electrical grounding, etc. Consider the fuel gas supply, nitrogen, steam, instrument air, and full FFG systems. One potential trouble spot is any fuel gas valves / regulators that drop the pressure significantly – consider possible corrosion on the downstream side.

The steam system for the flare provides assistance for the low pressure burners, normally only the first stage. Although not as large as a typical steam system for assisted elevated flares, the components must still be maintained: steam traps, control valves, manual valves, bypass minimum flow orifice, etc.

Another item to maintain while the flare is out of service are the strainers on the pilot fuel gas. Normally the fuel gas piping outside the fence is carbon steel, changing to stainless steel on the inside. Just outside the fence, there is a Y-type strainer used to protect the fuel nozzle at the mixer which is physically at the pilot. These strainers should be opened and cleaned of any debris when possible, ensuring proper fuel flow to the pilots.

The main manifold should have a drain pot for collecting any condensation from the flare header. Check all the instrumentation around that pot. Ensure there are no solids that have collected which would restrict its proper operation.

INSIDE THE FENCE CONSIDERATIONS

The maintenance / repair of the components inside the fence can only occur when the flare is out of service and fully isolated. The activities for these are outlined below.

The pilots and the ignition system components are the most common items to be maintained. In many cases, the most effective use of time is to replace the pilots during the outage. Useful pilot life can range from 10 to 12 years. After that, several of the components start to deform / crack, and are more susceptible to flame-out.

Replacement of all pilots is not inexpensive. MPGFs can easily have 40 pilots or more within the system. An intermediate solution can be to plan to replace half of the pilots during one outage, and the other half the next outage. Once personnel are allowed inside, an evaluation of the pilots would be made, and the worst half replaced.

For the existing pilots that will remain, the minimum action plan should be to clean the pilot mixer area and fuel nozzle. In some cases, insects have been known to infiltrate the piping between the mixer and pilot. A pipe brush can be used to clean this area.

As previously noted, the fuel gas supply piping inside the fence is normally constructed of stainless steel pipe. Proper installations will allow for some flexibility / growth of the piping. Inspect this pipe for any distortion / cracking, and repair as needed. Before reconnecting any fuel piping to the pilot, it should be swept out with high velocity air to remove debris and/or water.

The electronic ignitor element can be damaged over time, so these should be examined. It is more common to have damage with this system's wiring. The intense heat within the fence can cause connections to become loose or corroded, or the wiring insulation to break down. The wiring is contained within protected conduit, which should be examined for breaks / distortion. Even with undetected damage, it might be more effective to change all the wiring to ensure reliable life during the next cycle.

As noted above, the thermocouples can be changed from outside the fence. In some cases, the SS tubing that contains the thermocouples is damaged / distorted, causing the thermocouple to be wedged in place. While inside the fence, this tubing should be examined and repaired as needed. Note that this tubing is a special type, polished tubing with specially machined connectors. Consult with the OEM for details before replacing the components.

The runners are protected by either SS shielding or by a layer of rocks. The use of shielding allows for inspection of the runners, riser connections, the pipe shoes, foundations, utility piping, and instrumentation conduit. The shielding should be inspected for any missing bolting used to hold it in place. Discoloration is expected, but distortion which would block the ability for the various shield plates to slide over each other is not acceptable.

Figure 3 – Example of runner shields in good condition

Figure 4 – Example of damaged runner shields

If the runners are protected by rocks, any areas that do not have the minimum covering thickness (typically 12") should be repaired. This is especially true should there be exposed runner piping. The original supplier drawings should show the expected minimum thickness.

Figure 5 – MPGF runners protected with rocks

Figure 6 – Damaged runner exposed due to less than adequate rock covering

The runner pipe shoes should be inspected for their condition, checking for distortion and ensuring the ability to slide not be hindered by rocks or other debris. Foundations and bolting should be checked as well to ensure no major cracking / failures. Any nuts should be left with a proper gap to ensure the ability of the sliding design.

The ground covering inside the fence is rocks. Experience has shown that in many facilities, the rocks have been overheated and pulverized to a powder consistency. In most of these cases, the rocks used were limestone gravel. As the rocks break down, it affects the ability of the air to flow to cool the ground as well as provided a higher reflective surface. Should the ground covering be powdery, new rocks should be brought in to improve the surface. Usually, this new covering can be placed over the existing one. Proper rocks for the ground and concrete covering should be igneous granite (or equivalent) and 1-1/2" to 2" diameter size with no fines.

Figure 7 – Pulverized rock ground cover

The fence itself is not normally an item requiring maintenance. A flare outage is an opportunity to inspect of the bolting and panels from the inside. The majority of the panels used in the facilities are galvanized carbon steel, with the other option of stainless steel. In some cases, the galvanizing has been disrupted due to high heat for extended periods, or from too small of an interior footprint. In general, this is not a cause of concern as the panels will still perform the intended purpose of blocking the radiant heat. The panels are slotted at the connection points, and the bolting should remain slightly loose to allow for thermal expansion as needed. Note that the fence panels normally bolt from inside out and cannot be changed during operation. Any foundations / column bases inside the main fence should be covered by rocks for heat protection.

Figure 8 – Fence Panel Damage

The next consideration are the burners / risers / connections to the flare runners in the flare field. These are items that are exposed to the radiant heat from flames inside the fence containment.

In general, burners are very robust and will last the life of the flare. These burners have the proper metallurgy to handle high temperatures. When the burners are in service, the flame will be above the burner. Gas flowing through the burner will help keep the metal temperature below any destructive level. Any burners that can operate at lower pressure (usually just the stage 1 burners) may see a little more flame impingement at low flows and higher winds. However, these are typically steam assisted. The use of steam in the design is for smoke control, yet it will also help keep the metal at a reasonable temperature.

Gas risers which hold the burners are also built to handle the high temperature conditions. The weakest part of this system is the connection of the riser to the runners. These are normally O-let type connections.

Regarding inspections of these areas, the high pressure burner tips / risers / connections should be randomly spot checked (5 to 10 percent) for potential damage / cracks using typical NDE type inspection methods. All burners in low pressure service should all be examined.

Any flare knockout drums / liquid seal drums should be inspected at this point since the flare system is out of service. Generally, this involves opening of the vessel manway and visually inspecting without entering the equipment; however, vessel entry must be considered as a possibility. The related instrumentation / automated valving can be checked / recalibrated. Liquid seal drums have internals. Ensure those internals are in their proper places and the downcomer piping / distributor is not corroded.

IMPROVEMENTS

Take advantage of the flare system outage to install / implement improvements to the system. Potential improvements to consider with the OEM:

- Staging sequence – which valves open when
- Control system – local / DCS / dedicated safety system (i.e., Triconex)
- Improvements to safety bypass devices

PRE-TURNAROUND PLANNING

The planning for the flare outage / work to be performed is a key element for a successful event. The following are some key questions / considerations for this planning, and must be answered based on the specific outage taking place and company policies:

- The condition of the operating unit during the outage
  Will the unit be totally hydrocarbon free? Is there a need for a small temporary flare to be available to handle potential venting during the main flare outage? If yes, where will this be set up and what utilities are required?
- Consider where this work will fit into the overall turnaround schedule
  Flares are typically the last equipment to be removed from operation service, so much of the other turnaround work will be underway already when the flare is ready.
- Isolation of the unit from the flare system
  What is available to isolate the main flare header within the process unit from the flare system? Will this require all pressure relief valves and all connections to the flare header to be isolated with a Lock-out / Tag-out system? For ethylene production units, this represents several hundred connection points. Assigning a focal point to coordinate this effort will help ensure things go well. Check / mark the tag points on P&ID's. Field walk to verify these points.
  The alternative to this full isolation method is to isolate each runner with the manual valve that is already installed. The use of a slip blind on the downstream side of the manual valve gives that runner full isolation from the system. In this case, the upstream drums / header / manifold would still be "active".
- Schedule of activities
  Each activity time allotment must be set into the flare schedule. In many cases, the work can be completed using a day-shift only, and that should be the starting point.

Should the overall schedule require the use of the night-shift slot, extra equipment is required to provide lighting.

- Work permits / safety

  A key decision to be made is whether work inside the fence will be considered to be a confined space entry. OSHA defines confined space as:
  "*Many workplaces contain areas that are considered "confined spaces" because while they are not necessarily designed for people, they are large enough for workers to enter and perform certain jobs. A confined space also has limited or restricted means for entry or exit and is not designed for continuous occupancy. Confined spaces include, but are not limited to, tanks, vessels, silos, storage bins, hoppers, vaults, pits, manholes, tunnels, equipment housings, ductwork, pipelines, etc.*" [3]
  While the fence area does have limited egress, the design of the fence is such that it will provide adequate ventilation for any gas that could enter the space. No special devices are needed for this. As with any work area, the sources of energy must be isolated. Since there will need to be some equipment inside, normally one or more of the fence panel sections are removed, providing larger and easier exit areas. With these considerations, the unit may choose to treat this space differently only during the turnaround.

- Equipment to be used

  Since all the work is at / near grade, no large equipment / long reach cranes / man baskets will be required. Plan for access and staging for the equipment (track hoe, small crane(s), manlifts, vehicles) to include mats, pads, etc.

- Facilities

  Location and considerations for workers and staff. Flares are often far from the operating areas, which means many man hours can be consumed by traveling. Consider temporary facilities being stationed near working area - Port-o-lets, offices, tool trailers, parts storage, and/or break areas.

TURNAROUND WORK

Once access is granted for inside, one or two sections of fence panel should be removed using a small crane to allow for the other equipment access inside. The building of any planned scaffolding would be the next step performed.

Inspection of the pilots and burners can be done via ladders or a man-lift. Full inspection of the fence panels would require a man-lift.

[3] Occupational Safety and Health Administration web page, https://www.osha.gov/confined-spaces, covered by 29 CFR 1926, Subpart AA standard, May 4, 2015.

Full decontamination of a flare piping system is impractical but may not be necessary. A decision whether to perform this work depends on the isolation plan. Clearing of the header of hydrocarbons would be done once the process connection points are isolated, then the header can be cleared using a pressure / depressure (also called pad / depad) method utilizing nitrogen. This can be done since there are isolation valves at the MPGF runner inlets. This will help, but not fully guarantee isolated pockets are cleared of vapor hydrocarbons. Depad to either the atmosphere or to a temporary flare system.

Caution for working inside the flare fence. The ground is rock aggregate, making for rough and uneven surfaces. Any hot work (grinding / welding) should be handled carefully since this equipment has been in hydrocarbon service. The inside pipe walls can still contain contamination build-up, such as rust, desiccant, and catalyst fines.

Take the opportunity to blow out the pilot supply and steam piping to ensure no debris / water has been left in the piping.

COMMISSIONING / START-UP

The goal should be that all components are functioning as they were at the initial installation / start-up. Having the flare system still isolated allows for access inside to do final testing / repair as required. Items to consider once the planned work is complete:

- All foundations and ground cover in good shape with the rock covering.
- All tools / equipment removed from inside the fence.
- No exposed wires or other material that could be damaged with heat.
- All fence panels replaced.
- Ensure thermocouples are completely inserted to the pilots. Test by placing heat on the pilot head and checking the reading.
- Manually spark the electronic ignition and listen at each pilot to hear the spark.
- Remove personnel.
- Commission fuel gas system / ignite pilots.
    - Consider using both electronic and FFG systems to confirm both are fully functional.
- Troubleshoot any pilots that will not ignite. Shut down the fuel gas system and allow personnel back inside the fence. Note that the main flare system should still be isolated at this point.
- Once pilots are all burning, commission the steam system. Open the steam control valve 100 percent to confirm that the steam will not extinguish the pilots for those burners.

The main flare system is now ready for commissioning and placing back in service. Before opening to the burners, ensure that the oxygen level is acceptable in the header system. API Standard 537[4] recommends the oxygen to be below 6 mole percent.

SUMMARY

Pre-planning is the primary key to a successful execution of maintenance for the MPGF. Some of the work can be done without a shutdown of the flare system. Inspection from a distance is necessary since personnel is not allowed inside the fence with the flare in service. Work with the OEM to assist in planning. Utilizing checklists will help ensure proper planning and execution.

[4] API Standard 537, "Flare Details for Petroleum, Petrochemical, and Natural Gas Industries, *American Petroleum Institute,* 3rd Edition, Addendum 1, March 2020, pp. 61-62

AIChE PAPER NUMBER 5c

A BREAKTHROUGH INNOVATION IN REFRACTORY CERAMIC FIBER INSULATION: REDEFINING THE EFFICIENCY AND THERMAL PERFORMANCE OF PETROCHEMICAL CRACKING FURNACES

John Becht
Global Application Engineering Manager
Alkegen

Prepared for Presentation at the 2022 Spring National Meeting
San Antonio, TX, April 10 - 14, 2022

A BREAKTHROUGH INNOVATION IN REFRACTORY CERAMIC FIBER INSULATION: REDEFINING THE EFFICIENCY AND THERMAL PERFORMANCE OF PETROCHEMICAL CRACKING FURNACES

John Becht
Global Application Engineering Manager
Alkegen

Abstract:

Operators and designers of Ethylene Cracking Furnaces are under increasing pressure to improve efficiencies and yields, while generating lower emissions and consuming less fuel to run their furnaces. Selection of the optimal furnace lining insulation is a key factor to achieving the end user's goals of lower energy consumption, lower operational costs, improved safety, and increased production throughput of the ethylene production plant. Heat management solutions for ethylene cracking furnaces have evolved over time from traditional solutions such as insulating castables and refractory brick, to lower cost and more energy efficient refractory ceramic fiber and microporous solutions.

Recently, a step change in refractory ceramic fiber manufacturing technology has been achieved which redefines the insulation design and heat management capabilities imaginable for ethylene cracking furnaces. This technical paper discusses the newly developed refractory ceramic fiber in detail, along with illustrative examples of how its enhanced fiber properties of higher fiber index and lower shot content will allow end users to reduce energy consumption, heat loss, heat storage, and the associated insulation investment costs. Several case studies are presented which will contrast the enhanced efficiency and thermal performance benefits of the new refractory ceramic fiber material versus existing strategies for thermally insulating ethylene cracking furnaces.

Introduction

As air pollution and greenhouse gas emissions continue to threaten the livelihood of the world we live in today, industries worldwide are constantly being challenged to limit their fuel consumption and mitigate their carbon footprint. For most governments, companies, and countries, it is not economically feasible to simply halt production in order to abide by newer and more stringent emission control regulations. With that being said, it is critical for companies to continue to innovate and improve upon the traditional technologies that are used to efficiently operate today. Until new production processes are discovered that are more environmentally friendly, companies need to utilize more energy efficient solutions that coincide with the standard operating and production processes currently being used today.

The Petrochemical industry lies at the center of this discussion. The industry as a whole is a major contributor to almost all the materials and products that are used daily, so there is not an option to shut down production. In general, the production throughput of a petrochemical manufacturing plant is highly dependent on the efficiency of the furnace system used for operation. The furnace system and methodology used for production typically operate under extremely high furnace temperatures and energy input. This is entirely true for petrochemical cracking furnaces, in which this technical paper is focused around.

Given that petrochemical cracking furnaces operate at such high temperatures, they require innovative heat management solutions in order to reduce energy loss and cold face temperatures (also known as shell temperatures). Heat management systems include insulation linings consisting of dense brick, insulating firebrick, castables, microporous, and high temperature insulating wool depending on the furnace operating conditions and environment.

More recently, high temperature insulating wool has been used in the radiant walls and arches of petrochemical cracking furnaces. Radiant wall and arch insulation linings often consist of refractory ceramic fiber materials which are lightweight, resilient, durable, and optimal thermal performers. A newly innovated refractory ceramic fiber is redefining the thermal performance of insulation linings for cracking furnaces. The enhanced fiber properties of higher fiber index and lower shot content provide a flexible, lightweight, robust insulation lining that reduces the heat loss, cold face temperatures, and energy consumption of petrochemical cracking furnaces.

Methods & Materials

A petrochemical cracking furnace is the main unit to produce the fundamental raw materials such as ethylene and propylene in the petrochemical industry[2]. Ethylene is arguably one of the most important chemicals in the petrochemical industry, and the conventional ethylene production process is energy intensive and a heavy emitter of carbon dioxide[6]. There are a variety of cracking furnaces used in the petrochemical industry depending on which stage in the process the furnace is found and the feedstock being used. Some common examples and names to describe these cracking furnace systems are "ethylene cracking furnace", "pyrolysis furnace", "naphtha cracking furnace", and "steam cracker"[3]. In all cases, operation of the petrochemical cracking furnace system is the main engineering direction to improve upon the economic performance, and to meet technical and environmental requirements[2]. The two primary feedstocks for ethylene production are naphtha and gas (ethane, propane, butane, etc.), where the first step in the production of ethylene is to take the feedstock and crack it into ethylene and other various products inside the radiant coils of the cracking furnace[4]. This process is called pyrolysis. Pyrolysis is understood as the thermal cracking of petroleum hydrocarbons with steam, also referred to as steam cracking.

The configuration and conditions of a petrochemical cracking furnace can vary from plant to plant, but general specifications are put in place to ensure proper requirements for the design, materials, fabrication, and inspection are met. The American Petroleum Institute (API) has established a standard for fired heaters, API Standard 560 "Fired Heaters for General Refinery Service"[1]. For original equipment manufacturers, all efforts are made to best apply these standards to petrochemical cracking furnaces within reason. In the field, it is often hard for cracking furnaces to meet cold face shell temperatures as specified by API 560 due to the heightened operating temperatures seen in the radiant sections of a cracking furnace. Depending on the petrochemical cracking furnace operating conditions, design constraints, and specifications required, it is important to understand all the insulation materials available and their different characteristics in order to make the proper material selection and engineered lining design.

For any high temperature insulation lining, the main objective is to minimize heat losses and cold face temperatures in order to reduce energy consumption and to create a safe cold face temperature for the furnace to operate over its lifetime. The refractory lining system provides thermal resistance to retain heat and protect the casing from high temperature damage[1]. Operators and contractors have successfully been using a variety of different insulation materials and methodologies to insulate petrochemical cracking furnaces over the years. Until the discovery and invention of refractory ceramic fiber by Charles McMullen in 1942, heat treatment processes and applications predominantly all consisted of dense brick refractory (Figure 1) and castable lining solutions in order to operate at high temperatures. These solutions often require special masonry skills for installation, and are typically very heavy with minimal thermal efficiency and high heat storage properties. The dense insulation lining solutions provided adequate insulation performance, but lacked the superior advantages that high temperature insulation wool provides end users.

Figure 1. Brick Refractory.

Since the 1960's, high temperature insulation wools like refractory ceramic fiber changed the way insulation linings were approached for a wide variety of industrial applications. Refractory ceramic fiber in comparison to dense refractories provides greater thermal efficiency, lighter weight, lower heat storage, thermal shock resistance, resiliency, and ease of installation. The intangible benefits of refractory ceramic fiber were quickly realized by the petrochemical industry and have been the standard insulation lining solution for the radiant walls and arches of petrochemical cracking furnaces. Most petrochemical cracking furnaces now consist of refractory ceramic fiber modules and back-up blanket (Figure 2) as an insulation lining solution to satisfy the thermal performance specifications outlined by API 560. Refractory ceramic fibers can be made via a "blowing" or "spinning" manufacturing process, and there are slight differences between their performances depending on the process utilized.

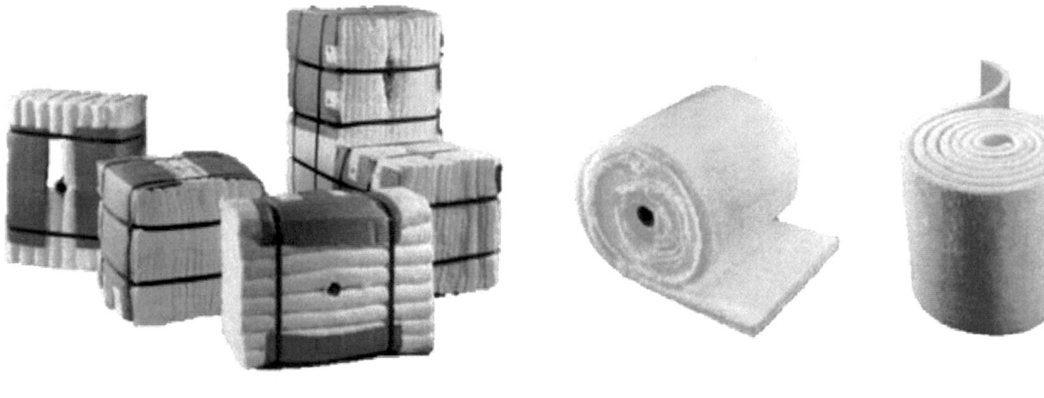

Figure 2. Refractory ceramic fiber modules and blanket.

Depending on how harsh the furnace conditions and environment is, sometimes a microporous back-up board can be incorporated into the insulation lining design in conjunction with refractory ceramic fiber to provide enhanced thermal performance and reduce heat loss and cold face temperatures. Microporous insulations (Figure 3) are often referred to as "super insulators", because they have inherently low thermal conductivity values. Although microporous insulation boasts enhanced thermal performance, they have limited high temperature use capabilities. They also typically come at an extremely heightened raw material cost and are very labor intensive to install.

Figure 3. Microporous insulation.

A breakthrough in refractory ceramic fiber manufacturing technology is redefining the thermal performance properties imaginable for refractory ceramic fiber insulation linings. In July 2020, technological advancements in the manufacturing process of bulk refractory ceramic fibers have created a refractory ceramic fiber with a higher fiber index and lower shot content. This can be seen in the scanning electron microscope (SEM) image shown to the right in Figure 4. This innovative material boasts a high fiber index refractory ceramic fiber that has exceptional insulating characteristics, reduced shot content and shot size, along with improved handling and tensile strength in comparison to the traditional refractory ceramic fiber that has been used since the 1960's. "Shot" can be understood as the non-fiber bi-product created during the manufacturing process, where grains of material were not turned into fiber. Shot is considered to not be an efficient resistor of thermal radiation in comparison to fiber.

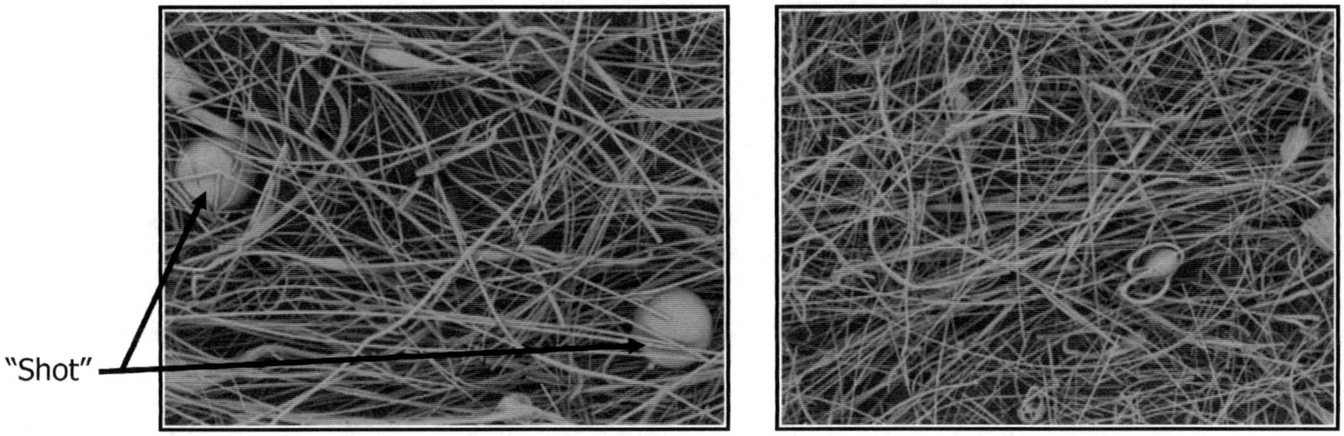

Figure 4. SEM Images of Traditional Refractory Ceramic Fiber (on left) & Newly Innovated High Fiber Index Refractory Ceramic Fiber (on right).

The result of this particular fiber makeup is a material with very low thermal conductivity values. The thermal conductivity of a material can be understood as a materials inherent ability to transfer or conduct heat under a steady state temperature gradient. The thermal conductivity curve of the high fiber index refractory ceramic fiber in comparison to traditional refractory ceramic fiber is shown below in Figure 5.

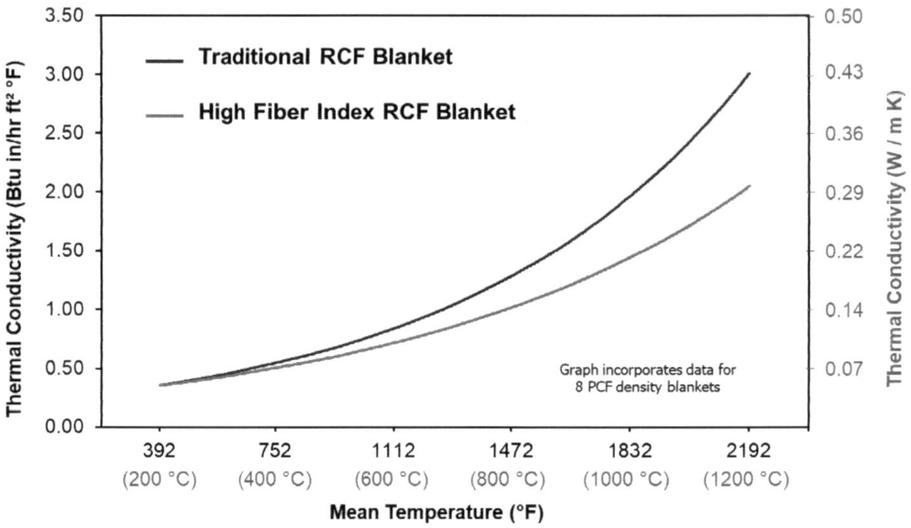

Figure 5. Thermal conductivity curve of traditional & newly innovated refractory ceramic fibers.

In addition to its superior insulating properties, the higher fiber index refractory ceramic fiber possesses additional improvements to traditional refractory ceramic fibers in areas like tensile strength, improved service life, and a "cleaner" and "softer" feel. These additional improvements are primarily due to its high fiber index. Fiber index is a material property that describes the proportion of fiber and shot content by weight. This property is expressed as a weight percentage, and traditional refractory ceramic fiber manufactured by blowing or spinning typically has a fiber index of ≈ 45-55%. The actual number of shot particles is small compared to the number of fibers, but in terms of weight the amount of shot and fiber are nearly equal. The high fiber index refractory ceramic fiber has very little shot based on weight, with a typical fiber index of ≈ 65%. By increasing the number of fibers, the tensile strengths of the high fiber index refractory ceramic fiber blanket are roughly 20% greater in comparison to traditional refractory ceramic fiber blankets. This is primarily due to a greater generation and interlinking of the fibers upon the needling stages of the manufacturing process. This creates a strong blanket, which is more difficult to pull apart and is better suited to withstand the rigors of installation. Along with the tensile strength enhancement, a longer service life in high vibration – high temperature applications can be expected given that there are more fibers per unit mass, and less large shot and shot content. Lastly, given the lower shot content and smaller shot size, the blanket itself has a smoother, "softer" and "cleaner" feel. This is convenient during installation, where contractors and installers can expect less dust, irritation, and more comfortable handling. Overall, with less shot and more fiber pound for pound, the higher fiber index refractory ceramic fiber offers improved handling for safer, quicker and easier installation, as well as increased durability and tensile strength that allows it to perform better and longer even in the most extreme conditions.

Case Study

To truly understand the added benefit of the high fiber index refractory ceramic fiber material, an assortment of insulation lining designs will be analyzed for their thermal performance properties. For illustrative purposes, the insulation lining scenarios included in this study are described in the Table 1 through Table 4 below. The lining examples chosen represent some of the more commonly used insulation lining solutions used in petrochemical cracking furnaces today, along with the newly innovated material. When it comes to the refractory ceramic fiber used, "AS" stands for alumina-silica, and "AZS" stands for alumina-zirconia-silica chemistry.

Insulation Lining Scenarios

Table 1. Traditional Spun Refractory Ceramic Fiber

Insulation Lining Details				
Item Color	Construction	Material	Thickness	Density
	Hot Face Material	AZS Folded Modules, Spun RCF	12" (305mm)	12 lb/ft³ (192 Kg/m³)
	Back-up Material	AS Blanket, Spun RCF	1" (25mm)	8 lb/ft³ (128 Kg/m³)

Table 2. Monolithic Blown Refractory Ceramic Fiber

Insulation Lining Details				
Item Color	Construction	Material	Thickness	Density
	Hot Face Material	AZS Monolithic Modules, Blown RCF	12" (305mm)	15 lb/ft³ (240 Kg/m³)
	Back-up Material	AS Blanket, Spun RCF	1" (25mm)	8 lb/ft³ (128 Kg/m³)

Table 3. Monolithic Blown Refractory Ceramic Fiber & Microporous

Insulation Lining Details				
Item Color	Construction	Material	Thickness	Density
	Hot Face Material	AZS Monolithic Modules, Blown RCF	12" (305mm)	15 lb/ft³ (240 Kg/m³)
	Back-up Material	Microporous Board	1" (25mm)	14 lb/ft³ (230 Kg/m³)

Table 4. High Fiber Index Spun Refractory Ceramic Fiber

Insulation Lining Details				
Item Color	Construction	Material	Thickness	Density
	Hot Face Material	AZS Folded Modules, High Fiber Index Spun RCF	12" (305mm)	12 lb/ft³ (192 Kg/m³)
	Back-up Material	AS Blanket, High Fiber Index Spun RCF	1" (25mm)	8 lb/ft³ (128 Kg/m³)

The application conditions chosen for this case study are representative of the uppermost operating temperatures seen in the radiant section of a petrochemical cracking furnace, along with the typical environmental conditions as outlined by API 560 Section 11.2, "Refractory Lining System Selection Specifications"[1]. For the purposes of this case study, we will focus on the radiant wall insulation of a petrochemical cracking furnace.

Application Conditions

Location / Orientation:	Wall
Total Insulation Lining Thickness:	13in (330mm)
Hot Face Temperature:	2300°F (1260°C)
Design Classification Temperature:	2600°F (1427°C)
Ambient Temperature:	80°F (27°C)
Wind Speed:	0 ft/s (0 m/s)
Emissivity:	0.9
Cold Face Temperature Requirement:	180°F (82°C)

Definitions:

Hot Face Temperature: This is the operating temperature of the unit, and the temperature that the insulation lining will see at its hot face. Typical ethylene cracking furnaces operate at approximately 2200°F (1205°C), and occasionally higher[8].

Design Classification Temperature: This is an adjusted classification temperature for the insulation used based upon the refractory lining system selection specifications outlined in API 560 11.2.1.a. which state, "Design hot-face temperature shall be the calculated hot-face temperature plus 300°F (165°C), based on the maximum flue-gas temperature expected for all operating modes with an ambient temperature of 80°F (27°C) with zero wind velocity."

Ambient Temperature: The temperature of the environment surrounding the unit. This can vary depending on the unit location. Examples are room temperature, or warmer and cooler temperatures for outdoor conditions.

Wind Speed: This is the wind (if any) present outside on the "shell" or "cold face" of the unit, which can affect the thermal transfer. This is not on the interior of the unit, where the hot face is present.

Emissivity: Defined as the ability of a surface to emit (radiate) energy, and is a material specific property. This is taken into account at the shell surface, where the emittance properties affect the thermal transfer at the cold face.

Cold Face Temperature Requirement: This is the cold face temperature specification for the unit based upon the refractory lining system selection specifications in API 560 11.2.2.a. which states, "The temperature of the outside casing of the radiant and convection sections along with associated ducts, fans, air preheater and SCR shall not exceed 180°F (82°C) at an ambient temperature of 80°F (27°C) with zero wind velocity."

Results & Discussion

The insulation lining scenarios were analyzed using a heat transfer computer model calculation based on ASTM C680-14 standard[5]. Information presented here on the cold face temperature, heat loss, and heat storage should be used only as a reasonable approximation in order to compare various refractory linings. Other important but variable site conditions can affect these results but are not normally considered in these heat flow calculations. Some conditions may include non-steady state operation, high gas flow and pressure, anchorage, low emissivity paint, external steelwork and shrouding. The thermal conductivity values for each individual material used in this case study can be found in the appendix for reference.

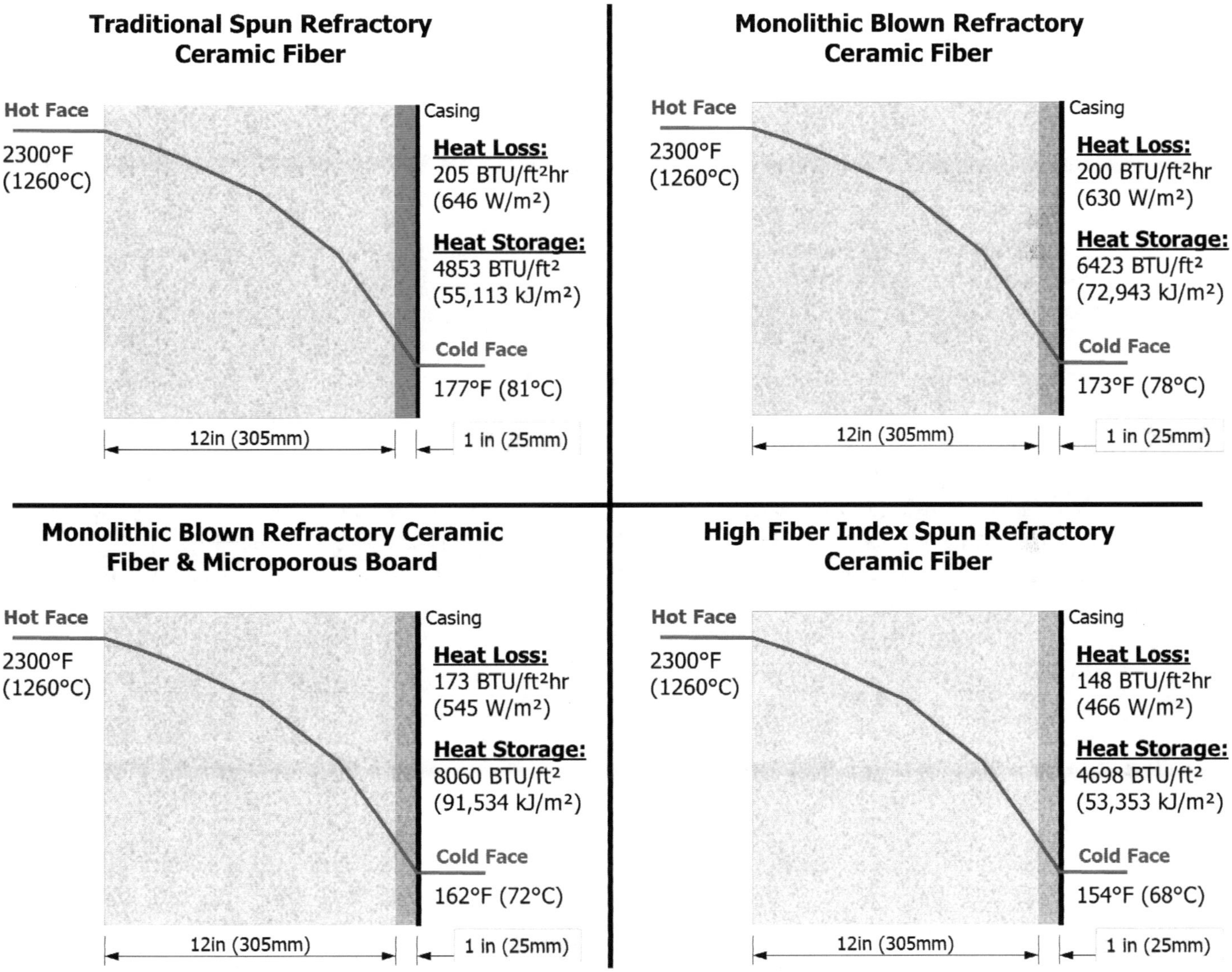

Figure 6. Thermal performance summary of insulation lining scenarios evaluated in case study.

How does this enhance the efficiency and thermal performance of a petrochemical cracking furnace? The superior insulating properties of the high fiber index refractory ceramic fiber allow operators of petrochemical cracking furnaces to reduce their overall energy consumption, emissions, save money, and improve safety. To quantify the benefits described, an example scenario is outlined below and the benefits are analyzed based off the thermal performance results of the insulation linings evaluated in the case study.

The following assumptions for an example petrochemical cracking furnace are shown below:

Furnace Cycle:	Continuous
Fuel Cost:	\$5.00 per 10^6 BTU
Cycles per year:	4
Cycle time:	2,016 hours (assuming 4 shutdowns per year that last a week long)
Lining Area:	5,000 ft^2 (465 m^2)

To understand the energy consumption, emissions, and overall energy losses of any insulation lining, the heat loss, heat storage, and available heat are the three main factors that need to be addressed in each approximation.

Heat Loss: This is the amount of energy per area, per hour that escapes through the insulation lining while the unit is operating. This varies with the furnace hot face temperature and insulation lining used, and will occur during the entire furnace cycle.

Heat Storage: This is the quantity of energy per area in the insulation lining during operation. This stored energy occurs only once during the cycle and will be lost upon unit shut down.

Available Heat: Defined as the gross quantity of heat released within in the combustion process minus both the dry flue gas loss and the moisture loss. It represents the quantity of "available" heat remaining for useful purposes and to balance losses due to walls, openings, etc. Both the heat loss and heat storage values are divided by the respective available heat values to provide a true heat loss and heat storage for the unit.

Energy Consumption & Emissions

Increasing demands are being placed on the petrochemical industry to reduce emissions in line with local regulations and international agreements. This makes the high fiber index refractory ceramic fiber a natural choice when it comes to increasing thermal efficiency and honoring corporate social responsibility commitments to sustainability. By incorporating the high fiber index refractory ceramic fiber into a petrochemical cracking furnace insulation lining, the heat losses seen through the insulation lining can be minimized, and in return the CO_2 emissions. Mitigating the amount of heat loss through the insulation lining has a direct relationship with the energy consumption required to operate the furnace, as well as the overall carbon footprint. The overall annual energy losses through the insulation lining scenarios and the contribution to CO_2 emissions are shown in the table below.

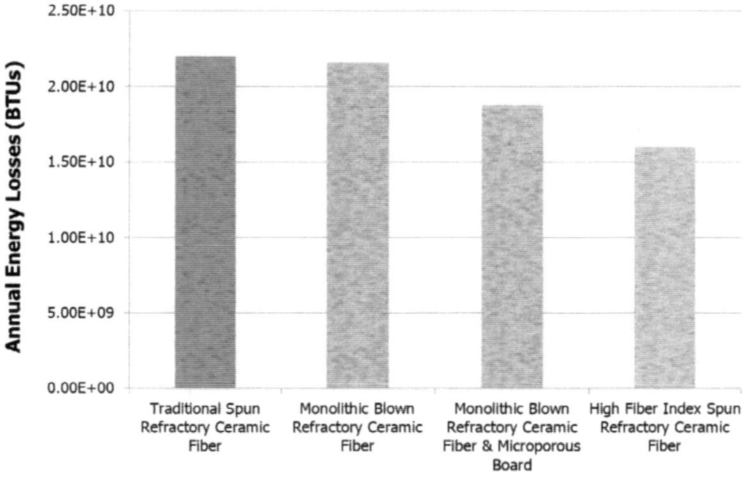

Figure 7. Insulation Lining Scenario

Insulation Lining Scenario	% Energy Consumption Reduction by using High Fiber Index Spun Refractory ceramic fiber
Traditional Spun Refractory Ceramic Fiber	38%
Monolithic Blown Refractory Ceramic Fiber	35%
Monolithic Blown Refractory Ceramic Fiber & Microporous Board	18%

Figure 7. Annual energy losses for each insulation lining scenario and energy consumption reduction when using high fiber index refractory ceramic fiber in comparison to other linings.

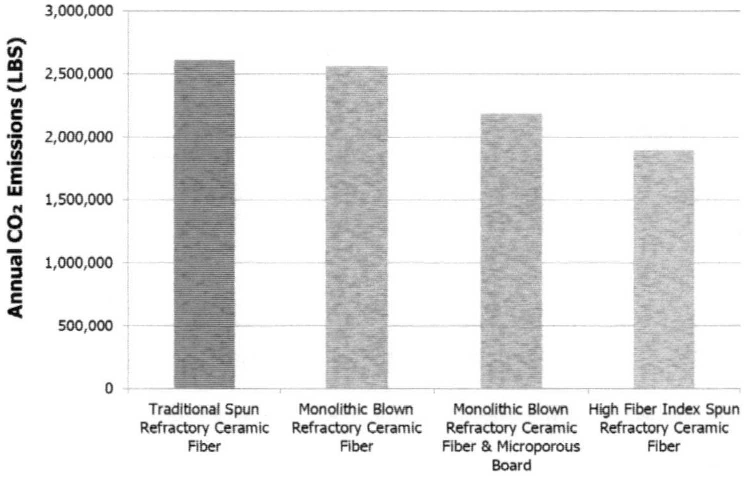

Figure 8. Insulation Lining Scenario

Insulation Lining Scenario	% CO_2 Emission Reduction by using High Fiber Index Spun Refractory ceramic fiber
Traditional Spun Refractory Ceramic Fiber	38%
Monolithic Blown Refractory Ceramic Fiber	35%
Monolithic Blown Refractory Ceramic Fiber & Microporous Board	18%

Figure 8. Annual CO_2 emissions for each insulation lining scenario and emission reduction when using high fiber index refractory ceramic fiber in comparison to other linings.

Energy Cost Savings

With rising worldwide energy prices, the reduction in energy consumption has a direct relationship with energy cost savings to the end user. In Table 5 below, the total annual energy costs sustained for each lining scenario are summarized, along with the possible energy savings benefit of using the high fiber index spun refractory ceramic fiber in comparison to the other insulation lining scenarios. The predicted savings assume adequate furnace controls to compensate burner openings for reduced fuel requirement. If the proper fuel / air ratios cannot be maintained at lower flow rates, savings may be reduced. Actual savings will depend on furnace operations, controls, refractory condition, and other such variables.

Table 5. Annual costs due to energy losses (heat loss, heat storage, available heat) of each insulation lining scenario and possible energy savings when using the high fiber index refractory ceramic fiber in comparison to the other linings.

Insulation Lining Scenario	Total Annual Energy Losses ($USD)	Potential Annual Energy Savings using High Fiber Index RCF ($USD)	Predicted Energy Savings using High Fiber Index RCF over 15-Year Service Life ($USD)
Traditional Spun Refractory Ceramic Fiber	$110,035	$30,281	$454,215
Monolithic Blown Refractory Ceramic Fiber	$107,796	$28,041	$420,615
Monolithic Blown Refractory Ceramic Fiber & Microporous Board	$93,902	$14,148	$212,220
High Fiber Index Spun Refractory Ceramic Fiber	$79,754	N/A	N/A

NOTE: Total annual energy losses are dependent on the size of the petrochemical cracking furnace and will vary proportionally with size, in particular the area of the hot face insulation lining. The predicted energy savings over the example 15-year service life shown above are assuming a constant fuel cost over the entire time period. Modern ethylene plants typically have 5-10 cracking furnaces[7].

Cold Face Temperatures

The cold face temperature is extremely important to maintain the overall structural integrity of a petrochemical cracking furnace, along with the health and safety of the operators nearby. The objective of any insulation lining is to create a safe cold face temperature for the furnace to successfully operate over its lifetime. The added thermal performance qualities of the high fiber index refractory ceramic fiber allow for lower cold face temperatures to be met, creating safer workplace and operation conditions. For ease of review, the cold face temperatures of the insulation lining scenarios are shown below in Figure. 9.

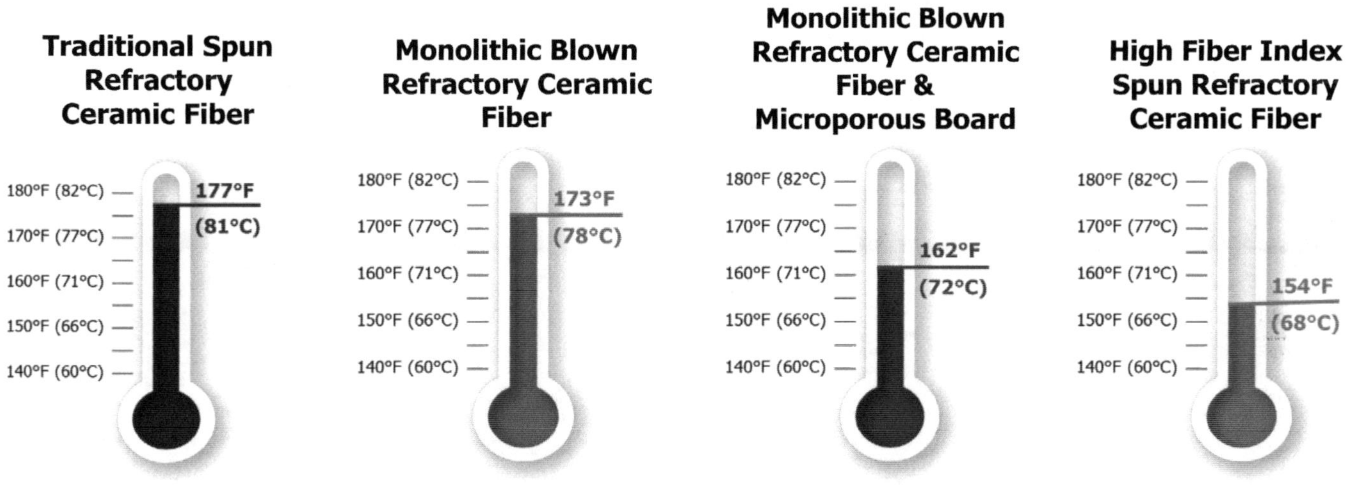

Figure 9. Cold face temperatures of all insulation lining scenarios.

Others

Raw Material Cost Savings

It is no longer required to use microporous insulation solutions to minimize energy consumption and cold face temperatures. Microporous insulation is significantly more expensive than refractory ceramic fiber, so considerable raw material cost savings can be achieved by incorporating the high fiber index refractory ceramic fiber in its place to satisfy enhanced thermal performance.

Installation Costs

No special masonry skills are required to install refractory ceramic fiber. Also, using a blanket back-up material in comparison to a microporous board back-up solution saves substantial time on installation and its complexity. Now that high fiber index refractory ceramic fiber performs better than microporous back-up thermal solutions, there is no need to consider the extensive installation steps necessary for a microporous board back-up solution.

Conclusion

The breakthrough innovation in refractory ceramic fiber technology is redefining the thermal performance of insulation linings used in petrochemical cracking furnaces. Operators now have the ability to further reduce their overall energy consumption, emissions, save money on energy cost, and improve safety while using a refractory ceramic fiber solution. The high fiber index and low shot content properties of the innovated refractory ceramic fiber material provides multiple benefits to end users because of its superior insulating properties, increased durability and tensile strength, and improved handling. In order for the petrochemical industry to continuously operate, step change improvements that reduce energy consumption, emissions, and boost efficiencies are required to make production a safer and more environmentally friendly process. This will depend on the help and collaboration with specialty materials leaders that enable customers to reduce fossil fuel consumption, save energy, and live greener. Innovative specialty insulation materials dedicated to sustainability and human health will enable the petrochemical industry to tackle some of the critical problems our society and environment face today.

References

1. American Petroleum Institute. API Standard 560. *Fired Heaters for General Refinery Service.* Fifth Edition. February 2016.

2. Jin, Y., Li, J., Du, W., & Qian, F. (2015). Integrated Operation and Cyclic Scheduling Optimization for an Ethylene Cracking Furnaces System. *Industrial & Engineering Chemistry Research, 54* (15), 3844–3854.

3. Rosemount Analytical & Emerson Process Management. *Improving Ethylene Plant Cracking Furnace Performance with Process Gas Chromatography.* Application Note, Chemical.

4. Fisher & Emerson Process Management. *Ethylene Production.* Chapter 1, page 1-15.

5. ASTM International. ASTM C680-14. *Standard Practice for Estimate of the Heat Gain or Loss and the Surface Temperatures of Insulated Flat, Cylindrical, and Spherical Systems by Use of Computer Programs.* Page 1-52.

6. Woodley, C. (2021, September 20). *Producing ethylene through a more environmentally safe process.* College of Engineering and Computing.

7. Tiziano Faravelli, Flavio Manenti, & Ranzi, E. (2019). *Mathematical modelling of gas-phase complex reaction systems: pyrolysis and combustion.* Elsevier.

8. Colannino, J. & John Zink Company LLC. (2008, January). *Ethylene furnace heat flux correlations.* Digital Refining Processing, Operations & Maintenance.

Appendix

Table A1. Thermal conductivities used for Traditional Spun Refractory Ceramic Fiber insulation lining scenario.

AZS Folded Modules, Spun RCF (12 lb/ft³)	
Temperature (°F)	K-value (BTU-in/(hr-ft2-°F))
392	0.569
752	0.735
1112	0.971
1472	1.179
1832	1.525
2192	1.941

AS Blanket, Spun RCF (8 lb/ft³)	
Temperature (°F)	K-value (BTU-in/(hr-ft2-°F))
392	0.361
752	0.555
1112	0.833
1472	1.249
1832	1.943
2192	3.095

Table A2. Thermal conductivities used for Monolithic Blown Refractory Ceramic Fiber insulation lining scenario.

AZS Monolithic Modules, Blown RCF (15 lb/ft³)	
Temperature (°F)	K-value (BTU-in/(hr-ft2-°F))
500	0.490
1000	0.840
1500	1.430
2000	2.190

AS Blanket, Spun RCF (8 lb/ft³)	
Temperature (°F)	K-value (BTU-in/(hr-ft2-°F))
500	0.400
1000	0.780
1500	1.360
1832	1.870
2000	2.150

Table A3. Thermal conductivities used for Monolithic Blown Refractory Ceramic Fiber & Microporous Board insulation lining scenario.

AZS Monolithic Modules, Blown RCF (15 lb/ft³)	
Temperature (°F)	K-value (BTU-in/(hr-ft2-°F))
500	0.490
1000	0.840
1500	1.430
2000	2.190

Microporous Board (14 lb/ft³)	
Temperature (°F)	K-value (BTU-in/(hr-ft2-°F))
392	0.152
752	0.187
1112	0.236
1472	0.305

Table A4. Thermal conductivities used for High Fiber Index Spun Refractory Ceramic Fiber insulation lining scenario.

AZS Folded Modules, High Fiber Index Spun RCF (12 lb/ft³)	
Temperature (°F)	K-value (BTU-in/(hr-ft2-°F))
392	0.550
752	0.680
1112	0.840
1472	0.970
1832	1.210
2192	1.480

AS Blanket, High Fiber Index Spun RCF (8 lb/ft³)	
Temperature (°F)	K-value (BTU-in/(hr-ft2-°F))
392	0.350
752	0.510
1112	0.710
1472	1.010
1832	1.430
2192	2.070

639918 Furnace Refractory Roundtable Discussions

Abstract:

Thomas Garza, PE

Fixed Equipment Engineering Manager

Chevron Phillips Chemical, LP

Robert (Bob) Pritchett

Fired Equipment Engineer

LyondellBasell

Chinh Dang

Furnace Mechanical Manager

Technip Energies

Prepared for Presentation at the 2022 Spring Meeting

San Antonio, Texas, April 14-14, 2022

REFRACTORY ROUNDTABLE DISCUSSIONS

Thomas Garza, PE

Fixed Equipment Engineering Manager

Chevron Phillips Chemical, LP

Robert (Bob) Pritchett

Fired Equipment Engineer

LyondellBasell

Chinh Dang

Furnace Mechanical Manager

Technip Energies

Abstract:

Refractory lined systems for ethylene furnaces are engineered systems that consist of monolithic refractories, high-temperature firebrick, and refractory ceramic fiber (RCF) to withstand aggressive service conditions at high temperatures. Refractory lined systems for ethylene furnaces are heat-resistant materials that are designed to protect the furnace casing plate from high flue gas temperatures during operations, as well as radiate energy onto the radiant coil tubes from the process burners. Refractory lined systems are exposed to high flue gas temperatures such as 2300°F (1260°C) at the floor of the radiant box to 2100°F (1149°C) at the furnace arch or transition section, while the convection section refractory can be exposed to temperatures ranging from approximately 2100°F (1149°C) at the shock bank to 300°F (149°C) as the gas exits the Hydrocarbon Preheat tube bank. Flue gas temperatures within the transition section ducting and the furnace stack are approximately 300°F (149°C) and less as the gas exits the furnace stack.

Exposure to these high temperatures can cause failure of the refractory lining system due to thermal expansion, thermal stresses, and normal deterioration (cracking, erosion, shrinkage, etc.) over the life cycle of the refractory.

This Refractory Roundtable Forum will discuss the various types of refractories, design parameters (e.g., thermal resistance, thermal expansion, mechanical strength, anchor materials selection, etc.), preventative maintenance and inspection techniques and the various repair methods available.

640296 Application of Fiber-Embedded, Polymeric-Based Materials As Pump Shaft Bushings in Abrasive Services

Abstract:

Application of bushings manufactured out of polymeric-based materials that offer self-lubricity, temperature and chemical resistance in critical, clean-service pumping systems (e.g. boiler feed water) is well documented and continues to be adopted by various organizations.

An area of opportunity has existed for dirty-service applications. Typically, bushings in said applications will be fabricated of softer materials as a sacrificial component (e.g. brass bushings, cutlass bearings, etc.) and often have more generous clearances to allow passing of entrained debris and/or reduce shaft wear, while also affecting pump efficiency/performance and at times, pump rotor stability.

Use of polymeric-based materials designed for clean-services is not well suited for applications with solids/debris entrained in the service, as these will be eroded/worn in a relatively short time.

Fiber-embedded polymeric materials offer a solution by providing abrasion/wear resistance, related to the material properties of the fibers used, while also providing self-lubricity, chemical and temperature resistance seen in traditional, clean-service materials. Additionally, by allowing clearances to be brought in closer to (or meeting) API recommendations, pumping systems often benefit from improved performance and rotor stability.

This paper will address application of two types of fiber-embedded polymeric materials as shaft bushings in two vertical lift pumps, in dirty/slurry-type services

On the Application of Advanced Pressure Actuated Seals in Legacy Turbomachinery

Rich Armstrong
Turbine Consultant
rarmstrong.cmgtech@gmail.com

Clayton Grondahl
President, CMG Tech, LLC
cmgtech@earthlink.net

Rich Armstrong is an experienced power machinery business leader having worked at GE for 17 years as well as for a Defense Company for 12 years focused on Naval Ship propulsion systems. At GE he worked in the Power/Energy and Oil and Gas groups spending much of his career focused on the commercialization of new products. Similarly in the Defense sector Rich lead a business unit that was advancing the state of the art in power and propulsion. For the past 3 years Rich has been supporting CMG Tech focused on bring the Pressure activated leaf seal to market. Rich holds a Masters in Mechanical Engineering and started his professional career as a Naval Nuclear Test Engineer.

Clay Grondahl is president of CMG Tech, LLC that is committed to advanced seal technology development and its application. At GE Power Systems he provided engineering and project management for the modification and performance improvement of operating GE gas turbines with installation of new and advanced technology components including seals. Earlier in his GE career he designed gas turbine rotating components and contributed to several advanced technology projects at the GE Corporate Research and Development Center and in GE Power Systems. His career started in the US Navy where he served on nuclear submarines making Polaris patrols and qualifying as Chief Engineer for the onboard nuclear steam plant.

Abstract

Seal technology for reduced shaft leakage continues to be developed and utilized in turbomachinery for improved performance. This lecture is focused on the consideration of retrofitting legacy equipment with advanced shaft sealing to recover performance and extend efficient operation. Several sealing concepts benefit from system pressure to close operating seal clearance. Of these, the Pressure Actuated Leaf Seal (PALS) is particularly attractive for retrofit in legacy equipment where misalignment and distortion are present. Installation of fixed seals require larger clearance to avoid rubs where there is ovality. The lecture will discuss PALS design for out of round conditions in which material is provided for sacrificial leaf tip wear-in on initial startup, similar to an abradable seal. Subsequent operating clearance is non-contacting and clearance approaches rotor run-out for greatly reduced seal leakage. Startup and shut-down rubs are avoided by flexure of PALS leaves on their support in response to differential pressure. Leaf tip closure with pressure is designed to provide clearance passing through transient thermal distortion and critical speeds. PALS configuration options are flexible to accommodate available seal space. PALS are robust in that they can ride over surface discontinuities making them a strong candidate for blade tip shroud sealing.

Introduction…should I consider advanced seals in my legacy unit?

Seal technology for reduced shaft leakage continues to be developed and utilized in turbomachinery for improved performance. Legacy turbomachinery shaft seals often struggle to maintain performance due to rubs from:

- casing distortion
- misalignments
- increasing start-up and shutdown natural frequency displacement

The real world is tough on seals. A sample of 27 steam turbines from Encotech (2007) found clearances 2-4 times nominal design of 0.015 inches as shown in Figure 1. GE publication by Cofer, et al (1996) noted "The internal mid-span (N2 packing) and first three HP and IP stage diaphragm packings are susceptible to rubs during startup. … design clearance for these packings is 0.015 inch, but after five years operation, are typically 0.060 inch. For GE E-class gas turbines Johnston (2000) states that for compressor discharge high pressure packing "In practice, most operating units have clearances significantly higher (20 to 60 mils) than nominal".

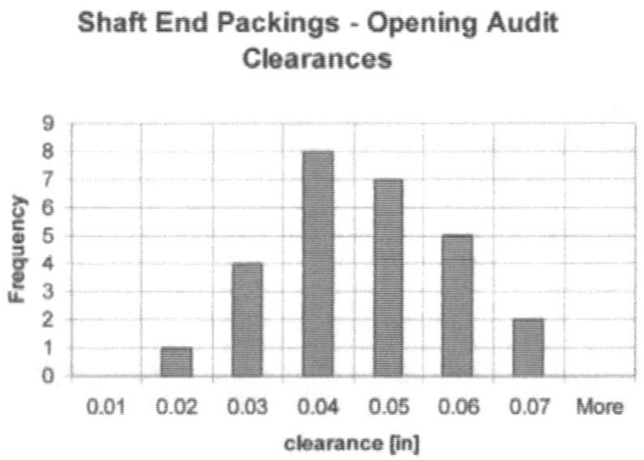

Figure 1 Legacy Steam Turbine Clearances

Advanced Seals typically will have physical clearances an order of magnitude less than Labryinth seals (e.g. 0.004 inches) but since labyrinth packing typically has numerous seal teeth, an "effective" clearance comparison in many applications will show that the advanced seals may have similar or are only slightly advantaged in effective clearance vs an "as designed" labyrinth as stated in Encotech (2007). The significant benefit coming from the advance seals is their ability to maintain these like new clearances despite real life conditions such as rubs that often occur during transient operations. It is notable that with increasing renewable power generation installed, many legacy units are being used in a more cyclic type application putting further stress on maintaining seal clearances.

Cofer (1996) stated that there was a 0.35% loss in unit heat rate for the 0.060 inch rubbed seal condition cited above vs 0.015 inch design seal clearance for that 500 MW reheat unit.

While most sources (e.g. Chupp (2006)) agree on the magnitude of these "real life/operating" vs "as designed" clearances, there is a significant range or variation when it comes to assessing the impact or benefits. It will be shown shortly that even using conservative benefits the paybacks associated with advance seals are significant.

The summary below was created from publicly available statements regarding the output and efficiency impacts of using advanced seals/retractable packing on legacy turbomachinery:

Retractable packing performance:	
Brandon "Retractable Packing" performance: (Power Engineering, August 2004)	Cites a NYSERDA study that found Brandon Retractable Packing "produced a 1% increase in turbine operating efficiency and a 2% increase in capacity". Also stated: "some users reported gains in capacity of 2-3%." more than 600 installations world wide.
TurboCare Inc "Retractable Packing" performance: (from 2001 web pages)	"typical efficiency improvements of 1 to 2% in heat rate and 2 to 3% in kW output". - 400 steam turbine installations ranging from 1,300HP to 1,100MW.
TurboCare Inc "Retractable Brush Packing" performance, i.e., retractable labyrinth seal packing with brush seals: (from 2006 web pages)	"up to 3% in unit efficiency and 4% in kW output". - over 600 (retractable packing) steam turbine installations.

Figure 2 Industry Statements on Advanced/Retractable Seal Performance

Another comparison can be taken from the Encotech (2007) report of 27 steam turbine units in which performance calculations of advanced seals versus real-life clearances is compared. Their methodology was based on a stage-by-stage, component-by-component analysis at constant unit steam flow. They calculated the performance impact of rubbed N2 packing and three HP stages of inter-stage packing in a 550 MW, 3500 psi, opposed flow steam turbine. Assuming those labyrinth seals are rubbed to a clearance of 0.060 inch vs. 0.015inch design clearance, the net performance benefit from a retrofit of labyrinth shaft packing with an Advanced Seal (PALS) in this steam turbine was calculated to produce ~ 0.5% improvement in heat rate and 1% in output.

A couple of points of caution when reading these above apparently wide-ranging performance benefits:

1. % efficiency is not the same as % heat rate nor points of efficiency. For example, a 1.5% efficiency change on a 33% efficient unit is that same as a 0.5% change in heat rate or points of efficiency.
2. Physical clearance data is usually taken at limited points of access and not in the fully constrained conditions of an operating unit and hence may not fully reflect the clearance in operation.

When it comes to evaluating the benefits, there are at least 2 components to the hard cost economic advantage:

- First is unit efficiency, meaning less fuel cost for the same output. As an example, a 0.5% improvement in heat rate on a 100 MW unit running at 10,000 BTU/kW-hr for 2000 hours/year would yield a fuel savings of **$20,000/yr** assuming a modest fuel cost of $2/BTU*10^6. In terms of net present value (NPV) at a typical 7% cost of capital this $20k/yr fuel savings results in an **NPV of over $1M** in benefit assuming a 7 year horizon. Another reference from a large Turbo machinery OEM stated that "A typical 10,000 hp compressor could save as much as $120,000/year in energy costs by upgrading to advanced seals". It should be noted that $20k/yr vs $120k/yr may not be so different understanding that the $20k/yr is based on a single seal/packing change out vs the $120k/yr is based on changing multiple seals in a given unit.

- The second benefit is the output increase. This is benefit could be extremely large or relatively unimportant depending on plant needs or power sale contracts. It seems appropriate to use the market average case of the typical capital cost of power plant. Assuming a conservative $500/kW and 1% output improvement this would yield an **NPV of $0.5M** in benefit for a single seal change out as described above.

Since each case should be evaluated on its own specifics, the point trying to be made is not one of precision, but rather, order of magnitude. Most advanced seals do have a cost premium vs the status quo, however, this premium is significantly less (5 to 10 times less) than the savings making a compelling case for consideration of advance seals on legacy turbomachinery where performance is typically not in an as designed condition.

Seal Technology Overview...which seal is best in my situation?

There are multiple types of seals associated with rotating equipment. Chupp (2006) presents a comprehensive summary of conventional and advanced dynamic sealing in turbomachinery at that time. As noted by Pedraza-Dalle, et al. (2019), recent development has been focused on seals that are non-contacting and adaptive that are designed to avoid rubbing under all operating conditions. For the purposes of this paper, we will focus on advanced clearance seals with technology readiness suitable for shaft and tip seal applications in turbomachinery and retrofit in legacy equipment including one non-contacting and compliant seal. The basic seal types selected meeting those criteria are briefly described below:

Labyrinth Seals are the most common type of seal where multiple clearance teeth/stages are used to step down the pressure. Tolerancing associated with manufacturing, thermal growth in operation, start-up shaft movement and vibration limit the tightness of these seals but they are widely used due to their simplicity and long life. They come in a variety of configurations shown in Figure 3 from Burcham (1979). Their use in retrofit of legacy equipment includes custom fit to accommodate observed wear patterns as well as application in combination with other advanced seals or enhancements such as the 'Guardian' ™ packing ring shown in (Figure 4) that is designed to protect labyrinth seal teeth during start-up transients.

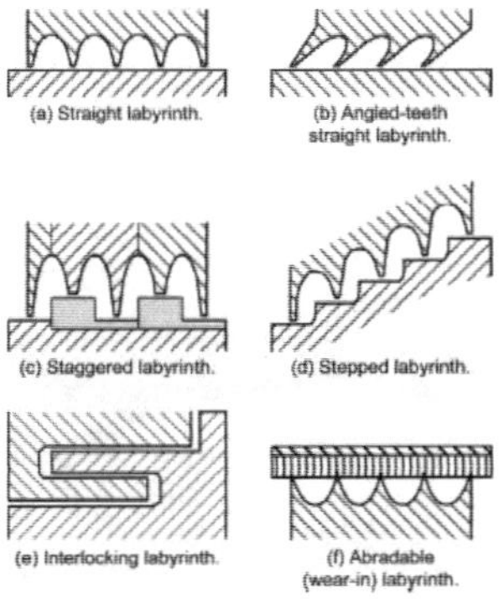

Figure 3 Labyrinth Seal Concepts

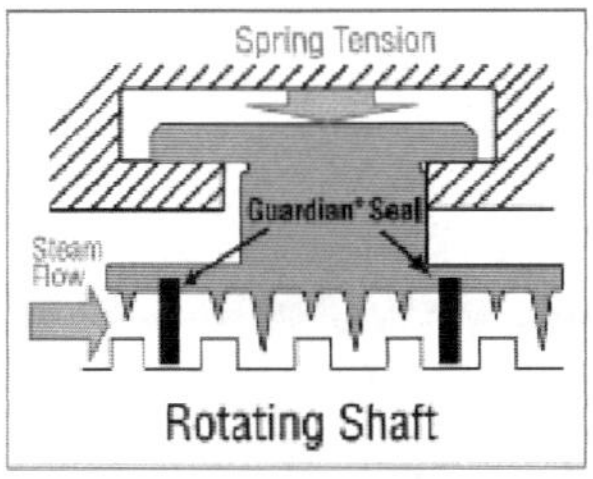

Figure 4 'Guardian' ™ Packing Ring (MD&A Turbines, 2021)

Abradable Labyrinth Seals, illustrated in Figure 3 (f), are labyrinth seals where a honeycomb or other abradable material is applied to the stator to create a sacrificial wear surface for rotating seal teeth to engage at a smaller operating clearance without risking their damage or deformation from metal-to-metal contact. Abradable sealing may also refer to thermal sprayed abradable material on gas turbine stator shroud surfaces for reduced clearance with unshrouded blades. In addition to performance gain from reduced operating clearance that practice can also correct turbine shell out of roundness.

Brush Seals are noted for being 'the first simple and practical alternative, to the finned labyrinth, for gas turbine air system seals', Ferguson (1988). Figure 5 is an axial view of a brush seal segment. They have provided substantial performance improvement over labyrinth seals due to reduced leakage of bristles in constant contact with the rotor that flexibly accommodate startup, shut down and other transient conditions. Brush seals are in common use in advanced turbo machinery applications because of their performance benefits and durability as described by Turnquist et al. (2005).

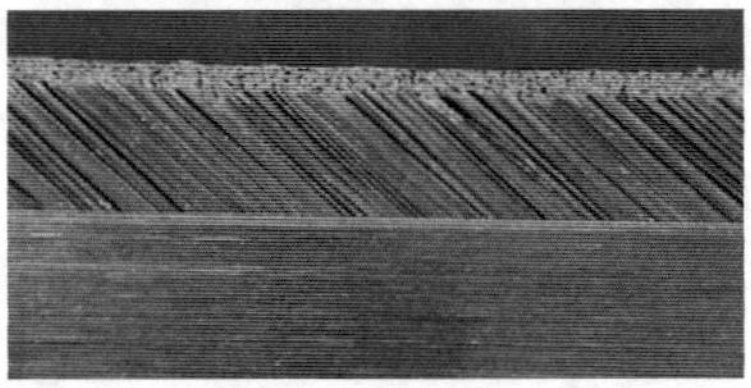

Figure 5 Brush Seal Bristle Pack Segment

Retractable Packing is another version of the labyrinth seal in which seal segments are spring loaded away from the rotor is depicted in Figures 6 and 6a. In the retracted position they have sufficient clearance to avoid start up rubs. As pressure increases to operating conditions the back side of the seal ring is pressurized to close the clearance with the rotor to close running design values. Since its inception in the 1980's retractable packing has also been used with brush seals. A key challenge with this approach is tolerancing to maintain the ability to overcome unplanned frictional forces. Retractable packing is typically applied in high pressure and intermediate pressure stages where there is sufficient pressure force to move the packing rings. Accurate diaphragm alignment is required for optimal performance as seal

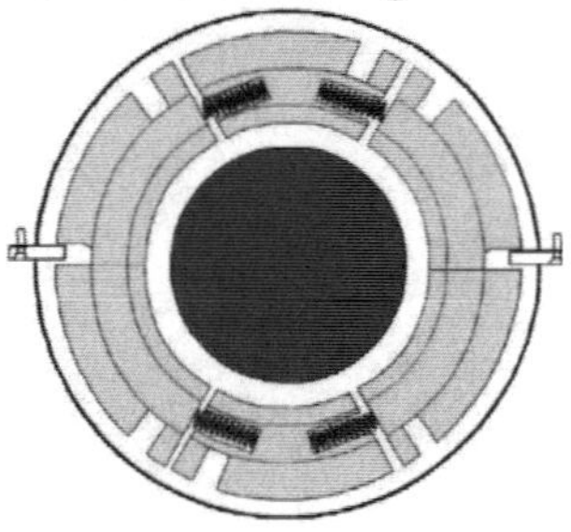

Figure 6 Retractable Packing Concept

Figure 6a Retractable Packing

damage can still occur at operating conditions if seals come in contact with rotor as noted by Little et al. (2001).

HALO Seals, the anacronym derived from Hydrostatic Advanced Low Leakage Seals, are a novel non-contacting, compliant seal. It is the invention of Justak (2011 and 2014). The seal is installed with a large radial clearance that reduces clearance under pressure (less than 5psid). The dynamic seal then self-adjusts to maintain a close clearance under all operating conditions, including dynamic rotor offsets. The seal is pressure balanced and in addition to low leakage the seal also provides rotor-dynamic damping as described by Justak (2013). The photo shown in Figure 7

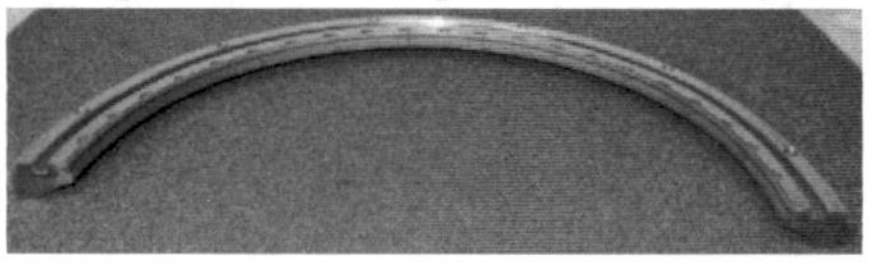

Figure 7 HALO seal segment

is also from Justak (2013). HALO development, testing at elevated temperature, 570F (300C), and application history is documented by Andres and Anderson (2014). They have been successfully applied in Dresser-Rand Datum compressors as described by Justak (2013) and Stiles (2016). The seal arrangement shown in Figure 7a is from Stiles (2016). The impact of seal clearance control has been validated to lower power input requirement by 2%.

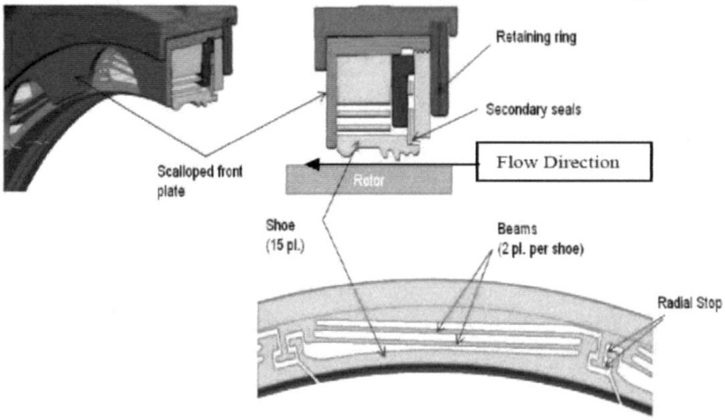

Figure 7a HALO seal arrangement

Pressure Activated Leaf Seal (PALS) shown in Figure 8 is one of the newest of the advanced seals being discussed. PALS uses differential pressure to bend thin sealing leaves in an arc to “engage” the shaft as pressure develops across the seal. Similar to retractable packing, PALS are designed to provide a large starting clearance to avoid startup rubs and then close to a small running clearance at operating conditions. PALS also incorporates an

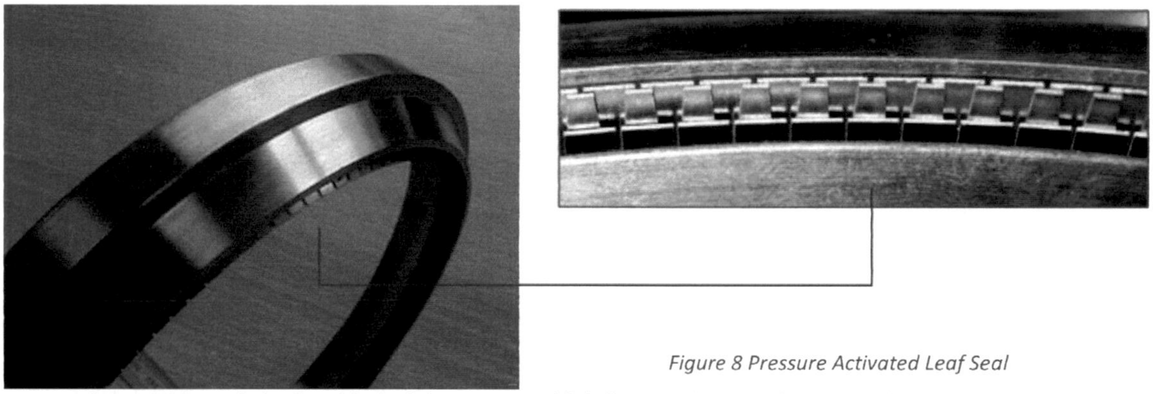

Figure 8 Pressure Activated Leaf Seal

aspect of abradable seals in that thin leaf tips may sacrificially wear into conformity with the rotor seal surface upon initial startup. This ‘wear-in’ capability mitigates the seal designers conflicted interest to provide as tight a dynamic seal as possible that does not rub due rotor dynamics, seal misalignment or stator runout. Subsequent to ‘wear-in’ PALS is a non-contacting seal and resiliently relaxes to a large clearance upon shut down.

Basic PALS construction is illustrated in Figure 9. It is composed of thin seal leaves, slotted and layered like shingles to block flow, a support member with a curved surface and backing ring joined with leaves and support to mount the assembled PALS. The seal is arranged circumferentially around the shaft with seal leaves facing upstream pressure. Seal leaves are elastically deflected in an axial direction by system differential pressure to close radially with the shaft, preferably above critical speeds and below operating pressure. When leaves are deflected into contact with the support member further seal closure is limited by the short cantilever leaf length remaining between the support ID and shaft. Leaves are designed and manufactured longer than anticipated eccentricities of the seal with respect to the operating rotor e.g., seal misalignment, run-out, wear into conformity with the rotor seal surface upon initial startup as noted above. PALS seal performance is independent of direction of rotation and rotor speed and can be designed for small or high operating differential pressure and nearly any seal diameter. PALS is the subject of several patents and extensive technology readiness testing at Cross Manufacturing is reported by Bowsher et al. (2014).

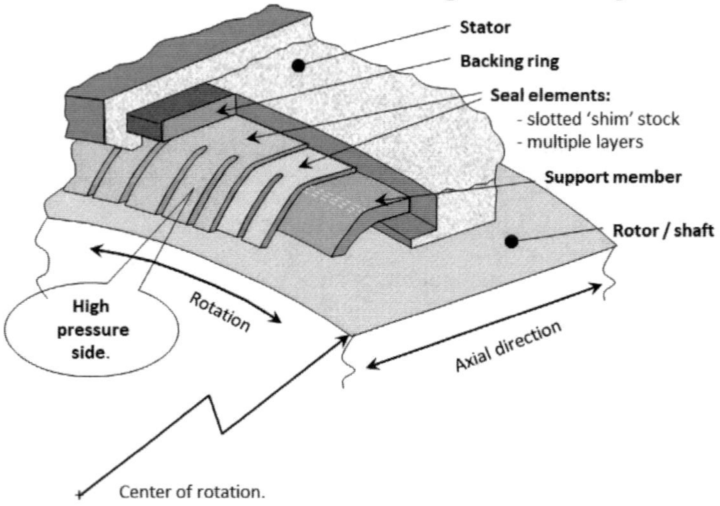

Figure 9 Pressure Activated Leaf Seal Illustrated

In terms of performance, testing has established that PALS is quite similar to a brush seal in effective clearance…essentially rotor run out creating the limit of effectiveness. An illustrative plot of seal closure versus pressure is shown in Figure 10 along with a prototypical seal section in Figure 11.

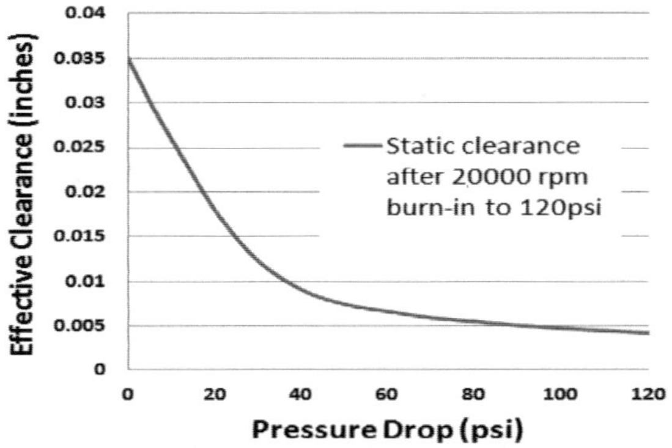

Figure 10 Effective Clearance of Pressure Activated Leaf Seal

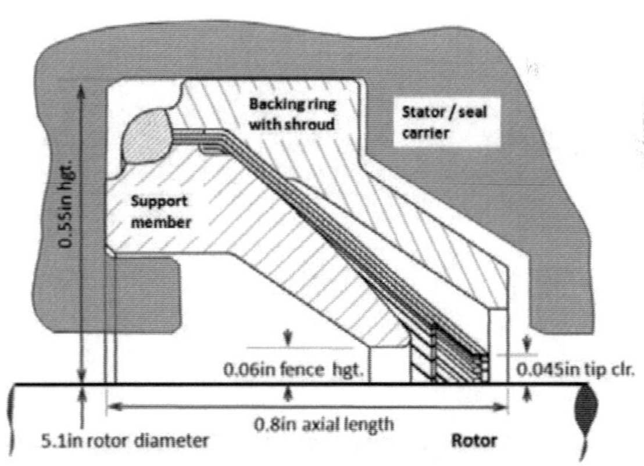

Figure 11 Pressure Activated Leaf Seal Section

Comparison of advanced and conventional shaft and blade tip seals

The below table and subsequent explanations are a qualitative comparison showing the advantages and weaknesses of the various technologies. A strength or weaknesses inherent to the design or operating experience for each seal type are indicated by an up arrow for positive design features and operating experience and a down arrow for negative design feature or unfavorable experience.

	Startup Rub Hazard	Operating Rub Tolerance	'Wear-in' Compliance	Low Leakage	Long Seal Life	High ΔP Capability	Tip Shroud Sealing	Axial Length	Rotor Dynamic Issues	Cost
Labyrinth Seals	↓	↓	↓	↓	↑	↑	⇑	↓	↓	↑
Abradable Labyrinth Seals	⇑	⇑	↓	⇑	↑	↑	↑	↓	↓	⇑
Brush Seals	↑	↑	↑	↑	⇑	⇑	↓	↑	↑	↓
Retractable Packing	↑	↓	↓	↓	↑	↑	↓	↓	↓	↑
HALO™ Seals	↑	↑	↑	↑	↑	⇑	⇑	↑	↑	↓
Pressure Actuated Leaf Seals	↑	⇑	↑	↑	↑	↑	↑	⇑	↑	⇑↑

Symbol key: ↓ Negative Vs other seals. ↑ Positive Vs other seals. ⇑ Somewhat favorable Vs other seals.

Table 1 Dynamic Seal Types Qualitative Comparison

Startup rub vulnerability:

Labyrinth seals	↓ often damaged by unplanned rubs compromising their effectiveness
Abradable seals	↑ designed to accommodate startup rubs but friction in retracting system can cause rubs
Brush seals	↑ bristles accommodate startup clearance changes by their deflection
Retractable	↑ typically designed to avoid start-up and shutdown rubs
HALO	↑ pressure differential activates seal to avoid start-up and shutdown rubs
PALS	↑ pressure differential activates seal to avoid start-up and shutdown rubs

Operating rub tolerance:

Labyrinth seals	↓ teeth are permanently deformed when rubbed
Abradable seals	↑ seal tooth damage can be tolerated but with possible performance loss
Brush seals	↑ bristles are designed to resiliently deflect on rubbing contact with acceptable wear
Retractable	↓ designed to resiliently deflect however mushroom effect does occur
HALO	↑ hydro-dynamic buffer inherently prevents metal to metal contact
PALS	↑ seal elements can resiliently deflect, with wear dependent on materials, pressure and speed for short durations without significant performance impacts

'Wear-in' seal compliance:

Labyrinth seals	↓ seal teeth are damaged if rubbed
Abradable seals	↓ 'wear-in' is tolerated but without seal clearance compliance
Brush seals	↑ bristles can resiliently deflect into compliance with the rotor
Retractable	↓ 'wear-in' is tolerated but without seal clearance compliance
HALO	↑ hydro-dynamic buffer inherently adjusts and therefore is wear in compliance is not necessary
PALS	↑ seal leaf tips worn into compliance with the rotor effectively eliminates misalignment of seal mounting components

Low leakage:

Labyrinth seals	↓ large seal clearance needed to avoid rubs or is created by unplanned rubs
Abradable seals	↑ effective seal clearance is reduced where there is adequate axial rotor movement between startup wear tracks and operating seal tooth path
Brush seals	↑ seal leakage ~10 - 20% of a 4 tooth labyrinth seal is often achieved
Retractable	↑ improved vs labyrinth but limited vs Brush or Pals due to limited avoidance capability and dependance on friction to activate
HALO	↑ non-contacting with small effective hydro-dynamic clearance and does not depend on friction
PALS	↑ non-contacting, shingled leaf seal members achieve small effective operating clearance… similar to brush seals and does not depend on friction

Long life:

Labyrinth seals	↑ typically have long life when operated without rubs.
Abradable seals	↑ long seal life unless there are significant unplanned rubs
Brush seals	↓ brush seals have reduced bristle contact wear and expected life has improved to 8+ years
Retractable	↑ typically have long life when operated without rubs
HALO	↑ long life expected due to non-contacting nature
PALS	↑ initial 'wear-in' of leaf tips provides for non-contacting under normal operating conditions; and resilient leaf deflection contribute to long life expectancy

High differential pressure capability:

Labyrinth seals	↑ operate dependably at high differential pressure
Abradable seals	↑ operate dependably at high differential pressure
Brush seals	↓ single stage designs limited to less than 400 psi differential pressure
Retractable	↓ HP same as Labyrinth but has challenges in LP due to overcoming stiction
HALO	↓ capability up to 660 psid
PALS	↑ available in high differential seal pressures (1000+ psi)

Tip Shroud Sealing:

Labyrinth seals	↑ traditional approach but significant rub hazard
Abradable seals	↑ addresses labyrinth challenges but adds cost and maintenance
Brush seals	↓ at risk of damage to bristles with interrupted shroud sealing surface
Retractable	↓ not applied due to low differential pressure and large diameters
HALO	↑ high cost due in part to large diameters
PALS	↑ capable of riding over shroud discontinuities and cost-effective at large diameters

Axial seal length:

Labyrinth seals	↓ inherently long because sealing is derived from small pressure drops across a succession of seal teeth. Seal-packing length can adversely affect bearing span
Abradable seals	↓ length is similar to other labyrinth seals
Brush seals	↑ seal with minimal axial length with a single pack of radial bristles
Retractable	↓ length is similar to other labyrinth seals
HALO	↑ similar to brush seal
PALS	↑ require less axial length than most labyrinths but more than brush seals

Rotor dynamic issues:

Labyrinth seals	↓ circumferential flow within seal cavities can incite rotor instabilities
Abradable seals	↓ similar to Labyrinth
Brush seals	↑ stabilizing effect on rotor dynamics though an improperly designed brush seal can induce thermal distortion of a rotor while starting up by extended friction heating
Retractable	↓ similar to Labyrinth
HALO	↑ a single plane of seal does not induce rotor instability
PALS	↑ a single plane of seal leaves do not induce rotor instability and friction heat input on startup is minimized by accelerated wear-in of thin leaf tips

Cost to manufacture:

Labyrinth seals	↑ typically not expensive to manufacture
Abradable seals	↑ features such as honeycomb rub strips and coated teeth add cost
Brush seals	↓ large quantity of small diameter bristles that must be oriented, closely packed and secured contribute to high manufacturing cost
Retractable	↑ some cost premium vs Labyrinth
HALO	↓ difficult to manufacture with complex EDM process
PALS	↑ leaves fabricated from sheet metal and the assembly of a small number of parts is expected to be cost effective in production

No one seal is "best" in all situations but due to the challenges in providing shaft and tip seals for legacy equipment advanced compliant type seals can offer significant advantages. The HALO and the Pressure Activated Leaf Seals (PALS) solve key challenges of other high performance advanced seals including:

- 'Wear-in' compliance: seals out-of-round shells*
- Startup rub avoidance: faster startup potential*
- Non-contacting running clearance ~ 0.004in
- Long life expectancy
- Low leakage performance gain
- Unrestricted\ design pressure
- No reverse rotation hazard
- Demonstrated shroud seal capability

*Often the most challenging issue in legacy equipment is casing distortion and start-up vibrations and in these situations the Pressure Activated Leaf is a solution worthy of serious consideration.

The HALO being advantaged in axial space and the PALS being advantage in high differential pressures, interrupted surfaces/tip sealing as well as manufacturing cost.

Applications for Pressure Activated Leaf Seals – Where are PALS being used or considered?

The PALS technology is being evaluated and tested for a broad range of other turbomachinery applications due to its ability to provide a non-contacting, cost effective solution where high performance/low leakage seals are needed. PALS most promising application areas include:

Machine Type	Locations	Comments
Compressors	Eye Seals, End Seals, Balance Pistons, Dry Gas Back-up	2021 application testing
Steam Turbines	Interstage Shaft, Shaft End Seals, Tip Shroud, N2 (HP/LP mid span) Shaft, Snout Ring	Evaluations for new units and retrofits, 3rd party testing
Gas Turbines	Compressor Discharge, Interstage Turbine Shaft, Tip Shrouds	2013 simulated shrouded turbine blade testing
Expanders	Shaft Seals, Dry Gas Seal Back-up	
Pumps	Eye Seals wear rings	2015 testing for water
Turbo Compressor	Various rotating locations	sCO2 in machine testing, Stapp (2018). Further 3rd party testing planned for 2021

Table 2 Applications for Pressure Activated Leaf Seals

To date PALS has been designed and tested to cover differential sealing pressures ranging up to ~2000 PSI with most of the testing done at ~100PSI. There is not a fundamental limit to the shaft diameter size range. Designs and testing have also been done as small as ~1 inch diameter and as large as ~7 inches. Testing has been done with air, CO2, water and steam. Temperature capability is similar to other seal technologies. There are multiple design variations and tuning parameters with which to work. In terms of differential pressure the most stright forward parameter is to adjust the tooth thinkness

As noted in the Turbo Compressor section of Table 3, the Peregrine Turbine Technologies sCO2 application required high differential pressure capability, ~ 2000 PSI, without large secondary flow performance charges. The seal would also need to avoid cross-coupled stiffness and potential rotor dynamic instability in sCO2 as well as a seal that could accommodate shaft displacement modes on startup and shutdown using soft radial foil bearings. Stapp (2018) provides additional discussion regarding sCO2 sealing. The solution was selection of Pressure Actuated Leaf Seals (PALS) that ran open at low speed and pressure and then closed down on the rotor at design speed.

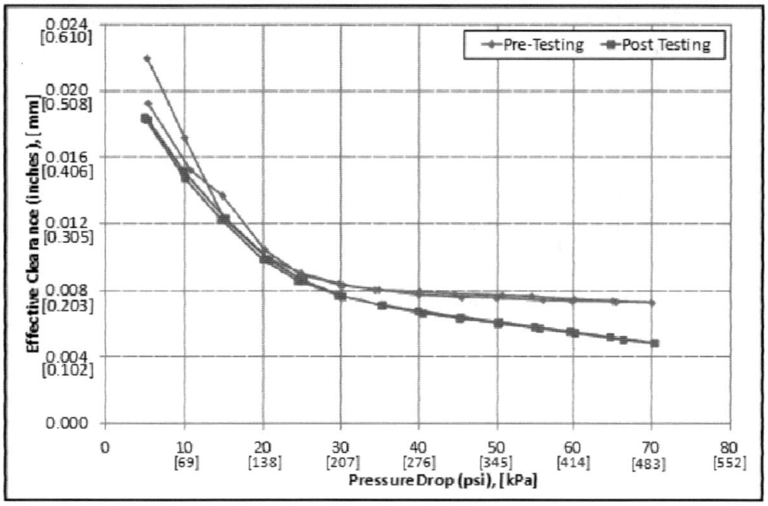

Figure 12 Tip Shroud Testing

Tip shroud sealing can be particularly expensive and challenging to maintain clearances in legacy units since most advance seals are costly at these larger diameters and are technically challenged in dealing with the interrupted surfaces and high speeds. The PALS technology is economically advantaged at large diameters and has shown significant promise in the testing done to date. The test results shown in Figure 12 from Bowsher et al. (2014) were taken before and after a 15 hour test using a dynamic rig with air and an interrupted surface to simulate a tip shroud.

PALS for Legacy – Should I consider using PALS in my machine?

In terms of dealing with casing distortion in legacy machines, the PALS design is well suited due to its relatively small tooth thickness which very quickly rubs off during initial start-up and hence conforms to the rotor from that point on in a non-contacting fashion. The below data and image were taken after a smooth rotor wear test at ambient temperature.

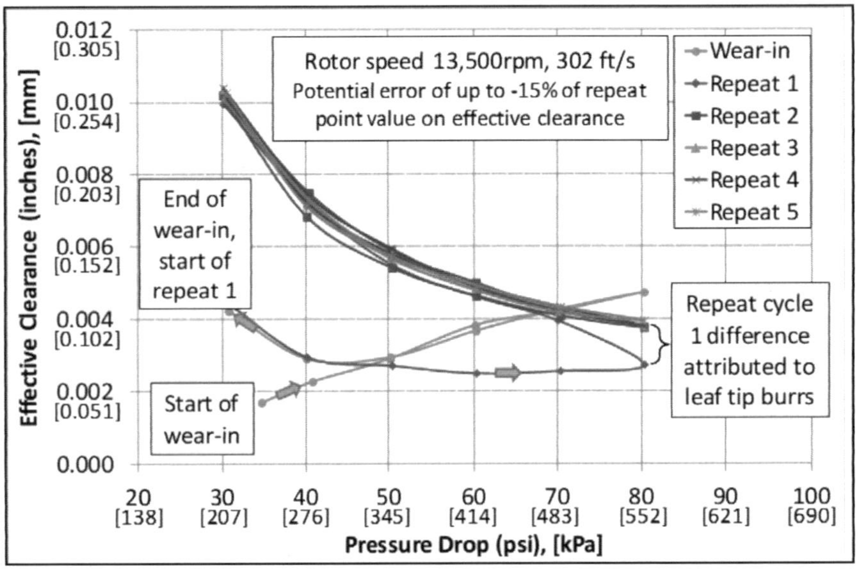

Figure 13 PALS Smooth Rotor Cyclic Testing

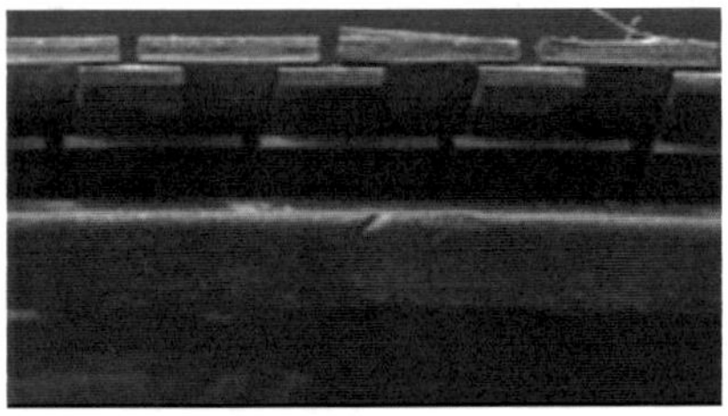

Figure 13a Post wear in PALS

Figure 13b Post wear in PALS

Conclusions from these tests:

- No significant rotor material loss or heating
- Consistent effective clearance after initial wear in
- Leaf tips wear to the rotor runout…minimal clearance
- Burrs form on both top and bottom leaves in both positive and negative axial direction

A major challenge in dealing with most any technology advancements in legacy equipment is the "space consideration". Brush seals are most advantaged in this factor since they typically take less space than a conventional labyrinth type seal and often are put in replace of only some of the labyrinth teeth giving a fall back to labyrinth type performance if there is an issue. PALS does take more space than a brush seal but has been manufactured and tested in 2 major orientations giving significant flexibility in terms or retrofitting into an existing machine as illustrated below.

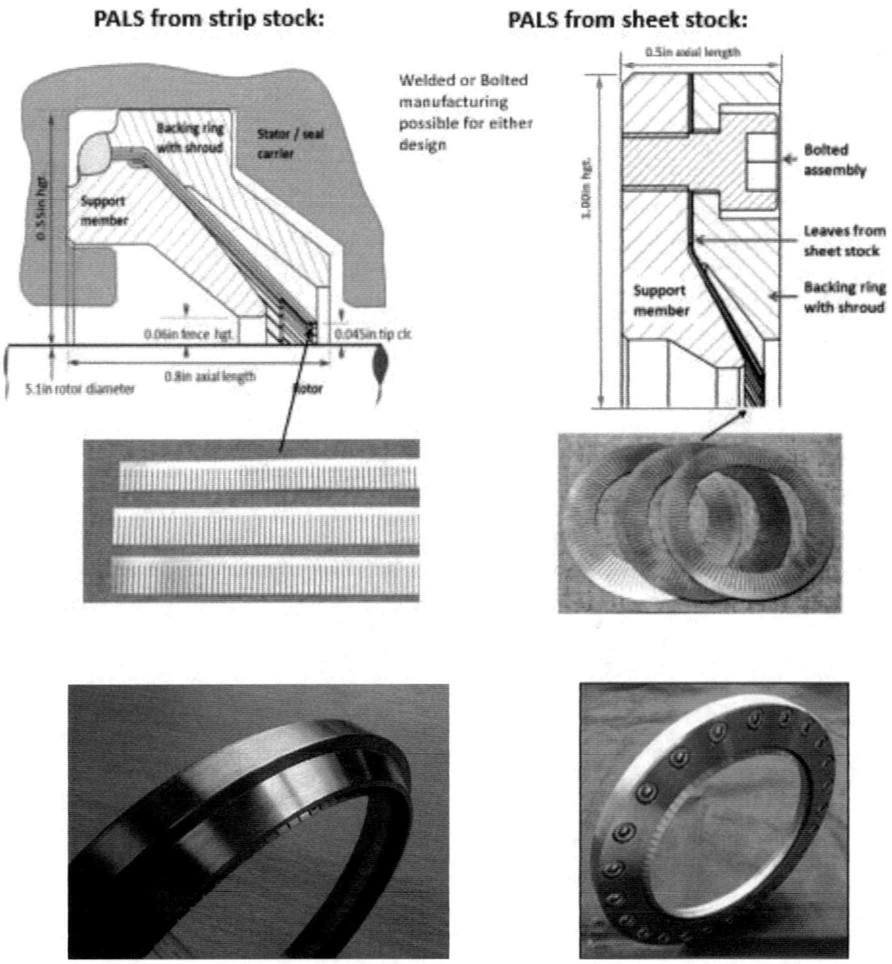

Figure 14 PALS Configurations

Summary and Next Steps

It is clear that advanced seals have a place in legacy turbo machinery due to their ability to extend high performance in many real-life conditions such as out of round casings and start-up rubs. It is also clear that when seals are not operating near their "as designed" conditions that the economics are highly attractive for advanced seals. The "best seal" depends on your specific application's needs. The pressure activated leaf seal (PALS), a cost-effective new comer to turbo machinery sealing, has some important advantages such as its 'wear-in' compliance with the shaft, high differential pressure capability and its non-contacting design during operation that make it a consideration in some of the more challenging applications.

ACKNOWLEDGMENTS

Cross Manufacturing for manufacture and technology readiness testing of Pressure Actuated Leaf Seals cited in this lecture.

NYSERDA, New York State Energy Research and Development Authority, is acknowledged for work sponsored under Agreement No. 9887 that demonstrated Pressure Actuated Leaf Seal feasibility.

Encotech, Inc. for their Performance Evaluation of Pressure Activated Leaf Seals in support of the NYSERDA project.

REFERENCES

Bowsher, A, Crudgington, P., Grondahl, C., Dudley, C., Kirk T., Pawlak, A., 2014, "Pressure Activated Leaf Seal Technology Readiness Testing", ASME-GT2014-27046.

Burcham, R. E., and Keller, R. B., Jr., (1978), "*Liquid Rocket Engine Turbopump Rotating-Shaft Seals*," NASA SP–8121; also cited in AIAA book (2014), "Turbine Aerodynamics, Heat Transfer, Materials, and Mechanics" Shih and Yang Editors, Chapter 3: "Turbomachinery Clearance Control," p 89.

Chupp, R. E., Hendricks, R. C., Lattime, S. B., and Steinetz, B. M., 2006, "*Sealing in Turbomachinery*," Journal of Propulsion Power, 22 (2), pp. 313-349.
Cofer, et al. (1996), "*Advances in Steam Path Technology*," GE Power Systems publication GER-3713E.

Encotech (2007), *"CMG Tech, LLC Pressure Actuated Leaf Seal Performance Evaluation",* New York State Energy Research and Development Authority, Agreement No. 9887.

Ferguson, J., 1988 "*Brushes as High Performance Gas Turbine Seals*," Proc. ASME 1988 International Gas Turbine and Aeroengine Congress and Exposition, American Society of Mechanical Engineers, pp. V002T002A006-V002T002A006.

Johnston, (2000),"*Performance and Reliability Improvements for Heavy Duty Gas Turbines*", GE Power Systems publication GER-3571H.

Justak, (2013). Hydrogen Compressor Seal Case Study - Utilizing HALO (Non-Contacting, Compliant) Inter-Stage, Impeller Eye, Buffer and Fail-Safe Seals. Texas A&M University. Turbomachinery Laboratories.

Justak, J. F., (2011), "Seal with stacked sealing elements": US patent 7896352 B2

Justak, J. F., (2014), "Self-adjusting non-contacting seal": US patent 8919781 B2

Little, Nathan T.; Sulda, Eric M.; Terezakis, Terry N., 2001, "Efficiency Improvement on Large Mechanical Drive Steam Turbines", Proceedings of the 30th Turbomachinery Symposium.

MD&A Turbines (2021), "*Steam Turbine Guardian Packing*", www.mdaturbines.com/parts/steam-turbine/guardian-packing/.

Pedraza-Dalle, et al. (2019), "*On the Development of a Pressure Actuated Leaf Seal for Turbomachinery Applications*" ASME Turbo Expo GT2019-90991.

San Andres, L and Anderson, A., (2014) "*An All-Metal Compliant Seal vs a Labyrinth Seal*", ASME Turbo Expo, Paper GT 2014-25572

San Andrés, L., and Anderson, A., 2015, "An All-Metal Compliant-Seal Versus a Labyrinth Seal: a Comparison of Gas Leakage at High Temperatures, "ASME J. Eng. Gas Turbines Power, 137 (May), pp. 052504. DOI: 10.1115/1.4028665 [ASME Paper GT2014-25572]

Stapp, D., 2018, SCO2 Symposium WHR Panel for Turbo Machinery Developments.

Stiles, David T.; Sandberg, Mark R.; Justak, John F.; Wilcox, Ed; Kuzdzal, Mark J.; Rohrs, Charles A.; Miller, Harry F. (2016). Application of Dynamic pressure-balanced Seals in a Multi-stage Centrifugal Compressor. Turbomachinery Laboratories, Texas A&M Engineering Experiment Station.

Turnquist, Norman A.; Bailey. Fredrick G.; Burnett, Mark E.; Rivas, Flor; Bowsher, Aaron; Crudgington, Peter, 2005, "Brush Seals for Improved Steam Turbine Performance". Presented at the Institution of Mechanical Engineers Retrofitting Steam Power Generation Plant Seminar, 20th September 2005

Title: Mechanical and Electrical considerations of electric motor/Drivers for large centrifugal compressors in Ethylene cracker applications.

Authors:

Kalyan Malnedi, is an application engineer for Elliott Group in Houston, Texas. He has more than 18 years of experience in centrifugal compressor research and development and selecting turbomachinery for various markets/applications including olefins markets. He holds MS in Aerospace engineering from Indian Institute of Technology -Bombay, Mumbai, India.

Shane Harvey, P.Eng., is the Manager of Americas Application Engineering for Elliott Group, a leader in supply of compressor and turbines for ethylene plants. He has more than 15 years of experience in sizing and selecting turbomachinery for the olefins market. He holds a BSc in mechanical engineering from university of Calgary and is a licensed Professional Engineer in the province of Alberta. He is the chair of the Rotating equipment subcommittee of the AIChE Ethylene Producers' Conference.

Axel Rauber, graduated from the Staatliche Studienakademie Baden-Württemberg, Mannheim, Germany in 2001 with a degree in electrical engineering – Diplom-Ingenieur (BA). In 2001 he started his professional career at ABB Medium Voltage Drives and is currently employed as a global product manager. He is responsible for VSI and LCI based products for high power applications.

Abstract:

Traditionally, large centrifugal compressor drivers used in Ethylene cracker applications are steam turbine drives. This has been a natural choice due to the availability of steam generated as part of the cracking process. For very large capacity plants, these steam turbine drivers are required to produce powers in excess of 100 MW. Recently, due to the clean energy initiatives, there seems to be a trend in the industry to apply electric motors for these ethylene cracking applications in lieu of steam turbine drivers. Large electric motors with VSD drivers brings its own challenges, limitations etc. In this paper, the authors will take a detail look at the mechanical challenges including but not limited to start-up demands and their impact on compressor selections, transient torsional requirements and also a detail review of electric motor/drive requirements, suitability of drive technology (LCI vs. VSI), experience trends, reliability and availability considerations. The paper also briefly outline the optional testing requirements including complete unit test and what extent of electrical and mechanical scope to be included.

1. Introduction:

The typical process to produce ethylene commercially is by thermal cracking of hydrocarbon feedstock in the presence of steam in a direct fired tubular furnace. Traditionally, steam turbine drivers are used to run the centrifugal compressors in this application due to the availability of steam generated as part of the cracking process. There seems to be a latest trend in the petrochemical industry to apply electric motors and drivers for centrifugal compressors used in ethylene cracker applications driven by various green energy initiatives, environmental regulations, requirements etc.

In this paper, the authors will address the electric motor selection, drive technology (LCI vs. VSI) considerations, impact of motor/drive technology on compressor design, the differences in compressor design considerations between steam turbine driver and electric motor driver for compressors used in ethylene crackers. In general, this discussion is also applicable to compressors used in applications such as PDH and LNG. Compressor type referenced and discussed in this paper is centrifugal compressor type. It is critical to optimize the compressor selection based on the driver type, application, service and all operating conditions including start-up, settle out, dry out conditions etc. [1]. The paper also address optional testing requirements with electrical motor drivers.

2. Ethylene Cracker:

To transform feed stocks such as ethane and naphtha into the desired olefins (Ethylene and Propylene) the feed stock is cracked by heating in the cracking furnaces. Feed stock is diluted with steam and heated to around 850degC. The now cracked gas is quickly quenched to stop the reaction. The Cracked/Charge/Process Gas Compressor provides the motive force to draw the cracked gas from the quench towers through a series of washes, knock outs, drying and finally increasing the pressure to allow separation. Cracked gas is sent to the separation side of the plant where large C3 and C2 refrigeration compressors provide the low temperature refrigeration to allow factional distillation of the products. Please see section-3 for additional information on Cracked/Charge/Process Gas Compressors, C3 and C2 compressors.

Please see below simplified process diagrams for ethylene cracking process (Figure 1), four stage cracked gas compressor (Figure 2) and propylene refrigeration system (Figure 3).

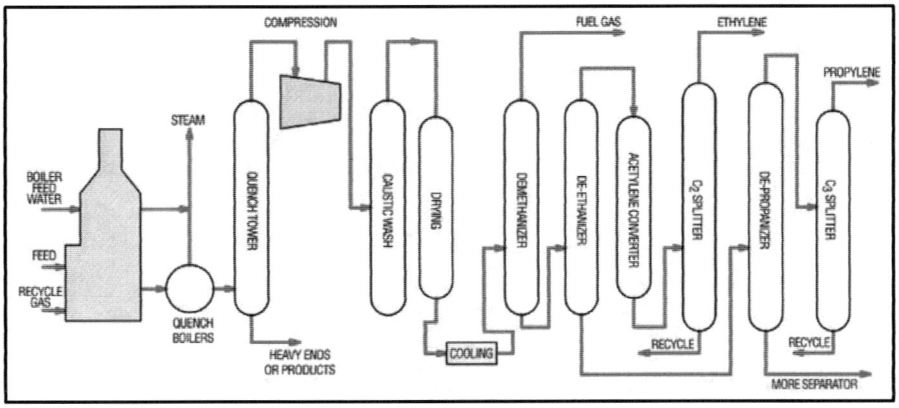

Figure 1: A simplified cross section of a centrifugal compressor

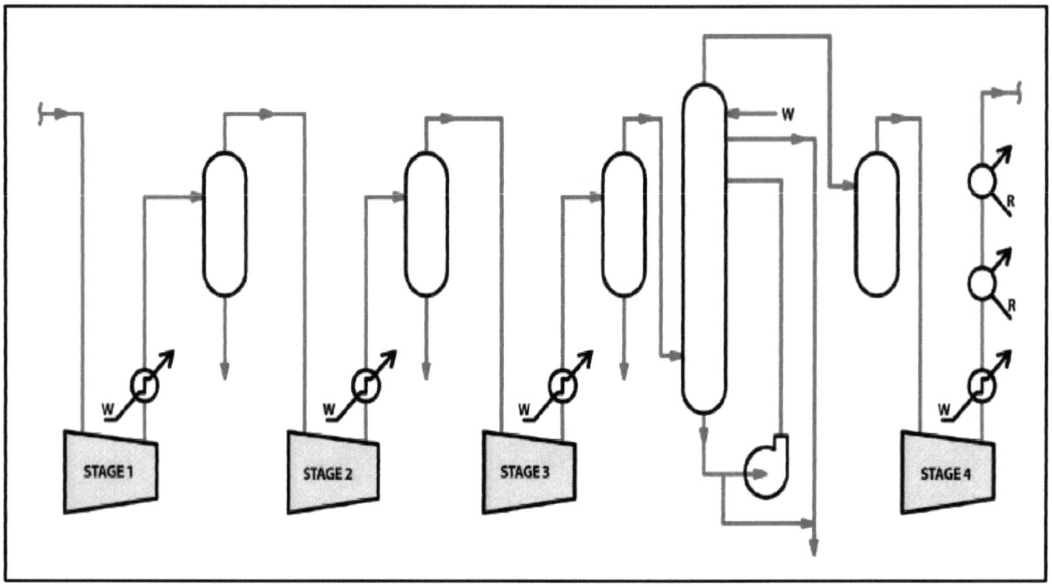

Figure 2: Four-stage Cracked Gas Compressor Train with Intercooling, Knock-out Drums, and Caustic Water Wash

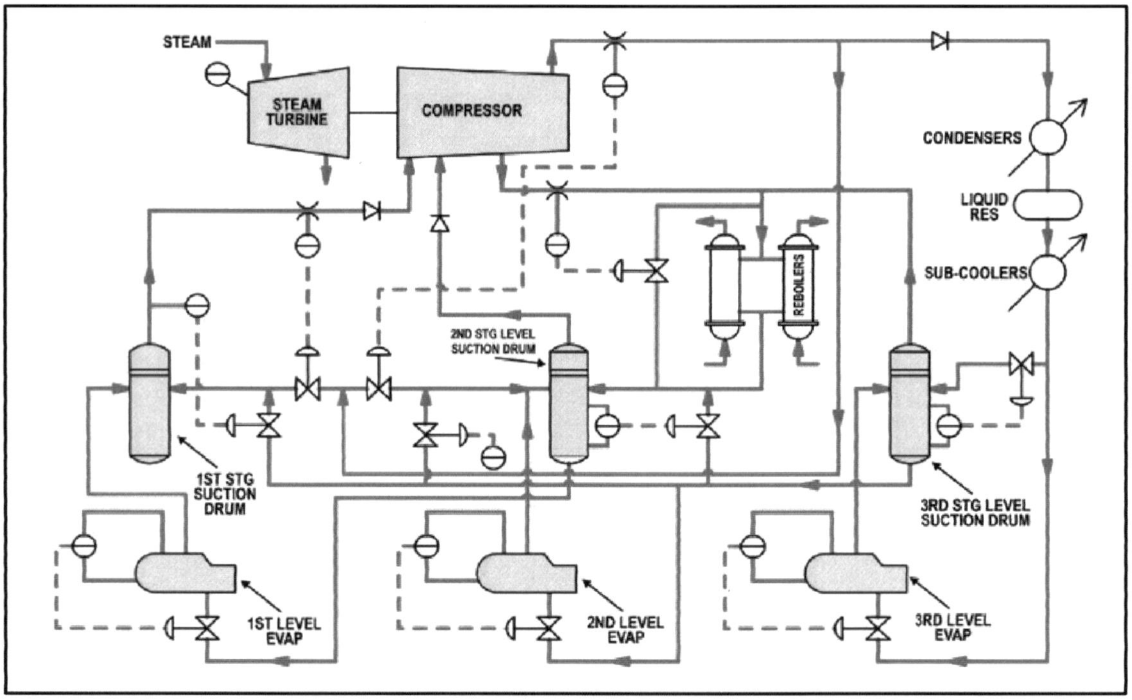

Figure 3: Typical Process Diagram of Propylene Refrigeration System

3. Centrifugal Compressors in ethylene Cracker applications:

Compressor type referenced and discussed in this paper is centrifugal compressor type. Centrifugal compressors are the dynamic type; compression is accomplished through the conversion of kinetic energy to static energy. A simplified cross-section of centrifugal compressor is shown below;

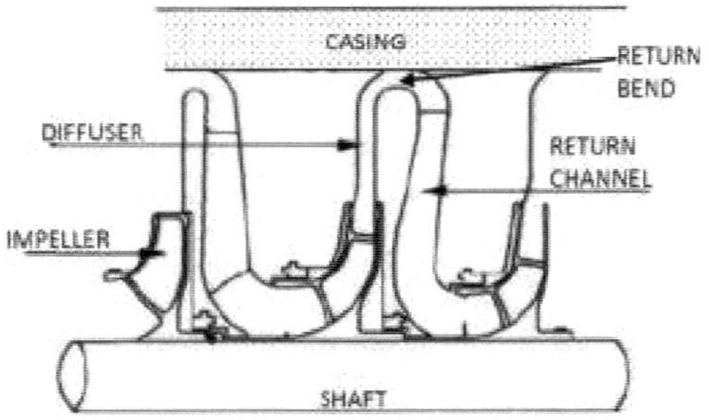

Figure 4: A simplified cross section of a centrifugal compressor

The two main characteristics of centrifugal compressors are volumetric flow through the unit and the head generated. Head generated inside a compressor is a function of the inlet gas MW. For the same suction and discharge pressure, a lighter MW gas requires a higher head than the heavier MW gas. Similarly, for a given mass flow and suction pressure a lighter MW gas will have a higher volumetric flow compared to a heavier MW gas.

There are three main services where centrifugal compressors are applied in ethylene applications [2]:

- Cracked/Charge/Process Gas Compressor
- C3 refrigeration compressor
- C2 refrigeration compressor/Binary refrigeration compressor

Cracked/Charge/Process Gas Compressor: These are high power compressors that draw gas from the cracking furnaces through various quench coolers and separation sections of the plant and pressurize the gas for further separation. They are characterized by low pressure, high mass flow rates, large volumetric flow rates and the need for multiple compressor sections to meet the required, typically two to five aerodynamic sections accommodated in two to three casings.

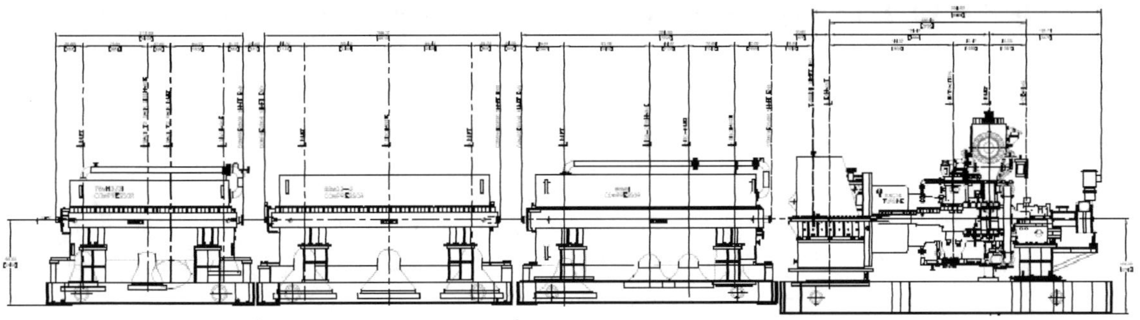

Figure 5: A typical five section three casings crack gas compressor train

C3 Refrigeration: Provides refrigeration duty around -40degF using or propylene or a mixture of two.

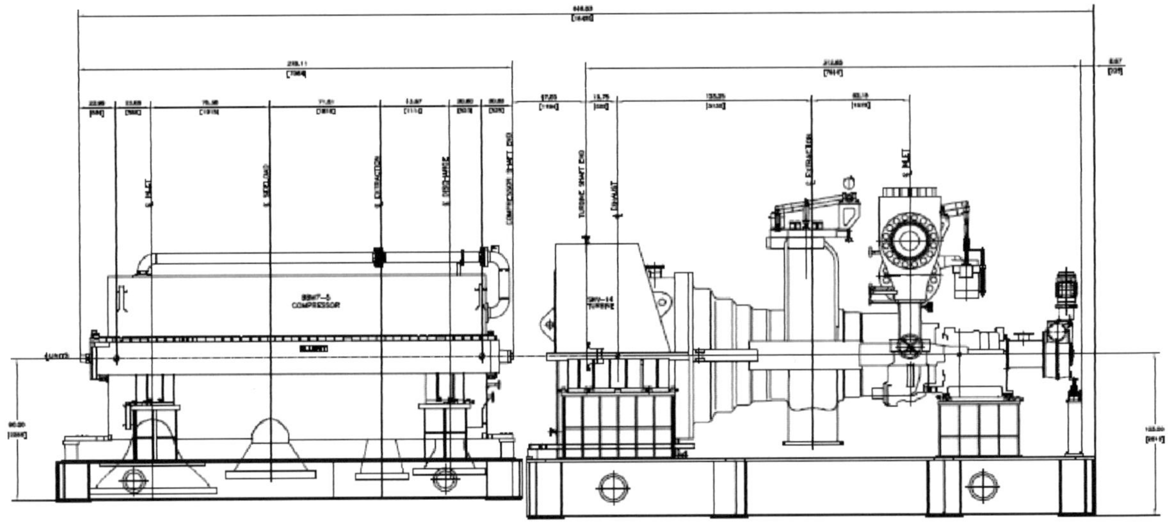

Figure 6: A typical three section single casing propylene refrigeration compressor train

C2 Refrigeration: Provides refrigeration duty around -150degF using ethylene as a refrigerant. Some process designs use a mixed-gas refrigerant for this service (Binary refrigeration)

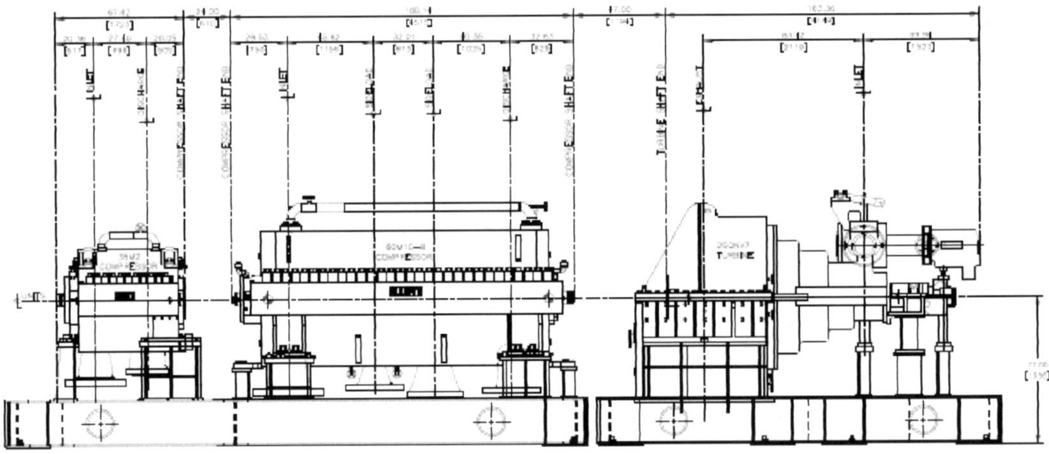

Figure 7: A typical five section two casings ethylene refrigeration compressor train

Refrigeration compressors are multi-sectional machines with sideloads or extractions depending on the process designs. Mismatches of the head required per section and large swings in volumetric flow rates between design and off design cases may impact overall compressor efficiency.

4. Drivers in Ethylene Cracker Applications:

Traditionally, large centrifugal compressor drivers used in Ethylene cracker applications are steam turbine drives due to the availability of steam generated as part of the cracking process. Recently, due to the clean energy initiatives, there seems to be a trend in the industry to apply electric motors for these ethylene cracking applications in lieu of steam turbine drivers.

Following is a list of major drivers used in Ethylene Cracker Applications:

1. Steam Turbine driver; with VHP or HP inlet steam depending on the service and power requirements. Typically, these are condensing type with HP or LP steam extraction
2. Electric Motor; 1X100% arrangement or a 2X50% arrangement for high powered applications
3. Drives: LCI or VSI drive, please see next section for additional discussion on the drive technology

When considering electric motor and driver, motor/drive commonality between various compressor trains is also an important consideration.

5. Electric Motors and Drivers:

5.1 Introduction

A variable speed drive system (VSDS) can be subdivided into components, which are shown in Figure 5.

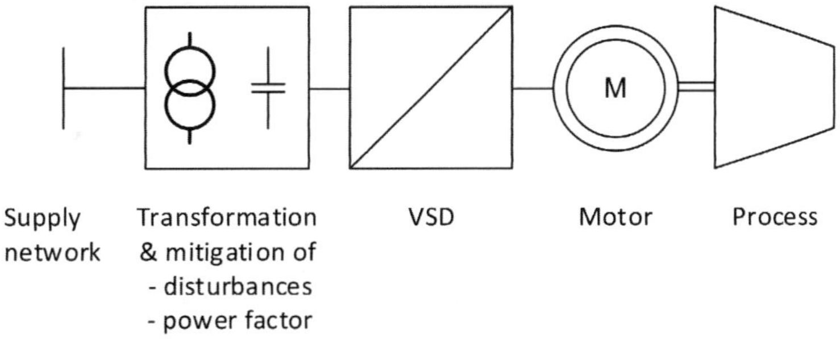

Figure 8: Variable speed drive system

- Supply network
  The supply network subcomponent consists not only of electrical parameters, but also includes the supply network power factor requirement and harmonic distortion limits.
- Transformation
  This subcomponent always includes a dedicated input transformer for VSDS driving large compressors. Depending on the requirements of the utility, a filter system may be needed.
- Variable speed drive
  In addition to safety, efficiency and cost, reliability is the most important selection criterion when choosing from a variety of solutions. For both technologies, voltage source inverters (VSI) and load-commutated inverters (LCI), high voltage, press-pack type semiconductors are used, which provide arc flash safety and lead to highest efficiency and reliability. Moreover, low-voltage

semiconductor solutions are available for VSI. For more information on the most used VSD system topologies see [3] and [4].

- Electrical motor

  For the discussed applications two types are used. Deploying a synchronous motor solution leads to best efficiency whereas an induction motor is simple with lower investment cost. The higher the power the bigger the advantage of synchronous motors not only due to experience and available references but also due to unprofitability and limited number of suppliers. Synchronous motors are preferred for the discussed applications, as the required shaft power is greater than 20 MW. Here a 0.5% higher efficiency can lead to enormous savings considering continuous operation over 20 years.

5.2 Comparison of LCI and VSI technology

The focus of this paper is on the integration of VSDS with rotating equipment. Nevertheless, other aspects must also be considered when selecting the technology. An active front-end VSI solution is not considered because regenerative operation is not required for the applications discussed. It would result in significant reliability and efficiency penalties due to the need for twice the number of active semiconductors.

- Network integration

  The network integration of a high power VSDS is not plug-and-play. Independent on the technology a VSD is a non-linear load, which injects current harmonics into the network. The current harmonics are reduced by using higher pulse number of the rectifier. Anyway, there are practical limits here. Due to tolerances on transformer phase shift a 24 or 36-pulse solution provides an optimum balanced with the complexity of the transformer and rectifier configuration. In general, it can be said that the smaller the difference between the primary and secondary voltage, the better the accuracy of the phase shift. In general, this means that the higher the secondary voltage, the smaller the tolerance, which implies a high primary voltage for high power drive applications, for example 22 kV, 33 kV or 110 kV.

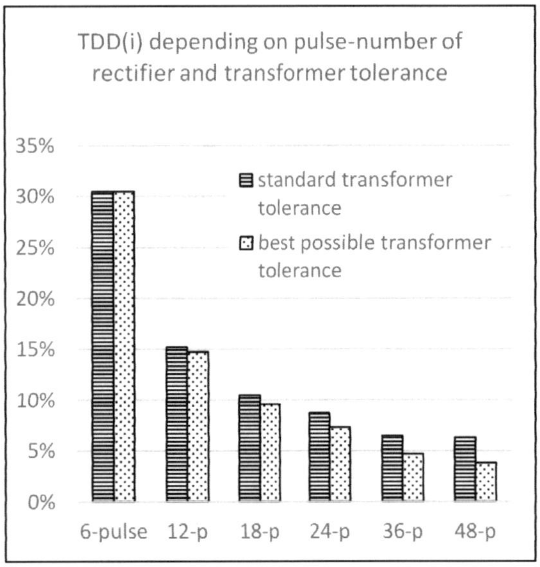

order [-]	f [Hz]		6-pulse	12-pulse	18-pulse	24-pulse	36-pulse	48-pulse
1	50	60						
5	250	300	100%	10%	15%	10%	15%	10%
7	350	420	100%	10%	15%	10%	15%	10%
11	550	660	100%	100%	15%	20%	15%	20%
13	650	780	100%	100%	15%	20%	15%	20%
17	850	1020	100%	10%	100%	10%	20%	10%
19	950	1140	100%	10%	100%	10%	20%	10%
23	1150	1380	100%	100%	15%	100%	15%	20%
25	1250	1500	100%	100%	15%	100%	15%	20%
29	1450	1740	100%	10%	15%	10%	15%	10%
31	1550	1860	100%	10%	15%	10%	15%	10%
35	1750	2100	100%	100%	100%	20%	100%	20%
37	1850	2220	100%	100%	100%	20%	100%	20%
41	2050	2460	100%	10%	15%	10%	15%	10%
43	2150	2580	100%	10%	15%	10%	15%	10%
47	2350	2820	100%	100%	15%	100%	15%	100%
49	2450	2940	100%	100%	15%	100%	15%	100%

Figure 9: Current harmonic distortion depending on rectifier pulse number and transformer tolerance
Figure 10: How current harmonics are considered depending on rectifier pulse number (rule-of-thumb).

Current harmonics of a 6-pulse rectifier are calculated as defined in IEC/TR 60146-1-2 [5]. Figure 7 shows, how the current harmonics are conservatively estimated based on the 6-pulse results (100%). Cancelled harmonics cannot be considered as non-existent, they must be assumed to be between 10 and 20% of their 6-pulse value due to transformer tolerances. Different values are used because

- Higher pulse numbers in the discussed power ranges, will lead to separate active parts of the transformer
- Higher pulse numbers require more complex transformer vector groups. For example, the simple star-delta configuration is used for a 12-pulse behavior, which leads to 10% of the 6-pulse value. Whereas higher pulse number require a zigzag or extended-delta connection

Inter-harmonics caused by the motor side operation of the VSD are neglected, which may not be sufficient depending on the VSD topology.

For LCI drives, the power factor is dependent on the operation point, which means a compensation is often required as explained in [6]. However, modern control methods show massive improvements compared to the classic way to design and control. More details can be found in [7].

If power factor compensation is required, a passive filter system connected to a tertiary winding of the input transformer will come with following advantages:

- A simple, lower pulse number rectifier can be used as the compensation system is also used for harmonic mitigation
- Robustness of the filter against changes of the network
- Flexibility in the design of the filter to consider the actual impedance of the network, since it is not sufficient to calculate the harmonic distortion of the voltage based on a purely inductive network. For example, a passive filter can be tuned to specific frequencies to minimize possible excitation of a parallel resonance by the VSD operation.

- Considerations related to the motor

Due to higher dv/dt connected to high switching frequency and common mode voltages at the motor terminals the motors are specially designed to the load requirements and VSD topology. DOL motors cannot be used for high power applications.

Due to higher current harmonic distortion and non-unity power factor, the LCI driven motors are bigger than the ones driven by VSI based VSDs. As the VSI inverter is self-commutated also induction or permanent magnet motors can be used, whereas the load commutated inverter (LCI) needs the back-EMF and therefore cannot run an induction motor.

- Considerations related to pulsating torques

As in explained in [6] torsional oscillation of the driven shaft has always been a concern with VSD for both LCI and VSI. Although the amplitude of such pulsating torques may be small compared to the driving torque, they can excite resonances when their frequency coincides with the natural frequency of the shafting (crossing). Literature is available for both LCI [8], [14], [15] and VSI [16], [17] and topology independent [18].

Demanding systems with high torsional stress on the shaft may require an advanced design process. Technical risk can be eliminated by designing mechanical and electrical components as a string. The output of such an electro-mechanical control system design is:

- Control philosophy (start-up, overspeed, etc)
- VSD parametrization and commissioning
- Detailed torsional data for the selected control philosophy, maybe models

State of the art tools to optimize large mechanical trains and mitigation measures like torsional resilient couplings are available. Modern modelling and simulation give new insight. Based on that, torsional optimized VSD control is developed. The following kind of mechanical excitations are to be considered:

- Direct excitation due to converter switching

  The converters are different in the way in which they produce harmonics and how they directly excite resonances. In general, VSI has smaller amplitudes of 6th and 12th harmonic than the LCI, but they are still present and must be dealt with. To reduce the pulsating torques for LCI driven equipment, a motor with two winding systems has been used since the beginning of VSD development. The pulsating torques can be of different types. The integer pulsating torques have frequencies, which are integer multiples of the machine frequency, and which are caused by the machine-side inverter The non-integer pulsating torques have frequencies that are non-integer multiples of the machine frequency. They are generated by the rectifier and are transferred via the DC link to the machine side. The frequency of the pulsating torques are predictable and can be presented in a Campbell diagram as shown in Figure 8, [6] and [14].

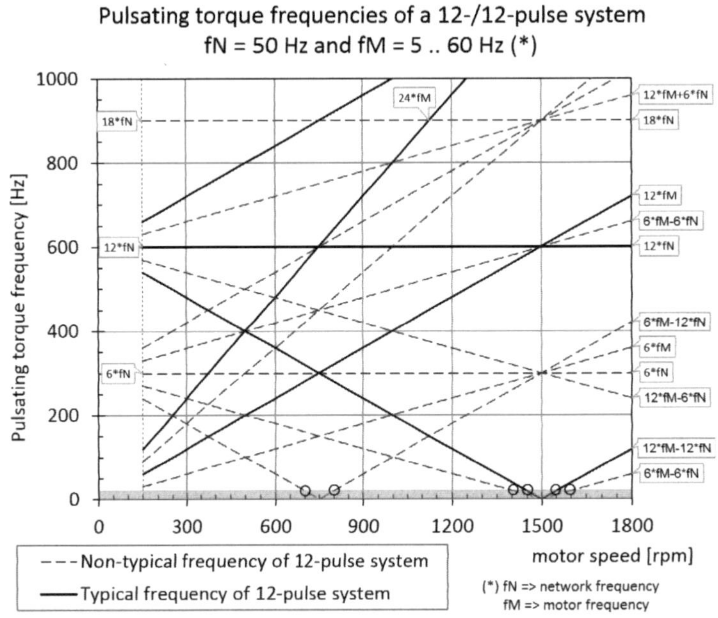

Figure 11: Frequency of the pulsating torques for a 50 Hz network and a 12-/12-pulse LCI driven 4-pole motor.

Different pulse numbers for the input and output converter can help when the line and motor frequency is identical. Anyway, it is recommended to choose the motor frequency to be different to the supply network frequencies as this avoids crossings with Eigen frequencies of the shaft train around the nominal operation point. The gray area in Fig. 11 indicates frequencies below 20 Hz, where the first mode of large compressors is expected. In the past, torsional interaction was avoided by skipping critical speeds with crossings (indicated with circles). Control algorithms (e.g. sequence control for the LCI) help to avoid crossings as well.

For VSI, a general statement cannot be given as the pulsating torques are heavily dependent on topology and control method.

- Closed loop stability

  The converters dynamic capabilities (i.e. their control bandwidth) are different. LCI has approximately 30 Hz linear bandwidth for a 2-pole / 60 Hz motor. The bandwidth of a VSI is much higher, which is good, but needs to be managed accordingly. To damp the first mode, a control bandwidth of 20 Hz is sufficient. VSI and LCI have different critical speeds, but both must be considered equally in a closed-loop. Active damping functions are available depending on suppliers. In general, as soon as an oscillation can be "seen" measured by the VSD, counteractions can be taken for example by acting on the torque reference.
- Strong control modulation: Control action can also be interpreted as pulsating torque. This phenomenon has not been fully addressed scientifically. It is independent of VSD topology and is caused by the control algorithm and settings. VSD suppliers can propose for example speed controller settings, which are considering the first TNF of the mechanical shaft train.

- VSDS reliability

A steam turbine is known to be very reliable. Reliability/availability figures of a steam turbine cannot be matched with a standard VSDS. The focus, when discussing the reliability of the VSDS is clearly on the VSD as it is the most complex component, independent on the technology. As explained in [15], Mean Time Between Failure (MTBF) and other reliability figures should be used with care. They are relevant when comparing different layouts in the same drive; however, they cannot be used to rank the reliability between different suppliers because of the different approaches that were taken to obtain the numbers. The design target must be to have a simple and robust topology because despite of many redundancy options, single-point-of-failures cannot be eliminated. Here the LCI has a clear advantage due to

- The simple topology is based on thyristor semiconductors and dc-link reactor, which are proven to be very robust and maintenance-free.
- The same bridge, topology and components are deployed on rectifier and inverter side.
- The scalability due to very simple series connection of thyristors, which is especially interesting for high power applications > 30 MW.

- No charging or discharging is required, which helps also during ride-through in case supply power is lost or disturbed.

Also, for very critical applications redundancy of electrical power components is only required for topologies based on low-voltage cells, to compensate the need of a large amount of power components (e.g. MMC). For other topologies, power semiconductor redundancy would lead to complex systems nullifying the advantage as anyway, single-point-of-failures cannot be avoided.
Independent on the topology a control unit is required. It consists of printed circuit boards including the main controller, measurement interface, sensors, the control panel and the control power supply units. Also, here many single-point-of-failures cannot be avoided.
In addition, the cooling of the VSD needs to be considered. Also, the reliability of the cooling unit is independent on the topology. For high power VSD of the size discussed in this paper water-cooled solution have been and will be dominant because

- Minimum load on the HVAC of the electrical room
- Smaller VSD footprint
- Lower VSD cost

Even though the VSD or HVDC system suppliers have gathered > 40 years' experience with water-cooling of power electronics and the availability of a variety of redundancy options (e.g. pump, heat-exchanger, de-ionizer, sensors) failure modes like leakages cannot be avoided.
Further, high maturity of the VSD is the basis for high reliability respectively availability.
Another interesting finding, when categorizing the field failures is, that only about 25% of trips are related to the drive itself. Out of this 25%, 90% are related to the implementation and design of the converter and the remaining 10% are related to the MTBF of the components (i.e. statistical failures). This means, that a proper system integration is very important. This includes

- Supply network integration
- Study of electrical-mechanical interaction
- System protection

To cover the system integration part and minimize the risk, project specific "Failure Mode Analysis" together with the end-user, engineering contractor & equipment manufacture are proposed.
As stated in [15], a VSD with a requirement to operate for years without a shutdown, be it scheduled or un-scheduled needs a redundancy on the macro level, i.e. a complete VSD or redundant converter.

5.3 Comparison of the LCI and VSI system topology (40 MW example)

Today the maximum shaft power supported by a single VSD based on VSI technology is <40MW. This means, that for a 40MW+ application two converters and two transformers are required, that work in parallel. Due to its simplicity the upscaling of output power of an LCI is simple and achieved by increasing the number of series connected thyristors. There are no changes in the topology. 100MW output power with a single LCI have been reached in the 1990s already.

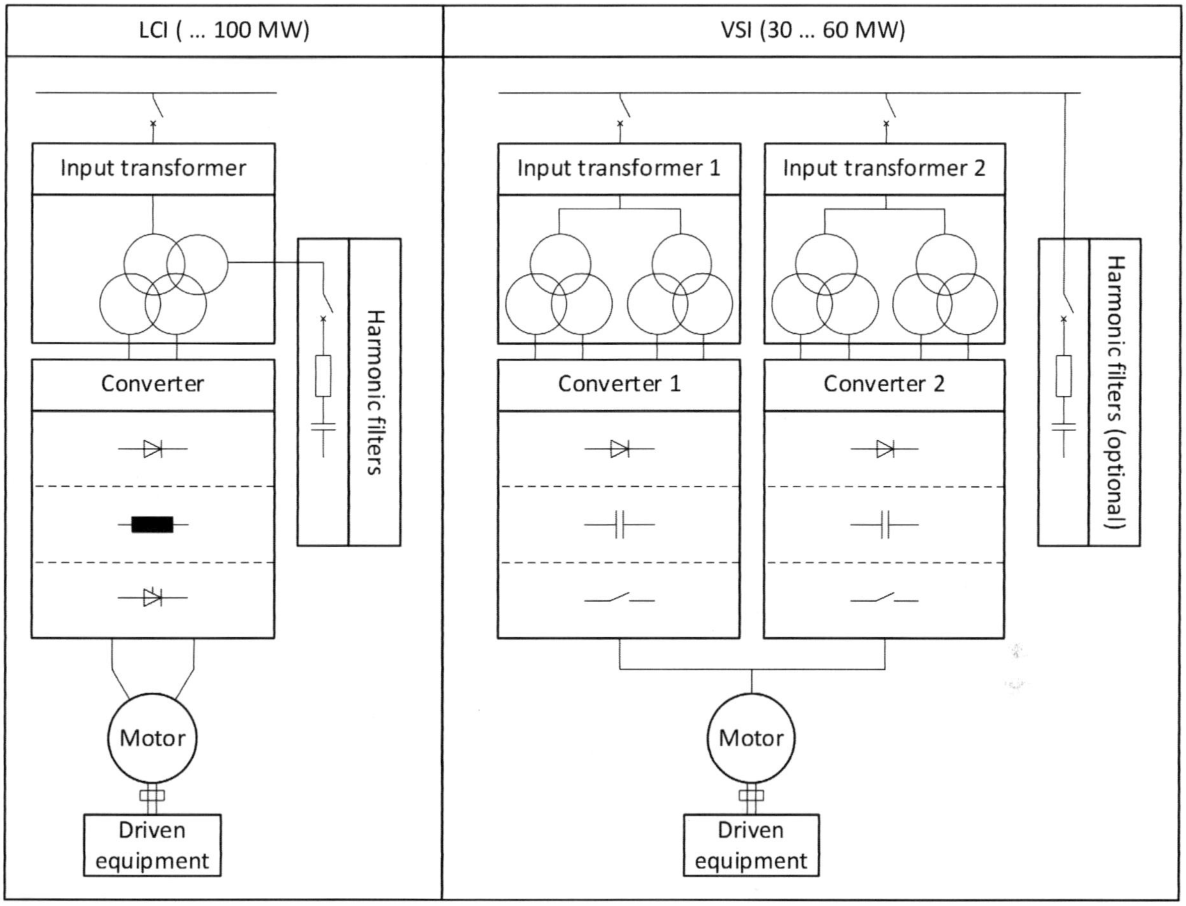

Figure 12: 42 MW electric motor/VSD driven compressor string, down nozzle mezzanine mounted

5.4 Comparison of physical size (40 MW example)

As discussed in [6], a general comparison of VSI and LCI is not possible. In addition, this comparison must consider whether components are installed indoors or outdoors. Anyway, a 40 MW example is presented in the table below.

Item	LCI system l x d x h [mm^3]	VSI system l x d x h [mm^3]	Comment
Input transformer	7000 x 5500 x 6000 (75 t)	**2x** 6500 x 4000 x 5000 (2x 45 t)	Outdoor, oil-immersed
Filter	**4x** 2000 x 2000 x 1500	n/a	Indoor, 4x filter branch
VSD	9000 x 2250 x 2740	**2x** 13000 x 1800 x 2700	Indoor (air-conditioned)
Motor	5510 x 3400 x 4500	5210 x 3400 x 4500	Water-cooled (IC81)

It is assumed, that the VSI solution does not require a harmonic filter for the mitigation of harmonic distortion.

Further, it must be noted, that the power density of LCI is unmatched, anyway a filter system for power factor compensation is usually needed. The numbers presented in the table above do not include the clearances to open the doors of the VSD compartments for example. The filter uses iron-core reactors for indoor installation. If it is installed outdoors air-core reactors are used, which increases the footprint significantly but could anyway be an interesting solution depending on the site-specific constraints. For this example, the difference for the motor is small. Nevertheless, it is possible that an LCI driven motor might require a bigger shaft height.

6. Impact of Driver selection on compressor design:

Below listed are some of the major difference in compressor design between steam turbine driver and electric motor driver.

6.1 Shaft end sizing: There is a significant impact on compressor shaft end sizing based on the driver. At times, this could results in going-up a compressor casing size. For motor driven applications in addition to the steady state excitations, transient torsional excitations also to be considered. Below are typical steady state and transient excitations to be considered based on the compression sting arrangement.

Steady State Excitations:

Centrifugal compressor and Steam Turbines: $1xf_{speed}$, $2xf_{speed}$, $N_{blade}xf_{speed}$

Gear: $1xf_{speed}$, $2xf_{speed}$, $N_{teeth}xf_{speed}$

Constant Speed Motor: $1xf_{line}$, $2xf_{line}$ (f_{line}=50 Hz or 60Hz) (Only applicable for DOL start)

VSD driven Motor: Nxf_{output} (Harmonics), $Nx(f_{line}-f_{output})$ (inter-Harmonics), N=6, 12, ...

Transient Excitations:

Electric fault of motor: 3 phase, 2 phase, single phase or ground, 2 phase to ground, $1xf_{line}$, $2xf_{line}$, decay in time

In addition to the above, driven string equipment inertia, short circuit air gas torque equations to be considered when sizing the shaft ends. In general, short circuit torque values are between 3 to 15 times full load rated torque. Synchronous motors will tend to be at the higher range of peak short circuit torque values.

Example of short circuit equation with torque multipliers:

Airgap torque equation - 3-phase short circuit

Te(t) = M0 e -t/t0 sin wt + M1 e -t/t1

M0 = 6.14 ; M1 = 1.15 ; t0 = 0.126 s ; t1 = 0.113 s ; w = 314 rad/s

Maximum value of torque **7.01** * TN, when t = 4.89 ms

Airgap torque equation - 2-phase short circuit

Te(t) = M0 e -t/t0 sin wt - M1 e -t/t1sin 2wt + M2 e -t/t2

M0 = 5.31 ; M1 = 2.67 ; M2 = 1.35 ; t0 = 0.185 s ; t1 = 0.635 s ; t2 = 0.343 ; w = 314 rad/s

Maximum value of torque **8.04** * TN, when t = 6.66 ms

TN = Rated Torque

All these information needs to be considered when designing the shaft end calculations.

Below table shows a comparison of compressor shaft end sizing between steam turbine driver and electric motor driver.

	Shaft end size (in), Hydraulic fit		
Driver Rating 42MW	**Compressor2**		**Compressor 1**
String Configuration	**Drive**	**Drive Thru**	**Drive**
Steam Turbine+ Compressor2+Compressor1	6.5	5.5	5.5
Synch. Motor w/VFD (LCI) +Gear+Compressor2+Compressor1	9	8.5	8.5
Synch. Motor w/VFD (VSI) +Gear+Compressor2+Compressor1	9	8.5	8.5
Synch. Motor (DOL) Gear+Compressor2+Compressor1	10.5	10.5	10.5

Figure 12: Driver configuration vs. Shaft end sizing

6.2 Starting torque requirements:

With Steam turbine driver, typically starting torque demands are not a major concern and a higher starting torque can be accommodated. With electric motor/drive, starting torque requirements becomes very important when sizing motor and drive and need to be thoroughly evaluated when finalizing the motor and drive sizing. Also, one more important design consideration is the lower speed, where torque demands may be higher, figure-14, example-2, in this case the thermal design of the motor is impacted.

Parameters to consider when calculating the starting torque requirements: String break-away torque, Settle-out pressure, suction and discharge piping volumes, driven equipment inertia, motor acceleration time, number of anti-surge loops in a multi section, surge margins etc. Each of these values can have a significant impact on the motor/drive size, starting characteristics.

Below are couple of examples of starting torque/load curves. First one is a variable torque from start to 100% speed. Second example is a linearly increasing torque up-to compressor minimum operating speed and constant torque from minimum operating speed to 100% speed, followed by constant power up-to maximum continuous speed (MCS). Each of these curves may impact the design of the motor and drive including overload capacity rating.

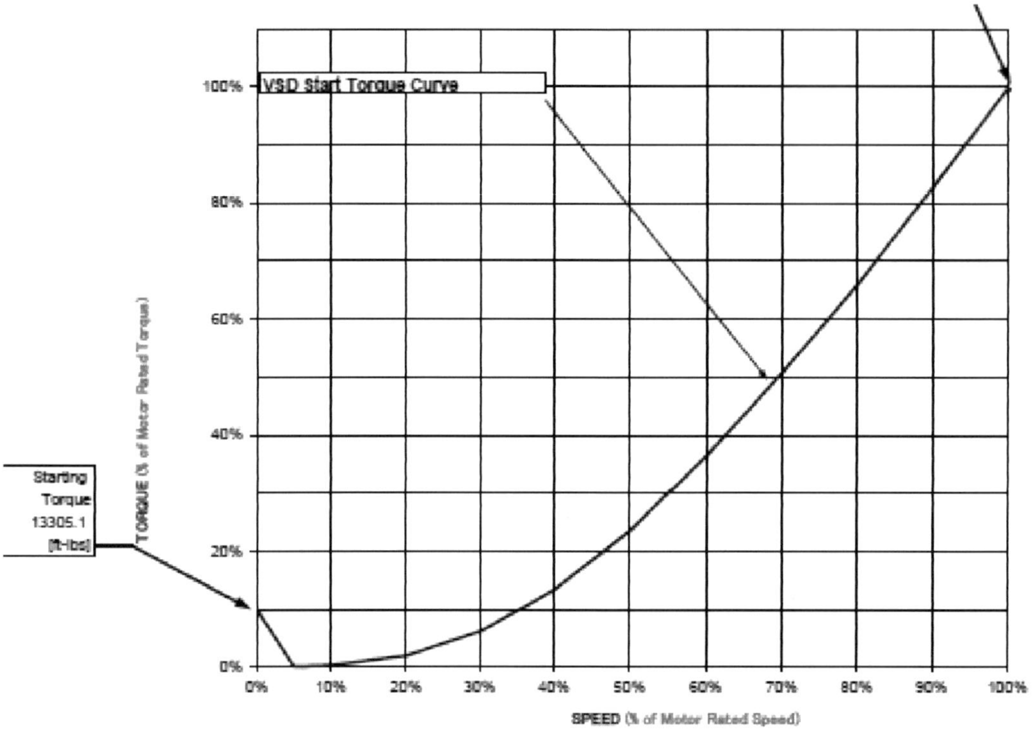

Figure 13: Example-1, Starting torque curve

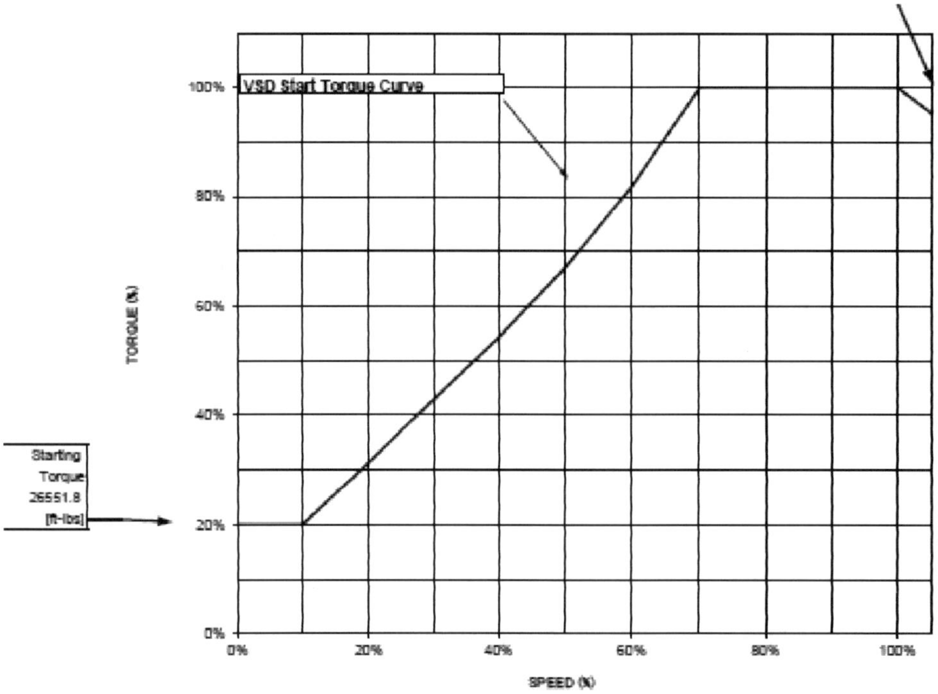

Figure 14: Example-2, Starting torque curve

6.3 Settle-out pressure requirements:

As mentioned in section 6.2 settle-out pressure plays an important role in starting torque requirements and properly sizing motor/drive requirements. A dynamic analysis is highly recommended in determining the starting torque requirements. Some of the considerations for high settle-out pressure/high starting torque applications.

1. Adding a manual suction throttle valve to reduce the suction piping volume. This valve will be used for start-up only
2. Vent/blow off process gas to reduce the settle out pressure
3. Increasing the motor/VSD overload power
4. Oversizing motor/drive rated power/torque
5. Increasing motor acceleration time, typical 120 sec with VSD
6. Fully system level dynamic simulation to optimize the starting torque requirements

6.4 Short circuit faults and Transient torsional requirements:

As discussed under section 6.1 short circuit faults and transient torsional requirements in determining the compressor shaft end size, coupling selections. Majority of the times, compressor OEMs may not have sufficient information available to perform steady state and/or transient torsional analyses prior to the order booking. There are lot of assumptions to be made during the pre-order and these need to be validated during the post order prior to motor/drive design and release to the manufacturing.

Also, when there is wide operating range, a dead-band may be required in the expanded speed range to avoid torsional excitation frequencies. Typically, this is applicable to the applications with speed range below 80% of the rated speed. Sometimes, this dead band can be up to 25%. Majority of the times, this is not an issue with steam turbine drivers as the torsional excitations can be addressed through spacer design changes and also the torsional excitations are not severe compared to electric drive/fault excitations.

In addition to the normal operating range, the plant operation may require running idle the compressor at a much lower speed, typically close to 25% of the rated speed. Such requirements needs to be considered when reviewing train torsional.

Some of the LCI driver OEMs can also offer VFD damping technology/active damping to suppress torsional excitations. Please see section-5 for additional information.

6.5 Coast down calculations:

Similar to steam turbine driver operation, coast down calculations and sizing of the rundown tank to be performed considering the full string inertia.

6.6 Additional considerations

a. Spare philosophy and drive commonality:

Based on the plant capacity and electrical load requirements a 2X50% train configuration or oversizing the motor/drive to match additional services may be considered. This allows the end user to have a common motor and driver spares. In addition to this having the same drive configuration between multiple compressor trains, allows the end user to consider a common hot standby spare, which can replace one of the fault drives without shutting down the compressor operation. The time to swap the standby VSD to be accounted in the compressor performance, drive and process control system design. This configuration provides additional redundancy to the plant operation.

b. Foot print comparisons steam turbine vs. motor driver

Please see below a comparison of 42MW olefins application with electric motor drive and steam turbine driver

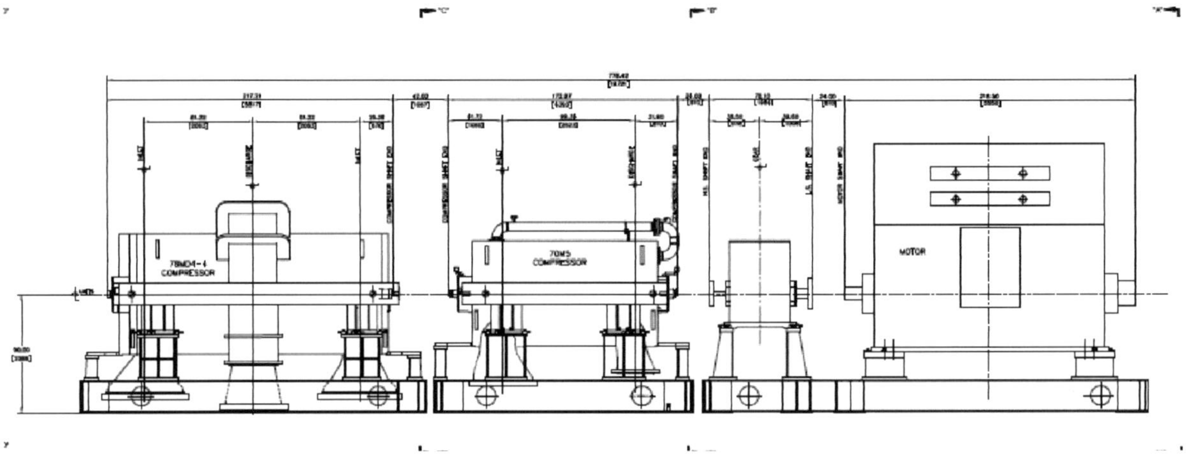

Figure 15: 42 MW electric motor/VSD driven compressor string, down nozzle mezzanine mounted

Approximate String dimensions: 67ft(L)X14.5ft(W)X16.9ft(H) (includes top half of the motor, can be removed for transportation)

Total String Weight: 753,500 lbs

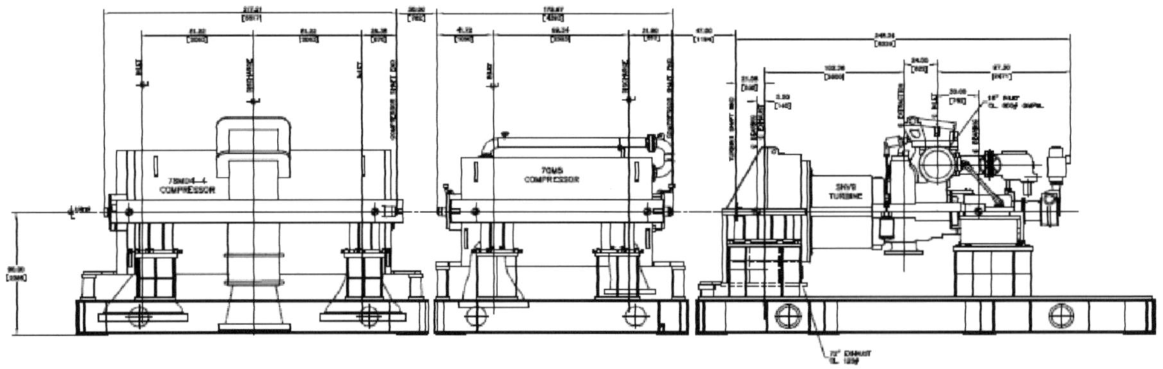

Figure 16: 42 MW Steam turbine driven compressor string, down nozzle mezzanine mounted

Approximate String dimensions: 66.5ft(L)X14.5ft(W)X13.5ft(H)

Total String Weight: 710,510 lbs

For motor/VSD driver, in addition to the above skid layouts, harmonic filter housing (for LCI drives) and power distribution center (e-house) dimensions and weights to be considered when reviewing the plot plans.

c. Grade vs mezzanine mounting:

Typically, compressor strings with steam turbine driver, especially condensing turbines are mezzanine mounted because of the ease of placement of auxiliary equipment like surface condenser, gland condenser etc. For electric motor driven strings, this opens-up the possibility of grade mounting vs. mezzanine mounting. Grade mounting/top nozzle configuration offers lower equipment installation cost, but may bring additional maintenance challenges especially if the compressors are horizontally split type, which is typical in ethylene applications. In a grade mounted horizontally split configuration, compressor main process nozzle connections need to be broken for maintenance. Alternatively, a barrel compressor can also be considered especially if the barrel compressor is not sandwiched in between two strings of equipment.

c. Strings with Gear Box:

Unlike, steam turbine driven strings, it is typical to have a Gear box in a motor driven compressor string. The combination of high peak torque, gear modes, and torsional natural frequency interference carries additional risk and to be properly evaluated prior to completion of Gear elements and couplings designs. The scope of this typically falls under the compressor OEM, who has the string responsibility. The outcome of this risk assessment also to be considered when duplicate low speed couplings are selected based on the duplicate motor design between multiple strings of equipment.

d. Turning Gear Operation:

Majority of the large steam turbine driven strings have a turning gear, used for slow roll of the turbine. In case of an electrical motor driven string a turning gear may not be necessary. For a larger/longer if compressor rotor bow is a concern a turning gear can be added to the string, either to the Gear Box (if present) or to the compressor. Also, a slow roll operation using VSD may be considered, if feasible this will avoid the need for a turning gear.

e. String testing:

At times, the end user may require a complete string test to verify the mechanical integrity of the full string. When such requirements are specified, it is very important to define the scope of the testing, equipment needs to be included, acceptance criteria etc. The string test can be as minimum as a unit mechanical string test with job motor, gear and compressors to full job equipment including input transformers, VSD, e-house, control panels etc. Also, some of the test requirements may include full load, full speed and full pressure tests. Depending on the scope of these tests a significant logistics, capital and schedule need to be considered.

7. Conclusions and Future trends:

As discussed in the paper, there seems to be a trend in applying large electric motor drivers to the Ethylene cracker applications, which predominantly uses steam turbine driver due to the availability of the steam. The authors addressed the mechanical and electrical consideration when applying large electrical motor drivers to the ethylene crackers. The authors also addressed the available drive technology, comparison of LCI vs. VSD technology, reliability, experience trends, compressor shaft-end sizing considerations, start-up torque requirements and impact on motor/driver sizing, considerations of short circuit faults, torsional excitations when designing the compression strings. The authors also discussed the spare philosophy, foot print comparison between steam turbine driven string and electric motor driven string, grade vs. mezzanine mounting, Gear box considerations and turning gear operation.

In general, these discussions/guidelines are applicable to all large electric motor driven compressor strings and not limited to the ethylene cracker applications. At the time of writing this paper, there seems to be very limited experience with motor/drivers over 60-70 MW especially in the oil and gas industry. However, if recent trends are any indication, we may see more growth/demand in using large electric motor drives beyond the above mentioned power rating.

8. Acknowledgements:

The authors thank Elliott Group and ABB for allowing them to publish this paper. The authors also thank their colleagues who contributed to the work cited in this paper.

9. References

[1] Kalyan Malnedi, Shane Harvey, "Challenges with Compressor Start-up/Dry out runs with dissimilar MW gases than the process gases" AIChE Spring Meeting, Ethylene Producers' conference, April 2019

[2] Shane Harvey, "Centrifugal Compressors in Ethylene Plants" AIChE Spring Meeting, Ethylene Producers' conference, April 2016

[3] IEC 61800-4 (First edition 2002-09) Adjustable speed electrical power drive systems – Part 4: General requirements – Rating specifications for a.c. power drive systems above 1 000 V a.c. and not exceeding 35 kV.

[4] A. Marzoughi, R. Burgos, D. Boroyevich, Y. Xue, "Design and Comparison of Cascaded H-bridge, Modular Multilevel Converter and 5-L Active Neutral Point Clamped Topologies for Drive Application" ECCE Paper.

[5] IEC/TR 60146-1-2 (Edition 4.0 2011-01) Semiconductor converters – General requirements and line commutated converters – Part 1-2: Application guide.

[6] A. Rauber, P. den Bakker, "Adjustable Speed Drive System Comparison VSI and LCI for High Power Applications", IEEE Paper No. 2018-PCIC-0346.

[7] T. Besselmann, P. Jörg, S. Van de moortel, "Power Factor Improvements for Load Commutated Inverters", Submitted to 2nd IEEE Conference on Control Technology and Applications, Copenhagen, Denmark, August 2018.

[8] A. Lenzi, P. Jörg, S. Falomi, D. Andreo, et al. , "Combined Torsional and Electromechanical Analysis of an LNG Compression Train with Variable Speed Drive System" Proceedings of the 38th Turbomachinery Symposium.

[9] P. Jörg, A. Lenzi, V. Depau, "Optmization Of Transient Behaviour Of Complex Turbocompressor Shaft Lines", Proceedings of the 2011 IEEE Industry Application Society Annual Meeting.

[10] T. Besselmann, S. Almér and P. Jörg, "Torque Harmonic Minimization for Load Commutated Inverters in Pulse Mode", IEEE Transactions on Industrial Electronics, vol. 65, no. 1, pp. 86-93, Jan. 2018.

[11] M. Kita, T. Hataya & Y. Tokimasa "Study of Rotordynamic Analysis Method that Considers Torsional and Lateral Coupled Vibrations in Compressor Trains with Gearbox", Proceedings of the 36th Turbomachinery Symposium 2007.

[12] Tanaka et al. , "Torsional-Lateral Coupled Vibration of Centrifugal Compressor System at Interharmonic Frequencies Related to Control in VSI".

[13] John A. Kocur & Maximilian G. Muench, "Impact Of Electrical Noise On The Torsional Response", Proceedings of the Forty First Turbomachinery Symposium September 24-27, 2012, Houston, Texas.

[14] L. Durantay, J.-M. Taillardat, J.-F.Pradurat, N. Velly, "State Of The Art for Full Electric Driven Refrigeration Compressors Solutions Using Adjustable Speed Drive: Which Combination of Technology Platforms Leads to the Best CAPEX & OPEX Solution up to 100 MW", PCIC North America 2017.

[15] P. Wikström, L. Terens, H. Kobi, "Reliability, Availability and Maintainability (RAM) of High Power Variable Speed Drive Systems (VSDS)", IEEE Paper No PCIC – 98-16.

EPC PAPER NUMBER 31a

UTILIZATION OF DYNAMIC SIMULATION FOR NEW STEAM CRACKER START-UP ACTIVITIES

James Kendall

Principal Engineer Process Sustainability

ExxonMobil Chemical Baytown

David Dutschmann

Principal Engineer Recovery

ExxonMobil Chemical Baytown

Jordan Bethea

Dynamic Simulation Engineer

ExxonMobil Research and Engineering Houston

Prepared for Presentation at the 2022 Spring National Meeting

San Antonio, Texas

April 11-14, 2022

EPC PAPER NUMBER 31a

UTILIZATION OF DYNAMIC SIMULATION FOR NEW STEAM CRACKER START-UP ACTIVITIES

James Kendall

Principal Engineer Process Sustainability

ExxonMobil Chemical Baytown

David Dutschmann

Principal Engineer Recovery

ExxonMobil Chemical Baytown

Jordan Bethea

Dynamic Simulation Engineer

ExxonMobil Research and Engineering Houston

Abstract:

Dynamic simulation as a training tool for increasing operator competency is not a new concept in the ethylene industry and continues to see advancement in both areas of application and use of new technology. Training often focuses on response to non-routine situations such as startup/shutdown and emergency events. Simulation software capability can be enhanced by DCS integration that connects the dynamic behavior of the model to controls and graphics that match the production system.

In addition to increasing operator competency, further value can be captured when applied to greenfield project start-ups. As with simulations of existing plants, superior console training accelerates operator proficiency in lieu of many years of experience. Dynamic simulations can provide greenfield steam crackers the opportunity to test procedures for clarity and completeness, perform DCS and safety logic application reviews, familiarize the operators with process behaviors ahead of the actual plant startup and even allow qualifying console operators on a new process flowsheet. First time startup procedures can also be crosschecked and validated in greater detail than possible with tabletop exercises. Further, in the area of process control engineering support, validating schematic operability, checking design of process control applications, and tuning controls pre-startup can be readily accomplished. Finally, focal areas such as techniques to minimize flaring and time to on spec can also be confirmed pre-startup.

In this paper, we will review how a dynamic simulation can be successfully implemented during a greenfield start-up by using general examples from industry and specific examples from recent ExxonMobil start-ups. Ultimately, the use of dynamic simulation can contribute significantly to the overall success and timeliness of a greenfield steam cracker startup.

Introduction

ExxonMobil's Baytown Olefins Plant started up a third train in 2018. With this addition the Baytown Olefins Plant became one of the largest ethylene plants in the world. A dynamic simulator was developed for console supervisor training. The simulator was complete well ahead of startup, facilitating multiple rounds of training.

In this paper we will describe how dynamic simulation can be successfully implemented during a greenfield start-up by using general examples from industry and specific examples from recent ExxonMobil startups. Ultimately, the use of dynamic simulation can contribute significantly to the overall success of a greenfield steam cracker startup.

Dynamic Simulation Basics

Operator Training Simulators (OTS) are high fidelity, custom models that cover the entire console, including DCS and safety systems. These offline applications are used to train operators on less frequent, higher impact scenarios such as startup, shutdown, loss of feed, and equipment failures. Value is derived from accelerating operator competency by increasing exposure to such situations without requiring many years of actual console operation.

The backbone of an OTS is the detailed process model that represents the dynamic behavior of a specific unit or portion of a unit. Since the first principle model is based on piping and instrumentation diagrams (P&IDs), a steady state heat and material balance, equipment specification sheets (valves, exchangers, pumps, etc.), and mechanical drawings (vessels, columns, reactors) for the unit, it reflects the actual unit heat and material balance, reaction kinetics, and utility interactions of the actual plant.

The high fidelity experience of the OTS is possible due to the detailed replica of the operating console which includes console hardware, operating schematics and interfaces, and control system database and applications. The behavior of the dynamic model is connected to the replicated console via a process called DCS integration. This allows for feedback of trainee actions to be sent to the model in real time, as reflected in Figure 1.

Control systems require electrical signals from instrumentation (e.g. 4-20 mA signals) which are converted and displayed to the operator as temperature, pressure, flow, etc. and integrated into the process control applications. The control system also sends electrical signals which move control handles, such as valve position.

An interface utility allows the control system to interact with the dynamic simulator instead of actual instrumentation by converting:

- Simulation variables into I/O for the control system
- Control system output into a process change in the dynamic simulator

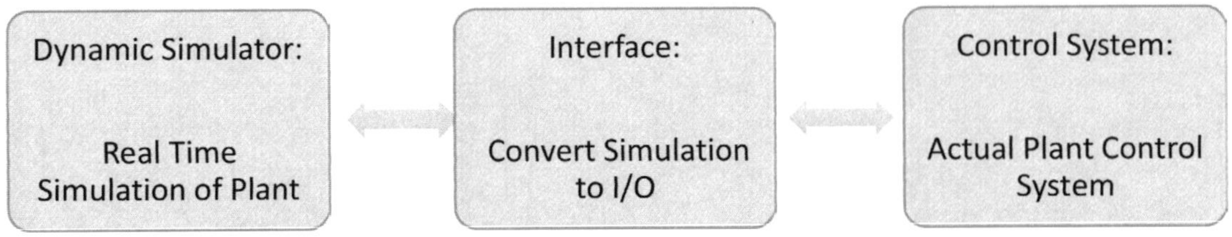

Figure 1. Interface between simulator and control system.

Matching Production Environment

Quantifying the value of training in a simulated production environment is difficult due to the lack of a counter-factual. The technical team believed the training yielded significant dividends in startup:

- Console Supervisors demonstrated familiarity with controls through high stress process swings in startup.
- Supervisors demonstrated good intuition for process limits and the consequence of process moves.
- Repeated execution of the startup procedure generated excellent familiarity with the startup framework and objectives of startup.
- Technical resources demonstrated familiarity with key process variables to watch to prevent machine trips.

Figure 2 shows the actual plant training room with stations that match the production environment. The match of the training environment with actual plant controls greatly enhances process training by giving operations the feel of real-world experience without creating process risk.

Figure 2. Training console

As with simulations of existing plants, superior console training accelerates operator proficiency in lieu of many years of experience. Training focused on startup, allowing operations to reduce flaring and time to on-spec in actual start-up.

Operations not only experienced the physical interface, but also gained experience with a copy of the operating schematics as reflected in Figure 3.

Production Environment Simulation Environment

Figure 3. Comparison of environments

Validating Control System

Prior to startup, the OTS enabled validating schematic operability, checking design of process control applications, and tuning controls pre-startup.

Particularly complex control applications, especially around major machinery, were validated with the dynamic simulator. In one case, the simulator revealed that a design mismatch between the major machinery control system and DCS controls could lead to the system to become caught in a loop where the DCS could no longer alter machine speed based on process input. Because the dynamic simulator revealed this control issue, engineers modified the system prior to startup.

Effective startup training also required tuning of control applications. This tuning was used for controls in startup instead of default tuning, facilitating better plant control in the first-time transient environment for the actual plant.

Grassroots Cracker Startup – Procedure Development

Dynamic simulations can provide greenfield steam crackers the opportunity to test procedures for clarity and completeness. Senior operations and senior technical reviewed procedures prior to dynamic simulation training. The OTS showed console supervisors could interpret the intent of some procedure steps differently and enabled correction prior to startup.

In addition, the simulator revealed several procedural steps where order had to be inverted to accomplish the intended objective. An early version of one procedure required the console supervisor to inventory a vessel with liquid when insufficient delta-pressure driving force was available because of vessel pressure at this phase in startup. An alternate approach was identified to add the liquid earlier in startup. The concept was fully developed and validated in the simulator before successful execution in startup.

In another example, a procedure required reducing feed to a process unit to a minimum value before proceeding to the next step. Executing this step in the OTS revealed that this action had an unanticipated consequence on a supporting system, eventually causing large process swings.

Implementation

Training all console supervisors on plant startup with the simulator was a significant achievement.

- Simulator demonstrated sufficient accuracy for effective training.
- All console supervisors experienced complete plant startup > 6 months ahead of actual plant start-up for the Baytown Cracker.
- All console supervisors experienced complete plant startup and other transient exercises in the two months prior to actual start-up.

Building a sufficiently accurate process simulator to achieve effective training required a significant commitment of manufacturing and process control resources. In many ways developing an OTS for a unit that has never run was more complex than for a running unit:

- Unit behavior in dynamic simulations is not known and intuitive for the trainer therefore significant process technical involvement is necessary to validate model behavior.
- Planned controls have never operated on unit. Simulator stability can be compromised by never before-tuned controls and therefore applications engineer involvement becomes critical even in early phases of simulator development.
- Actual startup training requires process engineer and applications engineer involvement just like a first ever startup of a unit.

The staffing plan for OTS development was consistent with meeting these challenges, requiring full time participation of process, technical, and process applications representations for several months.

Challenges

Between the start of dynamic simulator build and formal training, several key flowsheet changes and numerous P&ID changes occurred. The standard model build process would freeze P&IDs at the start of OTS development, but this is not useful or practical for a new plant in project development. Because the dynamic simulation team operated independent of the project team they were not aware of many of these changes until the initial model review.

In addition, for an operating plant a validated H&MB and operating data is available to the simulation team. For a non-existent plant, the best option is typically matching the model to the design H&MB. One key issue with this approach is overdesign:

- All equipment, especially exchangers, are tuned for design H&MB, but overdesigned in reality. As a result actual plant behavior and performance will not be accurate in the model.

- Design pressure drops built into the model can lead to parts of the model running at incorrect pressures depending on throughput.

All issues were successfully managed with reviews. Some key suggestions in scope identification for model development:

- Do not oversimplify the representation of key process equipment as this compromises the training utility.
- Carefully verify all model data in detail, line by line, before initial model reviews.
- The modeling team should avoid any assumptions design data (level tap locations, tan-tan dimensions, vessel elevations, etc.).
- If an assumption is made it should be clearly documented for review.

Because the intent of these models is for offline training only, risks are centered on lost opportunity costs resulting from diminished training experiences. For example, less than adequate ability to use the simulator for pre-startup training could potentially lead to multiple day startup delays. The adequacy of the training is also closely linked to the accuracy of the model behavior. For this reason, site engineers and operators are heavily involved during the creation of the OTS to approve of its quality. It is important to note that proper use of the simulator encourages conceptual learning rather than memorization of exact quantitative responses. Thus the end user needs practical understanding of model uncertainty or limitations and of how the process works.

Conclusions

Dynamic simulators were effectively used in training for the 2018 startup of the Baytown Olefins Plant's third train, contributing to the overall success and timeliness of the greenfield steam cracker startup. The simulation, manufacturing process, manufacturing technical, and control systems teams worked together to address identified issues to deliver an effective tool.

AICHE PAPER NUMBER 31B

PROMISING CHEMICAL TREATMENT APPROACH FOR 1,3-BUTADIENE FOULING CONTROL

Henrique Silva
LAM Regional Engineer
SUEZ - Water Technologies & Solutions

Tatiana Barbosa
Industry Subject Matter Expert
SUEZ - Water Technologies & Solutions

Fábio Rios
Olefins Process Engineer
Braskem

Williane Carneiro
Process Engineer Consultant
Braskem

Prepared for Presentation at the 2022 Spring Meeting,
April 10-14, 2022

PROMISING CHEMICAL TREATMENT APPROACH FOR 1,3-BUTADIENE FOULING CONTROL

Henrique Silva
LAM Regional Engineer
SUEZ - Water Techonologies & Solutions

Tatiana Barbosa
Industry Subject Matter Expert
SUEZ - Water Technologies & Solutions

Fábio Rios
Olefins Process Engineer
Braskem

Williane Carneiro
Process Engineer Consultant
Braskem

Abstract: At Braskem's Camaçari site, 1,3-butadiene is produced by extractive distillation with dimethylformamide (DMF) as the solvent. Since a longer run-length has been desired, the treatment of the extraction section has become more challenging, especially due to its historical issues with widespread polymerization at the bottoms of extractive towers. This paper recaps the general mechanisms involving fouling in the Butadiene extraction unit, followed by how SUEZ and Braskem are working together to control severe fouling with a promising chemical treatment program. The bottoms of extractive distillation towers operate at high temperatures severe enough to easily cause fouling of economizers, reboilers and tower chimney trays due to accelerated formation of rubber/sheet form of polybutadiene. The 4-Hydroxy-TEMPO chemical additive is commonly used in this system to inhibit this phenomenon, but for longer run-length, this approach has not proven to be adequately effective. To improve fouling inhibition efficiency, a new synergistic blend of 4-Hydroxy-TEMPO with a thermally selective activator has been used, covering critical areas of the process, showing improved results.

Introduction

By the end of this decade, the 1,3-butadiene market is expected to grow more than 30% as compared to 2022 due to the rising preference for this product in its several applications. Amid this scenario, 1,3-butadiene is also challenging to efficiently produce due to its very high reactivity and tendency for free-radical polymerization, producing high molecular weight polymer that can readily foul process equipment. Thus, one of the major concerns of reaching exceptional uptime levels in 1,3-butadiene plants is dealing with high fouling potential.

To make butadiene production more profitable, in the northeast of Brazil, Braskem's unit has been challenged, for the first time, to run a 6-year-campaign without any production issues and, consequently, critical fouling events must be avoided. This unit capacity, which started its operation in 1992, can deliver 118 kt of 1,3-butadiene per year. Historical data shows that one of the most common vulnerabilities in this unit is unscheduled shutdowns due to polybutadiene fouling in the bottoms of extractive distillation towers.

To minimize fouling and continuously run this unit for a new record run-length, Braskem and SUEZ have collaborated and managed several changes to the process control, maintenance, operating practices, and chemical treatment that are resulting in promising progress.

This paper discusses the change in the chemical treatment program and how positively it has impacted butadiene production results when combined with Braskem's good operating practices.

Brief Process Description

In Braskem, Q1 Unit II, the crude C4 is fed to in-series extractive distillation sections using dimethylformamide (DMF) as solvent. DMF can solubilize unsaturated compounds making butadiene distillation from other isomers easier. Hence, in the 1st extractive distillation tower (1st EDC), unsaturated and butadiene-alike components are removed in the bottom from lighter and saturated components in the presence of DMF. After the extract passes through the 1st stripper, the solvent is recovered. Then the butadiene and other isomers previously dissolved in DMF are sent to the 2nd extractive distillation tower (2nd EDC) to remove components heavier than 1,3-butadiene in the presence of DMF (Figure 1).

Most of, and purer, 1,3-butadiene stream leaves the top of the 2nd EDC and some amount of 1,3-butadiene is still dissolved in DMF along with the bottom stream which feeds a recovery column, bringing 1,3-butadiene back to the 2nd EDC. At T04, bottom products feed a 2nd stripper, and vinylacetylene (VAC) is removed from the process at the top. Both strippers provide hot solvent from their bottom streams, and solvent heat is recovered in some exchangers in the process. After leaving the bottom of the strippers, a small quantity of the solvent is sent to draw-off tar and other heavy contaminants by superheating the reclaimer (RCL). Other contaminants, lighter than DMF, such as dimers of butadiene, are removed by a distillation tower to purify the stream of solvent from the liquid distilled from the strippers.

The distilled stream from the 2nd EDC goes to the fractionation section, a double-series of simple distillation towers responsible for removing contaminant compounds that can be removed in the absence of DMF by simple distillation. The 1st fractionation column removes compounds lighter than 1,3-butadiene and the 2nd fractionation column, which receives the bottom product from the first, removes heavies in the bottom stream.

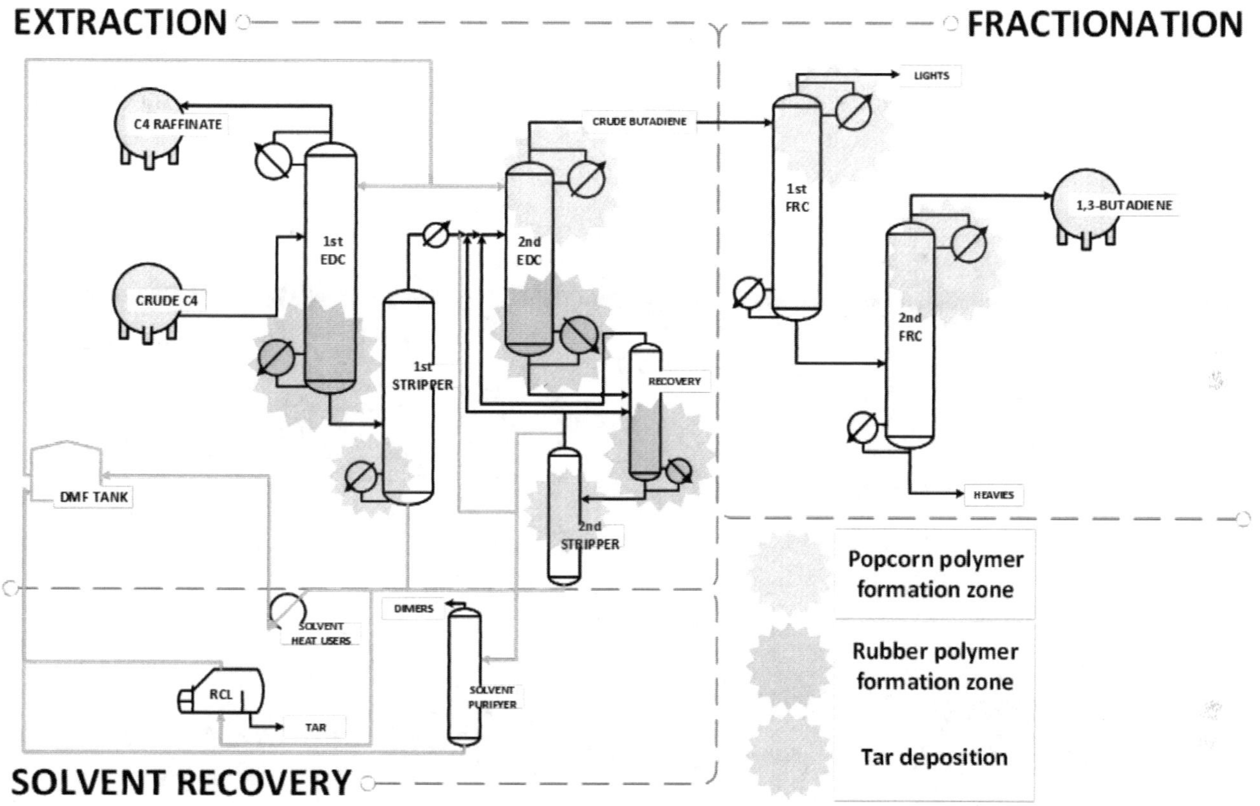

Figure 1. Simplified sketch of the butadiene plant.

Although popcorn polymer is usually a critical concern, rubber formation in hot areas of extraction and solvent loop are quite problematic, particularly if long runs (more than four years) are desired.

Historically, the Q1 Unit II unit has experienced a significant number of undesired shutdowns due to fouling in both extraction and fractionation areas. However, when oxygen intrusion was efficiently managed, popcorn formation could be avoided with the chemical treatment installed and already discussed in many discussion forums by butadiene and debutanizer producers. On the other hand, the rubber or sheet polymer formation mechanism is much more complex to deal with, especially because there is no way to avoid the high temperature demanded in the bottom of extraction towers and the accumulation tendency of deposit due to loop of solvent, characteristics inherent to this design.

Generally, the bottom of the 1st EDC and 2nd EDC is the most affected area by rubber fouling because of high temperature spots, about 130 °C that butadiene is exposed to. At such high temperatures, polymer inhibitor activity suffers a significant kinetic competition to polymerization induced by heat.

Fouling issues experienced in the extraction section

Usually seen in a rubber or sheet form fouled on extraction section, polybutadiene induced by high temperature can provoke a high molecular deposit strongly adherent to the metal surface resulting in a substantial increase of column pressure drop and several consequent troubles to extractive distillation, see Figures 2, 3 and 4.

Figure 2. Polybutadiene sheet adhered to a tray valve's surface.

The rubber/sheet form of polybutadiene is primarily induced by temperature but also can be catalyzed by other contaminants, such as oxygen and metals such as iron, for example, making the fouling rate even higher.

The mechanism responsible for the chain polymerization of butadiene is free-radical polymerization. Because of being a conjugated diolefin, 1,3-butadiene has a low activation energy enough to react under operating conditions of the production process.

One of the most significant challenges to dealing with heat-induced polymer in an extractive distillation column using DMF as solvent is to control fouling in condensing zone, that is, on vapor-liquid equilibrium interface and just below distillation trays. The last and hottest distillation trays as chimneys trays can also suffer catastrophic fouling if not controlled properly. Some chemical treatment strategies adopt dispersants as antifoulants to control fouling rate in this area, but in practice, it seems to help in the first years of run and after it accumulates even higher molecular weight residues that can foul unexpectedly, rendering cleaning maintenance even worse. Therefore, in long runs, dispersants could be a misleading choice. An additional complex matrix in the fouling with silicon-based antifoam and furfural degradation can bring more variables to the system.

Figure 3. Typical rubber formation on chimney trays and last tray seen below in the 1st EDC.

The Q1 Unit II last campaign started in 2012 did not achieve the desired goal. Premature cleaning was needed in both EDCs.

Figure 4. Q1 Unit II's 1st EDC obstructed valves from a bottom tray.

From the start of the run, after just over three years of continuous operation, pressure drop (dP) of 1st EDC increased by more than twice from start-up values, and an unscheduled shutdown for cleaning was needed (Figure 5).

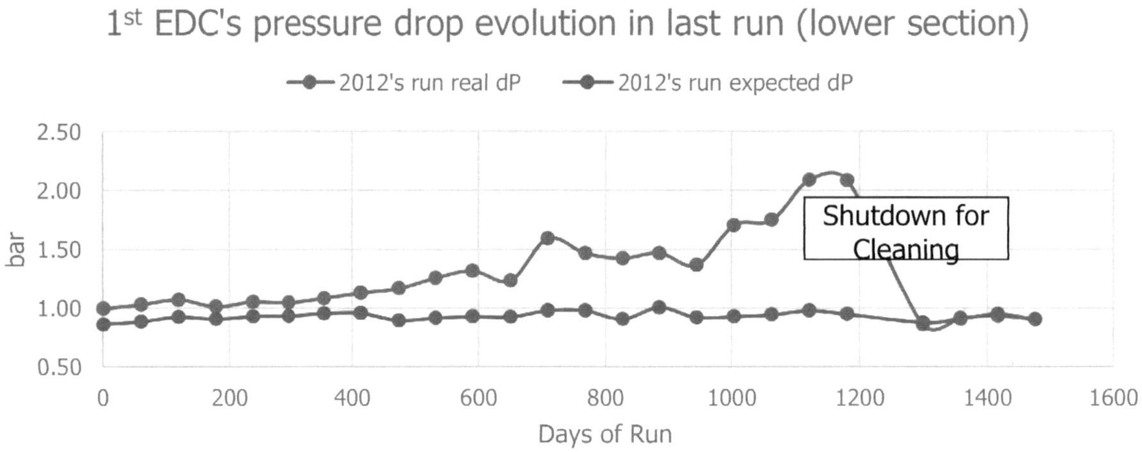

Figure 5. 1ST EDC´s real dP and expected dP by MRA

The expected dP is a calculated pressure drop behavior as a function of operating parameters using multiple regression analysis (MRA) at ideal conditions, normally right after the unit is started up. Hence, this curve can only see the initial conditions of the cycle, assumed at this time equipment is clean. Consequently, the expected dP represents column hydraulics without issues, including fouling effects.

On the other hand, real dP is the signal directly collected from the pressure drop indicator and shows an increase in response to plugged valves and restricted internal gas flow, so when too high, clean up of equipment is needed.

The difference calculated from the two dP curves above gives the differential dP, and a trend of positive values shows how much the pressure drop is higher than it should be at clean conditions. More than 1.0 bar of differential pressure drop was reached just before cleaning demand in the 1st EDC 2012's run case.

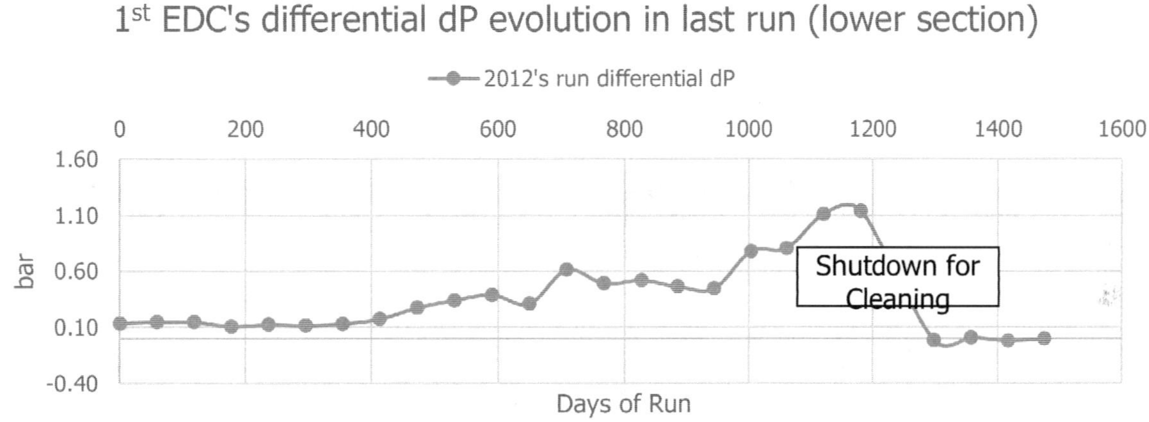

Figure 6. 1st EDC differential dP increase due to fouling

One of the important issues to notice in this campaign is that the 1st EDC real pressure drop at start-up was slightly higher than expected (about 0.14 bar), as seen in Figure 6, believed to be due to poor cleaning. Rigorous cleaning procedures at the intracycle cleaning occurred in 2015, which was a very important lesson adopted to guarantee the absence of polymers and seeds that can compromise the rest of campaign due to propagated polymerization in-chain reactions.

Since then, the rigorous cleaning has been considered one of the most impactful practices and deemed critical for the treatment program's success.

Some additional efforts directly related to the process that have been adopted to minimize fouling are:

- The original material of manhole's gaskets of distillation columns were replaced by a more torque resistant one. It helped to restrain the main source of oxygen intrusion in the plant;
- Hard control of the bottom temperature of EDCs to keep lowest possible values and minimize heat-induced fouling;

- Good operation of reclaimer to reach low tar levels, between 1 and 2 weight%, approach adopted by Braskem. It also helped reduce iron oxide;
- Air out under Nitrogen purge after cleaning filters and increased frequency of cleaning;
- Silicon residual analysis was started, which helped silicon dosing control;
- Investments in spare critical exchangers assisted the unit in avoiding intracycle shutdowns;
- Testing of nitrite residual in the solvent loop was started, which improved sodium nitrite dosing control.

Chemical treatment program updates

Besides starting a better control for nitrite residual, the chemical treatment program had other changes, such as dosages of antifoulants, which were optimized considering 1,3-butadiene concentration range in the feedstock. However, the philosophy to control rubber/sheet polymers gained a new aspect consisting of a new combination of chemicals in one product only, and never used in the plant before. SUEZ's new product consisted of an update to common 4-Hydroxy-TEMPO formulations combining the inhibitor with a thermally selective activator.

The coordinator is believed to make 4-Hydroxy-TEMPO more stable and effective in hotter areas, preventing it from being consumed unnecessarily in milder temperature zones. Thus, the combination gives 4-Hydroxy-TEMPO a level of selectivity to be available after reaching high temperatures, attacking free-radicals under elevated free-radical polymerization potential.

SUEZ believed that injecting the new antifoulant upstream of EDC's exchangers, in the hydrocarbon side, would provide better availability of the inhibitor to control fouling in critical zones. Figure 7 shows the scheme of the injection point.

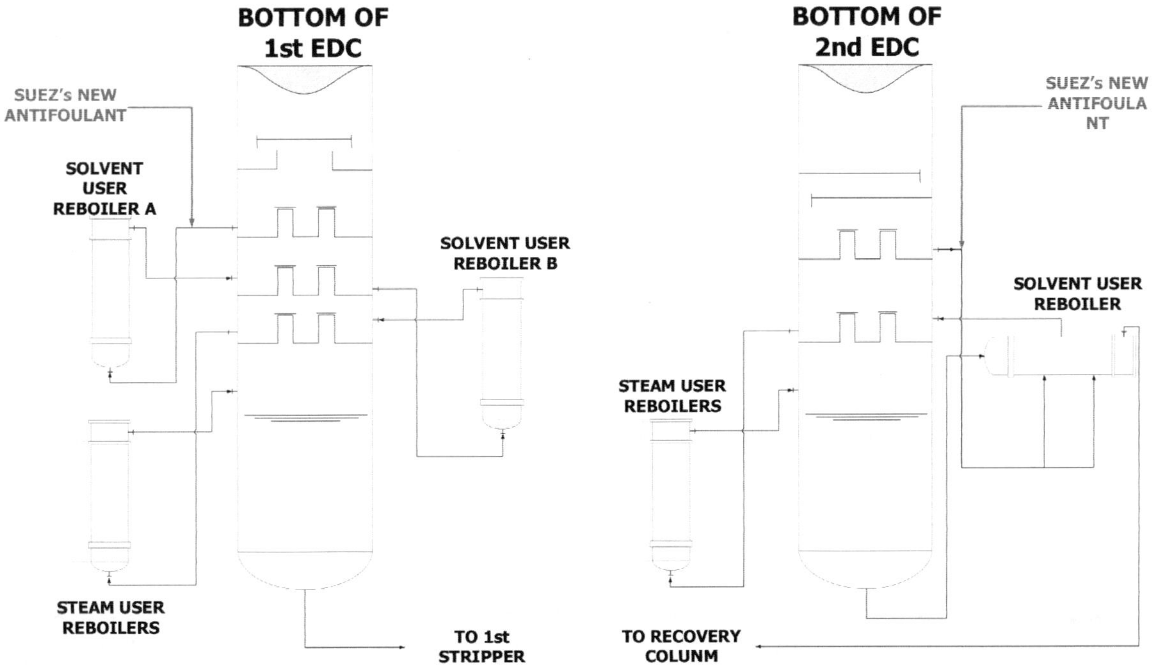

Figure 7. SUEZ's new antifoulant injection point in the bottom section of EDCs.

Other systems may also benefit from this new and improved chemical treatment approach. Most of the systems that use 4-Hydroxy-TEMPO in organic matter could upgrade to this new approach. Below are some possible systems that could be users of this technology:

- Light ends in ethylene plant-like deetanizer, depropanizer and debutanizers;
- Others 1,3-butadiene plant designs; and
- Isoprene extraction units.

Preliminary results of the current run

Since the new 6-year cycle was started in 2018, several positive results have already been seen under new strategies adopted. Compared to the past run, the 1st EDC has performed better, avoiding cleaning and showing no increase in differential dP, as shown in Figure 8.

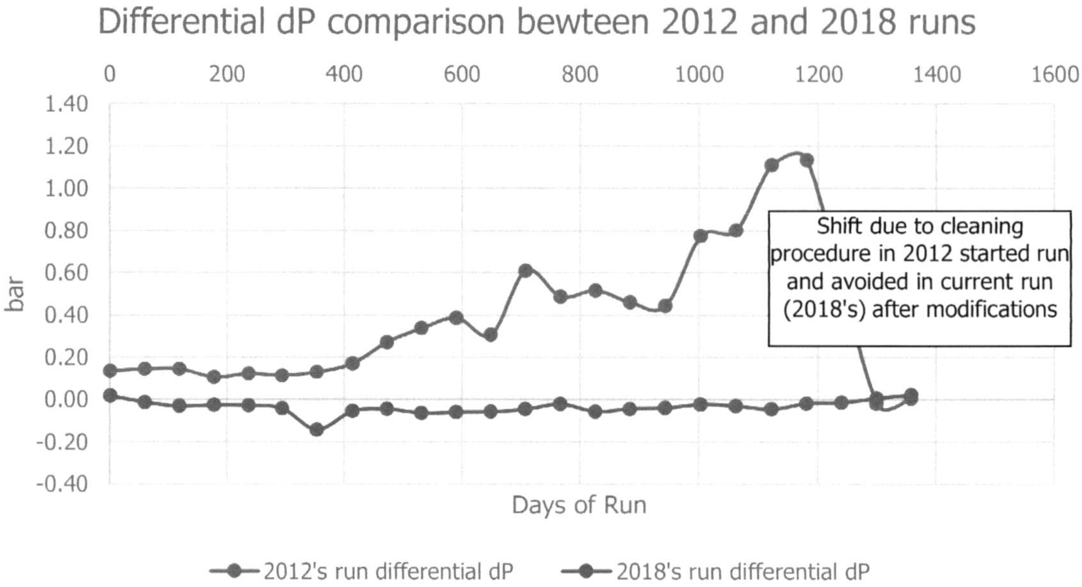

Figure 8. 1st EDC dP 2012 versus dp 2018 runs

It takes about 14 days to clean up this column (including shutdown and start-up procedures). Therefore, the impact of avoiding unscheduled shutdowns and subsequently ceased production is significant.

Because of a failure in the 2nd EDC pressure drop indicator, comparison between cycles is not available; if the run-length exceeds four years, that will be considered a success. Thus, the year 2022 will be proof this column run will overcome the last result.

As shown in Figure 9, the monitoring curves of pressure drop still give a promising expectation about increasing run length in 2nd EDC as well.

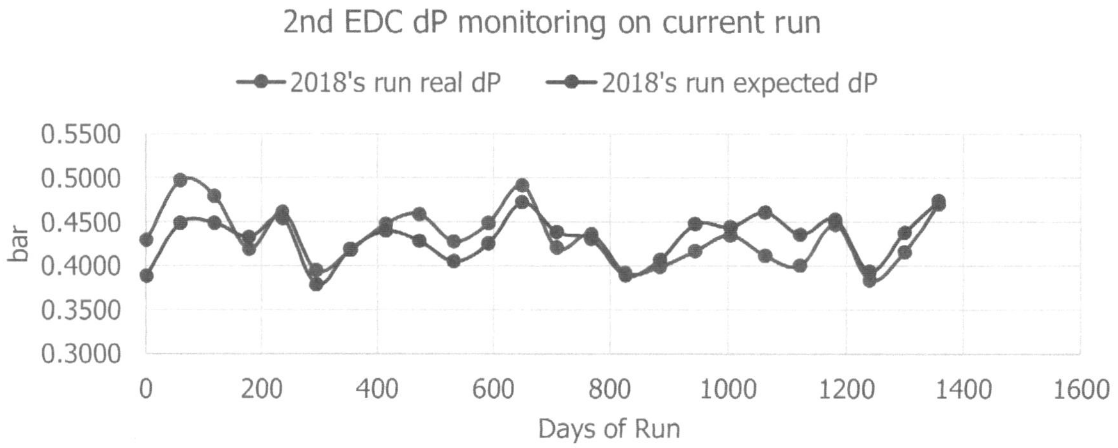

Figure 9. 2nd EDC dP trends on the current run.

Another notable impact is observed on the reboiler's run-length results, as shown in Figure 11. All critical equipment responded positively, exceeding the targeted campaign, with good condition for no hard cleaning, and most have far exceeded expectations, as shown in Figure 11.

Figure 10. 2nd EDC solvent user reboiler after 28 months of run.

Thousands of dollars have already been saved in cleaning procedures due to run-length increases. Additionally, the exchanger's cleaning has become easier with lower load adhered on the surface, and no damaged equipment by in-polymer volume expansion has been encountered.

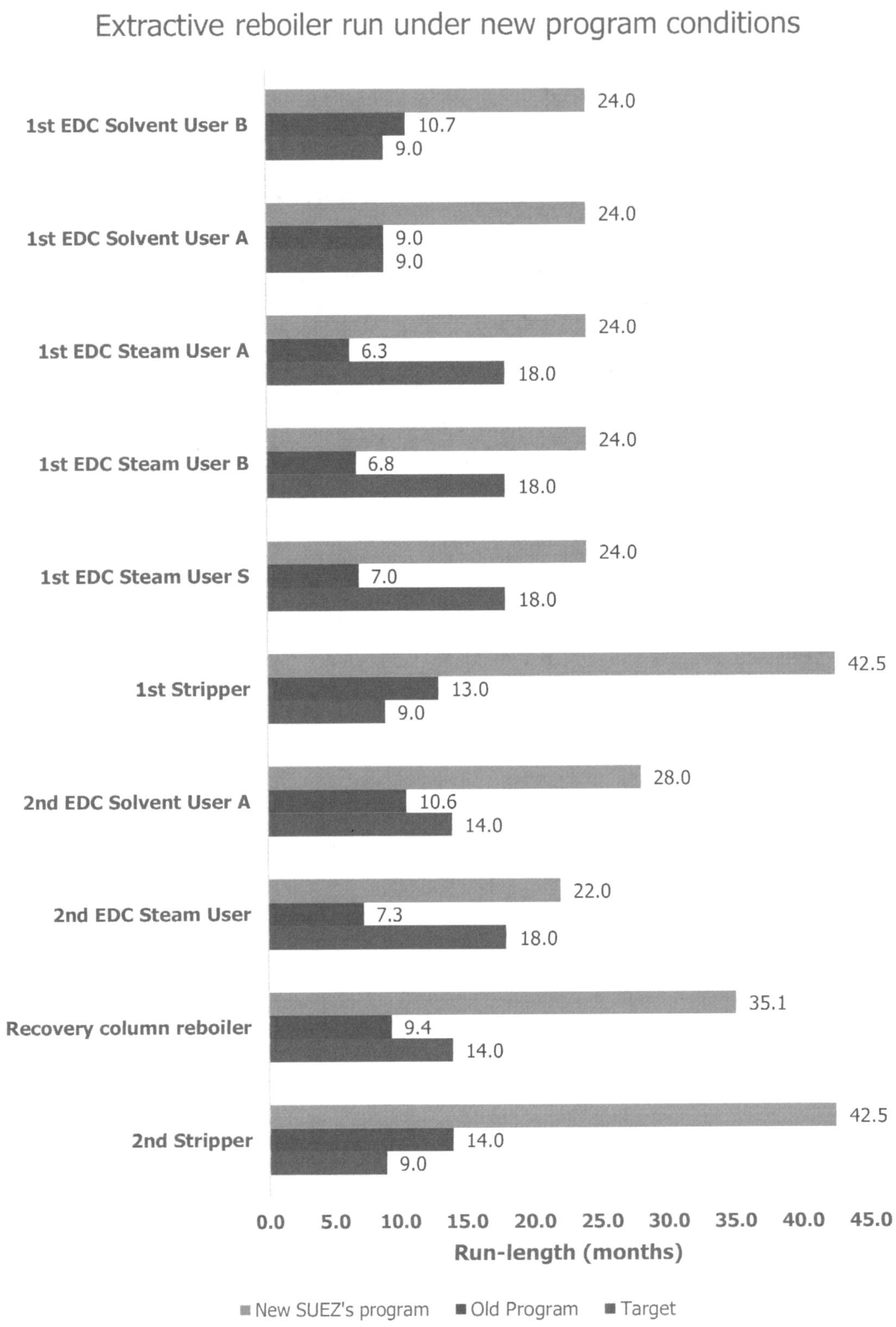

Figure 11. Reboiler's run length results.

Conclusions

Many improvements were adopted to control fouling in the extractive section of 1,3-butadiene. While the current unit cycle is heading into the 4th year of the campaign, there is no indicator to schedule a premature shutdown, and a 6-year-cycle is much more reasonable today than when this unit started up. Failures due to fouling seem to be in control. Historically, the plant that had many operating troubles due to free-radical polymerization, about 4-year run thus far without any hydraulic restriction, is already a major achievement. The newly adopted procedures have made a beneficial difference allowing SUEZ's chemical treatment program to work in the best possible with well-cleaned equipment and restricted level of contaminants. Also, it is clear, particularly by exchanger performance, that SUEZ's new antifoulant functions better than other common chemical treatments previously used in the extractive section. SUEZ's new antifoulant strategies could also be used in other systems such as light ends, Isoprene extraction, and different designs of butadiene units.

References

1. Robey, R., Wiese, H. & Morrel, C. (1944) "Stability of Butadiene". Esso Laboratories, Standard Oil Development Company, Elizabeth, N.
2. Joachim, G. & Eckhard Loser. (2005) "Butadiene". Wiley-VCH Verlag GmbH & Co. KGaA, Weinheim. 10.1002/14356007.a04_431.
3. Takao, S., Chigasaki-shi & Horaki, H. (1969) "Process for the separation of conjugated diolefins". The Japanese Geon Co.
4. Kirk-Othmer 5th ed. "Encyclopedia of Chemical Techonolgy". Vol 4. pages 365 to 384

31c

INCREASING BUTADIENE UNIT RUN LENGTH AND RELIABILITY WITH ADVANCED ANTIFOULANT TECHNOLOGY

Joice Boll
Global Technical Lead - Petrochemical
DORF KETAL

Américo Apóstolo
LATAM Business VP
DORF KETAL

Sudarshan Vijayaraghavan
SEA Sales Manager
DORF KETAL

Prepared for Presentation at the 2022 Spring National Meeting
San Antonio, Texas, April, 10-14, 2022

INCREASING BUTADIENE UNIT RUN LENGTH AND RELIABILITY WITH ADVANCED ANTIFOULANT TECHNOLOGY

Joice Boll
Global Technical Lead - Petrochemical
DORF KETAL

Américo Apóstolo
LATAM Business VP
DORF KETAL

Sudarshan Vijayaraghavan
SEA Sales Manager
DORF KETAL

Abstract: In recent years, ethylene producers have made significant progress to reduce costs and increase productivity by increasing the time between plant turnarounds. This is increasing the need for improved run-length and reliability of the butadiene section of the olefins complex. Fouling control in the extraction section of the butadiene process is one of the most important challenges limiting butadiene run-length and reliability. This paper discusses the fouling mechanisms in the extraction section and the challenges faced by butadiene producers to control fouling. Best practices and case studies are presented that show how improved fouling control can increase run-length and reduce maintenance and energy costs. One case history focuses on improving control of stripper column pressure drop. The second case history discusses how pump strainer cleaning frequency can be improved with better fouling control and how this improvement is correlated with performance of the recovery heat exchanger and stripper reboiler. In both cases, improvements in monitoring and control of fouling provided the olefins complex with better predictability of planned outages.

Introduction

There is an increasing need for improved run-length and reliability in the butadiene section of the olefins complex to match improvements in ethylene plant reliability. In the Butadiene unit, improved fouling control plays a crucial role in enhancing plant run-length, productivity, and reliability, and contributes to reducing maintenance costs, energy costs and carbon emissions.

In a butadiene plant, a major challenge to meeting targeted run-length is often related to fouling in the extraction section. Fouling in the extraction section due to polymerization is caused by a combination of factors, including monomer concentration, iron, oxygen, presence of C5s, low velocities, temperature, and residence time. Solvent degradation and the use of anti-foams increase the complexity of controlling the overall rate of fouling. Fouling is a non-linear process with potential for unexpected spikes in the rate of fouling. The lack of predictability increases the risk of an unexpected interruption in production and all costs involved with an unplanned shutdown.

This article describes the chemistry of fouling in the extraction section and the application of ACtify™ technology for improved fouling control. ACtify™ is the brand name used for the patented polymer inhibitor technology of DORF KETAL. It has been discovered that certain tertiary amines formulated with existing inhibitor technologies (like 4-hydroxy-tempo or 4HT) dramatically improves the speed and efficiency of polymer inhibition reactions[1].

ACtify™ technology has been used successfully in styrene units for several years[2,3], and recently has been shown to improve polymer control in the extraction section in butadiene. In the butadiene polymerization mechanism, ACtify™ technology enhances reaction speed of the inhibitor to reduce polymer formation and reduces dimer formation, resulting in lower 1,3-butadiene losses. The development of the first patented dimer inhibitor is testimony to the impact of ACtify™ technology on the speed and efficiency of inhibitor chemistries.

Case studies presented here demonstrate how improved monitoring techniques and advanced antifoulant technologies offer the butadiene producer a substantial improvement in fouling control. This offers better predictability of planned outages, improving productivity and reducing energy and maintenance costs and reducing risk of unbudgeted outages.

Butadiene Production Process

Along with ethylene and propylene, 1,3-butadiene (1,3-BD) is one of the most valuable monomers produced through the steam cracking of hydrocarbons. Once fractionated in the debutanizer of the olefins plant, the C4 cut containing a mixture of butanes, butenes and butadienes (commonly called crude C4s), is the feedstock to the butadiene unit. The concentration of 1,3-BD in the crude C4 stream is typically on the order of 40 – 60%, and balance of it is containing other C4 compounds along with C3 and C5 as contaminants (see table 1).

Table 1 – Usual composition for Crude C4.

Crude C4 Component	Usual Range (%)	Boiling Point (°C)
1-3 Butadiene	40 - 60	-4.4
Iso-Butene	5 - 30	-7
1-Butene	5 - 15	-6.9
Trans-2-Butene	2 - 7	0.9
Cis-2-Butene	1 - 5	3.7
n-Butane	4%	-1
Iso-Butane	0.6%	-11.7
Ethyl Acetylene	0.01 - 1.0%	8.08
Vinyl Acetylene	0.20 - 2.00%	5.01
1,2 Butadiene	0.20 - 0.50%	10.8

Due to the close boiling points and relative volatilities of butadiene and other C4 compounds, they cannot be commercially separated through a simple distillation. Selective solvents are used to extract the 1,3-BD from the rest of the C4s. There are several commercial designs of butadiene units using solvents such as N-methylpyrrolidone (NMP), dimethylformamide (DMF), or acetonitrile (ACN), to extract the 1,3-BD from the crude C4 mixture. After extraction, the 1,3-BD is removed from the solvent and distilled to a high purity product (see figure 1).

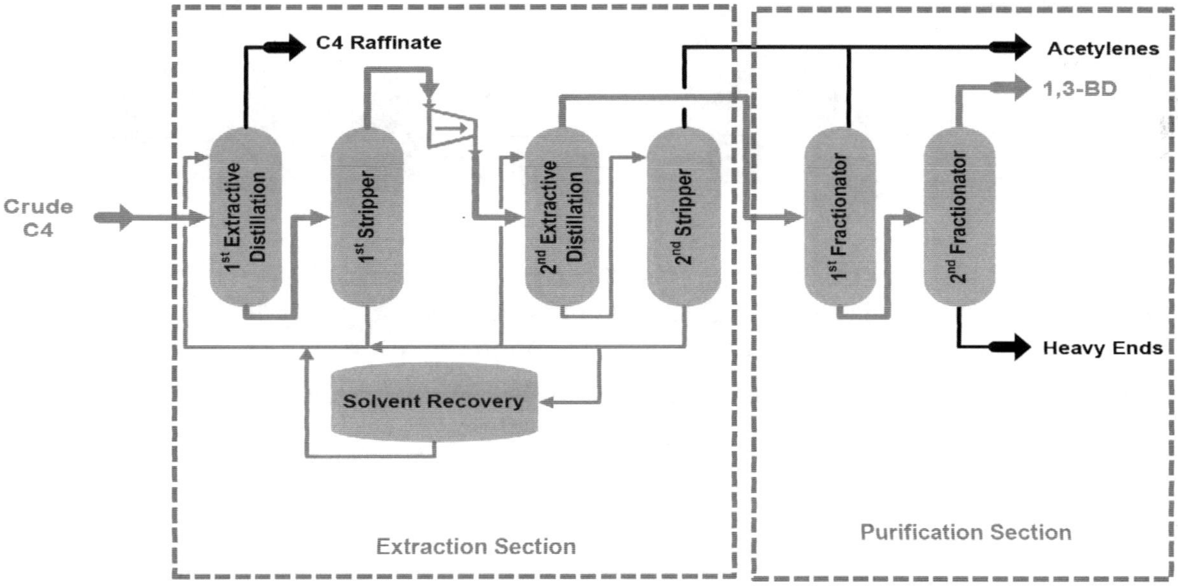

Figure 1 – Generic 1,3-butadiene extraction unit diagram.

In first extractive distillation column butanes and butenes, along with less soluble impurities in the solvent are removed from the overhead as C4 raffinate, and 1,3-BD containing stream (along with solvent) is sent to 1st stripper column. The 1st stripper column separates the main stream (relatively concentrated 1,3-BD) from solvent, sending it to a compressor. After compressing, this stream rich in 1,3-BD enters the second extractive distillation column.

The overhead stream of the second extractive distillation column concentrates 1,3-BD for downstream purification. The purification section removes acetylenes and other minor contaminants, thereby enhancing the 1,3-BD purity from 95-98% to at least 99.5% for the final product withdrawal. Solvent recovery unit is continuously working by receiving part of the solvent during the process & remove impurities such as water & heavies (degraded products, polymers & inorganics) from the solvent to meet solvent quality and specification.

Extraction Section Fouling Control

It is well known that both temperature and peroxide-initiated mechanisms contribute to the polymerization tendency of 1,3-butadiene. The temperature-initiated polymers are usually seen in the extraction section (figure 2a), and the peroxide-initiated polymers occur in the purification section (figure 2b – also known as popcorn polymer). Considering the temperatures and concentrations of 1,3-BD, the butadiene unit is prone to fouling in all sections.

Figure 2 (a)

Figure 2 (b)

The fouling seen in the extraction section is usually not related exclusively to the polybutadiene polymer, but also due to solvent degradation, iron catalysis, C5, oxygen contamination, and antifoam injection. Hence, fouling appearance can also vary significantly, as can be seen in Figure 3.

Apart from the foulant materials already described, the units using DMF as solvent can face challenges associated with solvent hydrolysis & slow degradation to formic acid and di-methyl-amine (DMA). Also, the injection of furfural (standard inhibitor indicated for DMF solvents) in excess in the unit generates furfural poly-condensate, which react with DMA to favor fouling.

The polymer formation in extraction area can happen through different mechanisms, including Diels-Alder dimerization, Aldol condensation, and free-radical polymerization. The free-radical mechanism is definitely the most predominant, being responsible for reduction in heat exchanger performance and increased pressure drop across column trays or packing. See Table 2 for more details on the mechanisms.

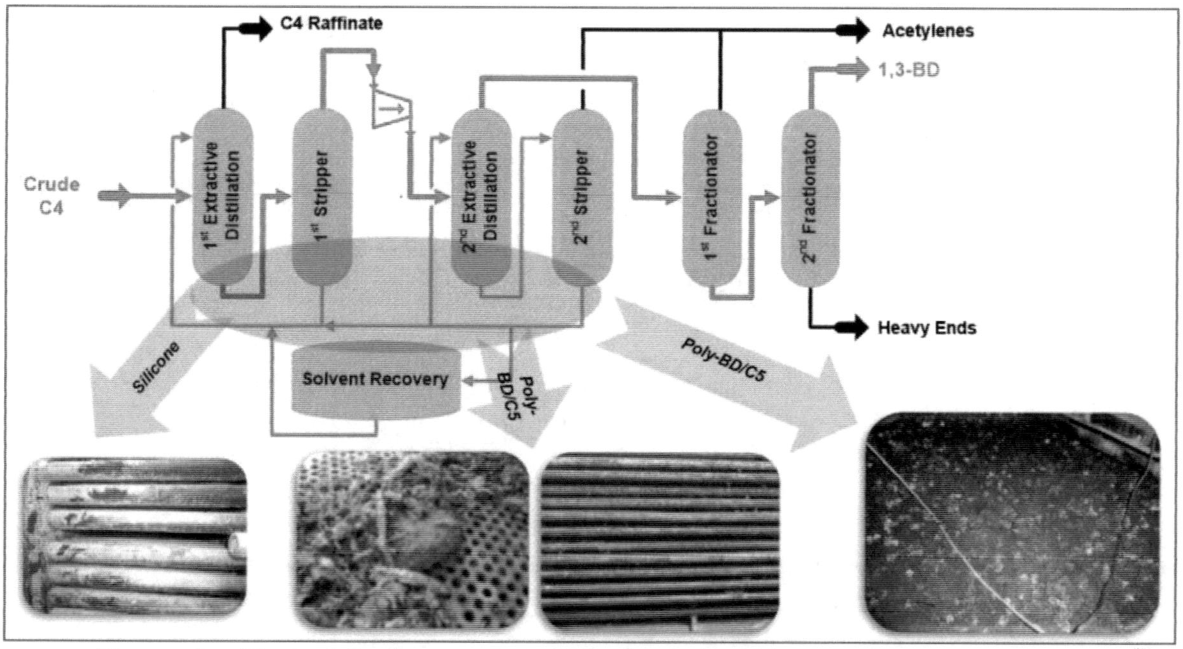

Figure 3 – Typical foulant materials find in a butadiene extraction section.

Table 2 – Reaction mechanisms seen in extraction section.

Diels-Alder Dimerization	**Aldol condensation**
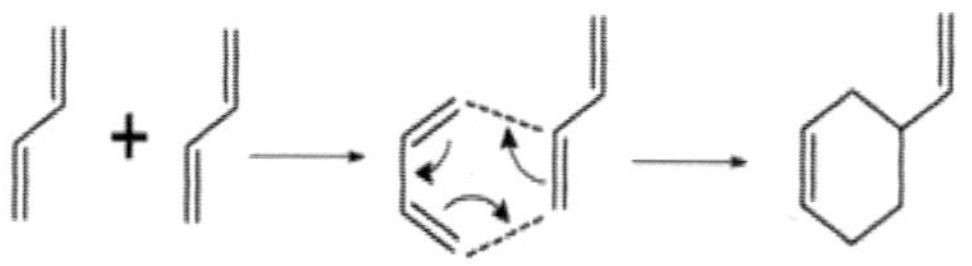	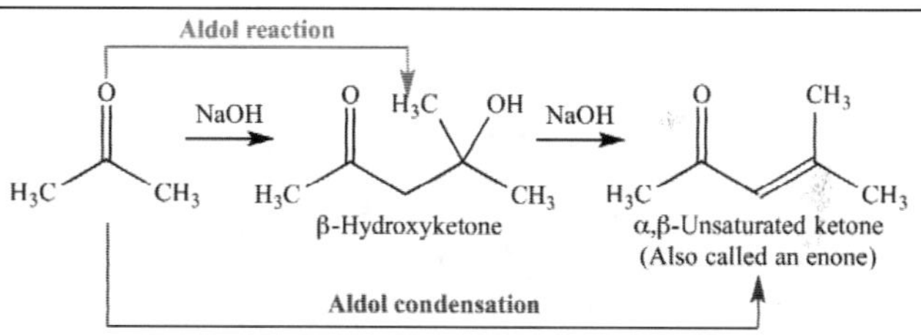
4-vinylcyclohexene (VCH) is obtained through Diels-Alder condensation process[4]. This is an endothermic reaction and only depends on temperature and residence time[5].	Carbonyl compounds can suffer aldol condensation in the presence of strong bases. In the butadiene unit, concentration of carbonyls in feed will be the critical factor to observe this mechanism.
Free-radical	
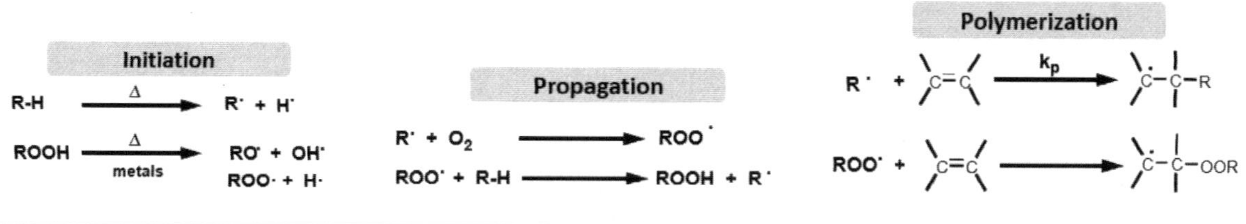	
Free radical formation is the most important contributor to fouling in this region. The radical substitution reaction occurs due to heat, high monomer concentration, and presence of peroxide and metals. The molecular weight of the component increases over time.	

Other factors that complicate fouling control include periods of low load conditions or efforts to push plant capacity above design by processing C4 feedstocks from other ethylene plants. Combination of these mechanisms promote challenges in fouling control in the solvent loop. In many cases, the default treatment recommended by licensors is not sufficient to meet challenging operating goals that were not part of the original design.

Best Practices for Extraction Section

Feed quality is a vital parameter for any unit, and in 1,3-BD production some contaminants in the feed can contribute in fouling. Table 3, based on DORF KETAL's experience, presents best practice for feed quality in butadiene to minimize the risk of fouling.

Table 3 – Crude C4 contaminants monitoring.

Feed Parameters	**UOM**	**Best Practice**
C5 content	ppm	< 500
Peroxides content	ppm	< 2.0
Carbonyl content	ppm	< 30

Solvent quality monitoring is also critical to extending butadiene unit run-length, guaranteeing that polymers, heavies, and/or degradation products are properly deconcentrated. Table 4 offers best practices for solvent quality.

Table 4 – Solvent quality monitoring parameters.

Solvent Parameters	**UOM**	**Best Practice**	
		DMF	**NMP**
Water	ppm/%	< 300	8 to 10
Iron	ppm	< 3	NA
Sodium Nitrite	ppm	1 - 2	200 – 300
Furfural	%	1.5 - 3.0	NA
Residue (Heavies)	%	< 2.0	< 0.2
Dimer	ppm	< 100 ppm	
C4	ppm	< 10 ppm	

Significant solvent degradation leads to high residue (heavies) in the system, increasing the tendency for deposits in low velocity zones. More dimer and C4 in the solvent increases potential for polymerization; reductions in these numbers help maintain healthy operation.

Silicon based antifoam is generally used by butadiene plants. High dosages of antifoam can lead to the deterioration in heat exchanger/reboilers run-length. Silicon is often found in deposits in low velocity zones. Once deposited, it cannot be easily removed, compromising heat transfer and causing early cleaning of heat exchangers. Antifoam injection should be monitored and reduced as much as possible. Typical injection rates are lower than 0.02 - 0.04 kg/ton BD.

Chemistry of polymer control

4-Hydroxy-Tempo (4HT) is a technology often used for free-radical polymer inhibition in ethylene, butadiene, and styrene production. The reaction rate (k1) in Figure 4 represents the speed of polymer inhibition. By activating the 4HT with ACtify™, inhibition speed increases (ka) significantly.

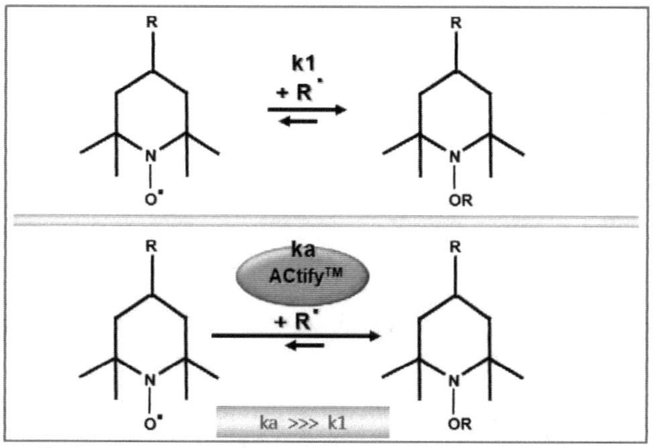

Figure 4 – Schematic representation of 4HT inhibition reaction.

The first reaction of two butadiene monomers in the solvent loop is the formation of butadiene dimer. It has long been theorized that dimer formation is not only formed by a condensation reaction, but partly the result of a free-radical step[5]. High temperatures in the solvent loop increase the rate of free radical and dimer formation. The speed of reaction of ACtify™ chemistry is fast enough to inhibit the free-radical mechanism of dimer formation. Slowing polymer formation earlier in the polymerization process gives ACtify™ a significant advantage over other options as the concentration of free radicals are lower, reducing inhibitor demand. Faster terminating reactions reduces the chemical required to control polymer formation and lowers the polymer molecular weight, reducing fouling tendency.

As shown in figure 5, the ACtify™ version of 4HT works better than regular 4HT, at less than half of dosage. The faster reaction speed of ACtify™ chemistry with free-radicals is lowering the concentration of free-radicals and polymer's molecular weights, reducing risk of deposition.

An additional benefit of ACtify™ technology is the potential to reduce NOx formation[6] in the ethylene furnace in the event there is a need to reprocess C4 streams from the butadiene unit to the ethylene furnace, since lower dosages can be used to control the polymerization process in the Butadiene unit.

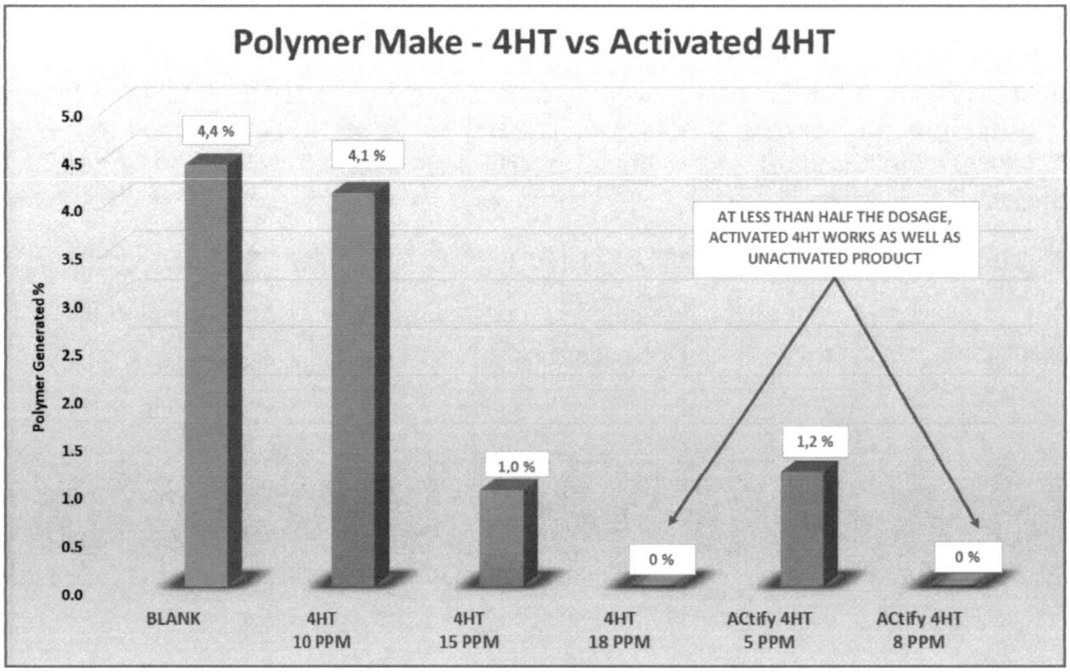

Figure 5 - Polymerization test with 4HT (Butadiene - 248F/120°C - 4 hrs) – Activated with ACtify™ and non-activated.

Patent of Butadiene Dimer Inhibition

Patent US-9598333 presents the technology for inhibiting butadiene dimers and polymers, where certain tertiary amines can dramatically improve existing inhibitor technologies. It confirms that dimer formation (via free-radical mechanism) are being inhibited due to faster efficacy of ACtify™ inhibitors.

Case Studies

Case study 1

This 100,000 tpy butadiene unit faced a bottle-neck situation due to severe fouling in the solvent stripper column. The flow corrected Delta P was increasing continuously since the first months of the campaign, with abrupt spikes by the end of run (Figure 6). Prior to implementation of the DORF KETAL treatment, the run-length reached up to 18 months with maximum throughput. Unscheduled shutdowns were necessary to overcome the bottleneck. With the use of ACtify™ technology, the plant has been able to achieve the targeted run-length of 36 months (green curve in Figure 6).

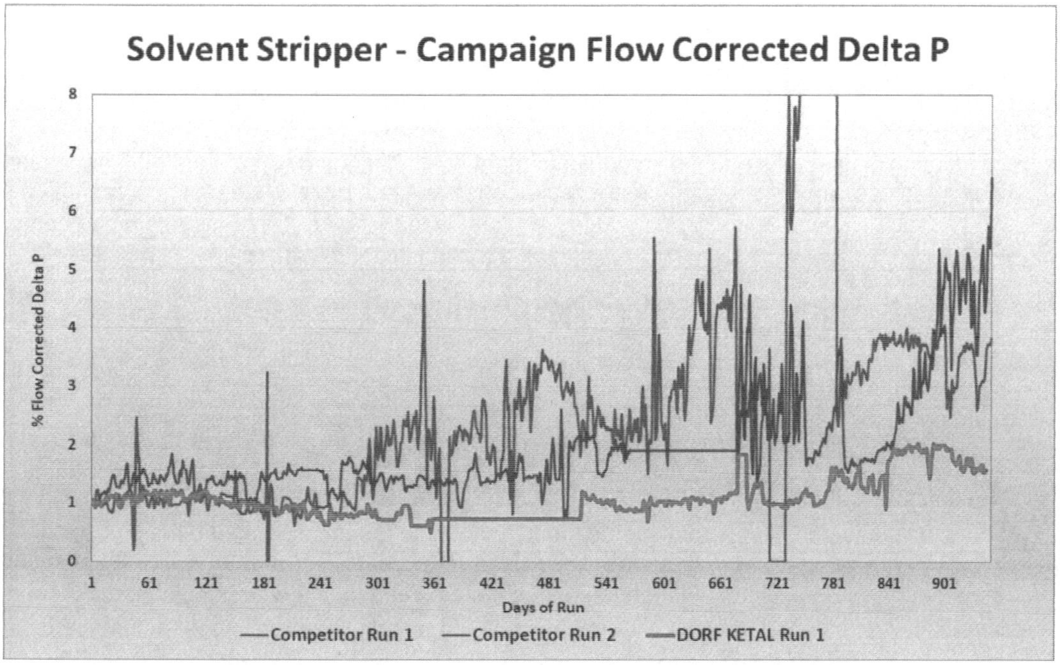

Figure 6 – Solvent stripper flow corrected delta P in different runs.

ACtify™ treatment also improved plant profitability by reducing maintenance and eergy costs by enhancing the efficiency of heat recovery system. Figure 7 shows comparative heat transfer performance.

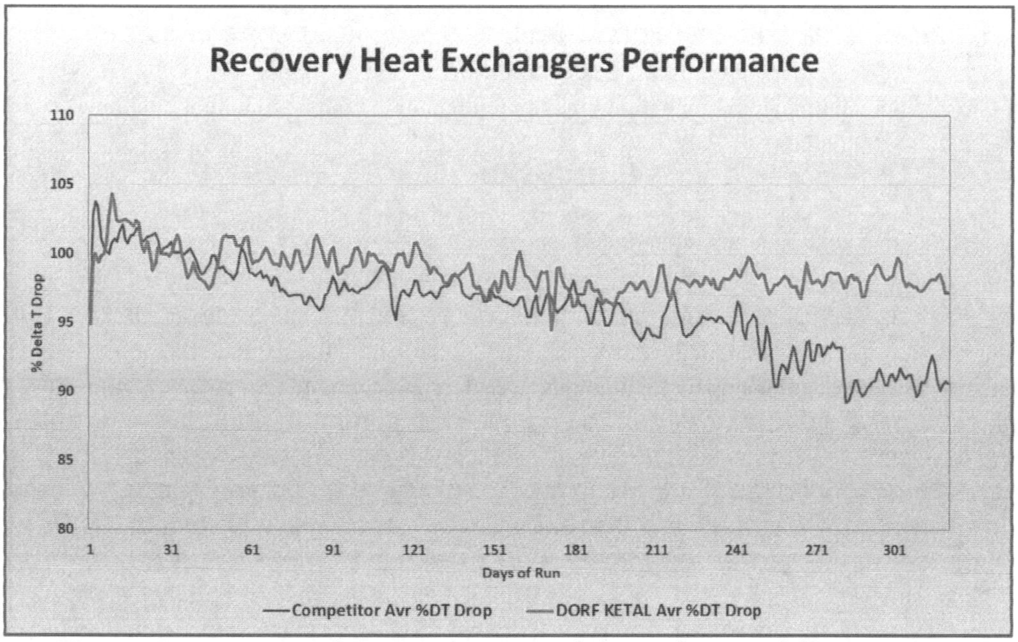

Figure 7 – Recovery heat exchangers delta T drop in different runs.

Case study 2

This 120,000 tpy butadiene unit was operating with licensor recommended anti-foam and solvent corrosion inhibitor. Within two years after start-up, the plant began experiencing an increase in strainer cleaning frequency and a significant reduction in heat transfer efficiency. Foulant analysis indicated the presence of butadiene polymers in the extraction section. Since the implementation of ACtify™ technology, the polymer formation and accumulation has dramatically reduced. Figure 8 shows how the time between strainer cleaning has increased.

The improvement in polymer control has stabilized the performance of heat exchanger and stripper reboiler, a clear indication that fouling is under control.

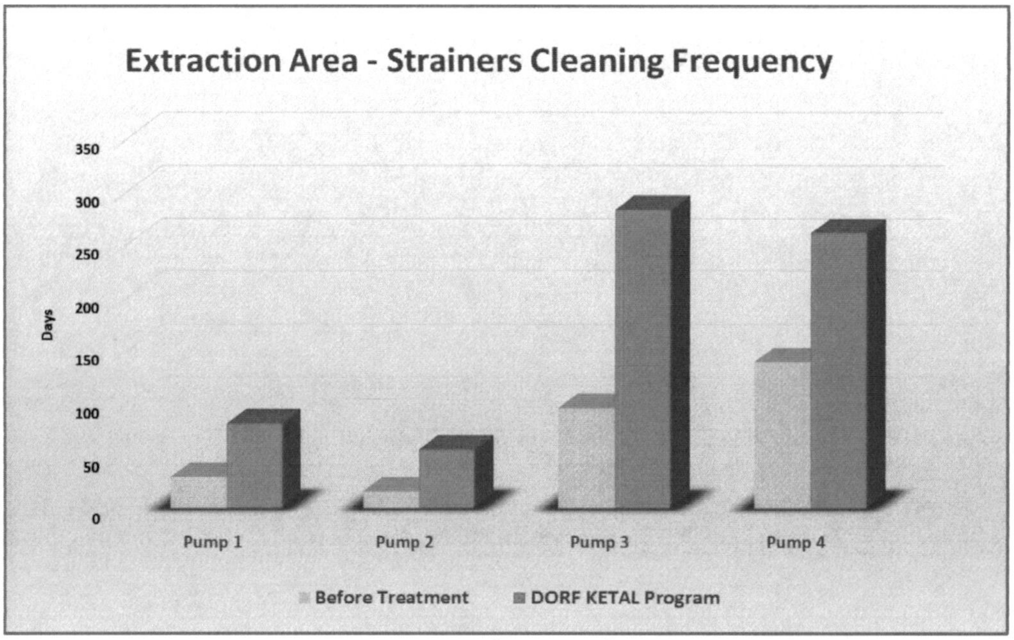

Figure 8 – Extraction area pump strainers cleaning frequency, before and after chemical treatment.

Conclusion

There is an increasing need for improved run-length and reliability of the butadiene section of the olefins complex. Fouling control in the extraction section of the butadiene process is one of the most important variables that can limit butadiene plant run-length and reliability. Fouling is a non-linear and difficult to predict process that creates operating risk.

Design of chemical treatment needs to consider the plant conditions, run-length targets, and presence of contaminants. As the case histories have shown, fouling control can be improved with an advanced antifoulant strategy. ACtify™ technology, a new generation treatment program, offers faster reaction kinetics with commonly available free-radical inhibitors. Faster termination of free radicals reduces dimer formation, subsequent formation of higher molecular weight polymers, and chemical demand. Faster fouling control reduces the risk of non-linear, uncontrolled fouling in the extraction section. This offers better predictability of planned outages, improving productivity and reducing energy and maintenance costs. The financial risk of unplanned outages is also reduced.

References

1. SUBRAMANIYAM, M., 'Additive Composition for control and inhibition of aliphatic monomers, and methods and preparation and use thereof', US 9598333, United States Patent and Trademark Office, (20 March 2017).
2. ONDYAK, J. NOLAND, J. and RAMASWAMY, P. N., 'Comparing DNBP Replacement Options in Styrene Production', *Hydrocarbon Processing*, (November 2017).
3. CHOKSI, J. et.al., 'A Study in Sustainable Styrene,' *Hydrocarbon Engineering*, (December 2018), pp. 21 - 24.
4. ALDEEB, A., ROGERS, W., and MANAAN, M., 'Evaluation of 1,3, butadiene dimerization and secondary reactions in the presence of oxygen', *Journal of Hazardous Materials*, (August 2014).
5. 'Butadiene Product Stewardship Manual', American Chemistry Council, (2019).
6. SUBRAMANIYAM, M., and KNIS, M, 'Quantifying the Risk of NOx Accumulation in the Ethylene Cold Box from the Use of Polymerization Inhibitors,' Paper 41C, Ethylene Producers Conference, (April 2021).

639905 Revamp of Yncc No. 2 Ethylene Plant

Abstract:

Yeochun NCC Co., Ltd (YNCC) Ethylene Plant No. 2 is located near Yeochun, Korea. The plant No. 2 ethylene technology was licensed by Technip Energies Process Technology (formerly Stone and Webster) in 1992. The original capacity of this plant was 350 KTA of polymer grade ethylene with the equipment designed for capacity of 450 KTA (based on 8,000 operating hours per year). YNCC#2 revamped plant on various occasions to 543 KTA. In 2009 Technip Energies Process Technology performed a feasibility study to identify the bottlenecks and required scope of modifications. The objective was to increase the operating capacity of the plant from 543 KTA to 859 KTA. In 2018, YNCC decided to implement 859 KTA expansion and contracted Technip Energies Process Technology to provide Basic Engineering Package (BEP) and Ethylene Technology Licensing for the planned capacity, as well as efficiency increase for YNCC No. 2 Ethylene Plant. This important plant revamp had many challenges due to the large capacity increase of ~160% from the current capacity.

This paper will discuss the various revamp techniques and construction challenges in this large capacity increase. Few examples are:

- Addition of two 192U furnaces and a gas turbine on a tight plot space.
- All three major compressors (Cracked Gas, Ethylene Refrigeration and Propylene Refrigeration) had a foot print replacement on the same foundation.
- High capacity trays were used in many towers.
- Parallel towers were added for few services where required capacity increase was beyond capacity limit of the high capacity trays.
- Efforts were done to minimize the parallel equipment addition due to the limited availability of the plot space. Many equipment were therefore replaced on the same foundation instead of adding a parallel equipment.

YNCC along with its construction partners Daelim and Hanwha successfully completed the construction in December 2020 and plant was started up and achieved the targeted ethylene production in January 2021.

SPRING22 +18TH GCPS
A Joint AIChE and CCPS Meeting
APRIL 10-14, 2022 • SAN ANTONIO, TEXAS

AIChE Paper 31e

Caustic Tower – Bench Market Study and New Analytical Method for Polymer Content

Steven Imbert
Senior Product Application Specialist Petrochemicals
SUEZ – Water Technologies & Solutions, Belgium

Joop Dees
Petrochemical Technology Leader EMEA
SUEZ – Water Technologies & Solutions, The Netherlands

Ruben Nackaerts
Lead Technologist
SUEZ – Water Technologies & Solutions, Belgium

David Hood
Senior Customer Application Engineer
SUEZ – Water Technologies & Solutions, USA

Prepared for AIChE 2022 Spring Meeting –
34th Ethylene Producers' Conference
San Antonio, Texas, April 10-14, 2022

Benchmarking Study and New Analytical Method for Polymer Content in Caustic Towers

ABSTRACT:

The process gas in ethylene units contains acid gases, like carbon dioxide and hydrogen sulfide, resulting from cracking process and coking/decoking reactions in the furnaces. The removal of these acid gases, to avoid downstream product contamination and/or other side-effects like catalyst poisoning, takes place in the caustic tower, which is integrated into the charge gas compressor.

A variety of other components, like aldehydes and ketones, are present as contaminants in the cracked gases and tend to start polymerizing under the basic conditions in the caustic tower to form aldol condensation polymer.

These reactive growing polymers can ultimately precipitate from the caustic solvent and deposit on the internals of the caustic tower, causing separation issues in the spent caustic drum and subsequently impacting the downstream spent caustic stripper, storage tank and wet air oxidizer.

The polymer formation can result in severe fouling, which can significantly impact the plant's overall efficiency and reliability (e.g., via throughput limitations or unscheduled unit shutdowns for cleaning activities).

A global benchmarking study and laboratory test work illustrate the importance of the caustic tower's operating conditions, such as temperature, caustic strength, and aldehyde content, on fouling tendency.

Finally, a new analytical method for quantifying polymer has been developed by a team of SUEZ experts. The method is being used to screen inhibitors in the lab and implemented at several ethylene crackers to monitor the fouling potential. The results have been compared with the classic analytical method (polymer extraction with solvent). Applying the new method reduces the analysis time from several hours to minutes, which enables more efficient fouling control of the unit. Additionally, the new method yields more repeatable and reliable results.

1.0 Process Overview

Cracked gases exiting the furnaces in ethylene plants contain acid gases, such as carbon dioxide and hydrogen sulfide. Carbon dioxide is present because of the coking/decoking reactions that typically occur in the cracking furnaces. The presence of hydrogen sulfide results from trace amounts of sulfur compounds in the feedstock and sulfur compounds injected into the furnaces to minimize coking and CO formation during the cracking reaction. These contaminants should be removed from the cracked gas to prevent any adverse effect on the acetylene converter catalyst and meet product specifications for polymer grade ethylene.

The acid gases are removed by one of the two following methods: (1) an alkanol amine system followed by a caustic wash, or (2) a caustic wash only. In the case where acid gas levels are relatively low, caustic washing alone provides an adequate and economical removal of these contaminants. Caustic washing systems are usually located between the third and fourth stage, or the fourth and fifth stage of the charge gas compressor.

The ethylene industry faces challenging times, with a strongly competitive environment and ever-changing feedstock prices and market conditions. Profitability is strongly affected by the geographical location of the unit and the type of feedstock processed. The type of feedstock also significantly affects the overall fouling tendency in the unit. For example, the presence of oxygenates in gas feeds can accelerate fouling in the caustic tower due to resulting elevated concentrations of aldehydes and ketones in the process gas.

2.0 Caustic Tower

The acid gases in the cracked gas are removed by countercurrent contact with circulating caustic solutions. The caustic system may consist of one or more towers, which are divided into several caustic sections of varying strengths with a water wash of the cracked gas before it leaves the system.

Typically, the caustic system has two or three caustic sections or loops. The first loop (called the bottom loop or weak caustic section) contains a circulating solution of caustic with 1-3% free caustic. The cracked gas enters the tower below this section while the spent caustic is removed via a skim system. To maintain a low caustic usage, a steam-cracking unit typically runs the caustic strength as low as possible in this section. Conversely, the concentration of aldehydes and ketones is at its highest, which promotes aldol condensation polymerization.

Above the weak section, the 2nd caustic circulation loop (called middle, but in some designs the strong caustic section) operates with a free caustic level, where This section may vary from 5-10%. If a plant has a third caustic stage, it will operate at the highest free caustic level of the three sections, and typically in the order of magnitude 8-10%.

The water wash section on top is used to avoid caustic carryover. Caustic carryover can result in a reduction of efficiency of the process gas compressor.

The make-up caustic is added to the strong section of the tower to maintain the required caustic concentrations. The plant receives concentrated caustic into the plant and dilutes it down to 10-20% for use in the caustic tower. The make-up caustic rate is roughly equal to the spent caustic rate. Figure 1 shows the typical configuration of a caustic scrubber.

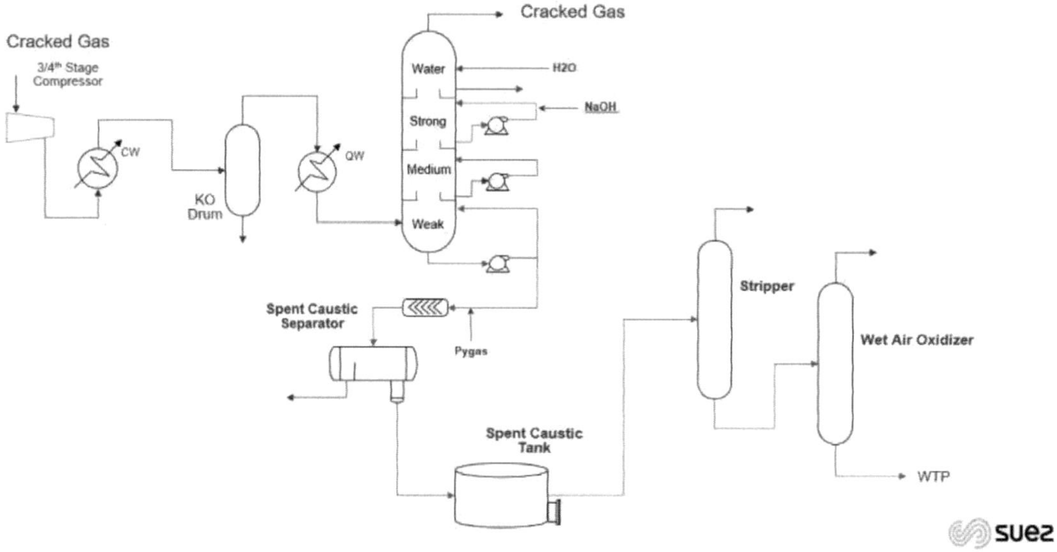

Figure 1: Typical configuration of a caustic scrubber

Generally speaking, caustic towers are operated in a temperature range between 30 – 50°C (~85 – 125 °F). This temperature is controlled by the inlet temperature of the cracked gas, e.g., a preheat-exchanger. A low temperature will increase the risk of hydrocarbon that condenses in the caustic tower, which can promote fouling and foaming in the caustic tower. Skimming the caustic is an effective way to remove the hydrocarbons. By controlling the temperature difference between the Top and Inlet of the preheater, condensation can be prevented or minimized.

On the other hand, a high caustic tower operating temperature in the order of magnitude 45 - 50°C (~120°F) should be avoided as the polymerization rate will increase significantly. Not only does the aldol condensation polymerization tend to increase, but the polymer formation also increases as a result of the free radical polymerization in case of the presence of unsaturated hydrocarbons. Operating the caustic column at relatively high temperatures will reduce the absorption of the acid gas.

The handling or the disposing of the spent caustic has become increasingly expensive as environmental regulations have put limitations on the plants. In response, many plants have been installing equipment to handle and treat the spent caustic locally at the wastewater treatment plant. To remove polymers, the spent caustic is washed with a hydrogenated pygasoline stream.
The spent caustic contains aromatic components, like benzene, which should be stripped out. Other approaches are used to neutralize the spent caustic and/or sulfide converting systems like wet air oxidation.

Monitoring the polymer formation while operating the caustic system is an essential step and should be done regularly to avoid operational issues, which count not only for the column itself, but also the downstream units processing the spent caustic.

The standard analytical method (extraction-evaporation) requires specialized equipment and additional time. The study concluded that the method has lower repeatability and doesn't represent the actual situation in relation to the used solvent. A new technique for polymer analysis has been developed, reducing the time and being more repeatable and representative for the real amount of polymer in the caustic system.

3.0 Areas of Concern

Fouling, emulsification and foaming are the three major problems encountered in the caustic system. These problems can result in extra operational costs (i.e., export of spent caustic), throughput limitation and plant shut down.

Fouling in the caustic system is typically characterized by an increasing dP over the tower section, and accordingly, a decrease in acid gas removal/efficiency or a build-up of solids in the spent caustic. Fouling is due mainly to aldol condensation reactions, but also, in some cases, due to free radical polymerization when operating at higher the temperatures.

Foaming is mainly characterized by a pressure fluctuation in the tower and by an increase in pH and level in the water wash section and a corresponding increase in caustic make-up requirements.

Emulsification may occur if the loading of hydrocarbons to the system is too high. This may be due to operating the tower too cold or an inadequate separation of the hydrocarbons in the compressor knock-out drum upstream.

4.0 Feedstock Contaminants

Contaminants in the feedstock, such as oxygen(ates), alcohols and ethers, can lead to the formation of constituents typically causing fouling problems. Compounds, such as ethanol and methanol found in naphtha and NGL feedstocks can be converted to aldehydes and ketones. For example, ketone hydrolyses to acetic anhydride and acetic acid. Ketones can also react with amine or caustic to form stable compounds, which cannot regenerate and increase caustic demand.

Aldehydes can lead to polymer formation via the well-known aldol condensation polymerization mechanism or pass through the column, affecting the downstream product quality (e.g., crude butadiene) and should be avoided.

Oxygen(ates) may be introduced from different locations via naphtha's, gas oil, refinery gas or flare gas recycle. Nowadays, feedstock is processed via plastic recycling technologies and form aldehydes and ketones.

5.0 Fouling Mechanism & Control

Organic fouling of amine and caustic systems is typically caused by aldol condensation polymerization, and less commonly, by free radical polymerization in the liquid phases.Free radical polymerization occurs according to the traditional initiation, propagation, and termination steps. The reaction involves dienes or unsaturated compounds, which have condensed from the cracked gas, and the polymerization can be catalyzed by metals and peroxides.

In the amine system, this mechanism may occur in the hotter areas of the process, such as the regenerator or stripper reboiler and the lean side of the lean/rich exchanger. Free radical polymerization products are darker in color and may be rubbery. The treatment program for free radical polymerization uses chain-stopper type antifoulant chemistry. This type of antifoulant terminates active free radicals.

The Aldol Condensation mechanism is the most prevalent mechanism causing fouling in acid gas removal systems, like the amine and caustic. Typical observations caused by Aldol Condensation polymerization, are:

- process caustic sample coloration typically orange, brownish-red to red
- red hydrocarbon liquid layer on top of the spent caustic
- granular orange to brown-orange deposits
- the presence of yellow tacky polymer in stage or skim samples

The red liquid is commonly referred to as "red oil." Further polymerization of the red oil yields tacky, yellow or orange polymer. The waste oil from the caustic tower is considered hazardous waste. The Aldol Condensation reaction occurs when aldehydes and ketones are absorbed in the caustic solution from the cracked gas.

The polymer formed will have a brownish/reddish color when chain length increases, starting from single molecules via oligomer to long chain molecules (polymers).

The reaction that takes place in a basic environment like the caustic system is:

$CH_3CHO + CH_3CHO \Leftrightarrow CH_3 - CHOH-CH_2 -CHO$

Various carbonyl compounds in the cracked gasses can react via the aldol condensation mechanism and form polymers. Cracked gas analysis has indicated that acetaldehyde is the most prevalent component and will most directly form the polymer and red oil. Other types of carbonyls typically found in the cracked gas include acetone, acrolein, propionaldehyde and formaldehyde.

The aldol condensation potential can be determined by analyzing the level of carbonyls in the gas, deposit analysis, sludge analysis, the caustic color on each loop, and by simulating the fouling with laboratory techniques. In steamcracking units, the carbonyl level in the cracked gas can be determined using gas-bombs for sampling purposes in combination with gas chromatography or acetaldehyde test tubes. Some units have installed online analyzers (in combination with H_2S/CO_2) at the inlet or outlet of the caustic tower to detect the presence of carbonyls like aldehydes.

Conducting laboratory test work with different aldehyde concentrations at different temperatures and caustic concentrations has shown the effect on the polymer formation taking place in the caustic system.

6.0 Fouling Control & Analysis

The new SUEZ Water Technologies & Solutions non-toxic, nitrogen-free aldol polymerization inhibitor program provides superior protection for both the caustic scrubber and the spent caustic system, reducing the polymer content significantly and fouling in these systems.

This technology approach can be used alone or in combination with other SUEZ's carbonyl scavengers, depending on the severity of the problem. The technology is a combination of functionalities:

- carbonyl scavenger – solubility increaser
- oxygen scavenger
- Optional: dispersant

The Generation 2 new carbonyl scavenger is an innovative technology and polymer solubilizing agent for caustic system. The carbonyl scavenger includes multiple hydroxyl groups on the incorporated carbohydrates and increases the solubility of the formed polymer.

The non-toxic aldol polymerization inhibitor hinders aldol polymer condensation by reacting to the carbonyl group, reducing polymer formation induced by peroxides, and dispersing polymers. It is a technology that enables the provision of full protection from the whole caustic system against polymer fouling.

7.0 New Monitoring Technology & Effectiveness of Anti-Foulant Technology

The performance of an antifoulant treatment on a caustic tower is often monitored by measuring the polymer content in the caustic samples. The polymer content is measured by an extraction of the caustic sample followed by evaporation of the organic solvent phase. As this method is not always available in the lab and is very time-consuming, an alternative filtration method is often being used, but with lower repeatability and presenting safety and health concerns.

A new analytical monitoring method has been developed, taking the following objectives into account:

- Safety
- Applicable in the field
- Repeatability
- Relatively fast
- Specific for aldol polymers
- Representing the current situation
- Applicable for the GEN 1 & 2 antifoulant technologies

The analytical results generated information in order to mitigate the fouling by controlling the operational settings and the dosage of antifoulant.

The method was developed in the SUEZ Center of Excellence lab based on synthetic samples. The method includes the extraction of aldol polymers in an organic solvent, followed by measuring the light absorbance of the solvent at a specified wavelength (based on the complete UV-VIS spectra). This absorbance at the specific wavelength indicates the extractable aldol polymer content in caustic samples.

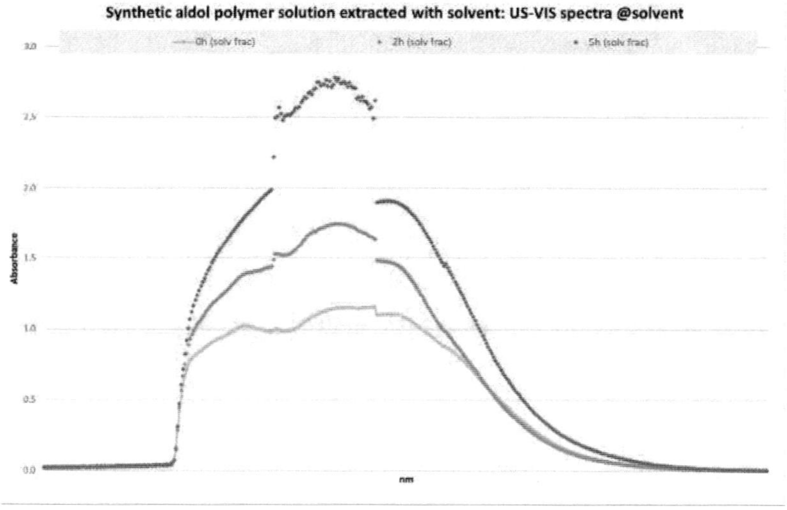

Figure 2: Synthetic aldol polymer solution extracted with solvent: US-VIS spectra @solvent

The correlation between the traditional analytical method and the extraction-absorbance method deployed for untreated synthetic samples has shown to be favorable.

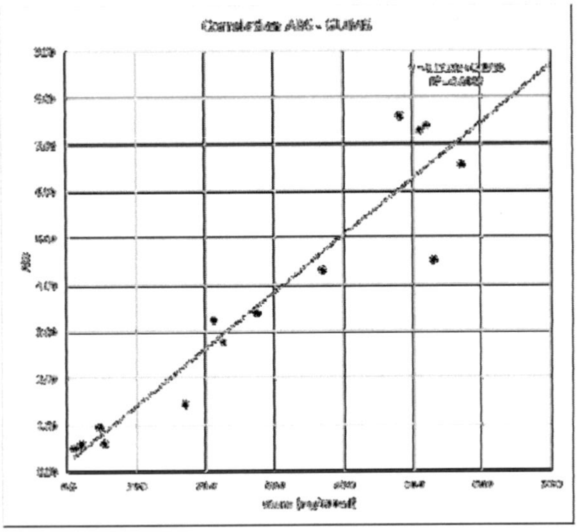

Figure 3: Correlation ABS - GUMS

The correlation is lower in the higher range as many polymers tend to precipitate out of the caustic solution.

Repeatability of the new extraction-absorbance method was shown to be very good (within these conditions):

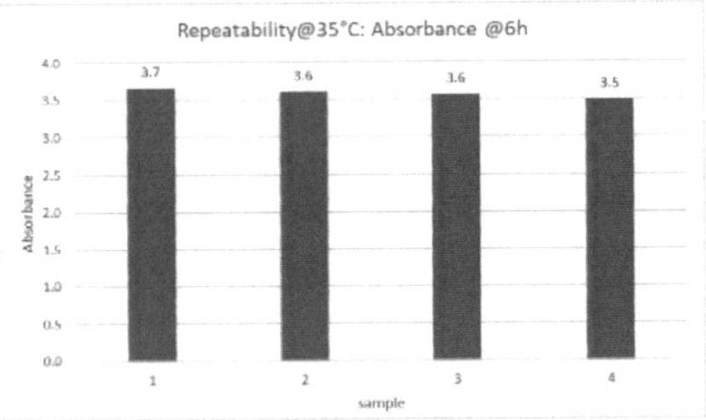

Figure 4: Repeatability @35°C – Absorbance @6h

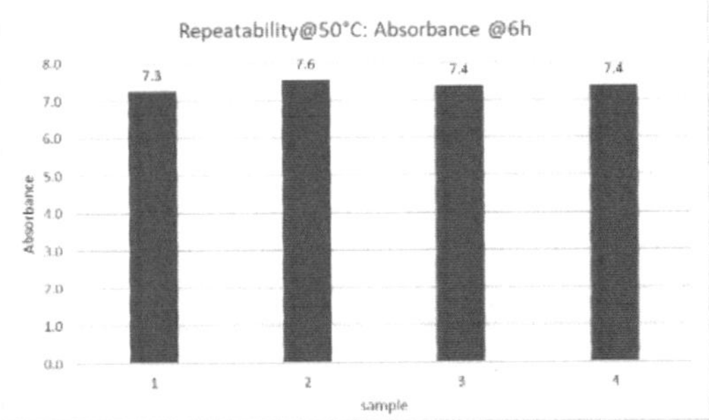

Figure 5: Repeatability @50°C – Absorbance @6h

Repeatability of the traditional method was shown to be good (within these conditions):

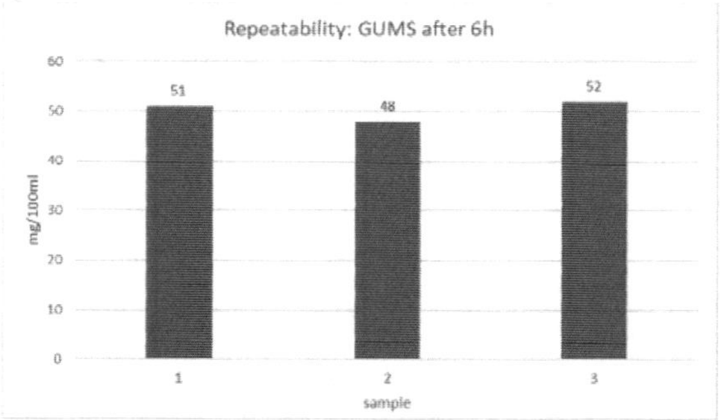

Figure 6: Repeatability – GUMS after 6h

Besides the new spectrophotometric method, other routine monitoring for a caustic tower should be included for performance-related criteria, like feedstocks being processed and the process conditions and characterization/quantification of the sludge.

8.0 Bench Market Study

Process conditions in caustic columns play an important role in affecting the polymerization reactions. Aldehyde concentration, temperature and caustic strength are the main contributors to the aldol condensation polymerization and red oil formation.

SUEZ conducted a market study, collecting the process parameters from more than 30 columns around the globe.

Condensation of cracked gas in the caustic tower is avoided by heating the cracked in preheat heat exchanger.

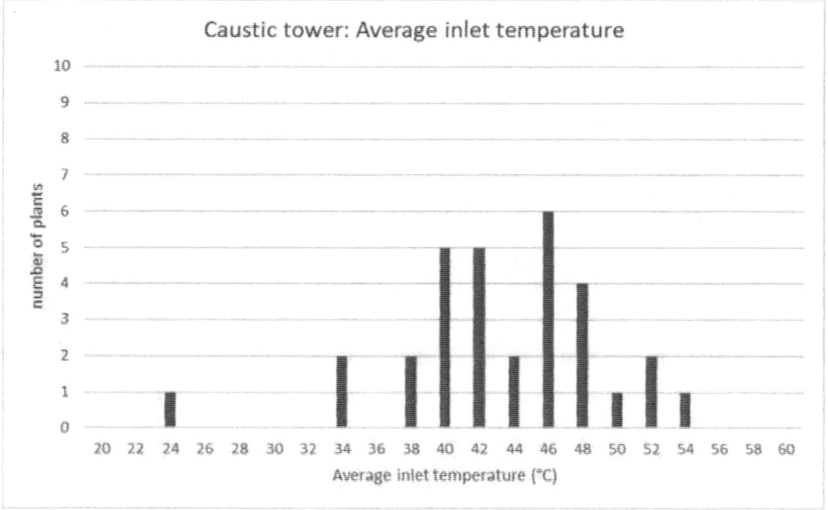

Figure 7: Caustic tower – Average inlet temperature

The study showed that the average caustic tower temperature is around 43°C.
This temperature is determined by:

- Outlet temperature of the stage of process gas compressor before the caustic tower
- Efficiency of the interstage cooler (cool water temperature - fouling)
- Operation setting (or capacity) of the preheat heat exchanger before the caustic tower

Remark: In exceptional cases, the preheat heat exchanger is replaced by a heater in the bottom loop circulation of the weak caustic.

It was surprising to observe that the column operating temperature was a function of the fouling tendency.

Seasonal and operation changes were found to have an impact on the temperatures at the caustic tower, and therefore, will affect the polymerization tendency.

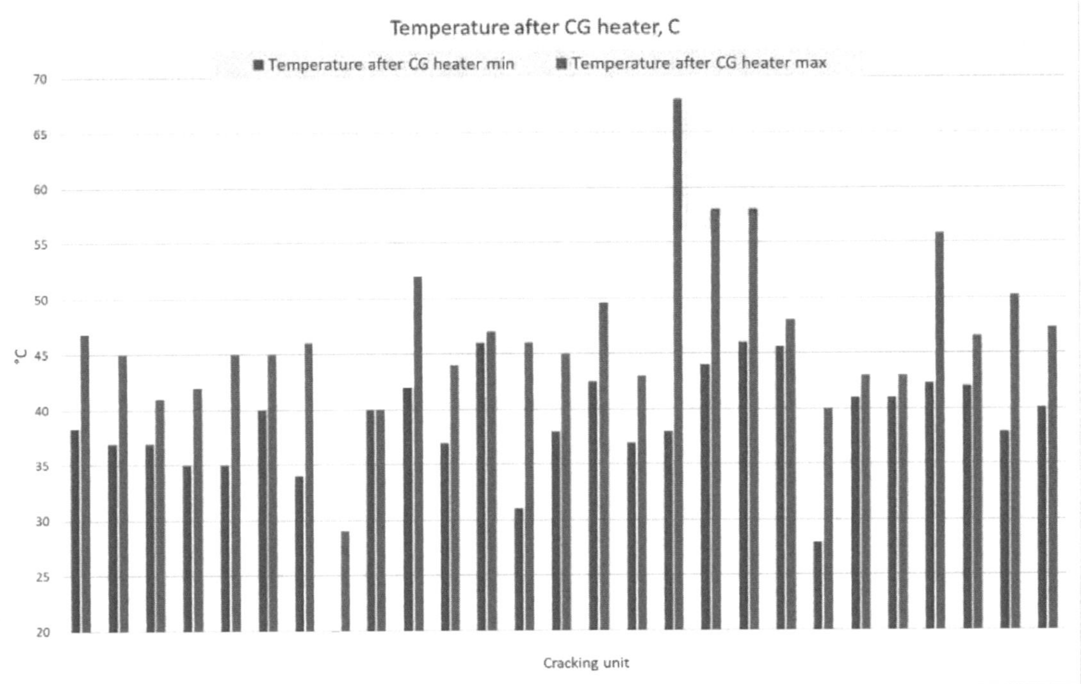

Figure 8: Temperature after CG heater, °C

The effect on the fouling tendency was simulated with laboratory experiments and proved the effect of temperature on the fouling tendency. Besides the temperature effect, other operating process parameters contributed to the fouling tendency in caustic columns and ranked:

- Acetaldehyde concentration
- Temperature
- Caustic strength
- Oxygen
- Residence time

9.0 Laboratory Simulation Experiments

As the effect from the process circumstances in caustic columns are significant, SUEZ conducted laboratory simulation experiments focusing on the polymer formation by varying the main contributors to the aldol condensation polymer formation:

A: Caustic Strength
B: Temperatures
C: Aldehyde Concentration

In caustic columns, all the sections have a different caustic strength affecting the polymerization tendency. On the other hand, the circulation loop with the lowest caustic strength (weak section) typically has the highest concentration of aldehydes.

A. Caustic Strength

Polymer formation in the function of caustic strength with synthetic samples with 500 ppm acetaldehyde:

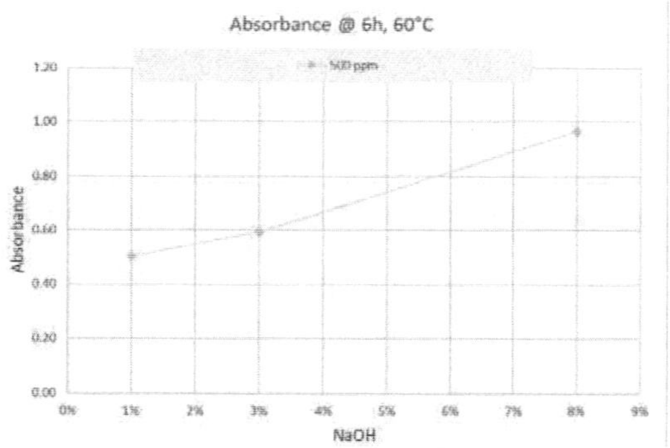

Figure 9: Absorbance @6h, 60°C

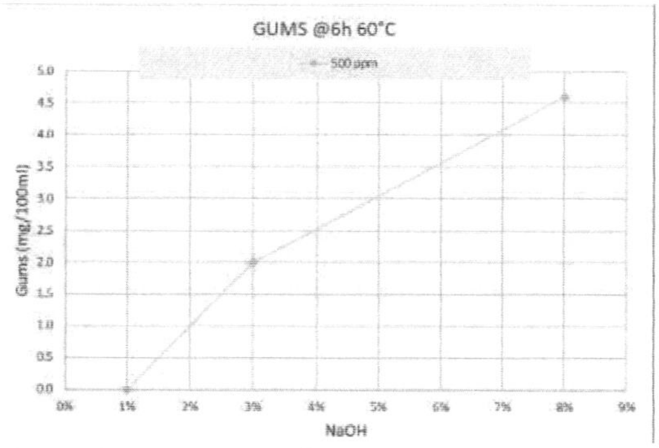

Figure 10: GUMS @6h, 60°C

Looking at the results of the solution of 500 ppm acetaldehyde, the gums and the absorbance increase with an increased NaOH strength. NaOH initiates the polymerization by making the aldehyde ionic.

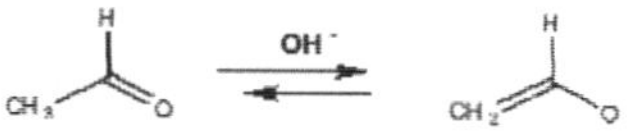

Working with an increased NaOH concentration (1% → 8%) will increase the concentration of OH-.

B. Temperature

Polymer formation in the function of temperature with synthetic samples with 2000 ppm acetaldehyde and 3% caustic.

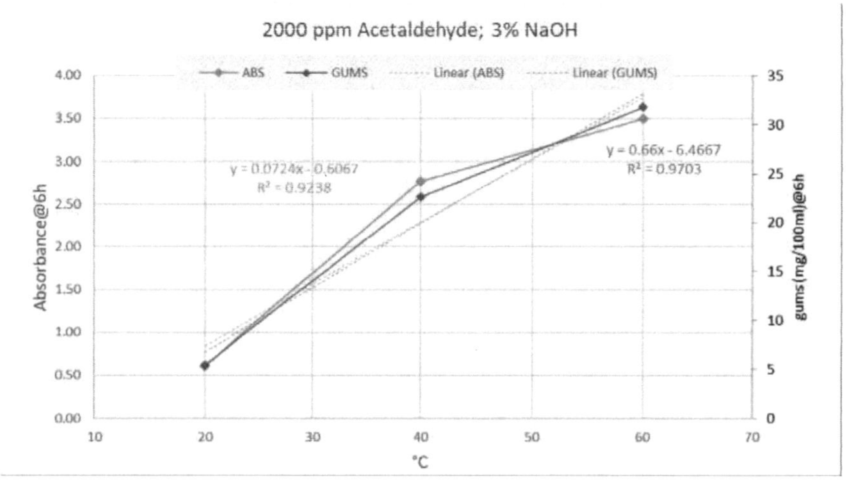

Figure 11: 2000 ppm Acetaldehyde, 30% NaOH

Based on the measured gums and absorbance, the temperature clearly accelerates the polymerization speed.

C. Acetaldehyde concentration

Polymer formation in the function of acetaldehyde content with synthetic samples with 1, 3 and 8% caustic.

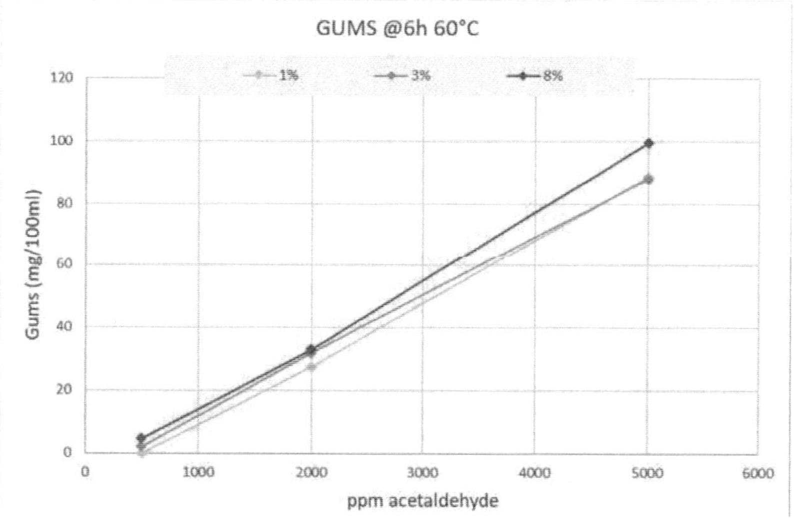

Figure 12: GUMS @6h, 60°C

Based on the gum analysis, the impact of the acetaldehyde is important.

Based on the results, we can conclude that these three parameters impact the polymerization reaction, and we can rank them in the following order of importance.

1. Acetaldehyde content
2. Temperature
3. Caustic strength

Of course, the residence time is also important, which is mainly impacted by the volumes in the sections and the makeup of water and/or fresh NaOH.

SUEZ has developed a tool, called CauTrate*, to estimate the polymer formation, which can be used to evaluate the impact of changing an operational parameter or of processing a certain feedstock.

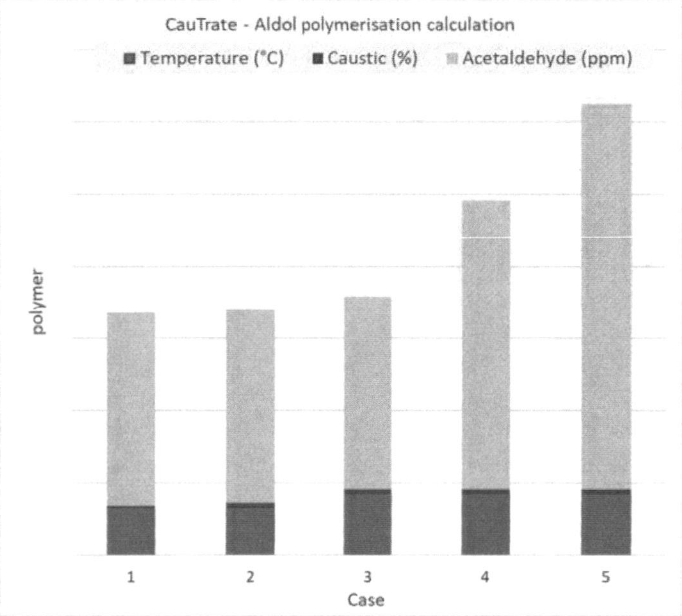

Figure 13: CauTrate – Aldol polymerization calculation

Case 1: Reference	
Case 2: Increase [caustic] in weak section	1 to 3%
Case 3: Increase column temperature	35 to 45°C
Case 4: Increase [acetaldehyde] in cracked gas	100 to 150 wtppm
Case 5: Decrease water make up	1.0 to 0.5 T/h

Together with the absorbance measurements on the different sections, SUEZ is well-positioned to control the Generation 2 antifoulant technology to mitigate the fouling.

10.0 Literature:

1. Aiche Paper Number32B, 2018 Conference

639321 CO_2 Reduction for Steam Cracking Furnaces By Full Hydrogen Firing

Abstract:

When firing fuel gas as generated within the cracker, the steam cracking furnaces emit about 90 to 95 percent of the plant's total CO_2 emissions. One option to reduce CO_2 emissions from the cracking furnaces is to increase the hydrogen content in the fuel gas. By switching to firing 98 vol.% hydrogen fuel gas, CO_2 emissions from the furnace stack can be reduced by about 95 percent for liquid crackers (typically 10 vol.% H_2 in fuel gas) to 85 percent for ethane crackers (typically 85 vol.% H_2 in fuel gas).

Due to the difference in fuel gas characteristics when moving to firing high hydrogen content fuel gas the flue gas mass flowrate decreases, the firebox efficiency increases, the heat load into the furnace convection section decreases and the adiabatic flame temperature increases. When revamping existing furnaces, next to the adequacy of the firing system, there will be consequences on the furnace run length and superheated high pressure steam production and the revamp strategy needs to be adopted accordingly.

This paper discusses the design considerations when firing high hydrogen content fuel gas. The effect on the cracking furnace performance such as run length and SHP steam production, will be further explored and the adequacy and considerations on the convection section will be discussed. In addition, the impact on the burner design will be touched upon. Technip Energies is developing and testing burners suitable for hydrogen firing. The results from both the large capacity LSV® bottom burner and side wall TSWB® burner firing full hydrogen are presented.

639294 Tackling Industrial Emissions through State and Federal Partnerships

Abstract:

Industry contributes significantly to U.S. greenhouse gas (GHG) emissions and tackling industrial emissions will therefore need to be a key element of U.S. climate policy. Furthermore, the resources that U.S. industry can bring to bear – including financial resources, technical knowledge, research capacity, and workforce skills – make it potentially a global leader for industrial decarbonization. Pioneering policy approaches, technical advances, and demonstration projects in the U.S. could help enable industrial transitions throughout the world. Realizing this vision will require concerted, collaborative efforts between the U.S. public and private sectors. A key to success will be establishing such partnerships at both federal and state levels of government.

639295 Life Cycle Thinking and the Energy Transition: The Path to Sustainability and a Circular Economy

Abstract:

Global energy and environmental challenges have spurred an increased focus on sustainable sources of energy while global investments in the green energy sector have surged. Low-carbon, "zero" emission technologies and the scientific advancements required to facilitate the energy transition—such as wind turbines, solar photovoltaic panels, and lithium-ion batteries for electric vehicles and energy storage systems—are pivotal advancements for the energy transition. Global energy production, rising concerns of climate adaptation, and the plastic waste crisis have increased focus on sustainable sources of energy, decarbonization efforts and waste reduction strategies such as advanced recycling of plastics. However, with a national and global focus on a climate plan that prioritizes clean energy, circular economy and sustainability, it is critical to integrate life cycle dimensions into future policies to identify, quantify, and assess the social, environmental, and economic implications of products, processes and technologies across their life cycles. Transforming the energy, plastics and waste management sectors and evolving to a circular economy will only be sustainable if systems-level impacts along the entire value chain are appropriately measured.

639296 Climate Change and the Persian Gulf Oil Producers

Abstract:

The increasing urgency of climate action is exacerbating uncertainty around future oil demand, a worrying prospect for Saudi Arabia and other governments dependent on oil export revenues. Persian Gulf producer countries are in a particularly tough situation. They stand to lose if climate action fails, due to a fast warming climate that could render their countries too hot for humans. They could also lose if climate action succeeds, given the level of government dependence on fossil fuel rents. This talk looks at the strategic, economic and political implications of climate change on the Persian Gulf oil producers, as well as the actions Gulf producers are taking to cope with the energy transition.

Green Ethylene by Steam Cracking of Renewable Feedstock

Omar Hamid, Renewables SME
Advisian (part of Worley group)

Abstract

Green ethylene is a nascent but growing industry. Companies such as Braskem have been producing bio-ethylene through the dehydrogenation of sugarcane-based ethanol since 2010. Global concern with climate change and support by consumers for decarbonizing the supply chain have created a demand that is expected to grow by more than 6% a year during the next decade. New and pending legislation have also encouraged this transition. Most global ethylene for polymer production is made by steam cracking fossil feedstocks. This paper explores the opportunities for processing renewable feedstock in existing steam cracker plants. The performance of processing hydrotreated vegetable oils, tallow, used cooking oils (UCO), renewable naphtha, renewable diesel, and Fischer-Tropsch (FT) wax is compared, and the facilities required to prepare renewable feedstock are discussed. The pathways offer a route to green ethylene by leveraging and extending the life of existing steam cracker assets while reducing net carbon emissions.

Background

Ethylene and propylene are the highest bulk chemicals produced globally at over 255 MT/yr [1]. These light olefins are reactive compounds that are widely used in the production of polymers to manufacture plastics. The final products from these monomers are ubiquitous and are found in toothbrushes, carrier bags, food packaging, mobile phones, computers, carpeting, clothing, and furniture.

Steam cracking plays a dominant role in olefin production. Roughly 95% of ethylene and 60% of propylene demand is met with stream cracking [2, 18]. In this process, hydrocarbon feedstocks, such as naphtha, ethane, etc. are converted to light olefins, such as ethylene and propylene, and other products. The balance of propylene comes primarily from refinery fluidized catalytic cracking (FCC) units or propane dehydrogenation (PDH) units. The global demand for ethylene and propylene is forecasted to continue to grow at 5.5% per year (2021-2029) [3]. General population growth leading to increased demand for processed food and beverages is considered a key driving factor.

The principal feedstock for steam crackers varies by region [1]. The low cost and availability of natural gas liquids (NGL) in the Middle East and United States make ethane the preferred and more profitable feedstock in those regions. Asian and European crackers prefer to process naphtha and liquid feedstock as these are sourced from farther locations and their logistics for transport are simpler.

Steam Cracking is Energy Intensive

Steam cracking is the most energy-consuming process in the chemical industry, representing approximately 8% of the sector's total global primary energy use. The steam cracking process accounted for approximately 180–200 million tons of CO_2 emissions by 2004 worldwide, as noted in Table 1, which compares US and European energy use and CO_2 emissions related to olefin production against global estimates. [4]

Table 1 Global Energy Use and CO_2 Emissions due to Olefin Production (2004)

Energy use and CO_2 emissions due to olefin production in the World, the US and Europe			
	World	**US**	**Europe [1]**
Total feedstocks (million tons)	**300**	**85**	**90**
Feedstocks (breakdown, wt%)	Naphtha 55 Ethane 30 LPG 10 Gas oil 5	Ethane 55 Naphtha 23 Propane 15 Gas oil 5	Naphtha 75 LPG 10 Gas oil 59 Ethane 5
Ethylene capacity	110-113	28-30	30-32 (75% in Western Europe)
Propylene capacity	53-55	16-17	17-18
Total energy use [2]	**2-3**	**0.5-0.6**	**0.7-0.8**
Total CO_2 emissions [3]	**180-200**	**43-45**	**53-55**

1. Enlarged EU and former Soviet Union
2. Fuel combustion and utilities included (exa Joules, 10^{18})
3. Fuel combustion, decoking and utilities included

The steam cracking reaction, which is highly endothermic, requires 700–1200 BTU per pound of hydrocarbon [5]. The capacities of individual heaters range from 20,000 to 200,000 tons/yr with some designed for a firing rate of 500 MMBTU/h. The high temperature required to initiate the formation of free radicals relegates the reaction to the radiant section of the furnace with only 35% to 50% of fired duty contributing to the reaction. This leads to CO_2 emissions of one to two tons per ton of ethylene production.

Decarbonization Pathways

The petrochemical industry will need to decarbonize to support the obligations of the Kyoto Protocol and COP26. Many major ethylene producers have announced their goals to achieve net zero carbon emissions by 2050.

The industry can reduce CO_2 emissions in a variety of ways [6, 10]. For instance, energy efficiency improvements would provide low investment opportunities to decarbonize, furnace efficiencies may be improved by adding air preheat, refrigeration systems may be optimized, and heater integration can be increased. The introduction of direct and indirect forms of carbon tax incentives could result in a greater degree of heat recovery than would have been considered for the design of most existing facilities. However, the fundamental nature of the endothermic pyrolysis reactions limits reduction by 25-30%.

The use of biomass-based fuels for firing offers a net zero carbon emissions. This approach, which does not reduce CO_2 emissions from the furnace stack is not considered to add CO_2 to atmosphere. Biomass absorbs atmospheric CO_2 when plants grow; therefore, any carbon emissions resulting from their combustion are considered carbon neutral.

Precombustion or post-combustion carbon capture and storage (CCS) offers pathways to deep decarbonization, provided CO_2 storage or pipelines are available in proximity. Precombustion carbon capture normally consists of a blue hydrogen plant where natural gas is reformed to hydrogen and fed as fuel to the furnaces. The CO_2 is captured from the intermediate syngas and sent for sequestration. Post-combustion carbon capture is accomplished by adding absorbers to the furnaces stacks to remove CO_2 from the flue gas. The low concentration of CO_2 combined with the presence of oxygen leads to efficiency challenges.

Companies are also investigating the electrification of cracking furnaces. Though a mature technology, electric heaters have been traditionally designed for low MW applications. Prior to the current emphasis on carbon emissions, the total cost of power for electric heating did not justify large electrical heaters in the petrochemical industry. Several companies are developing high MW electric furnaces for steam cracking, which must be powered with green (or similar) electricity to be able to decarbonize the stream cracker.

Switching feedstock offers another pathway to decarbonize ethylene plants. The pathway considers moving from fossil-based feedstock, i.e., NGLs and naphtha, to biomass feedstocks, e.g., ethanol, vegetable oils and biomass. Ethanol dehydrogenation was a popular method to produce green ethylene prior to the advent of large-scale steam crackers. This paper reviews the carbon reducing performance of hydrotreated vegetable oils and Fischer-Tropsch (FT) waxes to that of traditional steam crackers. Catalyst based cracking has received renewed interest with potential for reduced CO_2 emissions. Other methods to produce green ethylene, referenced in the literature, include furnaces fired by biogenic off-gas from the process, thereby yielding carbon neutral stack emissions.

Biomass Feedstock and Green Ethylene

Green ethylene, olefins produced from biomass-based feedstock, is chemically identical to fossil-based ethylene. Renewable biomass sources are considered as green feedstock. Materials that are naturally derived from growing plants, animal parts or manure can all be considered renewable. Examples of biomass sources include corn starch, soybean oil, ethanol, switchgrass, and landfill biogas. Renewable biomass feedstock can be in liquid, gas, or solid forms for conversion to olefins. The chemical characteristics of each determine the process complexity to produce ethylene.

Table 2 Typical composition of fatty acids in vegetable oils and animal fats

	Composition of Fatty Acids (% w/w)										
	12:0	14:0	16:0	18:0	18:1	18:2	18:3	20:0	22:0	22:1	24:0
Vegetable oils											
Cotton		0.6-1.5	22-26	2.1-5	14-21	47-58					
Peanut		0-0.5	6-14	1.9-6	36-67	13-43		1.32	2.52	1.0	1.23
Babassu	44-45	15-17	5.8-9.0	2.5-6	12-16	1.4-3					
Canola			1.2-6.0	1-2.5	52-67	16-31	6-14			1-2	
Coconut	44-51	13-21	7.5-10	1-3.5	5-8.2	1-2.6	0-0.2				
Rapeseed		0-1.5	1-6	0.5-3.5	8-60	9-23	1-13			5-56	
Sesame			7.2-9.2	5.8-7.7	35-46	35-48					
Sunflower			3.5-7.6	1.3-6.5	14-43	44-74					
Linseed			6	3.2-4	13-37	5-23	26-60				
Corn		0-0.3	7-16.5	1-3.3	20-43	39-62	1-13	0.24			
Olive		0-1.3	7-20	0.5-5	55-85	4-21					
Palm		0.6-2.4	32-46.3	4-6.3	36-53	6-12					
Poppy			12.6	4.0	22.3	60.2					
Jatropha Curcas		0-0.1	14-15.3	3.7-9.8	34-46	29-44	0.3				
Soybean			2.3-13.3	2.4-6	18-31	49-57	2-10			0.3	
Jatropha		0.5-1.4	12-17	5.0-9.5	37-63	19-41				0.3	
Animal Fats											
Lard	0.1	1.4	23.6	14.2	44.2	10.7	0.4				
Tallow	0.1	3-6	25-37	14-29	26-50	1-2.5	0.9				

This paper considers two types of green feedstock: triglyceride-based sources; and FT liquids from renewable sources. Triglycerides are the fat lipids that comprise most of vegetable oils and tallows. They consist of three fatty acids held together with a glycerol molecule. Vegetable oils from a variety of plants are used with some common sources being soybean oil, canola oil and palm oil. Use of triglycerides from waste streams are often favored to improve carbon intensity of the overall process. Such waste streams include used cooking oils (UCO), distillers corn oil (DCO) and tallow from rendering facilities. Typical triglyceride-based compositions in vegetable oil and animal fat feedstock are provided in Table 2. [7, 19]

FT liquids are produced by passing syngas (CO and H_2) across iron- or cobalt-based FT catalysts. The syngas can be produced by gasification of nearly any biomass source. Gasification and FT processes are generally considered for difficult-to-process lignin or cellulosic biomass that do not compete with food sources. A typical FT liquid yield is summarized in Table 3. [8]

Table 3 General comparison of high and low temperature FT syncrudes and conventional crude oil components

Components of Fischer-Tropsch Syncrudes (low and high temp) and Conventional Crude Oil			
Component	**High Temp FT**	**Low Temp FT**	**Crude Oil**
Alkanes	>10%	Major product	Major product
Cycloalkanes	<1%	<1%	Major product
Alkenes	Major product	>10%	None
Aromatics	5-10%	<1%	Major product
Oxygenates	5-15%	5-15%	<1% O (heavy)
Sulfur species	None	None	0.1-5% S
Nitrogen species	None	None	<1% N
Organometallics	Carboxylates	Carboxylates	Phorphyrins
Water	Major by-product	Major by-product	0-2%

Steam Cracker

Steam pyrolysis involves thermal cracking of hydrocarbons, which is a free radical mechanism involving initiation, propagation, and termination steps, as outlined in Figure 1 for an ethane molecule [9]. The complexity and number of reaction pathways increase with the molecular weight of the hydrocarbon.

Initiation $C_2H_6 \rightarrow CH_3^{\bullet} + CH_3^{\bullet}$

Propagation $CH_3^{\bullet} + C_2H_6 \rightarrow CH_4 + C_2H_5^{\bullet}$

$C_2H_5^{\bullet} \rightarrow C_2H_4 + H^{\bullet}$

$H^{\bullet} + C_2H_6 \rightarrow H_2 + C_2H_5^{\bullet}$

Termination $H^{\bullet} + H^{\bullet} \rightarrow H_2$

$CH_3^{\bullet} + H^{\bullet} \rightarrow CH_4$

$H^{\bullet} + C_2H_5^{\bullet} \rightarrow C_2H_6$

$C_2H_5^{\bullet} + CH_3^{\bullet} \rightarrow C_3H_8$

$C_2H_5^{\bullet} + C_2H_5^{\bullet} \rightarrow C_4H_{10}$

Figure 1 Thermal Cracking of Ethane: Initiation, Propagation and Termination

Paraffins, particularly normal paraffins, are good cracker feedstock. The initiation and propagation steps in reaction are most easily promoted in paraffins. This lends paraffins to yield higher selectivity towards ethylene and propylene production. In contrast, the bonds in carbon rings of naphthalene and aromatics are more difficult to break, which results in lower ethylene yields.

A steam cracker ethylene plant is divided into a hot section and a cold section, as illustrated in Figure 2. The hot section, which utilizes cracking furnaces and product quenching, consumes most of the energy required and emits CO_2 during fuel firing. Given that the reaction occurs only in the radiant section, heat recovery in the convection section is focused on feed preheat and steam generation.

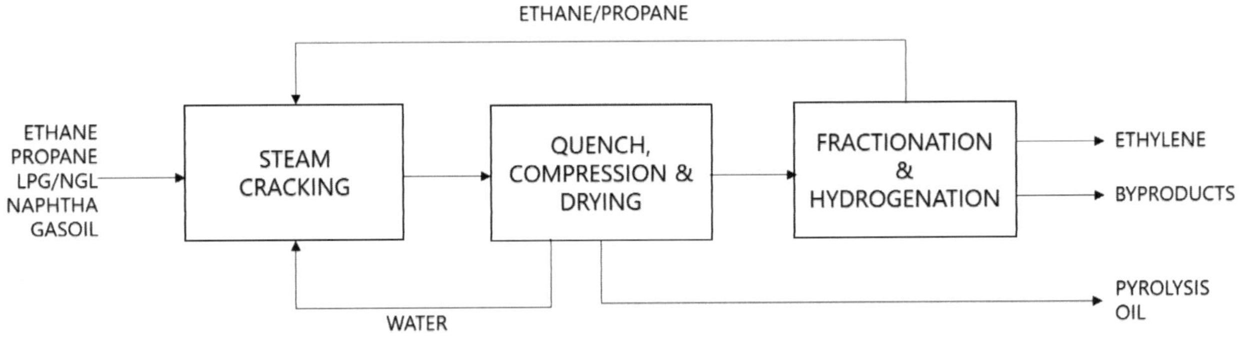

Figure 2 Steam Cracker Block Flow Diagram: Hot and Cold Sections

The cold section comprises a cryogenic fractionation section where ethylene and other products are purified. Byproducts are the major drivers for economics, and the cold section is optimized based on feedstock to produce high value products.

Biomass Feedstock Pathway – Hydroprocessing of Vegetable Oils, Tallow, UCO, POME and others

Triglyceride-based fats, oils and greases are relatively easy to convert to hydrocarbons. Sources such as vegetable oil, tallow, used cooling oil, distillers corn oil (DCO) and palm oil mill effluent (POME) can be hydrotreated to remove their oxygen content to be converted from carbohydrates to hydrocarbons. Triglyceride feedstocks consist of three fatty acid chains, predominantly C16 to C18 in length, that are bound by a propane backbone. Fatty acids carbon can be hydrodeoxygenation (HDO) or decarboxylated to produce single long-chain molecules, as shown in Figure 3. [20]

In the hydroprocessing unit, the triglycerides react with hydrogen at high pressures and temperatures. The backbone is removed as a propane molecule. Two competing reactions then take place to convert the remaining three fatty acid chains to hydrocarbons. Through HDO, hydrogen reacts exothermically with oxygen in the fatty acid to produce water and hydrocarbons. Decarboxylation entails hydrogen saturation of the carbon-oxygen double bonds to produce carbon dioxide molecules and hydrocarbons. HDO is the preferred pathway, because it minimizes the loss of biogenic carbon from the product; however, the hydrogen demand is high on the order of 2,400 to 2,800 SCF/BBL of fresh feed.

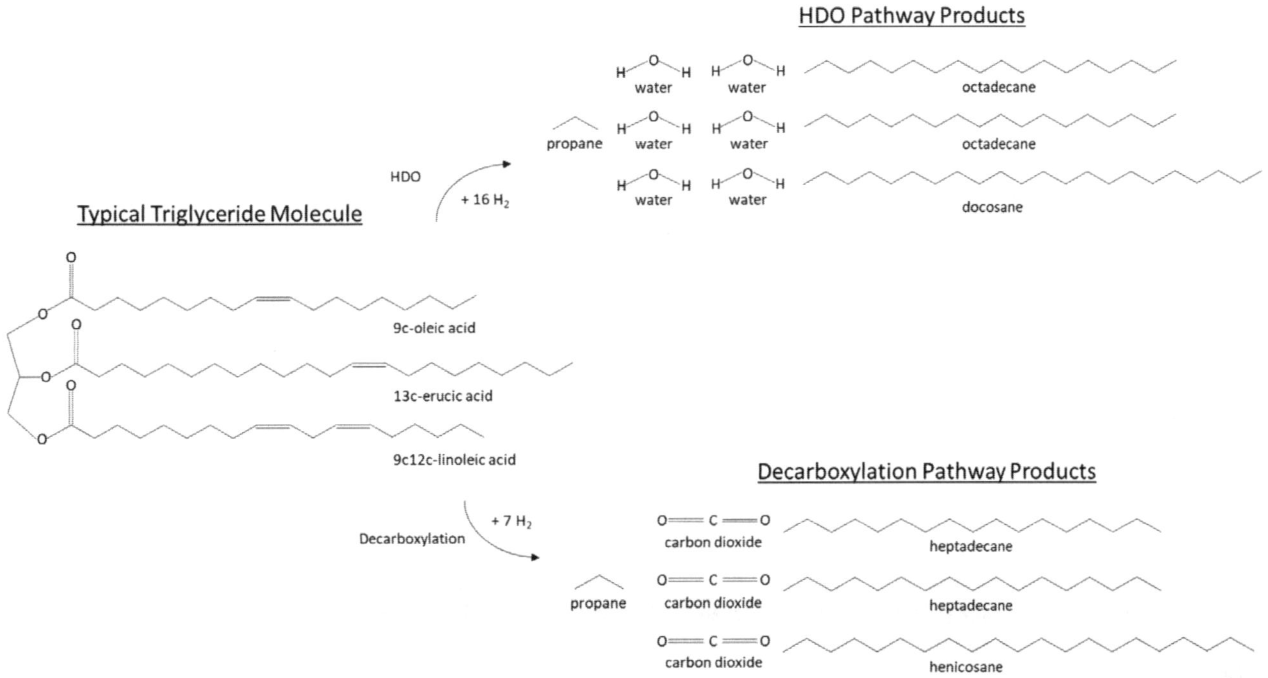

Figure 3 Conversion of Fatty Acid to Hydrocarbons through HDO & Decarboxylation

The resulting hydrocarbons are straight chain paraffins in the C16 to C18 range. Hydroprocessing of triglyceride-based oils is currently the leading pathway for production of renewable diesel and sustainable aviation fuel (SAF). The production of renewable diesel requires dewaxing to control the cold flow properties of the product. SAF production typically requires mild cracking. Normal paraffins are most selective to olefins production; consequently, neither dewaxing nor cracking is considered before feeding to a steam cracker.

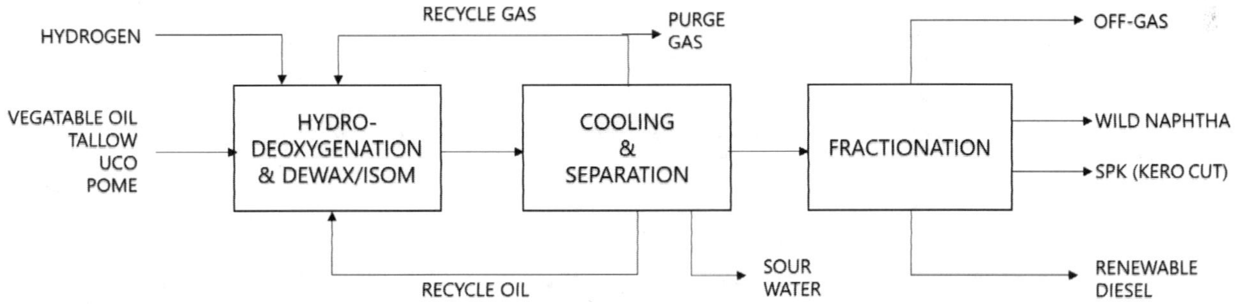

Figure 4 Hydroprocessing Block Flow Diagram

The hydropocessing unit receives fresh feed and makeup hydrogen, as illustrated in Figure 4, and the hydrogen rich gas is recycled to minimize fouling of the catalyst. This reaction is highly exothermic, and the product oil is recycled to the reactor as an inert fluid to control bed temperature rise. When producing fuels, the products of dewaxing or isomerization must be cooled and fractionated. The mature of this process has led to large-scale production for the renewable fuels market.

Dewaxing and isomerization are not considered for a stream cracker feed, and fractionation may not be necessary because nearly all products are paraffins suitable for cracking. The HDO wax product comprises molecular weights in the gas oil range. However, unlike fossil fuel sources, the material primarily composed of straight chain paraffins, and is low in naphthenic and aromatic compounds. Cracking is selective to olefins yields with low byproduct formation. The performance of a steam cracker is predicted with a HDO waxy product. The HDO waxy product composition is calculated using an in-house model assuming roughly 85% hydrodeoxygenation and 15% decarboxylation. The pyrolysis yields are estimated using a pseudo-mechanistic model based on available kinetic data [24, 25, 26] and adjusted to experimental yields published in literature [16, 17, 21, 22, 23,]. Table 4 provides a sample HDO product wax composition based on Soybean oil source and corresponding sample steam cracker yields.

TABLE 4 Steam Cracker Performance for Hydroprocessed Soybean Oil

Sample HDO Yield for Soybean Oil		Sample Cracker Yield	
Component	**wt%**	**Component**	**wt%**
CO2	1.2	H2	0.4
CO	0.5	CH4	10.0
H2O	10.6	C2H4	37.9
C1	0.1	C2H6	4.1
C2	0.2	C3H6	18.9
C3	4.9	C3H8	2.7
C4	0.3	C4	11.2
C5-C15	4.2	Pygas + BTX	13.6
C16	7.0	Pyfuel Oil	1.3
C17	11.2		
C18	58.9		
C19-C24	0.8		

Biomass Feedstock Pathway – Fischer-Tropsch Liquids from Woodchip, Algae or other Biomass sources

Biomass sources outside of the food supply chain are of interest for industrial uses. The use of vegetable and other edible oils for fuel or petrochemical applications is controversial in terms of weighing "food versus fuels" [11]. The Fischer-Tropsch (FT) process provides a pathway to convert non-food based lignan, cellulosic and algae sources, which are feedstocks that can be difficult to process but can be gasified then passed over a FT catalyst to produce synthetic hydrocarbons, as illustrated in Figure 5.

The FT process requires syngas, a combination of H_2 and CO, as its feedstock. Gasification is most commonly considered for conversion of woody and other biomass to syngas. The technology can convert any carbon-based feedstock to H_2 and CO. The process occurs generally at high temperatures and pressures in the oxygen (or air) and steam. Residue that cannot be gasified comes out as ash/slag. Sub-stochiometric oxygen ensures CO is favored of

CO_2 production. The methanation and water-gas shift reactions leads to hydrogen and methane make.

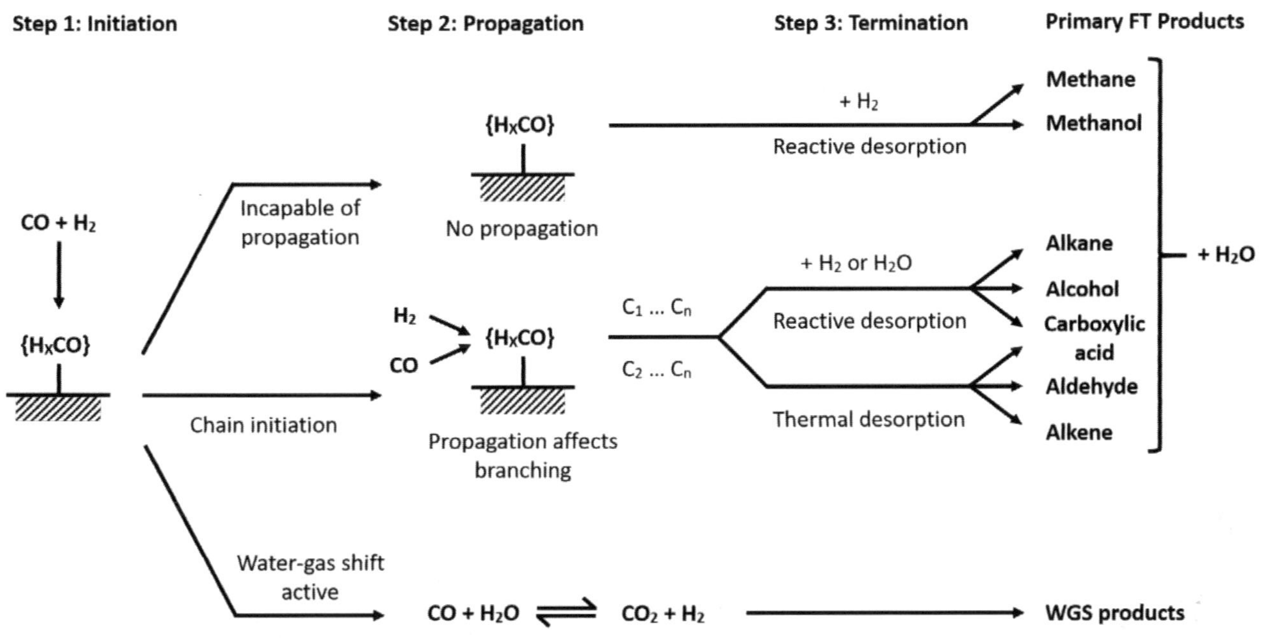

Figure 5 Fischer-Tropsch Catalysis and Products: Initiation, Propagation & Termination

The FT process converts syngas into longer chain hydrocarbons [12]. The mechanism involves polymerization and hydrogenation of CO through a chain reaction. The mechanism is not fully understood, but a proposed reaction network is suggested in Figure 6. The reaction is initiated with CO and H_2 being absorbed onto the catalyst. Additional CO and H_2 is absorbed to lengthen each chain in the propagation step. Reactive and thermal desorption are termination reactions resulting in a range of FT products.

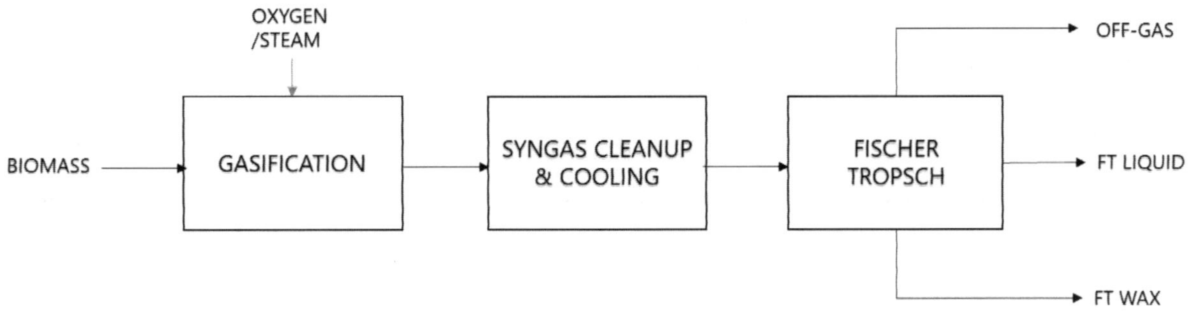

Figure 6 Fischer Tropsch Process Block Flow Diagram

The production of renewable hydrocarbons from biomass begins with gasification. Biomass is exposed to sub-stochiometric oxygen to produce H_2 and CO. This process forms small quantities of tar and other contaminants that are cleaned. The H_2:CO ratio is adjusted using a shift reaction, and the product is cooled before being fed to the FT synthesis loop. The FT waxy products are further processed to meet the specifications for fuels, lubricants, or other products. Gasification and FT processes are mature technologies, but relatively few plants have been built due to their complexity and cost. [13]

Table 5 Generic Compositions of the Main Types of Industrially Produced FT Syncrudes

Product Fraction	Carbon Range	Compound Class	Syncrude Composition (mass%)		
			Fe-HTFT	Fe-LTFT	Co-LTFT
Tail Gas	C_1	Alkane	12.7	4.3	5.6
	C_2	Alkene	5.6	1.0	0.1
		Alkane	4.5	1.0	1.0
LPG	C_3-C_4	Alkene	21.2	6.0	1.4
		Alkane	3.0	1.8	1.8
Naphtha	C_5-C_{10}	Alkene	25.8	7.7	7.8
		Alkane	4.3	3.3	12.0
		Aromatic	1.7	0.0	0.0
		Oxygenate	1.6	1.3	0.2
Distillate	C_{11}-C_{22}	Alkene	4.8	5.7	1.1
		Alkane	0.9	13.5	20.8
		Aromatic	0.8	0.0	0.0
		Oxygenate	0.5	0.3	0.0
Residue/Wax	$>C_{22}$	Alkene	1.6	0.7	0.0
		Alkane	0.4	49.2	44.6
		Aromatic	0.7	0.0	0.0
		Oxygenate	0.2	0.0	0.0
Aqueous Product	C_1-C_5	Alcohol	4.5	3.9	1.4
		Carbonyl	3.9	0.0	0.0
		Carboxylic acid	1.3	0.3	0.2

Hydrogenation and dewaxing are not considered for the stream cracker feed. The yields primarily comprise alkanes and alkenes with low quantities of oxygenated components, including alcohols, aldehydes, and ketones. The carbon chain lengths range from C1 to C40 with relatively even distribution. The concentrations of naphthenic and aromatic compounds are relatively low, and cracking is selective to olefin yields with low byproduct formation. The performance for a steam cracker is predicted with a FT waxy product based on a Cobalt based low temperature Fischer-Tropsch catalyst in Table 6.

The Opportunity

The industry demand for green ethylene is increasing. Anecdotal evidence suggests a premium is available for marketing ethylene with higher biogenic content [14]. Switching the feedstock of existing steam cracking assets to biomass feedstock provides an opportunity to extend the operational life or these units. Hydroprocessing of edible oils and gasification/FT conversion of cellulosic biomass can produce drop-in replacement feedstock for the crackers. The incremental addition of such feeds can reduce the carbon footprint with minor revamp to the plant. For greater substitution of feed, adjustments may be necessary to accommodate the liquid feedstock, which is high selectively toward olefins.

Table 6 Steam Cracker Performance for Co-LTFT Waxy Product

Sample Cracker Yield	
Component	**wt%**
H2	0.6
CH4	15
C2H4	35.3
C2H6	4.3
C3H6	17.6
C3H8	0.7
C4	10.1
Pygas + BTX	14.6
Pyfuel Oil	1.8

The global decarbonization effort has placed pressure on the petrochemical industry to reduce its net carbon emissions. Net-zero carbon emission may be achieved by firing the heater with biogenic off-gas. The cracker operation may be adjusted to generate sufficient off-gas to meet firing duty needs. Given that the source feedstock is biogenic, the resulting off-gas would be considered carbon neutral.

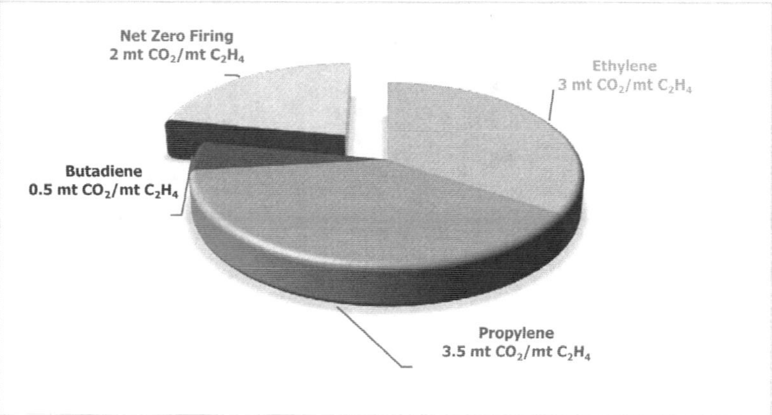

Figure 7 CO_2 Sequestration and Avoidance via Green Ethane Production

An underappreciated aspect of green ethylene is its potential to sequester carbon dioxide. The half-life of plastic degradation can range from 58 years (bottles) to 1200 years (pipes) [15]. The conversion of biomass to solid, stable and non-biodegradable plastics that can be are stored in landfills is a simple approach to the storage of carbon molecules that have been extracted from the atmosphere by vegetation. The yields based on HDO waxy product and FT wax feed suggest that over 5 tons of equivalent carbon dioxide is sequestered, and 2 tons of carbon dioxide is avoided for each ton of ethylene produced. Figure 7 provides the breakdown of the sequestering and avoidance of CO_2 based on this current review. These levels of carbon sequestration should motivate both industry and regulators to investigate pathways for ethylene plants to become carbon negative in support of the global decarbonization effort.

References

[1] The Future of Petrochemicals: Towards more sustainable plastics and fertilizers (2018) International Energy Agency (IEA) and Organisation for Economic Co-operation and Development (OECD). Accessed: https://iea.blob.core.windows.net/assets/bee4ef3a-8876-4566-98cf-7a130c013805/The_Future_of_Petrochemicals.pdf

[2] Dean, Christopher (2013) *Naphtha catalytic cracking for propylene production*, Digital Refining March 2013. Accessed: https://www.digitalrefining.com/article/1000787/naphtha-catalytic-cracking-for-propylene-production#.YcAhOGjMI2w

[3] Polaris Market Research (2021) Ethylene market share, size, trends, industry analysis report by feedstock, by application, by end-use, by region, segment forecast, 2021-2029. Accessed: https://www.polarismarketresearch.com/industry-analysis/ethylene-market

[4] Ren, Tao, Patel, Martin, Blok, Kornelis (2006) *Olefins from conventional and heavy feedstocks: Energy use in steam cracking and alternative processes*, Energy, Vol. 31, pgs 425-451. Accessed:
https://dspace.library.uu.nl/bitstream/handle/1874/21674/NWS-E-2006-3.pdf?sequence=1

[5] https://onlinelibrary.wiley.com/doi/10.1002/0471238961.0520082519211404.a01.pub3

[6] de Pee, Arnot, *et al.* (2018) Decarbonization of industrial sectors: the next frontier, McKinsey&Company, June 2018. Accessed:
https://www.mckinsey.com/~/media/mckinsey/business%20functions/sustainability/our%20insights/how%20industry%20can%20move%20toward%20a%20low%20carbon%20future/decarbonization-of-industrial-sectors-the-next-frontier.pdf

[7] Triantafyllidis, Kostas S. Lappas, Angelos A. Stöcker, Michael. (2013). *Role of Catalysis for the Sustainable Production of Bio-Fuels and Bio-Chemicals - 3.2.1 Microalgae.* Elsevier. Accessed:
https://app.knovel.com/hotlink/pdf/id:kt00BITF78/role-catalysis-sustainable/microalgae

[8] de Klerk, Arno and Furimsky, Edward (2010) *Catalysis in the Refining of Fischer-Tropsch Syncrude - 4.3 Comparison of Fischer-Tropsch Syncrude with Conventional Crude Oil.* Royal Society of Chemistry (RSC). Accessed: https://app.knovel.com/hotlink/pdf/id:kt00AAR5VB/catalysis-in-refining/comparison-fischer-tropsch

[9] Walzl, R. and Zimmermann, H. (2007) *Wiley Critical Content - Petroleum Technology, Volume 1-2 - 5.1 Ethylene from Pyrolysis of Hydrocarbons.* John Wiley & Sons. Access: https://app.knovel.com/hotlink/pdf/id:kt005GYY91/wiley-critical-content/ethylene-from-pyrolysis

[10] Saygen, Deger and Gielen, Dolf (2021) *Zero-Emission Pathway for the Global Chemical and Petrochemical Sector,* Energies 2021, 14, 3772. Accessed: https://www.mdpi.com/1996-1073/14/13/3772/pdf

[11] Rahul Gautam, Jagdeep K. Nayak, Achlesh Daverey, Uttam K. Ghosh, Chapter 1 - Emerging sustainable opportunities for waste to bioenergy: an overview, Editors: Chaudhery Mustansar Hussain, Sunpreet Singh, Lalit Goswami, Waste-to-Energy Approaches Towards Zero Waste, Elsevier, 022, pgs 1-55, ISBN 9780323853873.

[12] de Klerk, Arno (2013) Fischer-Tropsch Process, Kirk-Othmer Encyclopedia of Chemical Technology, January 2013. Accessed: https://doi.org/10.1002/0471238961.fiscdekl.a01

[13] de Klerk, Arno and Maitlis, Peter M (2013) What Can We Do with Fischer–Tropsch Products? Chapt 4 - Greener Fischer-Tropsch Processes for Fuels and Feedstocks (eds P.M. Maitlis and A. de Klerk) February 2013. Accessed: https://doi.org/10.1002/9783527656837.ch4

[14] Bio-based ethylene market size, share & trends analysis report by application, regional outlook, competitive strategies, and segment forecasts, 2019 to 2025. Grand View Research. Accessed: January 31, 2022 https://www.grandviewresearch.com/industry-analysis/bio-based-ethylene-market

[15] Benchaita, Tayeb (2013) Greenhouse Gas Emissions from New Petrochemical Plants: Background information paper for the elaboration of technical notes and guidelines for IDB projects, Inter-American Development Bank. Accessed: https://publications.iadb.org/publications/english/document/Greenhouse-Gas-Emissions-from-New-Petrochemical-Plants-Background-Information-Paper-for-the-Elaboration-of-Technical-Notes-and-Guidelines-for-IDB-Projects.pdf

[16] Wang, Feng & Xu, Yuanyuan & Ren, Jie & Li, Yong-Wang. (2010). Experimental investigation and modeling of steam cracking of Fischer–Tropsch naphtha for light olefins. Chemical Engineering and Processing: Process Intensification. 49. 51-58. 10.1016/j.cep.2009.11.005.

[17] Pyl, Steven P, Schietekat, Carl M, Reyniers, Maria-Françoise, Abhari, Ramin, Marin, Guy B, and Van Geem, Kevin M (2011) *Biomass to olefins: Cracking of renewable Naphtha*, Chemical Engineering Journal 176-177 (2011) pgs 178-187. Accessed: https://www.researchgate.net/profile/Ramin-Abhari/publication/251672664_Biomass_to_olefins_Cracking_of_renewable_naphtha/links/5c76f4eda6fdcc4715a142ce/Biomass-to-olefins-Cracking-of-renewable-naphtha.pdf

[18] Seddon, Duncan (2010) Petrochemical Economics: Technology selection in a carbon constrained world, Catalytic Science Series – Vol 8, Ed. Graham J. Hutchings. Accessed: https://vcmstudy.ir/wp-content/uploads/2016/06/VCMStudy-Duncan_SeddonPetrochemical-Economics.pdf

[19] Shahidi, Fereidoon. (2005). *Bailey's Industrial Oil and Fat Products, Volumes 1-6 (6th Edition).* John Wiley & Sons. Accessed: https://app.knovel.com/hotlink/toc/id:kpBIOFPVE1/baileys-industrial-oil/baileys-industrial-oil

[20] Rasmussen, Henrik, Gabrielsen, Jostein. (2020). *Production of Renewable Fuels: R100 and Co-processing.* 2020 AFPM Summit

[21] Murat M, Gholami Z, Šimek J, Rodríguez-Padrón D, Hidalgo-Herrador JM. (2021) Rendering Fat and Heavy Fischer-Tropsch Waxes Mixtures (0–100%) Fast Pyrolysis Tests for the Production of Ethylene and Propylene. Processes. 2021; 9(2):367. https://doi.org/10.3390/pr9020367

[22] Pyl, Steven P. (2012). *Biomass to Olefins: Steam Cracking of Renewable Feedstocks.* Methusalem Advisory Board Meeting. https://www.lct.ugent.be/sites/default/files/events/Methusalem%20Advisory%20Board%20Meeting%2019-06-2012%20Steven%20Pyl.pdf

[23] Steven P. Pyl, Thomas Dijkmans, Jinto M. Antonykutty, Marie-Françoise Reyniers, Ali Harlin, Kevin M. Van Geem, Guy B. Marin (2012) Wood-derived olefins by steam cracking of hydrodeoxygenated tall oils, Bioresource Technology, Volume 126. Pages 48-55, ISSN 0960-8524. https://doi.org/10.1016/j.biortech.2012.09.037.

[24] Pejpichestakul, W., Ranzi, E., Pelucchi, M., Frassoldati, A., Cuoci, A., Parente, A., Faravelli, T., Examination of a soot model in premixed laminar flames at fuel-rich conditions (2019) Proceedings of the Combustion Institute, 37 (1), pp. 1013-1021, DOI: 10.1016/j.proci.2018.06.104

[25] Ranzi, E., Frassoldati, A., Stagni, A., Pelucchi, M., Cuoci, A., Faravelli, T., Reduced kinetic schemes of complex reaction systems: Fossil and biomass-derived transportation fuels (2014) International Journal of Chemical Kinetics, 46 (9), pp. 512-542, DOI: 10.1002/kin.20867

[26] Ranzi, E., Cavallotti, C., Cuoci, A., Frassoldati, A., Pelucchi, M., Faravelli, T., New reaction classes in the kinetic modeling of low temperature oxidation of n-alkanes (2015) Combustion and Flame, 162 (5), pp. 1679-1691, DOI: 10.1016/j.combustflame.2014.11.030

640351 Hydro-PRT - the Differentiated Plastic Waste Recycling Technology from KBR -

Abstract:

Increasing global production and consumption of plastics is leading to accumulation of waste plastic in landfills and the environment, leading to societal, ecological and economic imbalances.

Advanced recycling technologies such as KBR's Hydro-PRT^{SM} process end-of-life and contaminated plastic waste, that can be further refined to produce virgin polymers and chemicals. KBR's Hydro-PRT is a differentiated, stable and scalable technology, based on a unique patented process using supercritical water to convert plastic waste into commercial products that can be used to produce new polymers delivering a truly circular economy.

The uniqueness of Hydro-PRT process offers following advantages:

High-product yields: no char or unwanted by-products, combined with the ability to control reaction conditions contribute to high liquid products yields.

Feedstock versatility: ability to process mixed composite polymers, flexible and rigid plastics and insensitivity to organic contaminants such as paper, card and biomass means the Hydro-PRT process can process a wider range of waste plastic feedstock than other competing technologies.

Scalability: direct heat transfer between the plastic and the heating medium (supercritical water) leads to an efficient heat transfer mechanism. Scale up is a simple process, not affected by heat transfer considerations.

Product stability: the water surrounding the plastic in the hydrothermal reactors acts as a suppressant to unwanted free radical reactions giving a higher yield of a far more consistent, stable product.

Product quality: the distillation of products allows for ready separation of fine solids and particulates into the heavy wax residue and water from the naphtha fraction. This provides high quality product oils (naphtha, distillate gas oil and heavy gas oil) with low solids and water content. Low chloride content, with most chlorides partitioning to the water phase rather than product oils.

Thermal efficiency: the process operates at lower temperatures requiring lower thermal energy input and the product gas can be reused in the process boiler to generate the supercritical steam providing a highly energy efficient process.

Operational flexibility: Hydro-PRT has a high degree of process control allowing products with different distillation ranges to be produced. Operators can bias products by operating parameters and blend of plastics in the feedstock.

Modular Offering: a modular offering of the technology brings cost and schedule benefits to any project.

Well-studied and demonstrated technology: the technology has been successfully demonstrated on pilot scale since 2014. A 20 thousand ton per year commercial scale plant utilizing Hydro-PRT is under design and construction by ReNew ELP in Wilton, UK. Using end-of-life plastic as feedstock, this plant will produce liquid hydrocarbon cracker feedstock alongside intermediate products. This plant will eventually process over 80 thousand tons of plastic waste feedstock per year and is scheduled to start-up in 2022.

Hydro-PRT provides clear environmental benefits:

- independent studies have demonstrated that advanced recycling technologies such as Hydro-PRT can reduce CO_2 emissions by 1.5 tons for every ton of plastic waste processed when compared to incineration.

- utilizing Hydro-PRT process to recycle plastics into new polymers leads to reduction in the use of fossil-sourced feedstocks.

- when recycling plastics waste via the Hydro-PRT process, minimal waste is produced -- impurities (colorants, additives, fillers etc.) in the plastic feedstock fall out into the heavier hydrocarbon feedstocks, which can be used in construction.

Benefits for business include reaching sustainability goals, developing sustainable infrastructure and reducing emissions and environmental footprint.

Hydro-PRT products resemble natural gas and oil with key applications ranging from making sustainable fuels to being refined into truly circular monomers, like ethylene.

The products, separated into fractions of choice based on their boiling points, are suitable for further processing in different types of crackers.

CIRCULAR ECONOMY WITH PLASTICS PYROLYSIS AND PURIFICATION

James Starr,

Pyrolysis & Purification Product Group Leader, Technip Energies

Veronique Reich,

Principal Process Engineer, Technip Energies

Audrey Tardy,

Plastics recycling Licensing Manager & Technologist, Technip Energies

Abstract: The world is hungry for a circular economy, eliminating our population's throwaway mentality for disposable resources in packaging, clothing, and other products. Social and political pressure is pushing further the need for quick evolution.

Mechanical recycling has been the bedrock of plastic recycling for many years. However, this route faces some shortcomings, such as its inability to handle low sorting quality plastic wastes, the downgrading issue caused by the degradation of plastic physical properties and the finite number of times a plastic good can be recycled.

Advanced chemical recycling can address these limitations. One of the fastest growing areas within advanced chemical recycling is pyrolysis of plastic rich waste. This route ensures "return to virgin plastic" (plastic monomer). Technip Energies is engaged in plastic circularity : "for each of our plastic technologies, we want to have a recycling technology". To achieve this vision a dedicated "Pyrolysis and Purification Product Group" has been set-up to focus on enabling feeding back plastic rich waste to steam crackers.

The presentation will use a case study to link the quantity of waste required to meet the coming legislation trends. It will explain the Technip Energies answer to support a plastic circular economy; a club of pyrolysis technologies to couple with Technip Energies' purification and feed treating technologies. As an illustration, the paper will showcase some integration solutions for both types of plastic pyrolysis products (plastic pyrolysis oil and plastic pyrolysis gas) for which some patent activities are on-going on the purification and recovery technology.

The Why of Plastic Recycling

2019 estimates suggest that of the roughly 410 million tonnes[6] [13] of the global polymer demand (PE, PP, PVC, PS, EPS, ABS, PET bottles and fibers, PU, engineering polymers), 240 million tonnes[6] enters the waste stream as post-consumers wastes (municipal and commercial), production scrap and agricultural films wastes for the commodity polymers (PE, PP, PET,PS and PVC).

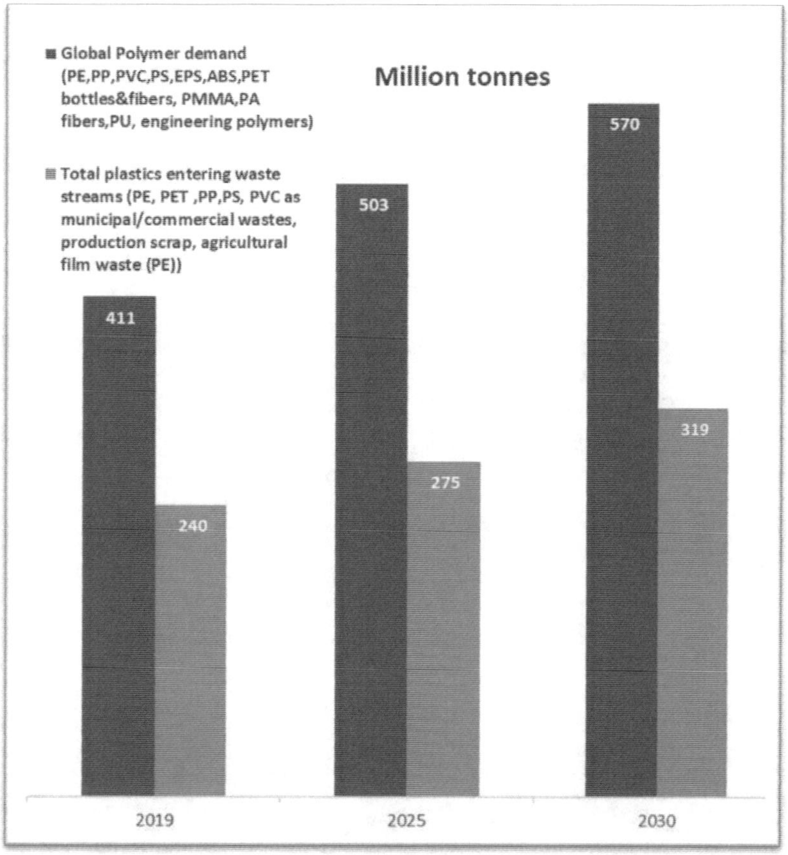

Figure 1: Global polymer demand and commodity polymers entering waste stream

Environmental studies report that the plastic wastes lead to severe environmental consequences. More than 40 percent of plastics are used only one time. Eight million tonnes per year of plastics enter the ocean representing 51 trillion microplastic particles. The social push to tackle the issue is huge and leads to emotional debates, which influence political policies and shape individual behaviors. In our linear extractive, production and consumption model, end-of-life wastes are seen as disposed material that are not needed anymore. This Take – Make – Waste model has been challenged and society has been moving into the circular economy.

Circular economy applies to all materials such as food, steel, cement, plastic and aluminum used to produce consumer goods. Acting on consumer goods is proposing solutions to reduce the 45 percent of the global GHG emissions caused by the production and consumption of these products[12]. The ELLEN MACARTHUR FOUNDATION defines the three principals of the circular economy as follow:

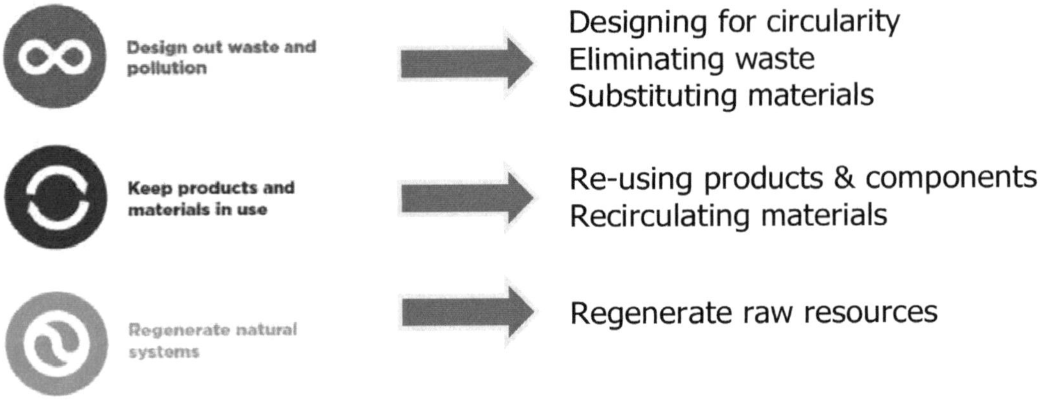

Figure 2: Circular Economy principles[12]

Governments are embracing the circular economy and have been implementing policies to drive the change. Plastic recycling is one of the key fundamentals of this circular economy. Europe is seen as the forerunner in tackling the plastic issue, leading the way on the "Reduce, Re-use and Recycle" hierarchy principle. Other nations are following its example and are implementing similar regulations.

To deeper understand how regulations are driving the plastic change, a worldwide overview is proposed:

Europe

Since 2015, the European commission has defined the first-of-a kind growth strategy roadmap for making the economy sustainable (the European green deal). Applied to the plastic industry, this strategy requires:

- The application of reduce, re-use, recycle options as much as possible;
- Mandatory increased requirements on recycled content in packaging, construction materials and vehicles;
- Member States shall take measures to prevent waste generation;
- Member States shall take measures to ensure that waste that has been separately collected and prepared for re-use and recycling is not incinerated;
- Building a circular business model;
- Increase investment into innovative technologies;
- All plastics packaging on EU market will be recyclable in 2030;
- No more usage of certain single-use plastics by end of 2021;
- Restriction on intentional use of microplastics.

The EU legislation framework is wide and key regulations driving the change are summarized in the following timetable:

Table 1: EU plastic regulations overview[9] [10]

	2020	2021	2023	2025	2029	2030	2035
Plastic bottles separate collection				77 %	90%		
Bio-waste separate & recycle at source			Dec.				
Plastic packaging recyclable by design						100 %	
Single-use plastic ban		100 %					
Separate waste collection + high quality recycling for plastics, paper, metal, glass		Dec.					
Separate waste collection + high quality recycling for textiles				Jan.			
Mandatory recycled content in Plastic bottles				25% PET only		30% Applying to all plastics	
Municipal waste recovery for recycle/reuse	50 %			55%		60%	65%
Landfill disposal						waste suitable for recycling recovery are not accepted in landfills	10% reduction of the amount of municipal waste landfilled
Plastic packaging recycling target				50 %		55 %	
Plastic packaging tax	**Italy/Spain (2022/2023)** 0,45 Euros per kilogram of single use plastic item on market **All members states** 0.80 € per kilogram of **non-recycled plastic packaging waste** (Weight of plastic packaging waste produced - Weight of plastic packaging waste recycled)						

Europe was also the first regional Pact to join the ELLEN MACARTHUR FOUNDATION global Plastics Pact Network.

United States

In August 2020,the US engaged into plastic circularity by implementing and signing the US Plastics Pact. It is unifying 95+ stakeholders across the entire plastics value chain. The pact provides the US 2025 roadmap on four targets to be addressed:

Figure 3: US plastic pact four-pillar actions

There aren't any plastic circularity related laws yet at national level but, at federal level policies have been implemented:

Table 2: US plastic regulations overview

Date	State	Category	Regulation
2007	Seattle, Washington	Regulation	Seattle's council passed a Zero Waste Strategy with a goal to recycle 70 percent of the state's waste by horizon 2025. The recyclable materials are plastics, glass, aluminum, tin cardboard and paper
Sept.2019	California	Regulation	Recycled content plastic bottle law : · 15 percent in 2022 · 25 percent in 2025 · 50 percent by 2030 with a penalty of 20 cents/pounds of post-consumer resin short of the target amount
2022	New Jersey	Regulation	Recycled content law : · 10 percent post consumer recycled content in rigid plastic containers in 2024 and 50 percent by 2036. · 15 percent postconsumer recycled content in plastic beverage in 2024 and 50 percent in 2045

United Kingdom

The United Kingdom has been following an approach of both taxation and giving guidance. The effective date of the plastic packing tax is 1st April 2022. The taxation system can be summarized as follows:

- Who is likely to be affected ?
  - Producers and importers of plastics packaging
  - Business consumers of producers and importers of plastic packaging
  - Consumers who buy goods in plastic packaging in the UK

  Exemption: Producers and importers of small amounts of plastic packaging as only manufacturing and importation of 10 or more tonnes of plastic packaging over a 12 month-period are taxed.
- Taxation principles
  - Applies to plastic packaging that does not contain at least 30 percent recycled plastics
  - £200 per tonne of plastic packaging

WRAP (Waste and Resources Action Program) set up the UK 2025 road map :

1. Eliminate problematic or unnecessary single-use packaging through redesign, innovation or alternative (reuse) delivery model
2. 100 percent of plastics packaging to be reusable, recyclable or compostable
3. 70 percent of plastics packaging effectively recycled or composted
4. 30 percent average recycled content across all plastic packaging

The UK has also joined the ELLEN MACARTHUR FOUNDATION global Plastics Pact Network.

China

China has reconsidered its imported waste policy and applied a plastic ban for low quality waste coming from overseas since 2018. Most of the low quality municipal solid wastes which were exported were indeed shipped to China. As an example, 60 percent of Europe's plastic waste was exported to China in 2018.

In September 2021, China also announced a five-year action plan with the following principles:

Table 3: Principles of China's plastics action plan

EPR (reduction)	Plan will encourage retailers and delivery firms to cut "unreasonable" plastic packaging
Incineration	Plan will boost its plastic recycling and incineration capabilities in the ranges of: - 580,000 tonnes in 2020 - raise waste urban incineration rates to around 800,000 tonnes per day by 2025
Bans (ultra-thin plastic)	China will ban the production of ultra-thin plastic bags nationwide as well as personal care products that contain plastic microbeads. Markets selling fresh produce will be included for 2025 target
Reduction (plastic tableware)	Objective 2025: the consumption of disposable plastic tableware in the catering and takeout area of cities above the prefecture level will be reduced by 30 percent
Alternatives promotion	Plan will also promote the use of alternative products such as bamboo, wood, paper and new biodegradable plastics
Preparing for re-use and recycle	Plan will boost recycling rates for agricultural plastic mulch to 85 percent

India

In October 2021, the Indian government announced the willingness to ban single-use plastics in 2022 and issued a draft nationwide proposal to mandate certain requirements for plastic carry-out bags and to ban certain single-use plastics.

India also set-up a Plastics Pact in Sept.2021 based on the following four pillars:

1. Eliminate unnecessary and problematic plastic packaging through redesign and innovation
2. Make/design plastic packaging to be reusable, recyclable or compostable
3. Effectively recycle (or compost) plastic packaging
4. Incorporate recycled content across all plastic packaging

Summary on Market Drivers

Europe's circular economy regulations framework strategy has opened the way toward plastic circularity. Regulation framework has moved globally. The players of the entire plastic chain are embracing the move such as brand owners making pledges which are beyond regulation (100 percent recycled content, shorter time frames) and plastics industry companies creating associations to coordinate their efforts. Plastic circular economy is one way to answer the global social push from Non-governmental Associations and individuals.

Plastic Recycling Technologies

Plastic recycling technologies can be classified in two main different process categories, namely the mechanical route and the chemical (or advanced) route that gathers depolymerization, dissolution and thermolysis, upon which pyrolysis based technologies are considered. The below scheme highlights how these technologies stands within the circular model:

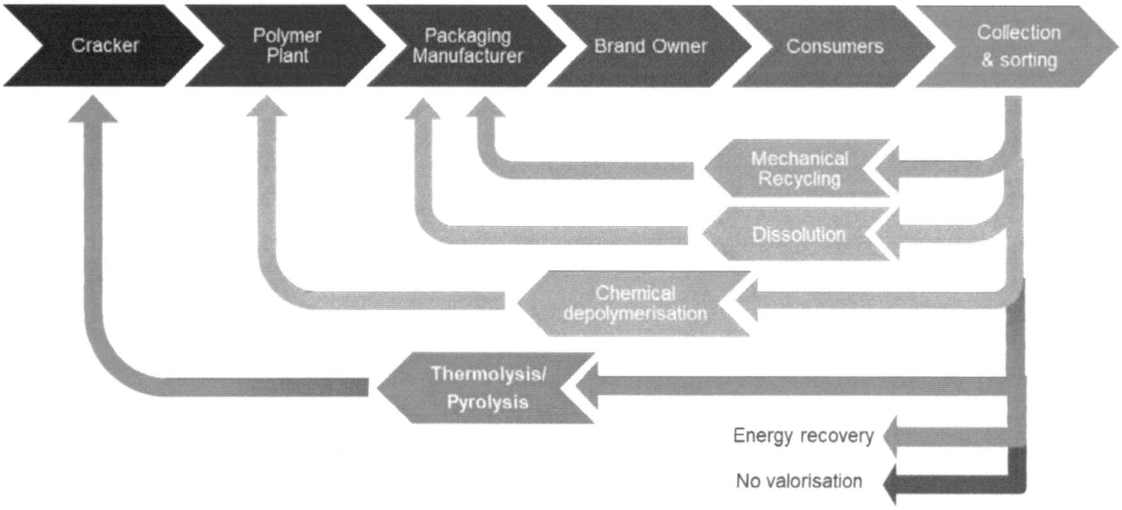

Figure 4: Plastic recycling routes

The mechanical recycling route is well-established and is a simple process with few unitary operations, leading to a low CO_2 footprint. Nevertheless, it requires a well sorted feedstock to be able to produce recycled plastic of an acceptable quality, and only certain types of polymers can be recycled this way. Intensive and efficient sorting methods have to be implemented for that purpose. The physical properties of the polymer are not entirely recovered and, over recycling cycles, degrade. Through mechanical recycling, plastic waste have a finite number of recycle times.

Advanced chemical recycling technologies are considered to be new, but, physical, thermodynamic and chemical principles are well-known. These technologies open the way to effectively answer plastic circularity from a consumer good-to-good point-of-view. From hard to recycle wastes, these technologies produce root monomers to be fed back to the petrochemical industry for the production of virgin-like recycled polymers having the same properties as the standard oil and gas ones. However, the processes can be complex depending on the technology and involve several unitary operations. Consequently, the CO_2 footprint shall be evaluated on a case-by-case base.

Among the chemical recycling methods, pyrolysis, a process involving common unitary operations is one of the fastest growing technologies. A case study is presented below to reflect how it can impact the plastic circularity business in 2040.

Pyrolysis Market Perspectives for 2040

It is expected that the overall worldwide plastic waste generated in 2040 will be in the range of 420 million tonnes per year[7] . The projection for how this waste will be managed is identified below[1] [5]:

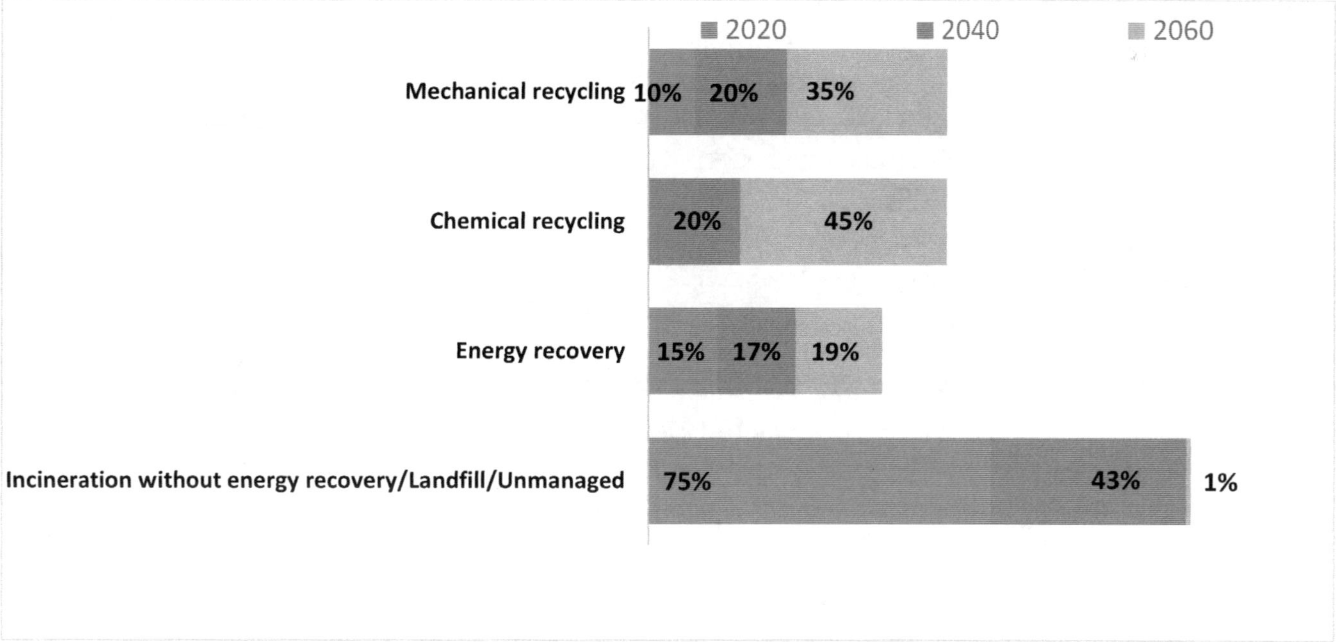

Figure 5: Plastic Waste Management projection from 2020 up to 2060

The expected chemical recycling rate in 2040 would be about 20 percent. Among them, pyrolysis is the only available technology that can treat polyolefins-based hard to recycle wastes (i.e., low sorted or ultimate plastic wastes). Depolymerization for instance, is currently mainly dedicated to PET waste treatment using sorted PET waste. Polyolefins represents roughly 45 percent of the global polymer demand[6]. The pyrolysis technology park is wide. The technology readiness level has increased over the last few years, and some technologies are ready for commercialization and industrialization while depolymerization has a slightly lower technology readiness level. Considering that the pyrolysis route will represent 75 percent of the advanced techniques in 2040 is thus a fair assumption. Therefore, the potential for plastic waste sent to pyrolysis can be estimated at about 63 million tonnes per year.

Considering that standard pyrolysis plant size will increase from the current 30,000-50,000 TPA (waste feed rate based) to 100,000 TPA and beyond, it would mean a huge pyrolysis market with circa 600 plants of 100,000 TPA operating in 2040 to meet the targets. The pyrolysis market is thus there, and the investments are expected to be boosted by states incentives and business commitment. To achieve the circularity there will be the need to integrate the pyrolysis products back to steam crackers. So not only do the pyrolysis products need to be present, but the steam crackers will need modifications to cope with the quantities of this new era feedstock.

How much ethylene from pyrolysis oil in 2040?

From the hereabove 63 million TPA of plastic waste that would potentially feed pyrolysis units and ultimately steam crackers, one can estimate the quantity of ethylene expected from this route. Considering typical yields from the pyrolysis unit, a heavy tail adjustment and typical ultimate ethylene yield achievable through steam cracking of the heavy cut, about 13 million TPA of ethylene would be produced. Our in-house projection for overall worldwide ethylene production by 2040 is about 300 million TPA (considering the 2021 production and a 2.3 percent increase per year). Therefore, the ethylene expected from pyrolysis products would represent roughly four percent of the overall expected ethylene production by 2040.

Meeting business commitments and regulation will require a matrix of solutions. Pyrolysis will be a strong player in the overall scheme with other advanced recycling techniques, such as depolymerization. Mechanical recycling will still be the linchpin for converting large quantities. The waste collection and sorting will need to alter and improve to support the future of this eco-system and so the investments will need to start now. Overall, the plastic recycling route will not remove the need for fossil-based feedstocks, but it is a large step to ridding our future generations of plastic as a waste. If a fossil-free society is the target, then it should also be supported by bio-sourced feedstocks like ethanol to help in reducing the reliance on oil.

Pyrolysis and Purification Solution including Pyro Club Approach

It is well known that waste has a demographic. Much like population densities and cultural orientations, the compositions will vary from country to country, or even region to region. Further, the different collection methods of both post-industrial and post-consumer waste exacerbates the differences. Similarly, steam crackers are unique; there may be similar flavors of steam cracker, but each will have its own idiosyncrasies.

There is also a large cultural difference between pyrolysis and waste management companies and the steam cracker owner/operator. Steam cracker owners are used to highly specified and regulated feeds that are traded based on known benchmarks and contaminant limits. Whereas pyrolysis technologies and waste management companies seldom know the complete composition of the feedstock they are processing, especially considering the stringent limits expected by steam cracker operators. Pyrolysis is by no-means alchemy and although it thermally decomposes hydrocarbons back to something that can be processed, any contaminants in the feedstock to the pyrolysis reactor will need to be removed or reduced ready for onward processing within a steam cracker.

There are two core principles to the approach of Technip Energies, firstly to have access to a range of pyrolysis technologies (our Club) to allow us to answer the challenge irrespective of the waste demographic and location of crackers and, secondly, to have purification technologies to act as the funnel between the pyrolysis technologies and the steam cracker.

Over the last 18 months, whilst others have focused on a single pyrolysis technology, we have been scouting a vast amount of pyrolysis technologies with the aim of building our club with a rich variety of technologies, each having its own uniqueness in the application to a steam cracker or demographic. Our large database was then down-selected by a process of due diligence and collaborative conversations to align on business intent. This is just part of the club membership process as the overall intent is to end in a formal relationship agreement to support the ongoing deployment of their, and our, technologies.

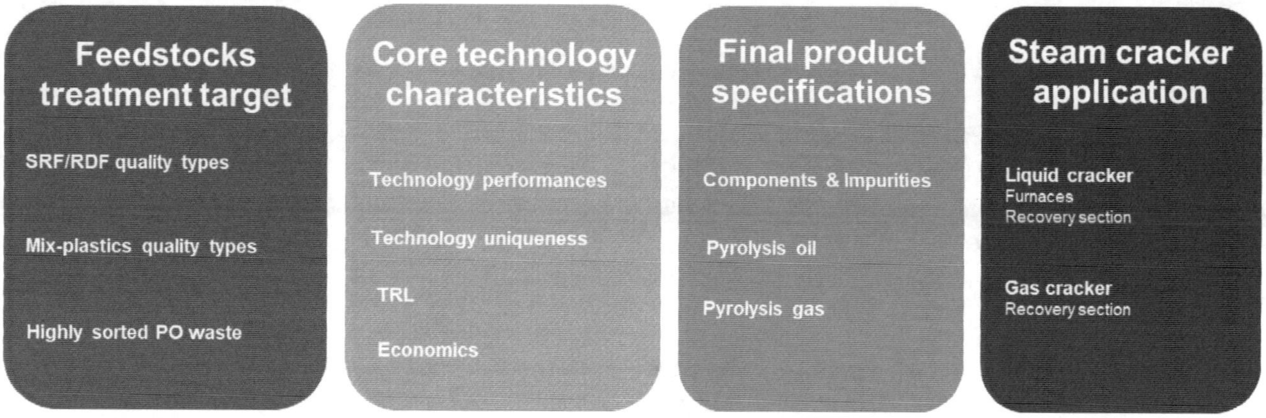

Figure 6: Four Segments of Differentiation for Pyrolysis Technologies

Some of the key aspects we were interested in were the types of waste the technologies had specialized in, the economics of a unit and the product quality. A further consideration was always in and around "deployability" of the technology; was it already in commercial use and was the team in the right mindset for licensing and the routine engineering of a licensed package?

With these technologies coming on board, we have four collaboration agreements to date, with at least another two going through due diligence. We believe we are reasonably placed to implement them anywhere in the world. These technologies then need to be integrated with a steam cracker.

Pyrolysis Products Integration within Steam Crackers

Two Types of Pyrolysis Products (oil and gas) with Similar Constraints

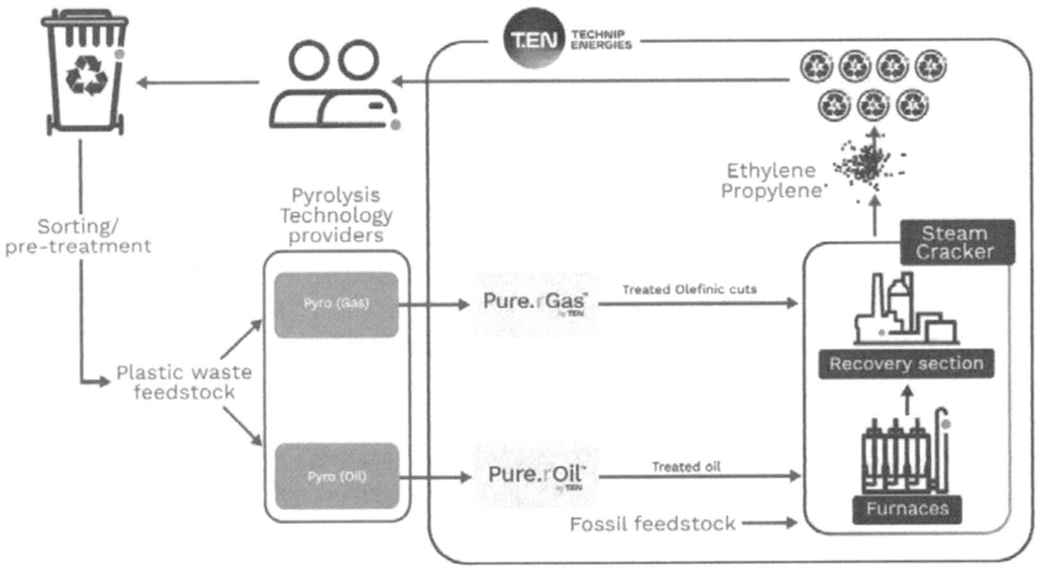

Figure 7: True circular economy thanks to pyrolysis products

Pyrolysis of plastic-rich waste ensures true circularity once the associated products are successfully processed in steam crackers, since it ensures virgin quality of recycled products. Depending on the pyrolysis technology applied, the main pyrolysis based product obtained can either be a gas or an oil. For both types, some similar challenges need to be mitigated before integrating them in reasonable quantities within steam crackers. These include the following:

- **Presence of contaminants :** Pyrolysis product quality is heavily dependent on the quality of the waste processed. Highly sorted, clean plastic waste will lend itself to high quality product output. At the opposite end, low quality waste will induce a low quality product. Contaminants are diverse and can cause a variety of problems in a steam cracker. Purification is therefore a must to ensure successful processing in these units.

- **Variability in the feedstock processed :** The quality of plastic waste processed fluctuates from place to place due to demographics/geography and/or over time due to seasonal variations in consumer habits. The purification needs to be resilient to these fluctuations

Purification Location

By essence the purification train of pyrolysis based gas products needs to be centralized with the cracker. Many synergies are possible between the two units in terms of utilities generation, effluent treatment, ... Even though the purification associated with oil product can be centralized with the pyrolysis unit or decentralized close to the cracker, most of the time it is also more efficient to consider a purification co-located with the cracker to benefit from integration aspects and to lower the overall CAPEX.

Pyrolysis Oil Purification for Steam Cracker Integration: Why ?

There is a large range of pyrolysis technologies which can produce pyrolysis oils from plastic waste. The range of technologies and wastes used as feed result in a wide range of pyrolysis oil qualities. Obviously, the operating parameters used within the pyrolysis reactor, like pressure, temperature, residence time or heating rate will induce variability in the produced pyrolysis oil. However, there are common trends in any pyrolysis based oil. The components can be classified in three main types :

- The attractive compounds, like paraffins and naphthenes, which provide good olefins yields through steam-cracking
- The "ugly" compounds like olefins and aromatics which have no or low olefin potential through steam-cracking, but on the reverse which will promote fuel-oil make. if some di/tri olefins or polyaromatics are present in the oil, this trend is further amplified.
- The contaminants and potential heavy oil tail that can jeopardize the operation of steam crackers with a variety of negative consequences.

Contaminants may affect both the furnace part and the recovery section of the steam cracker.

Oxygenated compounds would convert in the furnaces mainly to CO and H2O, but conversion is not complete and all kind of oxygenates (ketones, aldehydes, alcohols, ...) may remain, together with unconverted oxygen itself. These oxygenates are prone to promote organic acid formation, thus corrosion and fouling. The furnace run-length may be impacted and the final products may be contaminated.

Nitrogenated compounds can decompose to ammonia, which would then accumulate within the quench and dilution steam systems, which may pollute the ethylene final product and/or poison the palladium catalysts. Where NO containing compounds are fed to a cracker there might be the risk within the furnace to generate some NO. If some oxygen is also present, originated from some oxygenated compounds fed to furnace (and/or from other external stream fed to the cracker), then a potential NOx safety issue may arise within the cryogenic section of the plant. The NOx chemistry can be summarized as follows[2,3]:

$$2\,NO + O_2 \rightarrow 2\,NO_2$$
$$NO + NO_2 \rightarrow N_2O_3 \text{ (blue liquid)}$$
$$N_2O_3 + \text{Unsaturated hydrocarbons} \rightarrow \text{NOx gums}$$
$$NOx/N_2O_3 + NH_3 \rightarrow NH_4NOx \text{ salts}$$

Explosion incidents when some NOx gum/NOx salts are reheated have already been reported in Europe, the USA and Japan.

The halogenated compounds (especially chlorinated) can lead to furnace radiant coil corrosion. High amounts of salts and/or metals within pyrolysis-based oil would also affect the convection coils leading to increased corrosion and coking rate so leading to reduced furnace run-length. The recovery section may be impacted as well if the pyrolysis oil is polluted by some specific metals like mercury; it would lead to damage within the aluminum plate fin heat exchanger of the cryogenic section through metal embrittlement[4].

Finally, if untreated, the heavy tail in the oil feedstock can exceed the end point specification. Special attention is required to check that the proper vaporisation of the feed is ensured, that the furnace cross-over temperature remains adequate, and that fouling tendency within the convection section remains within acceptable limits.

Pyrolysis Oil Purification for Steam Cracker Integration : Technip Energies Answer

Technip Energies has developed a technology to condition the pyrolysis oil produced from plastic waste, with the ability to process as wide a range of oils as possible, while mitigating all the above discussed issues. This technology is based on existing referenced technologies. Hydrogenation for saturation of olefins and some contaminant removal is based on commonly available technologies from selected and qualified catalyst suppliers. The purification strategy comprises a dedicated multilayer contaminant removal scheme with several levels of safeguard for key components including chlorides, oxygenates, metals, etc.

It maintains the operability of the steam cracker unchanged by allowing:

- The mitigation of fouling and corrosion issues in the furnaces
- A final product on-specification
- Poison-free operation of ethylene plant catalysts

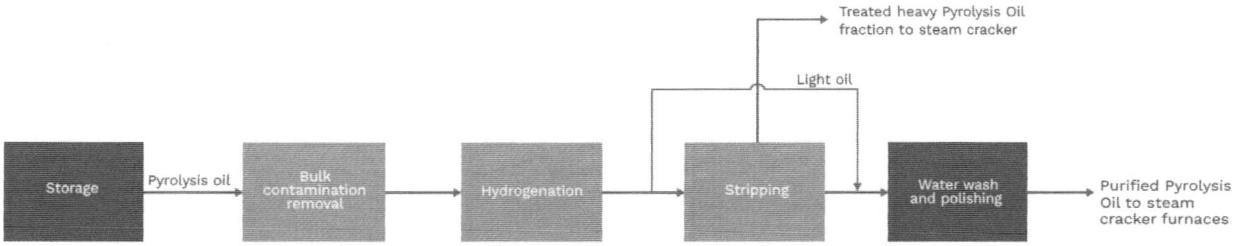

Figure 8: Pyrolysis Oil Purification Scheme

The principal sources of the technology used are:

- Technip Energies' Chloroff technology, developed and implemented many years ago for the clean-up of spent lube oil
- Hydrogenation technologies and catalysts tested on actual oil samples
- Technip Energies' quench technology, used on a large number of operating liquid cracking plants
- Technip Energies' Ripple Trays™ used as the tray of choice in many fouling systems in ethylene plants

Knowing that each cracker has some special features in terms of furnace set-up, etc, the design can be tailored on case-by-case basis.

Pyrolysis Gas Purification for Steam Cracker Integration : Why ?

Among all the advanced recycling techniques using the concept of pyrolysis of plastic waste in the absence of oxygen, the ones exporting product gas present some significant differentiators for integration with a steam cracker. Indeed, pyrolysis gas technologies offer a high light olefins content which allows a cracker integration directly within the recovery section of the plant. The by-pass of furnaces enables:

- An attractive carbon footprint
- A plastic based feedstock that can act as an olefins capacity adder and not only as fossil feedstock replacement (especially if steam cracker is not recovery section limited)
- Integration in various kinds of crackers including those that don't have furnaces able to crack liquid feedstock

However, integration of pyrolysis gas products issued from plastic waste into existing steam crackers is a challenge that needs to tackle the dual concern of being able to capture the high value compounds while managing the presence of contaminants.

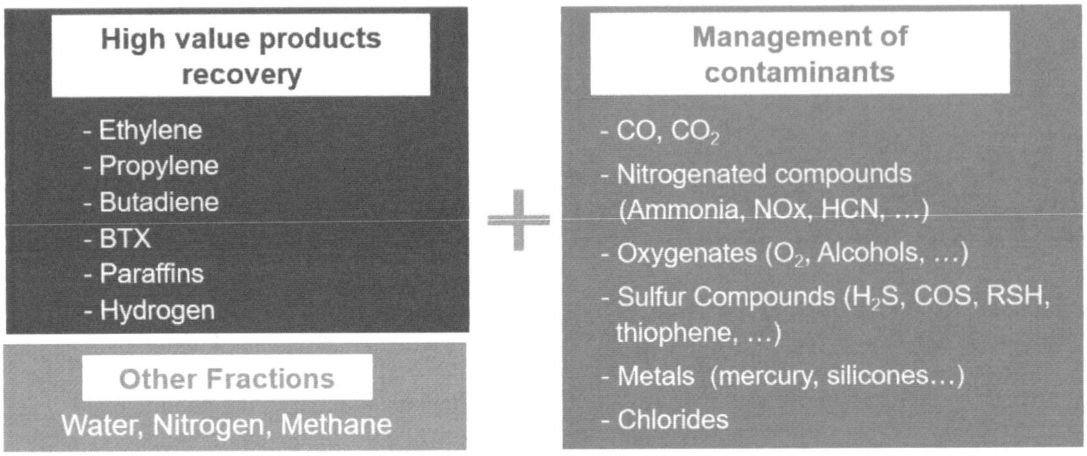

Figure 9: Pyrolysis Gas Integration within Steam Crackers : A Dual Concern

The high value compounds inherently present in the product gas are ethylene, propylene, butadiene and BTX. Paraffins will act as an olefins yield booster once recycled to extinction, together with the usual cracker recycles. Light end fraction is a mixture of methane, hydrogen, and nitrogen. At the same time, a range of various contaminants prone to induce negative effects on the steam cracker needs to be mitigated (NOx, CO, metals, chlorides, HCN, acid gases, oxygenates, sulfur compounds, ...).

The main reason for the purification requirement is that mixing this gaseous cut with the main cracked gas of an ethylene plant would present some safety issues, whatever the type of cracker:

- For front-end demethanizer scheme (back-end hydrogenation), integrating a stream containing NO, oxygenates, butadiene and ammonia would induce some safety issues within the cryogenic part of the plant (refer above discussion on NOx chemistry and associated incidents)
- For front-end deethanizer or depropanizer scheme (front-end hydrogenation), integrating a stream containing a lot of CO upstream the acetylene reactor is not an option, due to greater thermal runaway risks. In addition, the risk of exotherm is further accentuated because by its nature the pyrolysis product gas will undergo some CO fluctuations due to the variability of the waste being processed.

Obviously, the purification is also a must to:

- Mitigate fouling and corrosion issues
- Maintain the final products on-specification
- Keep the run-length of the hydrogenation catalyst unchanged

Overall, the pyrolysis gas purification presents very similar challenges to those encountered in Refinery Off Gas (ROG) treatment.

Pyrolysis Gas Purification for Steam Cracker Integration : Technip Energies answer

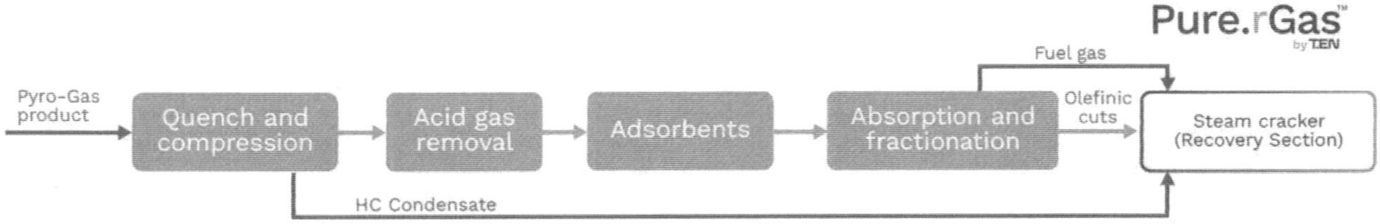

Figure 10: Pyrolysis Gas Purification Scheme

The scheme is a sequence of unitary steps already proven in industry at large scale like water wash, amine unit, caustic wash, drying and adsorption.

To ensure safe operation of the ethylene plant, a liquid absorption of light ends to fractionate the light components (including NO and CO) from the main olefinic streams is implemented. The process is based on a liquid absorption taking place in mild cryogenic conditions, purposely chosen to eliminate the risk of NOx gums/salts formation (which takes place especially below -100°C)[2,3].

By essence, the purification train is centralized within the ethylene plant that will then receive three types of purified products:

- Fuel gas fraction, which contains the light end like methane, hydrogen, nitrogen, CO, O_2, NO, ...
- Olefinic liquid fractions (one or several cuts) that contains ethylene, propylene, butadiene and C_5- paraffins
- Hydrocarbon condensate cut that contains the BTX fraction

Multiple synergies with the cracker can advantageously be optimized like utilities (including regeneration facilities, steam system and cooling water), the regeneration gas facilities, and the effluent treatment (spent caustic treatment, waste water treatment). Customized solutions to define the best integration scheme for a given ethylene plant are optimized on a case-by-case basis.

Summary and take-aways

The plastics recycling market is potentially huge and dynamic. It is pushed by several drivers such as legislation and company's ESG targets, as they seek to lower their carbon footprint attributed to the kind of feedstock used and business commitments to use recycled materials.

Overall optimization of our eco-system will consist of a mix of approaches to cover the 3R adage (Reduce, Re-use, Recycle) of the plastic circularity. Waste consolidation and a matrix of recycling routes (mechanical/depolymerization/chemical) together with bio-sourced feedstocks will be required.

Chemical recycling promises to provide an effective solution to answer plastic circularity from a consumer goods standpoint, where new plastics are produced from post-consumer wastes, without a degradation in product quality. Technip Energies is committed to supporting and defining solutions to make the circular economy a reality. We are building a pyrolysis technology providers club to be in position to address market pressure in a wide variety of situations and have developed purification flow schemes with patent applications pending.

For the benefit of society, end users, producers and technologies can support plastic circularity, but need to collaborate to shape the new environment. We can support companies with customized solutions that define the best integration scheme for a given ethylene plant, optimized on a case-by-case basis. The selected process schemes benefit from our expertise in ethylene technology licensing and our comprehensive ability to optimize the entire value chain from scouting of pyrolysis and waste to the implementation phase supported by our experience as an EPC contractor.

References

1. McKinsey & Company, 2018. "How plastics waste recycling could transform the chemical industry". Thomas Hundertmark, Mirjam Mayer, Chris McNally,Theo Jan Simons, Christof Witte
2. Dane Grenoble, EPC Spring 2012 meeting. History of NOx in Ethylene Industry
3. EPC 2007 Tutorial : NOx safety in Ethylene Plants
4. S. Mark Wilhelm, Alan MacArthur and Russ Kane, Methods to combat Liquid Metal embrittlement in cryogenic aluminium heat exchangers
5. Nexant – Technoeconomics – Energy&Chemicals- advances in Depolymerization technologies for Recycling – Tech 2021S10 - May 2021
6. AMI Consulting – The global mechanical recycling industry 2020 – May 2020
7. Based on AMI Consulting – The global mechanical recycling industry 2020 – May 2020 with an additional CAGR of 3% to reach 2040 year.
8. ICIS training – An introduction to plastics Recycling – March 2022
9. European Parliament website : https://www.europarl.europa.eu/portal/en
10. European commission website : https://ec.europa.eu/info/index_en
11. WRAP website : https://wrap.org.uk/
12. Completing the picture – How the circular economy tackles climate change – ELLEN MACARTHUR FOUNDATION – Material economics – 26 September 2019.
13. AMI Consulting – Global plastics industry-Seminar report 2021 – October 2021.

SPRING22
+18TH GCPS
A Joint AIChE and CCPS Meeting

Efficient Steps for Meeting Ethylene Process Sustainability Goals

Presenter/Principal Author:

Dr. Raghu Narayan,
Advisory Director, Ingenero Inc.
Houston, Texas
rnarayan@ingenero.com

Sameer Thorat,
Director: Analytics and Engineering, North America, Ingenero Inc.
Houston, Texas
sthorat@ingenero.com

Prepared for Presentation at
American Institute of Chemical Engineers
2022 Spring Meeting and 18th Global Congress on Process Safety
San Antonio, TX
April 10 – April 14, 2022

Efficient Steps for Meeting Ethylene Process Sustainability Goals

Presenter/Principal Author:
Dr. Raghu Narayan,
Advisory Director, Ingenero Inc.
Houston, Texas
rnarayan@ingenero.com

Sameer Thorat,
Director: Analytics and Engineering, North America, Ingenero Inc.
Houston, Texas
sthorat@ingenero.com

Keywords: Sustainability, Net-Zero, ESG, Ethylene Plant Performance, Tracking, Optimizing, Reducing Emissions, Smart Sustainability Dashboards

Abstract

With ESG becoming a 'core' performance goal, 'minimizing' emissions, maximizing yields, and minimizing waste generation are key to achieving the KPI targets of the Chemical and Refinery Process Operations. Ethylene plant operations and Refinery operations operate on a plan developed by the Optimization group, to accomplish the business and ESG targets.

Constantly tracking 'Plan vs Actual' is an extremely important undertaking. Concurrently, gathering all the data and transforming it to meaningful information becomes part of the unique memory, that can be utilized to establish capabilities and constraints, during the current 'run length of the asset (between turnarounds). Developing such markers and utilizing it to achieve 'best achievable performance' has become critical.

Ethylene industry is looking towards sustainable solutions, the new technology options would take time to be commercial and profitable process. The major influence on the ethylene manufacturing plants sustainability in next few years would be through solutions which would drive lower emissions in current operation through sustainable improvements.

Use-cases will be presented to discuss the opportunities and challenges of meeting sustainability goals through technology improvements, smart modifications, and utilization of power of data analytics through AI/ML.

1 Introduction

As we all are well aware, petrochemicals are an important building block of a huge range of products that underpin modern daily life and economic activity, globally. They are everywhere in our daily life and play a fundamental role in most of our daily activities and have become part of virtually every device we use to ensure our well-being, protect our health and security.

Olefins – ethylene, propylene, as well as a variety of C_4 olefins and others are critical intermediates in the manufacture of a vast range of products. As most of us know, ethylene and propylene are basic petrochemicals used to produce broad variety of polymeric materials, while the C_4 family of olefinic hydrocarbons are used to create a variety of products elastomers, that we use in our daily life.

The last twenty years has seen a significant production of olefins from Europe and North America to the Middle East and East Asia, driven by a growing population as well as a high rate of fast growth.

Since the Rio Earth Summit in 1992, economic development has attempted to proceed hand-in-hand with sustainable development, i.e., balancing economic, environmental, and societal responsibilities. The Paris Agreement adopted in December 2015 at the UNFCCC's Conference of the Parties (COP21) had proposed a pathway to low-carbon development by offering the signatory parties "the opportunity to present their (intended) nationally determined contributions" with an eye on minimizing 'negative' impact on social and economic development.

Several leading manufacturers of Olefins have presented their long-term plans to achieve their Environmental, Social and Governance (ESG) goals. We hear the term 'Goal is Net Zero'!

2 Goal of Net-Zero

- 2050 target – EU and many countries, 20% of world's largest *2,000 firms have Net-Zero targets*
- *61% of global emissions come from nations and companies that have Net-Zero targets*
- *68% of global GDP comes from nations and companies that have Net-Zero targets*

ESG is one of the core ethylene plant performance goals, with concerted efforts on carbon-footprint reduction over this decade and future, motivated by regulations and changing consumer demands and preferences. For example:

- Regulations
    - 19 European countries have implemented a carbon tax averaged to $42 per metric ton of carbon

- The carbon emission trading network across the EU reached a price of $68 per metric ton
- Consumer Demand
  - 6 in 10 consumers will change their shopping habits to reduce environmental impact
  - Public perception of company's sustainability (current emissions, targets, investment)

3 Facets of Sustainability

Sustainability has many facets:

- Safety, Employees, Assets and Community
- Energy Usage, e.g., Steam, Electricity
- Yield, effective conversion of feedstock
- Operational Reliability, Efficiency
- Water Usage and effluents
- Emissions, e.g., CO_x, NO_x
- Meeting all Environmental Regulations, Air, Water, and Solid Waste

It is easy to make bold 'declarative' statements. Most of us are well aware of the bold statement President John F Kennedy made on his inaugural address; "First, I believe that this nation should commit itself to achieving the goal, before this decade is out, of landing a man on the moon and returning him safely to the earth."

To make what President Kennedy stated a reality, it took a very determined effort and a constancy of purpose by a very dedicated group, for Neil Armstrong and Buzz Aldrin to land on the moon on July 20, 1969.

Such a determined effort is indeed required by the Ethylene Producers – a constant drive to do the following:

- Many of the Ethylene production assets are over decades old and the industry has to make the most of these assets, while adhering to the items listed above to achieve corporate ESG goals
- There is no one solution for the assets that are operated. Each asset has to be addressed as an 'individual patient'
- Lastly, as demand and consumer preferences change, plans for the future investments have to be constantly monitored.

4 Approach to Sustainability

Thus, "Sustainability" needs to be addressed in a stepwise approach for maximum gains:

- <u>Step 1</u> Tracking the Operation
- <u>Step 2</u> Optimizing
- <u>Step 3</u> Reducing emissions and waste

Tracking sustainability KPI's with production targets for each asset is the first step. We have all heard Peter Drucker's famous statement, "If you can't measure it, you can't improve it".

4.1 Tracking

With the explosive growth in sensor hardware (as well as soft sensors) and computational technology (Data Analytics), "Smart Sustainability Dashboards" with periodic reports for efficient monitoring and constantly tracking 'Plan vs Actual', are now readily available for tracking the performance of each asset. Such tracking also provides clearly understood GAPs. Definitive path-forward plans can be developed to close such GAPS.

Digitalization is indeed providing manufacturers gather all the available data and 'transforming' it to meaningful actionable information as well as 'tuning' the asset to achieve 'best achievable performance', which is critical.

Shown below is an example of a 'Tracking Report', focused on Energy Usage and Emissions.

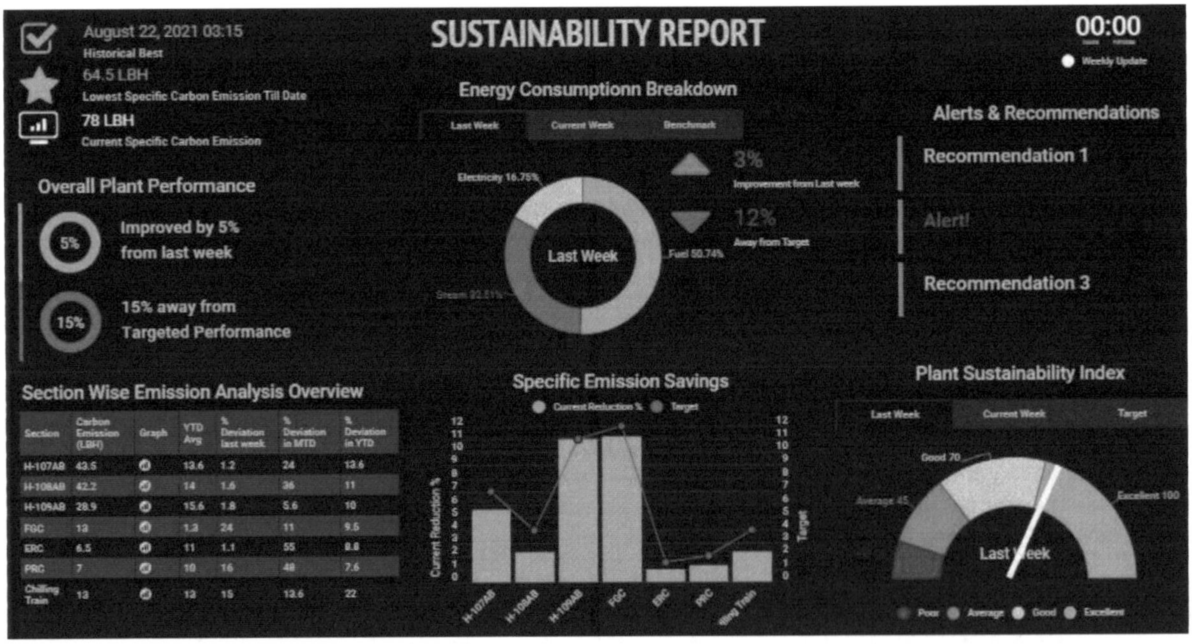

Figure 1. Customized Sustainability Dashboard

Figure 1 leverages live analytics as well as data visualization of emissions.

Auto-generated reports summarizing emissions, comparing energy usage and emissions over selected time periods provides real-time awareness of current operations.

4.2 Optimizing

While each asset will have unique constraints (e.g., feedstock flexibility), ethylene plants can apply digitalization and analytics to achieve sustainability goals, by being able to optimize its operations to achieve sustainability goals. Some of these are listed below:

- Special studies analyzing equipment and systems with the most room for improvement
- Customized solutions for critical systems, to identify and address underperformance and optimize energy consumption and emissions targets
- Appropriate use of digital twins to provide accurate predictions of impact of change on entire process
- Quantify realistic reductions associated with each change

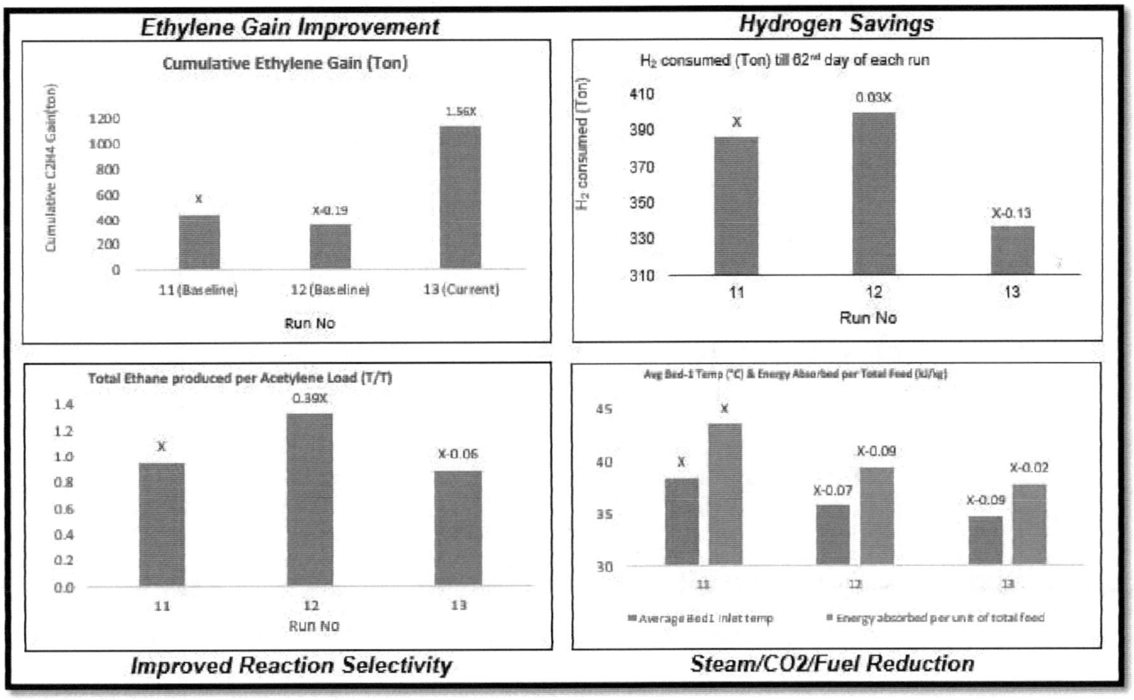

Figure 2. An example case study involving Acetylene Reactor Optimization

Realized Gains
732 T (185%) Ethylene
61 T (16%) Hydrogen
200 MW (9%) Energy
14 T (9%) Fuel Gas
475 T (25%) less CO2 Emission
717 T (22%) Ethane Recycling cost

Figure 3. Summary of Operational Gains

5 Case Studies

Another Case Study, that was conducted and implemented involved the propylene refrigeration circuit. The compressor was experiencing a high load due to hydraulic issues.

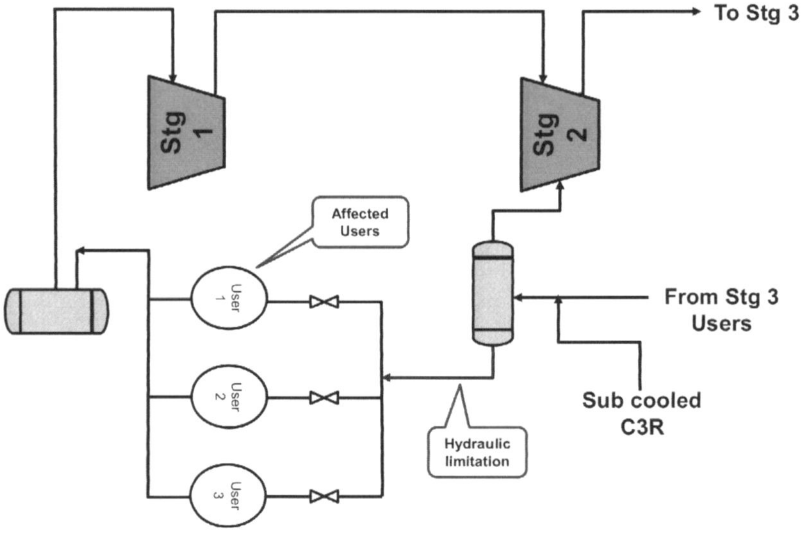

Figure 4. Illustration of the Propylene Refrigeration (C_3) Circuit

Upon detailed analysis of the above circuit, the following issues were discerned:

- Increased overall C3R circulation flows impacted flow from C3R LPFD to first stage exchangers
- Limited subcooling
- Increased flashing and higher pressure drop
- Critical condenser – limited by refrigerant flow

- Higher refrigerant flow requirement

Minimizing the pressure drop across critical loops reduced the power requirement of the Propylene Refrigeration compressor.

- Rerouted sub-cooled propylene directly to consumers thus overcoming hydraulic issues
- Reduced flashing across level valve
- Resolved flow limitation
- Pressure drops and internal flow and pressures decreased
- Substantially reduced Propylene refrigerant requirement

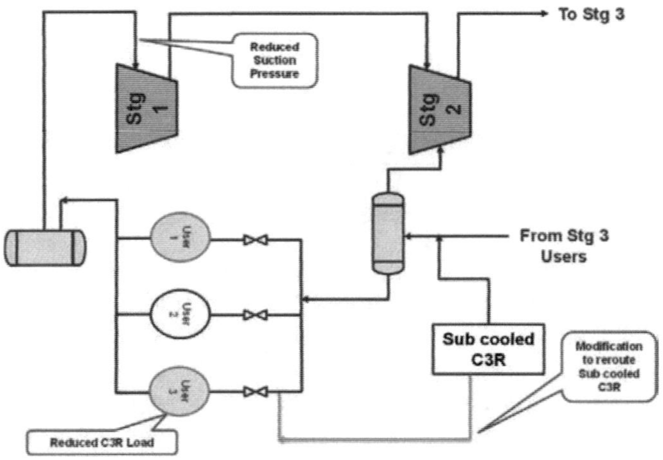

Figure 5. shows the modifications that were incorporated into the circuit.

Thus, overall Propylene compressor energy requirement was reduced by 5% to 7% over base case.

Similar Optimization examples have been developed for the Furnaces. Furnace hybrid models were utilized to optimize oxygen levels to minimize energy requirement without adversely affecting coking rates in the coil.

Real time recommendations to minimize stack oxygen, by optimizing Damper opening and Burner air registers have been developed.

Hybrid models also identified the best operating window considering, historical data and correlations between critical parameters, its impact on coking, current performance, and operating scenarios – feed and heat load, as well as known constraints and limitations.

These significantly improved furnace efficiency, furnace run length and reduction in fuel usage by 3% to 4% in energy and equivalent CO_2.

6 Conclusion

'Smart' tracking process powered by AI/ML can direct the asset to achieve best achievable performance, in a cost-effective manner, while meeting sustainability goals. As stated earlier, this effort takes a significant 'constancy of purpose', based on the life cycle of the asset. Additional benefits are a deeper understanding of the behavior of the asset, which could result in further studies for continuous improvement of the performance of the asset.

Finally, digitalization, done properly, helps gathering all the available data and transforming it to actionable meaningful information.

7 References:

1) Asen, E., 2021. *Carbon Taxes in Europe*. [online] Tax Foundation. Available at: <https://taxfoundation.org/carbon-taxes-in-europe-2021/> [Accessed February 2022].

2) Meredith S., 2021. *Why the world's largest carbon market is experiencing a boom like never before.* cnbc.com. Available at: <https://www.cnbc.com/2021/05/18/why-europes-carbon-market-is-experiencing-a-boom-like-never-before.html/> [Accessed February 2022]

3) Haller K, Lee J, Cheung J. *Meet The 2020 Consumers Driving Change*. IBM; 2020. Available at: <https://www.ibm.com/downloads/cas/EXK4XKX8/> [Accessed February 2022]

4) Net Zero Scorecard. eciu.net Available at: <https://eciu.net/netzerotracker/>. Published 2022. [Accessed February 2022]

5) 5. Graham J. Net-zero emissions targets adopted by one-fifth of world's largest companies. Reuters.com. https://www.reuters.com/article/us-global-climate-carbon-business-trfn/net-zero-emissions-targets-adopted-by-one-fifth-of-worlds-largest-companies-idUSKBN2BF2ZX. Published 2021. [Accessed February 2022]

6) Sustainability in Petrochemicals, IEAGHG Technical Report 2019-01 February 2019, IEA Greenhouse Gas R&D Programme

AIChE Paper Number: 639189

Paper 78d - A Practical Approach to Near-Term Decarbonization of the Ethylene Plant

Alex Brayton
Senior Process Engineer
Technip Energies
Houston, TX
alexander.brayton@technipenergies.com

Rishav Jain
Application Engineer
Mitsubishi Heavy Industries Compressor Corporation
Hiroshima, Japan
rishav.jain.xe@mhi.com

Keywords: ethylene, electrification, decarbonization, emission reduction, carbon management, energy transition

Abstract

With the petrochemical industry's increasing focus on reducing the carbon footprint of their respective plants, nearly all industry players worldwide want to achieve carbon neutrality as soon as possible. Hence, producers will need to employ transition technologies that are available today to decarbonize their respective plants. This study is mainly focused on exploring these transition technologies to reduce CO_2 emissions in a cost-effective manner and their consequent ethylene plant arrangements.

The paper starts by presenting a typical ethylene plant which will act as the 'base plant' for this case study. Then, a 'reduced emission' ethylene plant is introduced which employs a new cracking furnace design having the potential to reduce the fuel gas requirement by 25 percent. This, in turn, reduces the amount of carbon dioxide produced by the 'reduced emission' plant by 25 percent.

Since the amount of available VHP steam is less in the 'reduced emission' plant, all the compressor trains cannot be driven by steam turbine drivers alone. Two cases employing a combination of Steam Turbine compressor drivers and Electric Motor compressor drivers are analyzed and compared with the 'base plant'. Impact to the utility balances and CO_2 emissions of the studied plant configurations are summarized in order to provide a holistic picture of the 'reduced emission' plant.

Thereafter, the study evaluates the 'reduced emission' plant and 'base plant' on the parameters of costs such as piping, rotating equipment, new cracking furnace, etc. and maintenance/reliability. Finally, the paper is concluded by providing recommendations on how this 'reduced emission' configuration can be implemented by a petrochemical company in a practical and economical way.

1 Introduction

1.1 Motivation for Emission Reduction

Petrochemical feedstock makes up 12 percent [1] of global oil demand and with the petrochemical industry's focus on reducing the carbon footprint of their respective plants, nearly all industry players worldwide want to achieve carbon neutrality as soon as possible. While promising technologies to enable completely zero emissions are in various stages of development, it may take another ten to fifteen years before widely implemented on an industrial scale. Hence, producers need to employ transition technology. This study is mainly focused on exploring these transition technologies and their consequent plant arrangements in ethylene plants.

1.2 CO_2 Reduction Philosophy for the Ethylene Plant

The major source of direct CO_2 emissions in an ethylene plant is from firing fuel gas in the cracking furnaces and auxiliary steam boilers, if any steam import is required above the furnace steam generation. While we cannot reduce the firing associated to perform the cracking reactions, the cracking furnace can be redesigned to minimize firing associated with steam production. In addition to direct CO_2 emissions, an ethylene plant has indirect emission from imported electricity to drive pump drives and other equipment. To minimize indirect CO_2 emissions from the ethylene plant, electrical power from renewable and other low CO_2 sources should be maximized.

The above CO_2 reduction philosophy has the potential to reduce the CO_2 emissions of an ethylene plant by as much as 30 percent. This reduction can not only help end users to achieve their goals to improve their carbon footprint, but also, depending upon the region, end users can claim tax credits/benefits offered based on their plant's reduced carbon footprint.

As a consequence of the reduced steam production in the reduced emission ethylene plant, there is insufficient steam to drive all the major compressors as in a typical plant. Hence, there is an opportunity to electrify the major compressor drivers to utilize low CO_2 electrical power. The electrification of the chemical industry is underpinned by the availability of this low-carbon power generation with the expectation that the power sector will decarbonize at a faster pace.

The benefits of electrical motor drivers over steam turbine are lower CO_2 emissions, lower maintenance costs and simplicity of operation. Furthermore, since electric motors can be operated without requiring steam raising and warming of turbines, the electrification will allow end users faster start-up and turnarounds, with a potential to reduce overall flaring during these operations.

1.3 Study Scope

The focus of this study is to look at technologies available today to reduce the CO_2 emissions of an ethylene plants being built in near-term (before 2030). These are chiefly: 1) a novel reduced emission cracking furnace with lower steam production and 2) electrification of the main compressors with imported electricity from low CO_2 sources. Although outside the scope of this study, the role of both mature and emerging technologies such as carbon capture, electrification of furnaces, hydrogen firing in cracking furnaces, etc. in the long-term should not be ignored. These technologies are expected to play a critical role in the path to a "net zero" ethylene plant.

The following study explores the changes required to achieve a reduced emission ethylene plant with technologies available today and investigates the holistic impact to the plant including utility systems, CAPEX/OPEX, and CO_2 emissions. Two options are explored in which the major compressors are partially electrified in a reduced emission ethylene plant, and a cost analysis is performed to determine which option is optimal.

2 Base Case (Traditional Plant)

2.1 Configuration

A 12.3 MM lb/day (~2000 kTA) "world-scale" ethylene plant was selected as the study case to investigate impact of a low CO_2 design on the overall plant design. This plant utilizes ethane-based feedstock. The major compressors in the plant are the Cracked Gas Compressor (CGC), Propylene Refrigeration Compressor (PRC) and Ethylene Refrigeration Compressor (ERC) trains. In a typical ethylene plant, steam turbines drive all the compressors. Table 2.1 shows the compressor process power requirement for the base plant.

Table 2.1: Compressor trains of a typical ethylene plant

Steam Turbine Driver	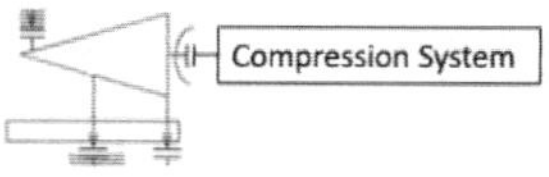	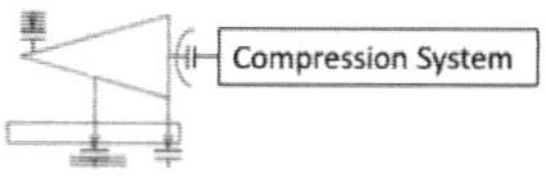	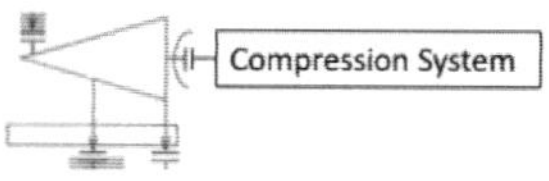
Train Name	**CGC**	**PRC**	**ERC**
Normal Power [MW (hp)]	72.6 (~97,300)	30 (~40,200)	34.1 (~45,700)
Turbine Rated Power [MW (hp)]	83.8 (~112,400)	33.4 (~44,800)	37.7 (~50,600)
Turbine Type	Extraction Condensing	Extraction Condensing	Extraction Condensing
Typical Size (L x W x H)	87 ft x 18 ft x 18 ft	33 ft x 13 ft x 13 ft	37 ft x 13 ft x 13 ft

2.1.1 Traditional Furnace Overview

Various ethylene furnace designs which utilize a range of feedstocks (such as naphtha, ethane, methanol, mixed, etc.), are being operated worldwide. While the details of these furnace configurations differ, the basic structure is the same in which the feedstock is cracked within radiant coils in the presence of steam. Transfer Line Exchangers (TLE) are then used to quench the cracked feedstock and boiler feed water is converted to high-pressure steam via absorbing the energy from quenched feedstock. In the furnace convection section, feedstock is preheated, vaporized, and superheated along with steam against hot flue gas and boiler feed water may be preheated in an economizer. TLEs are the primary source of high-pressure steam in an ethylene plant and any steam shortage is taken care of by additional boilers. Hence, steam is inherently available in an ethylene plant and any direct reduction in steam would also affect the production capacity of plant unless the furnace design itself is changed. In the traditional furnace configuration, large quantities of CO_2 are emitted from furnace stack. Figure 2.1 illustrates a simplified schematic of the cracking process.

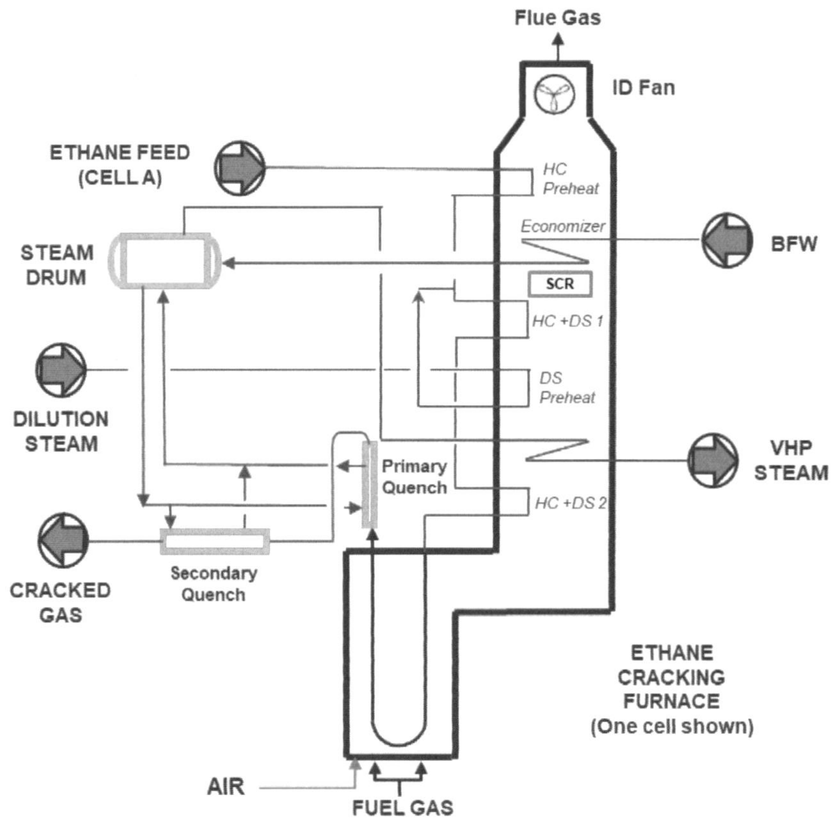

Fig 2.1: Schematic of conventional cracking furnace (Base Case)

2.1.2 Role of Steam

Steam plays an important role in the ethylene plant. In the traditional plant configuration, steam is typically used to drive major compressors and critical pump services to efficiently utilize the steam generated within the furnaces and to improve reliability in case of electrical power outage. Since steam is also utilized within the radiant section to dilute the hydrocarbon feedstock to favor ethylene production and reduce coke generation, complete elimination of steam is not feasible. Additional steam is still required for process heating in recovery section. However, any steam utilized to drive rotating equipment can be replaced by electrical power from low CO_2 sources such as wind, nuclear, or hydroelectric.

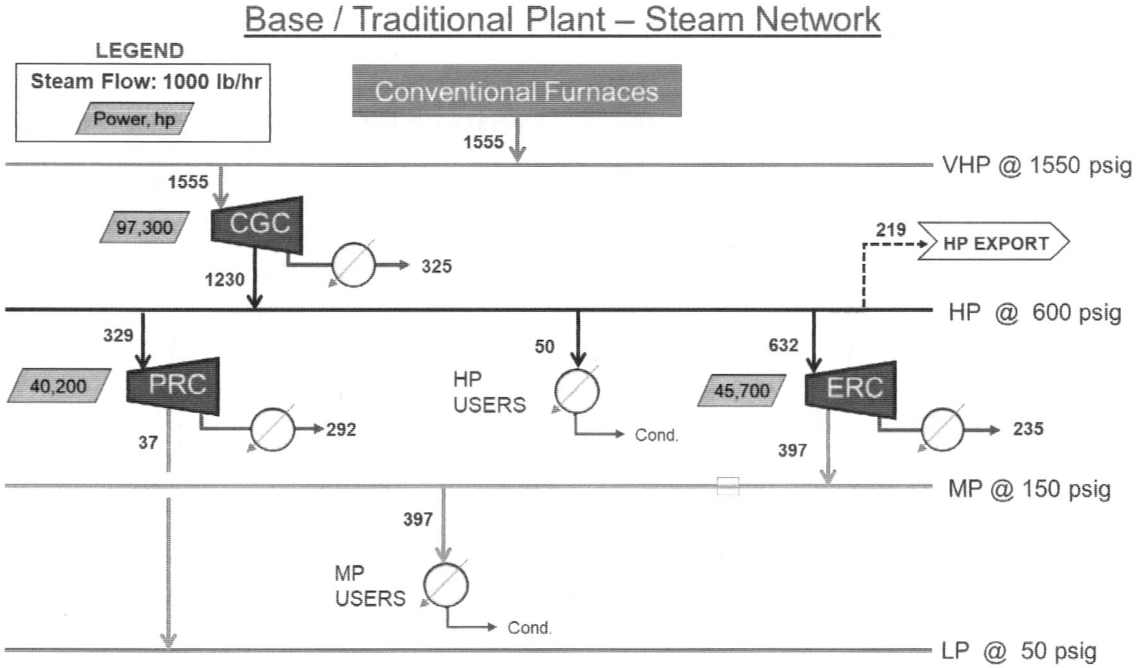

Fig. 2.2: Steam distribution network for a traditonal ethylene plant

Fig. 2.2 shows the steam distribution for the base plant configuration. Steam produced in the conventional furnaces is sufficient to satisfy process demand for normal operation without need for auxiliary boiler firing. Most of the steam is being used to drive the major compressor turbines. Additionally, MP steam and LP steam are utilized for miscellaneous process heating in the plant. Due to excess VHP steam production in traditional furnaces above process demand, HP steam is exported to OSBL for use in the petrochemical complex. While steam is utilized as efficiently as possible within a traditional cracker steam network, a significant portion of the produced steam is condensed against cooling water.

3 Electrification Option 1

3.1 Configuration

In this option, the conventional cracking furnaces will be replaced by a reduced emission furnace design with lower fuel gas firing and VHP steam production. Due to the reduction in the amount of VHP steam, not all compressors can be driven by the steam turbine driver, and some will be moved to motor drivers. Furthermore, the steam that is still produced in the reduced emission furnace still needs to be utilized efficiently, so some compressor drivers may need to be left on steam.

For Option 1, the Cracked Gas Compressor is selected for electrification, since it is the largest driver. Hence, the plant would maximize the emission reduction benefit from using low carbon electricity to drive the compressor. However, the power requirements for the CGC pushes the maximum reference for electrical drives. To stay within a reasonable motor power (< 50 MW (67,051 hp)) for ethylene applications, the CGC has been split into two trains for this study.

3.1.1 Reduced Emission Furnace Overview

Previously, Technip Energies has presented on a Low-emission Cracking Furnace which can produce same cracking with a lower overall fired duty and steam production (AIChE Paper Number: 617345)[2]. While the Low Emission Furnace has potential savings of 30 percent reduction in fuel gas and 60 percent reduction in VHP steam production, some development and process optimization are still required to achieve this target. For the purposes of this study, we selected demonstrated technology that is available today with a slightly lower value of 25 percent reduction in fuel gas consumption and a 40 percent reduction in VHP steam production. To distinguish between the Low Emission Furnace design, we will refer to the furnace design considered in this study as the "reduced emission furnace" design.

In the reduced emission design, the furnace has been redesigned to achieve lower fuel gas firing and less steam generation, while still maintaining the heat required for cracking. In order to achieve this goal, the BFW economizer, which is used to preheat BFW, is removed from the convection section to minimize steam production. The convection section duty is utilized to pre-heat combustion air against hot flue gas. In the reduced emission furnace, the heat from furnace effluent quenching is used to raise some saturated steam in the primary quench exchanger (albeit less than in the conventional furnace), while the secondary quench exchanger is utilized for final feedstock superheating, instead of steam generation as in the conventional design.

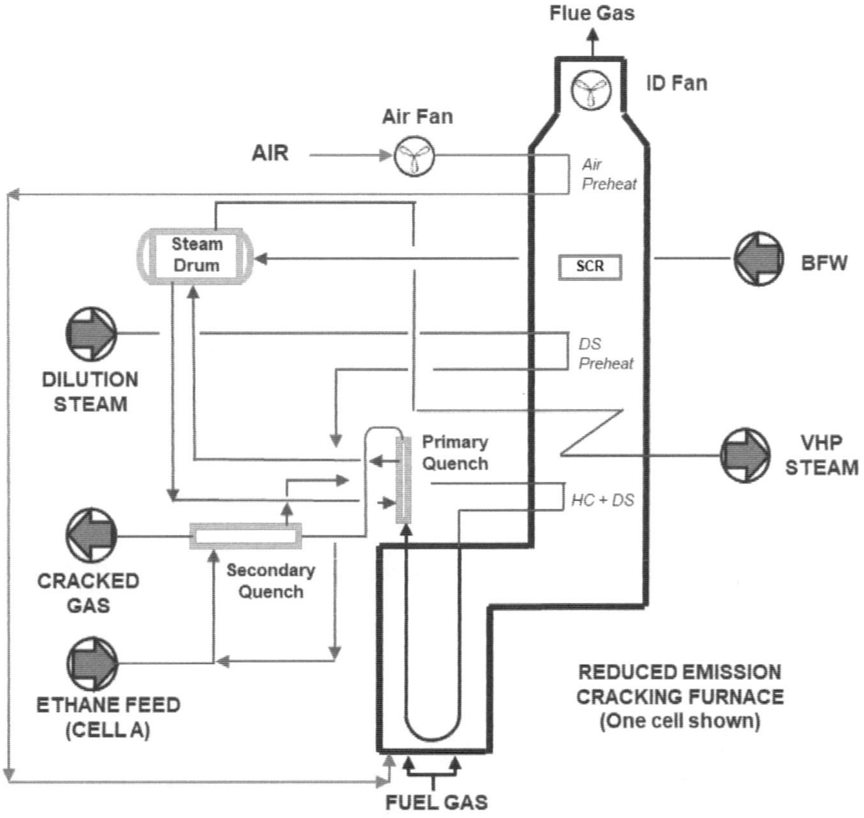

Fig. 3.1: Schematic of reduced emission cracking furnace (Option 1 & 2)

3.1.2 CGC Motor Arrangement (Split)

The CGC train is divided into two nearly identical trains. Note that it may be possible to reduce the cost of this option by adjusting the CGC process balance to achieve identical drivers, however, this is a subject for future study. Table 3.1 summarizes the two motor drivers for the Cracked Gas Compressor trains. An advantage of splitting the CGC drivers is the HP train can be run more efficiently at a higher speed when compared to a single driver.

Table 3.1: Specifications of Motor Drivers for CGC trains

Motor Driver	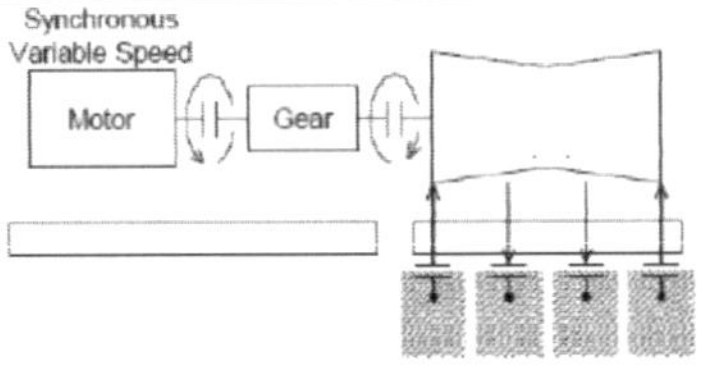	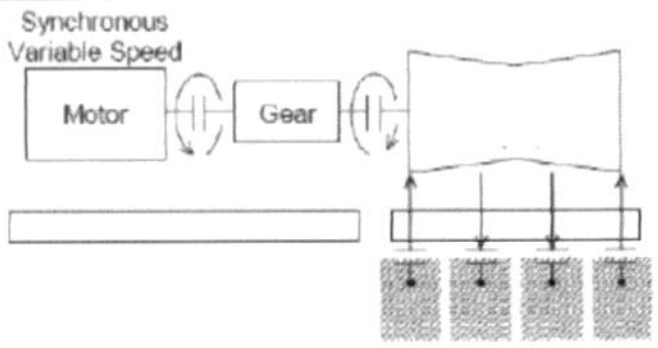
Train Name	**CGC - LP**	**CGC - HP**
Rated Power [MW (hp)]	41.2 (55,250)	44.3 (59,400)
Motor Type	2 Pole	2 Pole
VFD Type	LCI or VSI	LCI or VSI
Typical Size (L x W x H)	70 ft x 20 ft x 20 ft	60 ft x 18 ft x 18 ft

3.2 Impact to Plant Utility Balances

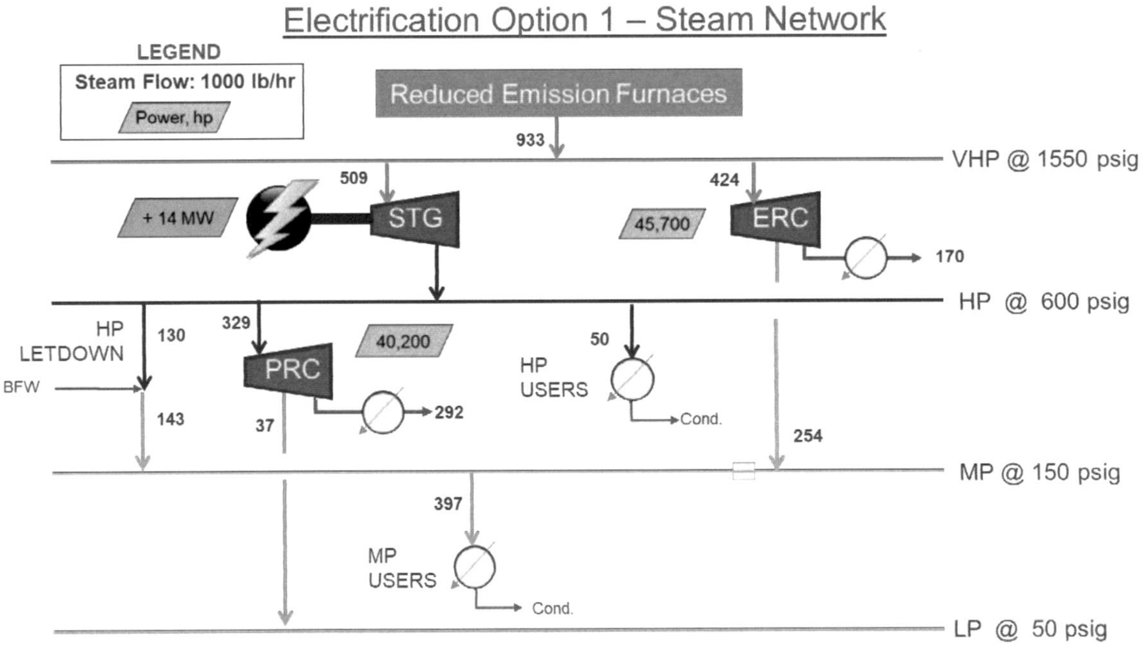

Fig. 3.2: Steam distribution of Option 1 ethylene plant

Fig. 3.2 shows the steam distribution network for Electrification Option 1 with the CGC moved to an electric driver. The traditional cracking furnaces have been replaced with a reduced emission furnace design with a 40 percent reduction in VHP steam production. The ERC steam turbine is moved to VHP level to utilize a portion of the VHP steam generated by reduced emission furnaces. Additionally, a back-pressure STG train is installed to utilize the remaining VHP steam to generate electricity, which will offset the power import requirement from OSBL. Without consideration of the STG, steam cannot be utilized efficiently in network, as removal of the CGC turbine reduces the flexibility of the steam network. Even with the added STG train in Option 1, a significant amount of HP steam is required to be letdown and desuperheated to satisfy MP process demand. There is a possibility to utilize the HP Letdown steam by adding another STG train, which could generate an estimated four MW of additional power. However, the design would not be entirely practical given the limited amount of steam energy available. Note that MP and LP steam demand for process heaters remains unchanged. As a result of lower steam availability in the Low CO_2 plant design, HP steam is no longer exported to OSBL. For a true comparison to base option, additional firing in HP boiler in OSBL needs to be considered.

In addition to the steam network, electrification of the ethylene plant also has implications for other utility systems. These impacts are discussed below, and a comparison table is presented in Section 6 which compares all three options.

- **Fuel Gas System:** In the base option, furnace fuel gas is supplied internally by cracker residue gas without need for any external fuel gas supply. Some fuel gas will be exported to OSBL based on the fuel gas balance in the base case. In the electrification option, the reduced emission furnace design can readily achieve a 25 percent reduction in furnace firing duty. Some additional OSBL firing is considered to raise the HP steam export (no longer can be exported), which lowers the net reduction in firing. With less fuel gas demand for furnace firing, hydrogen-rich residue gas can be exported from the ethylene plant. For Electrification Option 1, the increment in cracker fuel gas export is approximately 17,900 lb/hr as compared to the base plant. Note this export value considers the HP boiler firing to supply HP Steam export to OSBL.

- **Cooling Water System:** As a result of the plant electrification, the CGC surface condenser has been eliminated and the surface condenser duty for the ERC and PRC have been reduced. The CW demand of the plant decreases by ~14 percent as a result of the lower total surface condenser duty.

- **Electrical System:** As expected, the electrical system is the most dramatically impacted. Some power will be produced by the STG train utilizing the remaining VHP steam. However, an increase in power import of nearly 72 MW (i.e. 41 + 45 – 14) will be required to drive the CGC train. With the exception of the added compressor driver, ISBL power consumption for typical process users will remain approximately the same as the base case.

3.3 CAPEX & OPEX

Next, a high-level cost analysis for Electrification Option 1 is compared to traditional ethylene plant, both in terms of Capital Expenditure (CAPEX) and Operating Expenditure (OPEX). Cost considerations are summarized below in turn. A cost comparison for all options is presented later in Section 6.

CAPEX considerations:

- **Cracking Furnaces:** The conclusion from an internal cost estimate is that the cost of the reduced emission furnace is similar to that of the conventional furnace. Hence, there is no significant impact to the furnace cost as a result of the reduced emission design.
- **Rotating Equipment & Auxiliary:** The major CAPEX consideration for electrification is associated with the change-out of the steam turbine system with electric motor system. Cost analysis conducted herein includes the major electrical auxiliaries associated with the electrical motor, including the VFD system, harmonic filter, transformers, switchgears, and additional E-house space for the VFD, transformers, etc. Note the cost analysis estimate is based on using LCI type VFD drive.
  For Electrification Option 1, the total equipment cost associated with the CGC on electrical motor is more than double of the conventional plant with CGC on steam turbine. This significant increase in cost is associated with splitting the CGC into two trains. Additionally, an STG train is added for this option, which is associated with a higher cost. Furthermore, the cost of the ERC steam turbine is 10 percent more expensive than the base option as a result of it being moved to VHP steam.
- **Electrical Cabling for Motor & VFD:** The cost of electrical cabling for the motor and VFD system has been included. The estimate basis assumes the compressor/motor house is approximately 1500 feet away from the ethylene plant substation, which is assumed to house the motor VFD drives and auxiliaries. Some additional allowance has been assumed for interconnection cables between transformers and VFD, etc. Note that the electrical cabling from the complex main substation located in OSBL to the ethylene plant substation has not been considered, which could be quite significant depending on the distance and power demand of ethylene plant.
- **Steam / Cooling Water Piping:** Cost savings from the reduction in CW and steam piping in the electrification case has been included. The header sizes have been reduced down by approximately one NPS as a result of the lower steam production and CW circulation. The steam piping and CW piping associated with the CGC turbine and surface condenser have been eliminated.

The change in OPEX of the reduced emission plant can be calculated by the cost increase of the electricity import, less the revenue generated by exporting excess fuel gas.

Assuming a price of \$2.5 / MM Btu for the fuel gas export, an estimated \$12.2 MM USD / yr can be generated by selling excess cracker fuel gas (based on 8,400 operating hours per year).

Alternatively, the excess fuel gas could be utilized to generate electricity or steam reformed to produce hydrogen, which is outside the scope of this study.

Assuming a price of 5¢ / kWh for import electricity, the increase in OPEX due to higher electricity import in Option 1 is $30.2 MM USD / yr. Note that the assumed electrical price is based on the industrial rate for conventional electrical power as representative of US Gulf Coast, in absence of sufficient data for a power price from low CO_2 sources. Hence, this analysis assumes a competitive rate for low CO_2 electrical power, which will depend on future technology development or government subsidy.

Hence, the overall OPEX increase as compared to the base case cracker is $18 MM USD / yr.

It is noted that even with an assumed competitive price for renewable energy, the overall OPEX in the electrification case is expected to increase significantly when compared to the conventional ethylene plant. It is important to note that OPEX of the electrification option is highly dependent on the price of the electricity, which can vary from region to region.

3.4 CO_2 Emissions Reduction

Replacement of the traditional cracking furnaces with the reduced emission design offers a direct emission reduction of 25 percent as a result of the lower furnace firing. Consideration of the electrical power and the steam export in the base option will offset the net CO_2 emission reduction to 15 percent. The overall reduction will be dependent on CO_2 footprint of the electrical import, but for this study we will assume an electrical import from a low emission source such as wind or hydroelectric (based on 26 ton CO_2 /GWh[3] [11.8 lb CO_2 / kWh] including life-cycle emissions).

3.5 Maintenance / Operability / Stability

Table 3.2: Maintenance comparison between steam turbine and electric motor

Items	Steam Turbine Driver		Motor Driver	
Major Inspection (Casing Disassembly)	Good	Period: 8 years Workday: 12 days	Better	Period: >8 years Workday: 10 days
Operability	Good	Multiple Operations required to start-up	Better	Automatic Startup with One Push Start

Motor drivers are easier to maintain as compared to steam turbine drivers. However, since the equipment count in Option 1 is higher as compared to traditional plant, the maintenance work will increase in Option 1.

4 Electrification Option 2

4.1 Configuration

As mentioned above, Option 1 has several shortcomings, which contribute to higher plant CAPEX and OPEX. First, the CG compressor was split into two trains to keep the motor size in reasonable range, which results in a significant cost increase as compared to typical steam turbine. Secondly, the steam produced in the reduced emission furnaces cannot all be efficiently used with the CGC on motor. Hence, some letdown of HP steam is required, which reduces the plant efficiency and making it more expensive to operate.

Hence, we explored a second configuration for reduced emission plant to see if shortcomings of Option 1 could be improved. In Option 2, the CGC train will be driven by steam turbine and the other two trains will be driven by motor drivers.

4.1.1 PRC and ERC Arrangement:

Table 4.1 summarizes the motor drivers for the PRC and ERC trains. The power requirements for the PRC and ERC are within a comfortable range for a single motor.

Table 4.1: Specifications of Motor Drivers for the PRC and ERC trains

Motor Driver	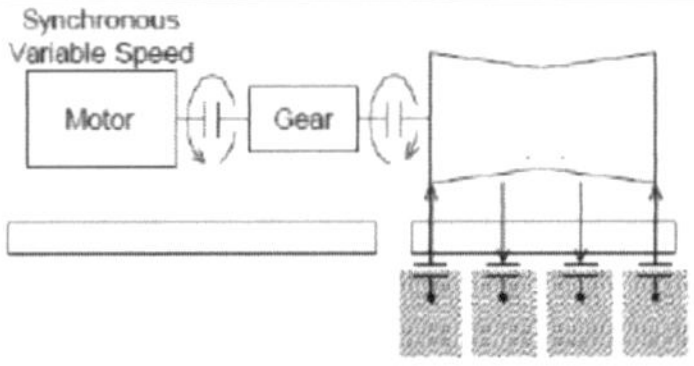	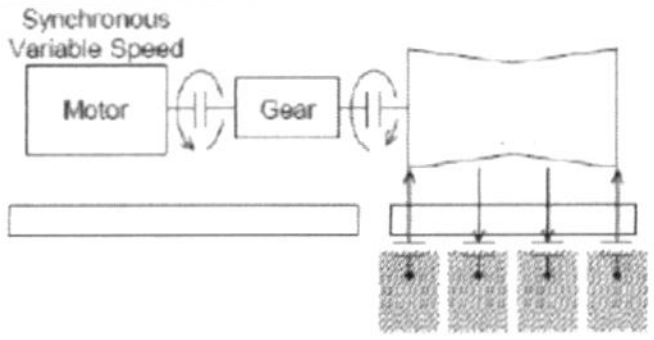
Train Name	PRC	ERC
Rated Power [MW (hp)]	34.1 (45,730)	38.5 (51,630)
Motor Type	2 Pole	2 Pole
VFD Type	LCI or VSI	LCI or VSI
Typical Size (L x W x H)	60 ft x 18 ft x 18 ft	60 ft x 18 ft x 18 ft

4.2 *Impact to Plant Utility Balances*

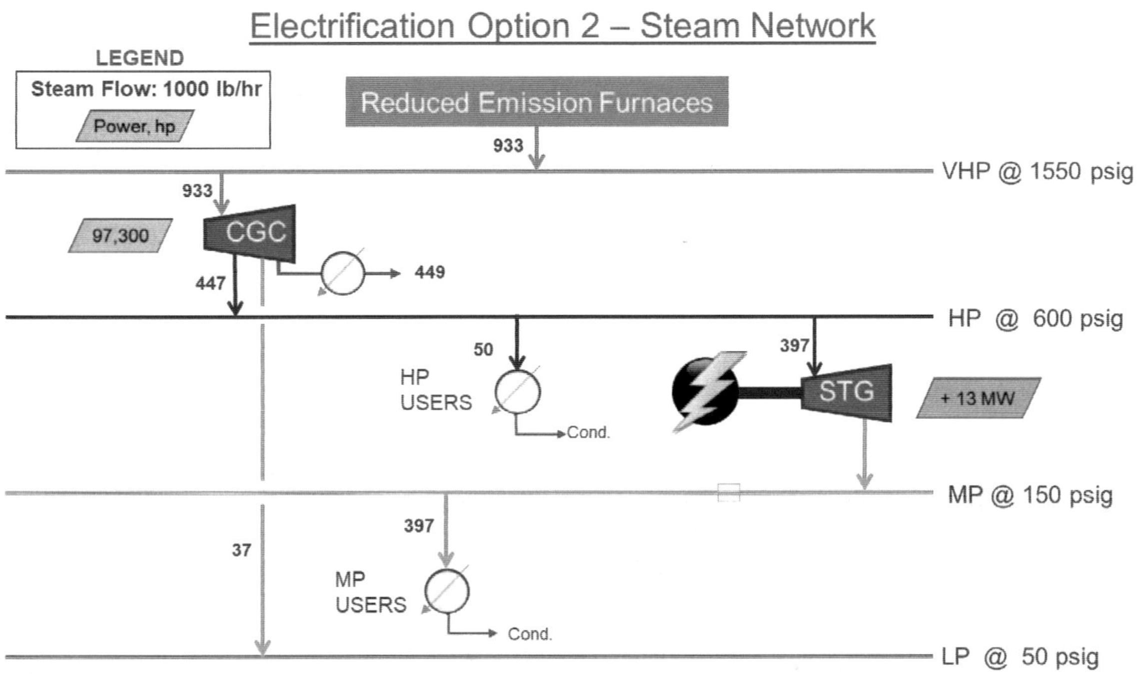

Fig. 4.2: Steam distribution of Option 2 ethylene plant

Fig. 4.2 shows the steam distribution network for Electrification Option 2 with PRC and ERC trains moved to electric motor drives. As in Option 1, the traditional cracking furnaces have been replaced with reduced emission furnaces with 40 percent reduction in VHP steam. In this configuration, CGC is driven by an extracting-condensing turbine with dual extractions from HP and LP levels. Additionally, a back-pressure STG train is also installed to utilize the remaining HP steam to generate electricity. This helps to improve the efficient use of steam in the network. Furthermore, no HP letdown is required in Option 2, which is an improvement from Option 1. The impact to the other utility systems is discussed below. In addition to the steam network, electrification of the ethylene plant also has implications for other utility systems. These impacts are discussed below, and a comparison table is presented in Section 6 which compares all three options.

- **Fuel Gas System:** Since the furnace fuel gas firing and OSBL firing is unimpacted by switching the compressor drivers, the fuel gas export is the same as Option 1.

- **Cooling Water System:** As a result of the plant electrification, both the PRC and ERC surface condensers have been eliminated. However, the surface condenser duty for the CGC has been increased. Hence, the CW demand of the plant decreases by ~16 percent for Option 2, which is similar to reduction estimated for Option 1.

- **Electrical System:** The electrical import requirements for Option 2 decreases as compared to Option 1 due to the lower power requirements of the PRC and ERC trains. For option 2, the net power import increases by approximately 60 MW (i.e. 34.1 + 38.5 - 13) is required.

4.3 CAPEX & OPEX

Cost considerations for Option 2 are summarized below.

CAPEX considerations:

- **Cracking Furnaces:** Same as Option 1.
- **Rotating Equipment & Auxiliary:** The relative increase in equipment cost associated with the PRC and ERC trains on electrical motor is less than associated with the CGC train. However, the PRC and ERC trains on electrical drives are still around 30 to 50 percent more expensive than conventional steam turbine drives. Additionally, the cost of the STG train needs to be considered, which was added to increase efficiency of steam network.
- **Electrical Cabling for Motor & VFD:** Cost of electrical cabling for the motor and VFD system is estimated on same basis as described in Option 1. Since the combined electrical power of the PRC and ERC is less than the split CGC trains, the cost of electrical cabling has reduced slightly for Option 2.
- **Steam / Cooling Water Piping:** The reduction in CW piping and steam piping increases when compared to Option 1, since two steam turbine and surface condensers have been eliminated in Option 2.

Since the fuel gas balance does not change as compared to Option 1, the OPEX reduction associated with higher fuel gas export remains at $12.2 MM USD / yr. The increase in OPEX due to electricity import in Option 2 is $23.7 MM USD / yr.

Hence, the overall OPEX increase as compared to the base case cracker is $11.5 MM USD / yr. Again, we find that the OPEX has increased due to higher electricity import.

However, Option 2 is cheaper than Option 1 in terms of OPEX and CAPEX.

4.4 CO_2 Emissions Reduction

The direct CO_2 emissions from Option 2 are similar to Option 1, since fuel gas firing remains the same. Although the power import in Option 2 decreases as compared to Option 1, the impact to overall carbon footprint is low, as the study assumes low CO_2 electrical power. Hence, the net CO_2 emission reduction considering electrical power is similar to Option 1.

4.5 Maintenance / Operability / Stability

Table 4.2: Maintenance comparison between steam turbine and electric motor

Items	Steam Turbine Driver		Motor Driver	
Major Inspection (Casing Disassembly)	Good	Period: 8 years Workday: 12 days	Better	Period: >8 years Workday: 10 days
Operability	Good	Multiple Operations required to start-up	Better	Automatic Startup with One Push Start

Motor drivers are easier to maintain as compared to steam turbine drivers. However, since the number of equipment in Option 2 is higher as compared to a traditional plant, the maintenance work will increase in Option 2. However, the number of equipment in Option 2 is lower than in Option 1, so this is the better option in terms of maintenance requirements.

5 Electrification Challenges/Limitations:

There are some points which need to be addressed in case of motor drivers:

- Availability of low CO_2 electrical power. Use of conventional electrical sources may serve to increase the overall CO_2 footprint of the reduced emission plant when considering the indirect emission of the electrical power.

- If a plant uses renewable sources of power from external sources and the external source is not fully controlled by the plant management, it may decrease the overall availability and reliability level of the entire plant. Blackouts, brownouts, and other impacts to external power grid are all challenges that will have a greater impact to the ethylene plant with electrical motor drivers.

- Customer should have a contract with grid to ensure uninterrupted power supply, because in case load shedding occurs, depending upon the amount of load shedding, the customer may have to shut down the plant. This is a big disadvantage for an electric system which needs to be addressed before switching to motor drivers.

These scenarios are some of the cases which need to be analyzed while designing a motor driver plant and appropriate risk action plans must be prepared for them.

6 Summary

The following table summarize the plant energy, utility, and emission impact for all three options:

Parameter	**Units**	**Base Option / Traditional plant**	**Electrification Option 1**	**Electrification Option 2**
Furnace Firing	MM Btu/hr (% of base)	3539 (100%)	2654 (75%)	2654 (75%)
HP Boiler Firing in OSBL (Note 1)	MM Btu/hr	-	302	302
Cracker Fuel Gas Export	M lb/hr	5.3	17.9	17.9
VHP Steam Production	M lb/hr (% of base)	1555 (100%)	933 (60%)	933 (60%)
HP Steam Import (Note 2)	M lb/hr	-219	-219	-219
Cooling Water Circulation	M GPM	164	141	136
Electrical Power Import (Note 3)	MW	8	80	68
CO_2 Emissions from Furnace	% of Base Plant	100 %	75 %	75 %
Specific Carbon Emissions (Note 4)	% of Base Plant	100 %	85%	85%

Notes:

1. In order to make a fair comparison, HP boiler firing is considered for the electrification case to satisfy the HP steam export in the traditional plant. It is captured in the specific carbon emission reduction.
2. Steam export is shown as negative flow.
3. Electrical power import = ISBL user + motor drivers – Power generated from STG. ISBL user is assumed to be the same for all options.
4. Considers both direct and indirect emission sources including furnace emissions, electrical power import, steam import/export, etc. CO_2 footprint for electrical import considers a low emission source.

The following comparison table presents a CAPEX summary in relative terms of the base case:

Equipment Cost	Base Option / Traditional plant	Electrification Option 1	Electrification Option 2
Cracking Furnace	1.00	1.00	1.00
CGC Train	1.00	2.10	1.00
PRC Train	1.00	1.00	1.30
ERC Train	1.00	1.10	1.50
STG Train	-	1.00	1.00
Steam/CW Piping	Base case	-1.0 delta	-1.5 delta
Electrical Cabling	Base case	1.0 delta	0.88 delta

Based on the above analysis, the following results can be concluded:

Parameter	Base Option / Traditional plant	Electrification Option 1	Electrification Option 2
Number of Equipment (Drivers)	● (3 turbines)	× (3 turbines + 2 motor with VSDs + 1 generator set)	▲ (2 turbines + 2 motor with VSDs + 1 generator set)
CAPEX	●	× (10% higher than Base Plant)	▲ (5% higher than Base Plant)
OPEX	● (Base case)	× ($18 MM /year increase)	▲ ($11.5 MM /year increase)
CO_2 Emissions	× (Base case)	● (15% reduction)	● (15% reduction)

Parameter	Base Option / Traditional plant	Electrification Option 1	Electrification Option 2
Maintenance	●	▲	▲
Operability	●	▲	●
Stability of operation in case of load shedding	●	▲	▲

● Good (Less amount is required),

▲ Moderate (Moderate amount is required),

× Bad (High amount is required)

It can be seen that different types of plant arrangements are advantageous in different conditions. If our main target is to reduce the CO_2 emissions, the following conclusions can be made:

- As ethylene producers look for opportunities to reduce the carbon footprint of their new and existing assets in the near-term, plant electrification will be a major consideration.

- While the processing scheme of the ethylene plant remains similar to that of the conventional plant, the reduced emission cracker design will change the energy distribution of the ethylene plant, resulting in a significant impact to the major utility systems.

- Power generation within ethylene plant via steam turbine generators may have to be considered to make efficient use of the steam that is produced in the reduced emission design.

- On an overall comparison, Option 2 comes out to be better plant as compared to Option 1. However, as industrial applications and experiences with large electrical motors sufficient to drive the CGC machine (> 80 MW) continues to develop, the cost associated with Option 1 could decrease considerably.

- The optimal plant configuration will result in balancing several key aspects to ensure a cost-effective and energy efficient plant design:
    - Quantity and reliability of low carbon electrical power
    - Steam availability from furnaces
    - Overall utility demand across the petrochemical complex

- Above strategies for reducing furnace firing and driving major compressors with electric motors are proven technologies and available today.

ABBREVATIONS

CAPEX	Capital expenditure
CGC	Cracked Gas Compressor
CO_2	Carbon dioxide
CW	Cooling Water
ERC	Ethylene Refrigeration Compressor
HP	High Pressure
hp	Horsepower
ISBL	Inside Battery Limits
kTA	Thousand Metric Ton per Annum
LCI	Load Commutated Inverter
M	Thousand
MM	Million
MSTA	Million Short Ton per Annum
MW	megawatt
NPS	Nominal Pipe Size
OPEX	Operating expenditure
OSBL	Outside Battery Limits
PRC	Propylene Refrigeration Compressor
STG	Steam Turbine Generator
TLE	Transfer Line Exchanger
VHP	Very High Pressure
VSI	Voltage Source Inverter

REFERENCES

1. IEA. *The Future of Petrochemicals. (2018).* https://www.iea.org/reports/the-future-of-petrochemicals . Accessed 30 November 2021.
2. Oud, P. van Boekel, T. Low Emission Cracking Furnace. *Conference Proceedings of The 33rd Ethylene Producers Conference.* April 2021. Dallas, TX. https://plan.core-apps.com/aiche_spring21/event/27ce62c4ce7e90ead9385e489be0c911. Accessed 10 January 2022.
3. *Comparison of Lifecycle Greenhouse Gas Emissions of Various Electricity Generation Sources*. (2011). World Nuclear Association Report .
4. TMEIC. *Variable Frequency Drives – a Comparison of VSI versus LCI Systems. (2013).* https://www.tmeic.com/Repository/Media/Comparison%20of%20VSI%20versus%20LCI%20Systems%20FINAL.pdf. Accessed 10 February 2022.

AIChE Paper Number: 78f

Propane Dehydrogenation and Steam Cracking Heuristic Approaches to Energy Efficient Plant Integration

Torben Hoefel
Product Management Sustainable Olefin Technologies
Linde GmbH, Engineering Division

Richard Koller
Process Design Engineer
Linde GmbH, Engineering Division

Prepared for Presentation at the
2022 Spring Meeting and 18th Global Congress on Process Safety
San Antonio, Texas, April 12, 2022

Propane Dehydrogenation and Steam Cracking
Heuristic Approaches to Energy Efficient Plant Integration

Torben Hoefel

Product Management Sustainable Olefin Technologies

Linde GmbH, Engineering Division

Richard Koller

Process Design Engineer

Linde GmbH, Engineering Division

Abstract

The integration of propane dehydrogenation (PDH) and steam cracking by the combination of the olefin product recovery has been already studied in the early 1980s. While thermal cracking of ethane as well as heavier hydrocarbon feedstocks is still by far the main source of ethylene, catalytic dehydrogenation of propane to propylene has gained increasing market share in the recent past due to the highly attractive propylene selectivity. As both processes yield products of a similar nature, integration of both recoveries may appear simple and advantageous from total equipment point of view.

A more detailed analysis of bulk composition, trace components and operation requirements of each technology shows that combining of product recovery may result in a significantly more complex separation task compared to the individual separations. Rather than raising synergies in separation, both effluents bear the risk of "contaminating" each other. Moreover, process integration always adds complexity and possible undesired interdependencies.

This paper presents a more detailed study of integration potentials for PDH and steam cracking. The most important differences between cracked gases and PDH effluents are described. Related characteristics of different PDH technologies and their impact on integration scenarios are analyzed on a high level.

Some promising scenarios are derived, which focus on synergetic potentials of PDH and steam cracker integration, while possible risks from detrimental effects are avoided or minimized.

Finally, the paper outlines some properties from an existing plant, where a highly customized small-scale cracker and ethylene recovery process is integrated into a large-scale PDH. The design includes a substantially simplified cold section, which applies pressure swing adsorption technology, omitting the requirement of very low temperatures and matching or even exceeding olefin recoveries of conventional low temperature processes.

1. Motivation

Ethylene and propylene are two widely applied commodity chemicals [1,2]. While thermal cracking is still by far the main source of ethylene, catalytic dehydrogenation of propane to propylene has gained increasing market share in the recent past.

Propylene used to be mainly obtained as a by-product from steam cracking and refinery catalytic cracking processes until the early 2000's [2]. However, to provide ethylene and ethylene-based products at a low price, gaseous feeds, especially ethane, are increasingly used in the steam cracking processes. Especially in the US, shale gas significantly contributes to the low costs of gaseous cracker feedstock [3]. The shift from the traditional naphtha feedstock to lighter fractions leads to a reduction of the co-production of propylene [4], further favoring the introduction of on-purpose propylene production processes like propane dehydrogenation (PDH).

Furthermore, the intensifying attention towards sustainable production processes will likely favor processes like ethane cracking and propane dehydrogenation due to high selectivity towards high value chemicals (HVC) and low carbon losses towards fuel-grade products. To visualize this advantage, Figure 1 shows where two processes for the production of ethylene and propylene are compared, adopted from [5,6].

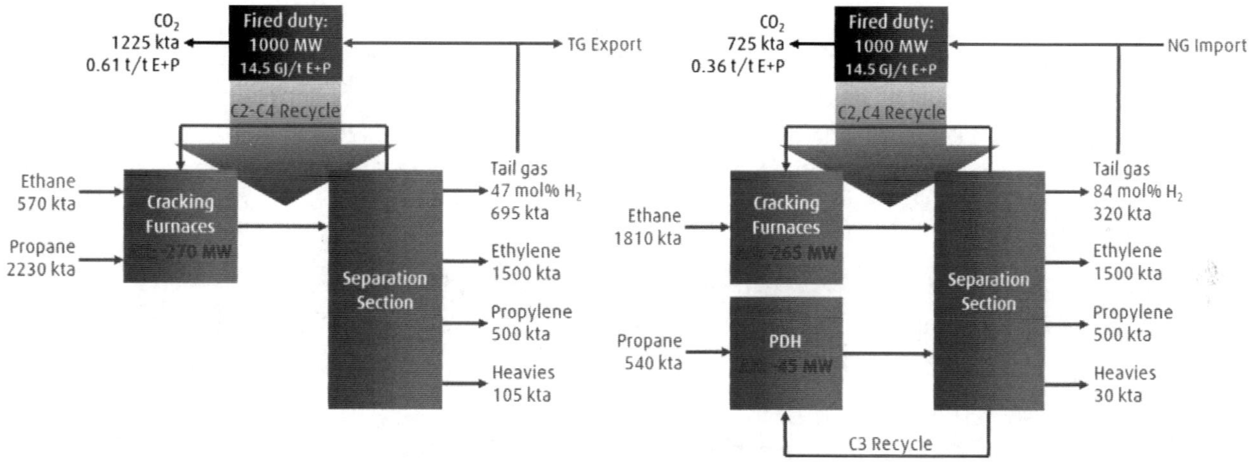

Figure 1: *Two different process alternatives for the generation of 1500 kta ethylene and 500 kta propylene. The first process utilizes propane cracking with some ethane cracking for generation of propylene and ethylene. It is assumed that all C2-C4 components are recycled to minimize heavy byproducts and save feedstock. The methane-rich tailgas is utilized to provide the required thermal energy, which is assumed to be 1000 MW. Some tailgas is exported. The second process utilizes ethane cracking and propane dehydrogenation for the generation of the same amounts of ethylene and propylene. The hydrogen-rich tailgas and some imported natural gas are utilized to provide the required thermal energy, which is again assumed to be 1000 MW.*

Similarly to [5,6], the example indicate that significantly less feedstock is required for the combination of ethane cracking and propane dehydrogenation (approximately 16 wt%) and the feedstock is significantly shifted to ethane. At the same time, ethane cracking and propane dehydrogenation generates more hydrogen, less methane and less heavy byproducts.

Although the required heat of reaction is somewhat higher for the combination of ethane cracking and propane dehydrogenation (approximately 310 MW instead of 270 MW), it is furthermore claimed that both processes require a similar amount of thermal energy / fired heat, which is estimated to be approximately 1000 MW or 14.5 GJ per ton of ethylene plus propylene (electrical energy and energy import/export not considered). Assuming this comparable energy efficiency of both processes, the combination of ethane cracking and propane dehydrogenation will yield however significantly lower CO_2 emissions, which results from the significantly decreased carbon intensity of the applied fuel.

This reduced carbon intensity also results from the above outlined selectivity advantage and significantly reduced methane concentration in the tail gas, which is typically utilized to balance the thermal demand of the process. The example indicated in figure 1 indicates a difference of 500 kta CO_2 emissions, which is a 40% reduction of CO_2 emissions of the combination of ethane cracking and propane dehydrogenation compared to propane/ethane cracking.

As outlined above, the CO_2 emission factor [t CO_2 / t HVC] is driven by the two aspects energy efficiency [GJ/ t HVC] and fuel carbon intensity [t CO_2 / GJ]. The low fuel carbon intensity is one significant advantage of highly selective / carbon efficient processes like ethane cracking and propane dehydrogenation. Depending on the detailed process design and other constraints, the energy efficiency of an olefin generation process can greatly vary, e.g., between 11-20 GJ/t HVC. However, unpublished studies have shown that highly efficient processes like ethane cracking and propane dehydrogenation may also yield similar or even improved energy efficiency compared to propane/ethane cracking, despite the somewhat higher heat of reaction required. Some of the underlying reasons are:

- Lower separation effort due to smaller amounts of byproducts
- Lower reaction temperatures for propylene generation, requiring reduced effort for heating/cooling, which yields opportunities for reduced energy shift to other heat sinks including atmosphere ("exergy loss")
- "Feed/effluent" exchanger are rather common in PDH reactor units, also facilitated by the lower reaction temperature requirement. Such exchanger reduces the need to shift heat to "waste heat boiler", again reducing the degree of exergy loss and enabling machine electrification.
- Some PDH units do not require water as diluent. High water dilution is a main contributor to energy shift/loss to steam and/or atmosphere, due to phase change close to ambient conditions.
- Some PDH technologies use intermittent fired heater as the main heat source. These might be simpler to electrify compared to cracking furnaces.

The two processes were further compared in an economic benchmarking, with the following arbitrary main assumptions:

- Ethane: 250 USD/t
- Propane: 360 USD/t
- Ethylene: 950 USD/t
- Propylene: 900 USD/t
- Fuel Import/Export: 13 USD/MWh
- CO_2 Emission Cost: 0-160 USD per ton of direct CO_2 emission

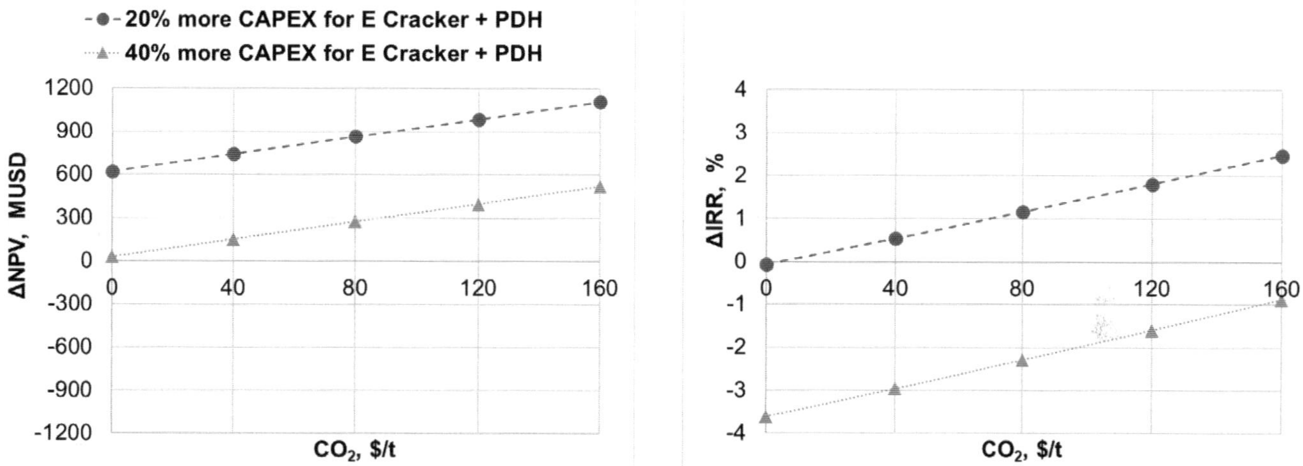

Figure 2: *Economic comparison of propane/ethane cracking with the combination of ethane cracking and propane dehydrogenation. The diagrams show the improvement in net present value (NPV, left diagram) and internal rate or return (IRR, right diagram) by the combination of ethane cracking and propane dehydrogenation for 20% more CAPEX (blue circles) and 40% more CAPEX (orange triangles), compared to propane/ethane cracking. The differences are shown as a function of the direct CO_2 emission cost.*

Figure 2 visualizes the results of the economic comparison in terms of net present value (NPV) and internal rate of return (IRR) difference between the combination of ethane cracking and propane dehydrogenation compared to propane/ethane cracking. It was arbitrarily assumed that the combination of ethane cracking and propane dehydrogenation requires either 20% more or 40% more CAPEX. The results indicate that in terms of NPV, the combination of ethane cracking and propane dehydrogenation likely results in superior characteristics for all investigated CO_2 prices, even in the case of 40% more CAPEX. Of course, the NPV difference is lowest with 0 $/t CO_2 emission cost. With increasing CO_2 cost however, the combination of the two highly selective process not only benefit from the carbon efficiency saving on feedstock but further improve their advantage due to lower cost associated with CO_2 emissions.

However, considering the CAPEX sensitive IRR, the results show that significantly increased CAPEX (e.g. 40%) possibly caused by the requirement of installing two plants will likely result in a lower IRR of ethane cracking and propane dehydrogenation compared to propane/ethane cracking, even in case the CO_2 prices are on the high side. Thus, if the economic target is the optimization of the IRR, one could possibly come to the conclusion to choose propane/ethane cracking, despite the superior carbon efficiency and CO_2 emission factor of ethane cracking and propane dehydrogenation.

It is thus concluded that a smart and CAPEX saving plant integration of ethane cracking and propane dehydrogenation must be targeted to reach optimized results for process performance and all economic key figures simultaneously.

2. Design Aspects

Due to the similarity of feedstocks and products, the separation process integration potential of propane dehydrogenation and steam cracking has already been recognized in the 1980's [7]. For a very early commercial PDH unit, some aspects of PDH/steam-cracker integration have been reported [8,9].

However, since then, according to the knowledge of the authors there hasn't been a single commercial project until recently, where these processes were integrated into a single plant design. This might have several reasons. In the early stage of propane dehydrogenation commercialization before 2000, where the business potential of PDH was not yet well recognized, a dedicated plant design for the PDH was naturally the preferred choice in order to generate the first commercial references. However, also between 2000 and 2020, where PDH evolved to a fully commercial and well referenced technology, companies apparently hesitated to go for integrated designs.

This is presumably due to the reason that a detailed analysis also yields a lot of differences. First, the bulk component concentrations of the product gas may greatly vary. Some typical compositions of cracking and PDH product gases are shown in Table 1.

Table 1: *Comparison of typical average concentrations of dry net product gas from different cracking processes and PDH. Once through yields are displayed, all concentrations are in wt-%. Dilution media like steam or hydrogen are not included.*

Component	Ethane Cracking	Naphtha Cracking	PDH
$H_2 + CO$	4.0	1.0	1.5
CH_4	3.7	15.4	1.4
C_2H_2	0.5	0.7	0
C_2H_4	52.5	31.0	0.3
C_2H_6	34.8	3.4	2.5
$C_3H_6 + C_3H_4$	1.1	16.2	34
C_3H_8	0.1	0.4	60
C4	2.2	9.6	0.3
Pyrolysis Gasoline	0.9	19.3	
Pyrolysis Fuel Oil	0.2	3.0	

The concentrations of different components can vary greatly when comparing gas (ethane) cracking, cracking of heavy feedstocks (e.g. naphtha) and dehydrogenation of propane. While the light (C1-) fraction of ethane cracking and PDH shows similar ratios of methane and hydrogen, the methane concentration for naphtha cracker is significantly higher, possibly imposing challenges on the cold section (Demethanizer) design when combining product gases from PDH and liquid feedstock cracking. Naturally, the C2 content for PDH product gases is very low, which implies that de-ethanization would be less efficient in case both product gases are mixed upstream. Of course, the C3 and in particular propane concentration is significantly higher in PDH product gases and one has to trade-off combined separation of both C3 fractions with a possibly somewhat increased energy requirement. Finally, PDH product gases yield extremely low C4+ components compared to cracking, and one may have to accept the disadvantages caused by the integrated C4+ processing.

Second, other process characteristics may even play a more important role, when it comes to integration of PDH and steam cracking, which is summarized in Table 2, adopted from [6,10].

Table 2: *Integration-relevant process aspects of different PDH technologies and comparison to steam cracking.*

Process	Moving-bed	Fixed-bed, cyclic vessel operation	Fixed-bed, reformer type	Fluid-bed	Steam Cracking
Licensor	UOP	Lummus Technology	ThyssenKrupp Industrial Solutions, (Linde)	Dow, KBR	Several, including Linde
Compressor suction pressure	atmospheric	vacuum	increased	atmospheric	atmospheric
Compressor discharge pressure	10-16 bar a				32-40 bar a
Hydrogen	from reaction and as diluent for coke control	from reaction	from reaction	from reaction	from reaction
Water	maybe saturated from quench	maybe saturated from quench	as diluent	maybe saturated from quench	as diluent
Trace Components	may contain coke, catalyst fines, CO_2, H_2S, Cl, O_2 (possibly in case of oxidative dehydrogenation), oxygenates (possibly generated during steam reforming), various contaminants from feed impurities				coke, CO_2, H_2S, and various other possible (incl. NOx, dosing chemicals, COS, Hg, ...)

Propane dehydrogenation may be considered as a rare and remarkable example on how so many fundamentally different (reactor) process designs are simultaneously brought to full-scale commercialization. Any of the commercialized PDH technologies has its specific needs for the separation section design and thus, any integration scenario with steam cracking may require a tailored approach. For example, the operation pressures of the reactors and as a consequence the product gas compressor suction pressures noticeably vary depending on the applied PDH technology. Also, the required discharge pressure of the product gas compressor is typically very much lower for the PDH compared to steam cracking. Any water content or water dilution of a PDH technology will require a dedicated design regarding any possibly required quench and drying sections. Furthermore, some PDH technologies utilize hydrogen as dilution medium, which favors the separation of such at a front-end. Most importantly, further (trace) components, which might be specific for a certain PDH technology and/or steam cracking, require particular attention, as thoughtless integration schemes would result in the "contamination" of product streams and/or recycle fractions to the individual reaction sections. Depending on the applied reactor technology, PDH product gases may or may not contain components like chlorine, catalyst fines, oxygenates or even oxygen (in case of certain oxidative dehydrogenation processes), just to name a few.

Third, further important aspects regarding integration include differences in operation, turnaround cycle and start-up requirements. One has to pay particular attention in the detailed process design in order to ensure maximum degree of freedom in operability, possibly even including the option of individual plant section operation while the other plant section is off-line. This very much depends on the respective capacities of the individual section as well as site- and client-specific boundary constraints and targets.

Summarizing, differences in bulk composition, trace components and other process characteristics require a sophisticated integration concept. The integration must be tailored depending on the applied PDH technology and cracker type (gas or liquid feedstock). Furthermore, the target ethylene / propylene capacity and the available feedstocks play a key role in the choice of the integration concept („which of the two plants is smaller and is integrated into the larger one"). The authors believe that the integration of the front-end separation (hot section, compression, trace removal) appears in general challenging. However, the tail-end separation and utilities (distillations, refrigerant loops, hydrogenations, steam, other utilities) bear promising integration potentials both at small and at large-scale.

3. Examples of Integration Scenarios

The following chapter outlines some promising integration scenarios.

The first scenario is in particular suitable for a front-end Deethanizer design but can be also applied for any cracker separation sequence [11]. For the PDH section, a PDH technology with hydrogen dilution or any other PDH technology could be applied, which is designed in terms of a front-end light fraction removal sequence. The two processes are outlined in Figure 3.

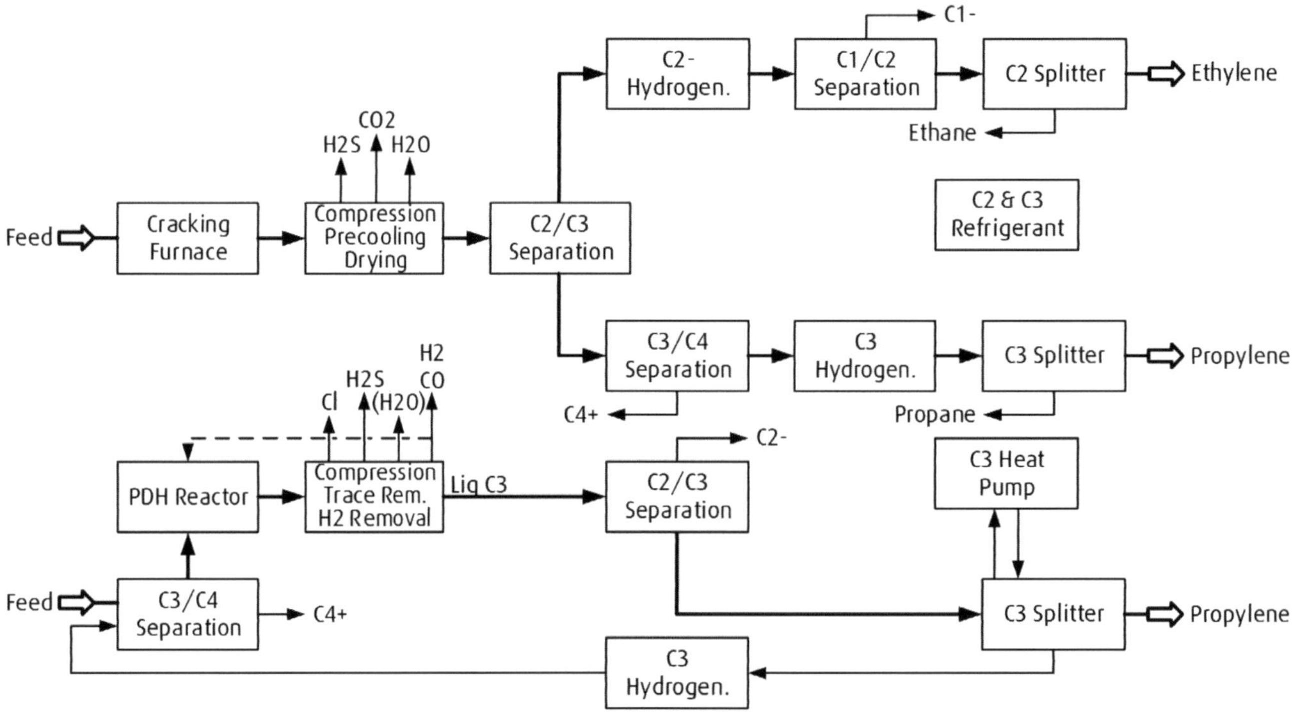

Figure 3: *Schematic process flow of a steam cracker with front-end Deethanizer (top section) and a propane dehydrogenation with front-end hydrogen and light fraction removal (bottom section).*

The authors of [11] found it of particular advantage for the integration with respect to CAPEX, OPEX and operability, when it is carried out after a first (partial) separation step for each product gas, where at least some of the C1- fraction of the cracked gas is already removed and at least some of the hydrogen from the PDH product gas is already removed. This originates from the logic that the high amount of hydrogen originating from the PDH product gas can be separated from C3 with rather simple effort and the effort would disproportionally increase, if the whole PDH product gas would be mixed with the whole cracked gas fraction.

While the "soft" hydrogen removal from the PDH product gas may be carried out in an isolated individual process unit, the C1- fraction separation of the cracked gas is conveniently carried out in a process unit, where the PDH cracked gas is already integrated. One embodiment of such

approach is sketched in

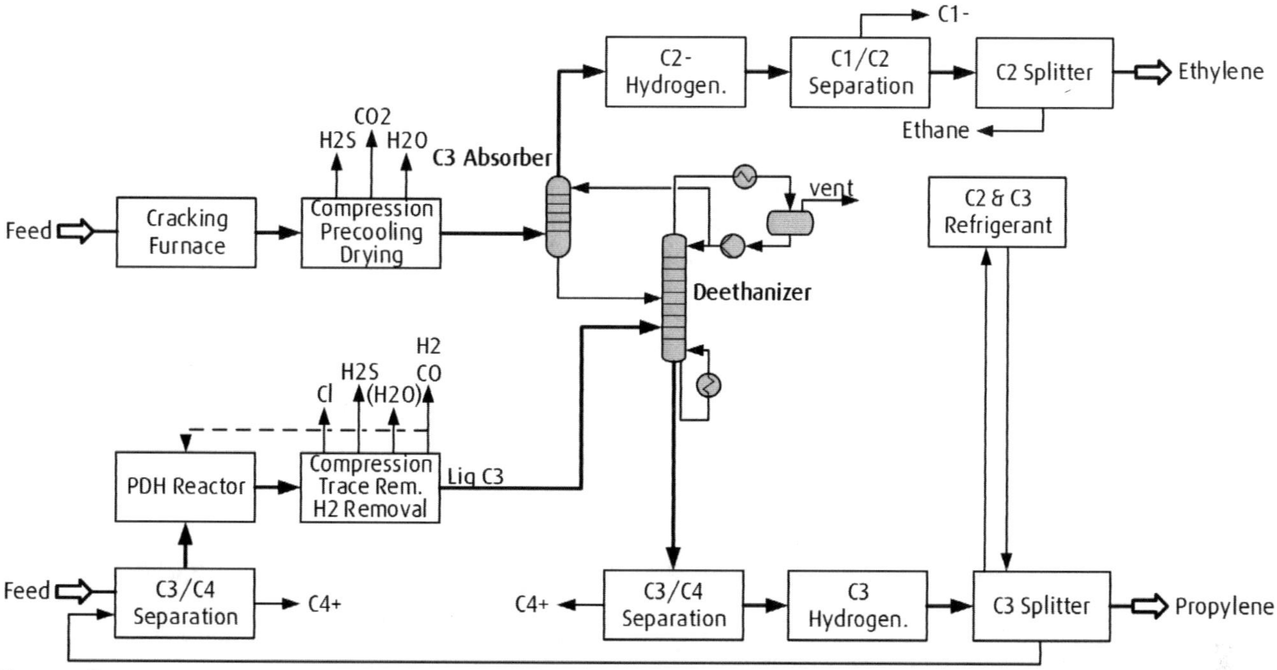

Figure **4**.

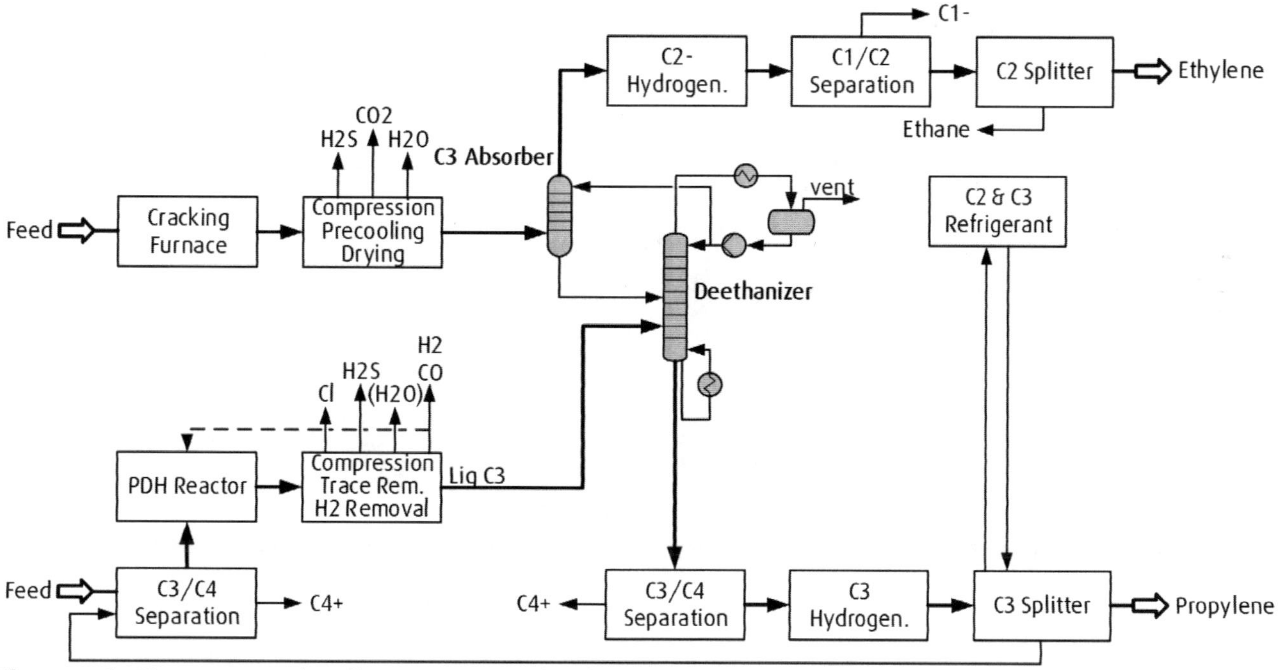

Figure 4: First *example of an integrated solution. Concurrently required units like Deethanizer, C3 Hydrogenation, C3 Splitter and refrigerant/heat pumps are integrated.*

A C3-rich and hydrogen depleted fraction originating from the PDH product gas is for example sent to a lower section of a Deethanizer column, where the majority of the C1- fraction (or C2- fraction) originating from cracked gas is already separated. By doing so, a Deethanizer design can be realized, which is very close to the design of a conventional steam cracker. Also, the required utilities of such Deethanizer section do not increase disproportionally, as the hydrocarbons originating from the cracked gas and the PDH product gas are sent to just the right sections of the Deethanizer for most efficient purification. Another large integration potential is within the C3 splitter section, where the propylene product can be generated from a single separation column. As such separation systems are typically designed as heat pumps, the heat pump stage of the C3 splitter can be conveniently integrated into a C3 refrigerant cycle. Such integrated multi-purpose compression systems (multi-level refrigerant supply, heat pump and product delivery combined in one machine) are well referenced designs within the C2 and C3 sections of certain olefin plant licensors [12].

It was found that such integration scenario is suitable for small or medium sized PDH sections combined with medium or large sized crackers. Combination of PDH and ethane cracking is particularly interesting, where the total propylene capacity is lower than 750-800 kta, allowing a single train design of the C3 splitter. The product gases of ethane cracking in particular and PDH furthermore fit together in the suggested integration example due to more comparable conditions and concentrations regarding C4 fraction, propane content and MAPD content. Special attention has to be paid however towards propylene, MAPD and C4 paraffin/olefin spec of the propane recycle. While cracking furnaces are comparably robust regarding these impurities, the specifications for certain PDH technologies could be extremely sensitive. Such aspects need to be covered for example regarding the design of the C3 hydrogenation and possibly two C3/C4 separation steps will be still required, as depicted in

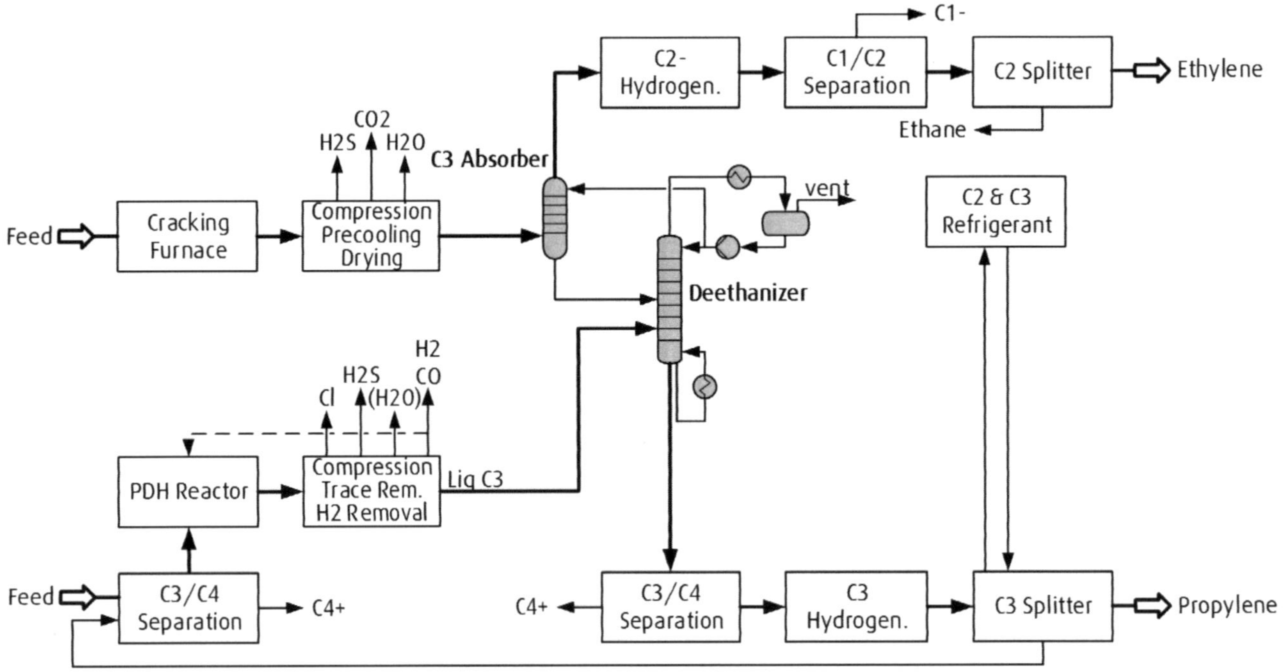

Figure **4**.

It was concluded that this integration scenario is likely suitable for any PDH technology, as any technology applies a certain combination of cold partial C3 condensation and distillation for the removal of hydrogen, methane and C2 components. While the distillation section (the de-ethanization) would be part of the integrated separation, the PDH section itself would just require a simple partial condensation step, which could for example be designed similar to the one depicted in Figure 5.

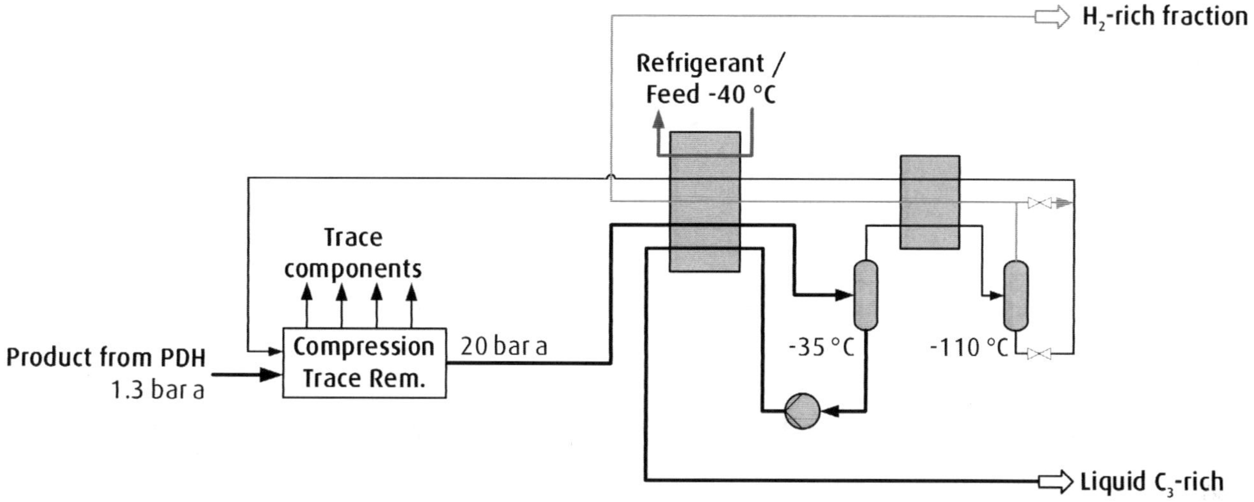

Figure 5: *Example of a "soft" hydrogen removal, likely suitable for any PDH technology.*

In a second example, in case the PDH section contributes predominantly to the total olefin capacity and only a small ethylene capacity is required, the flow scheme can be tailored accordingly, as shown in Figure 6. The process flow is now rather modeled after the original process flow of the PDH. Again, the PDH effluent is depleted in hydrogen and gets into contact with the cracked gas at a lower section of a Deethainzer column, where the light fraction from the cracked gas is already depleted, as described in [11].

Furthermore, a tailored concept is chosen for the small-scale ethylene recovery section [13], which is shown in more detail in Figure 7. Large-scale cracker separation sections exhibit highly energy integrated and complex solutions for the demanding task of C1- separation requiring a remarkable amount of static and rotating equipment for heat exchange, separation and refrigerant generation. In the modified process scheme the liquid olefin fraction generation and purification is still carried out by a simple and compact low temperature section enabling a process design with temperature requirements not lower than e.g. -70 °C. The demanding light fraction purification and olefin product recovery is instead carried out by a pressure swing adsorption at ambient temperatures and a high-pressure light fraction product is retrieved.

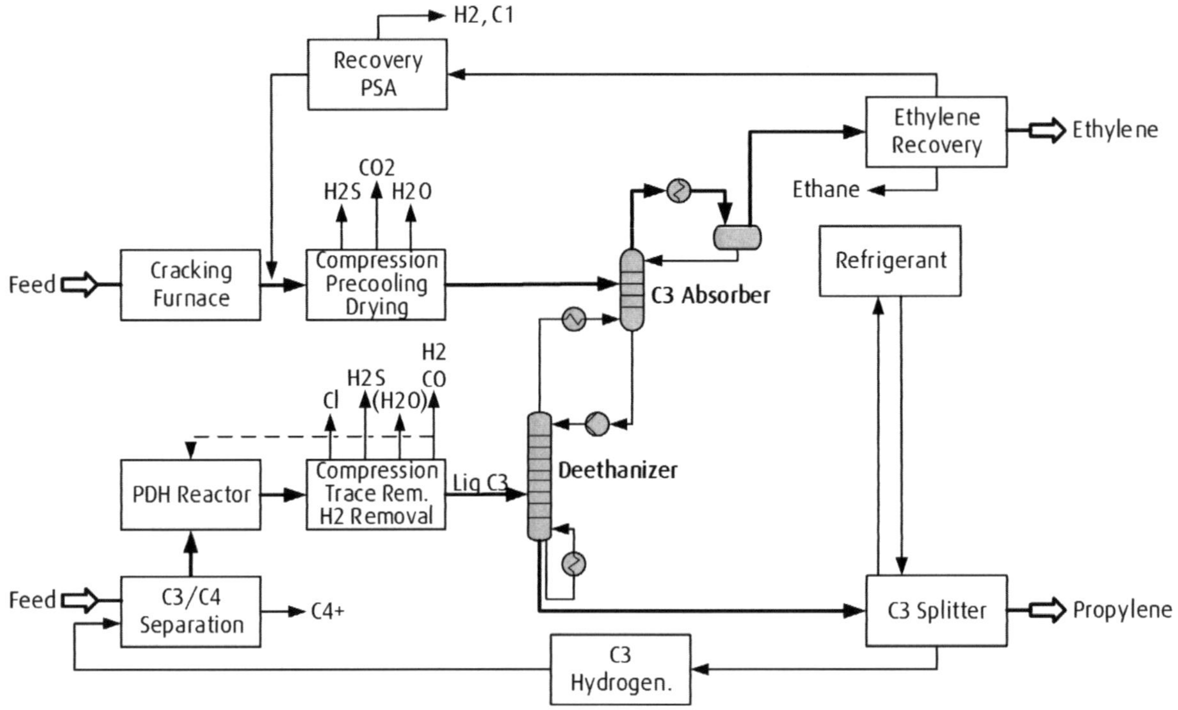

Figure 6: *Second example of an integrated solution, suitable for small-scale ethylene capacities. Ethylene recovery includes C2- hydrogenation.*

The modified process design is well suited for small- and medium-scale olefin recovery plants, where the application of equipment intensive low temperature generation would be of significant capital cost disadvantage. Moreover, the process design comes with some further interesting properties:

- The simplified low temperature section design allows all equipment to be located inside a coldbox, allowing prefabrication of the whole Low Temperature Olefin Recovery in only two units (Cold Box and Recovery PSA).
- No rotating equipment like compressors, expanders or pumps are required in the cold section.
- The PSA does not require a dedicated compressor and the cracked gas compressor is utilized instead.
- The concept enables very high olefin recovery rates from light fraction matching or even exceeding these of conventional low temperature processes.
- The light fraction is produced at high cracked gas pressure, which enables a direct high-pressure hydrogen recovery by utilizing a 2nd PSA.

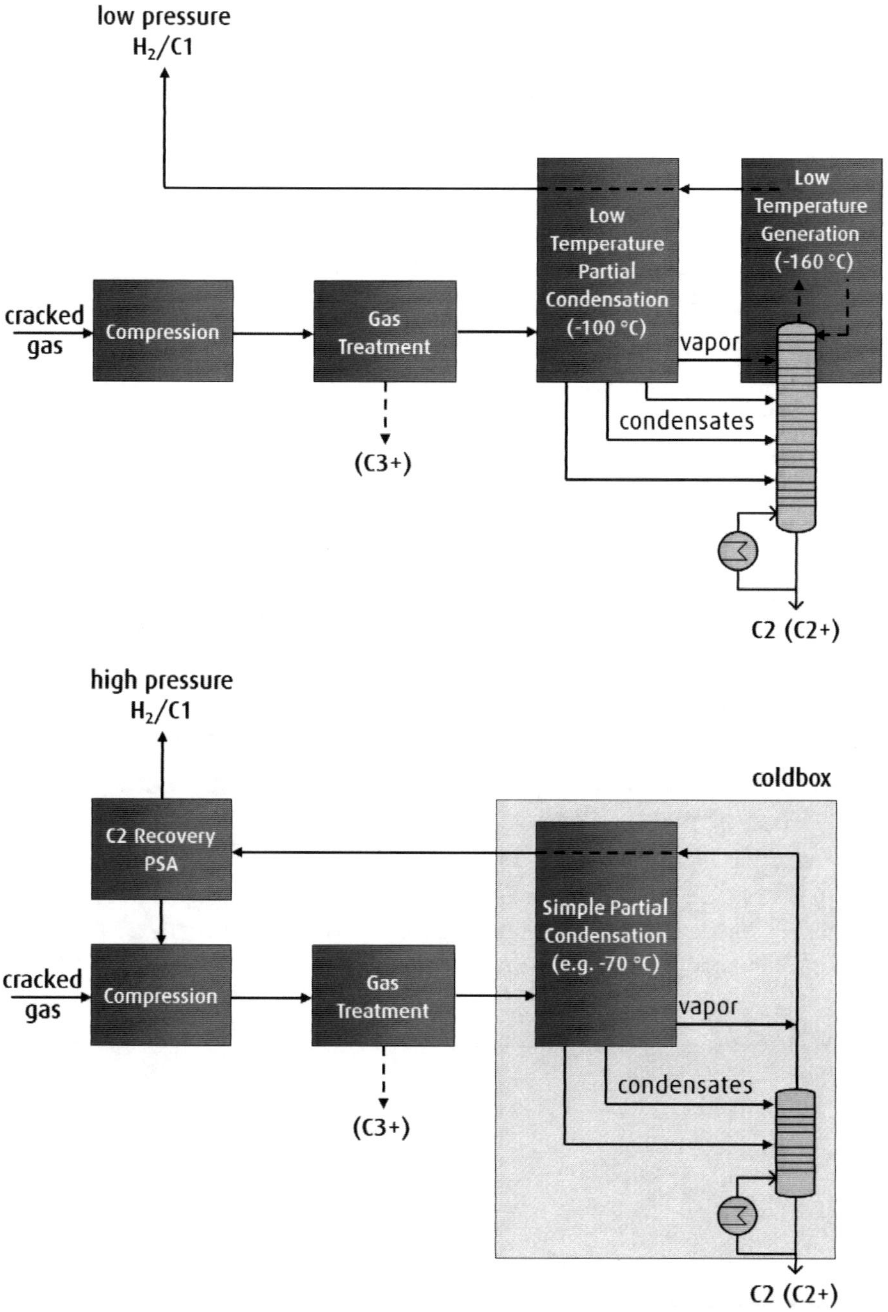

Figure 7: *Top section: A conventional large-scale C1-separation in an ethylene plant requires very low temperatures and a complex setup of highly integrated equipment. Bottom section: Tailored C1- separation for a small-scale ethylene processing unit, applying a simple cold section in combination with a pressure swing absorption.*

4. Concept Verification in an Existing Plant

A cracker and integrated separation section design was carried out by Linde recently, where a small-scale ethane/propane cracker was integrated into a large-scale PDH separation train. The design was similar to the second example outlined in the previous chapter and covers a total propylene capacity of 600 kta and a total ethylene capacity of 55 kta.

The design includes a fully shared Deethanizer system, the PDH C2- fraction is processed in a common ethylene recovery section and the cracker C3+ fraction is processed in a common C3 processing section. The C2 and C3 refrigerant systems are fully shared, including C2 refrigerant utilized for the PDH section (coldbox) and C3 splitter heat pump integrated into a C3 refrigerant compressor, which has a total of four process stages. Utility systems for steam, cooling water and others are fully shared, as both plant sections are fully integrated within a single plot.

Due to the low capacity, the C2 recovery section utilizes a compact process concept similar to that shown in the bottom section of Figure 7. The cracker comprises furthermore a "one furnace" design and the single C2/C3 cracking furnace is equipped with special aluminum alloy coils allowing very long runtimes longer than 120 days. Of course, the plant is designed for continued operation and propylene production while the furnace is in decoking mode.

The plant start-up was carried out in 2021 and the PDH sections including propylene recovery were successfully operated independently of the cracker section, allowing earlier start-up without the cracker sections and the ethylene recovery. Later, cracker and ethylene recovery start-up were conducted in an accelerated manner, benefitting from the PDH process units, which were already in operation. First operating experience indicated that cracker process units can be faded in and out smoothly (e.g. for the purpose of decoking the single furnace or running the cracker in "campaign style" as per ethylene demand).

5. Summary and Conclusions

It was concluded that a high process carbon efficiency and thus low CO_2 emission intensity of the fuel results in favorable process economics of PDH and ethane cracker combination. Precondition is however that an efficient and CAPEX saving plant integration of ethane cracking and propane dehydrogenation is achieved.

Differences in bulk composition, trace components and other process characteristics require an elaborate integration concept and the integration must be tailored depending on PDH technology and cracker type (gas or liquid feedstock). Generally summarized, the tail-end separation (distillations, refrigerant loops, hydrogenations) and utility sections have the most promising integration potentials. The target ethylene / propylene capacity and the available feedstocks play a key role. In addition to the process aspects highlighted on a block flow diagram level, there may be further energy integration opportunities, including for example steam and/or quench water. While the process design must satisfy the individual requirements of each plant section for example in terms of turndown, start-up, and turnaround cycle, the integration furthermore bears opportunities in material take-off reduction, as an integrated plant likely consumes less plot and construction material compared to two individual plants.

First start-up and operation experience from an existing plant have proven the feasibility and advantages of PDH and steam cracking integration.

6. References

1. H. Zimmermann, R. Walzl (2012), "Ethylene" in "Ullmann's Encyclopedia of Industrial Chemistry", Wiley-VCH Verlag GmbH & Co. KGaA, Weinheim
2. H. Zimmermann (2013), "Propene" in "Ullmann's Encyclopedia of Industrial Chemistry", Wiley-VCH Verlag GmbH & Co. KGaA, Weinheim
3. G. Ondrey (2014), "Making Propylene On-Purpose". Chemical Engineering, 121(1):13-16.
4. D. Lippe (2010), "Ethylene, propylene prices swing; growth in feedstock ethane continues". Oil & Gas Journal, 108(27):53–58.
5. M. Pretz et al. (2016), "Dow Fluidized Catalytic Dehydrogenation (FCDh): The Future of On-Purpose Propylene Production". AIChE Spring Meeting and 12th Global Congress on Process Safety, Houston, Texas. Paper Number 192d.
6. M. Pretz et al. (2017), "Shaping the future of on-purpose propylene production". Hydrocarbon Processing.
7. D.J. Phillips, J.L. Glazer (1984), "Process for the production of ethylene and propylene". United States Patent US 4,458,096 A
8. P. Chaiyavech (1991), "Experience in operating an ethane feed steam-cracker and propane dehydro complex". 2nd International Conference Gas Utilization: impact and Contribution towards Industrial Development in SE Asia, Jakarta, Indonesia, August 21-22.
9. P. Chaiyavech (2002), "Commercialization of the World's First Oleflex Unit". The Journal of the Royal Institute of Thailand, 27(3): 664-672.
10. S. Kogan et al. (1998), "Selective Dehydrogenation of Propane using an Improved Dehydrogenation Catalyst". 10th Annual Ethylene Producers' Conference, New Orleans, Louisiana. Paper Number 16a.
11. T. Höfel et al. (2021), "Process and plant for producing propylene by combination of propane dehydration and steam cracking with Pre-separation steps in both processes for removing hydrogen and methane". European Patent EP 3 652 138 B1.
12. H. Linde (1985), „Gewinnung von Ethylen" in „Tieftemperaturtechnik", Springer
13. T. Höfel, J. Kunkel (2022), "Process and plant for obtaining hydrocarbons". International Patent Application WO 2022/013424 A1.

AIChE Paper Number: 79a

Plastic waste contaminants: a deal-breaker for steam cracking?

Marvin Kusenberg
ir.
Ghent University, Ghent, Belgium

Azd Zayoud
Dr. ir.
Ghent University, Ghent, Belgium

Martijn Roosen
ir.
Ghent University, Ghent, Belgium

Thi Hang Dao
Dr.
Ghent University, Ghent, Belgium

Mehrdad Seifali Abbas-Abadi
Dr. ir.
Ghent University, Ghent, Belgium

Andreas Eschenbacher
Dr. ir.
Ghent University, Ghent, Belgium

Steven De Meester
Prof. Dr. ir.
Ghent University, Ghent, Belgium

Kevin M. Van Geem
Prof. Dr. ir.
Ghent University, Ghent, Belgium

Prepared for Presentation at the 2022 Spring National Meeting
San Antonio, Texas, April 10 – April 14, 2022

AIChE Paper Number: 79a

Plastic waste contaminants: a deal-breaker for steam cracking?

Marvin Kusenberg
ir.
Ghent University, Ghent, Belgium

Azd Zayoud
Dr. ir.
Ghent University, Ghent, Belgium

Martijn Roosen
ir.
Ghent University, Ghent, Belgium

Thi Hang Dao
Dr.
Ghent University, Ghent, Belgium

Mehrdad Seifali Abbas-Abadi
Dr. ir.
Ghent University, Ghent, Belgium

Andreas Eschenbacher
Dr. ir.
Ghent University, Ghent, Belgium

Steven De Meester
Prof. Dr. ir.
Ghent University, Ghent, Belgium

Kevin M. Van Geem
Prof. Dr. ir.
Ghent University, Ghent, Belgium

Abstract:

Pyrolysis of plastic packaging waste is a promising recycling route to tackle the ever-growing pollution of environments and ecosystems. At the same time, by using waste-derived feedstocks for production processes of base chemicals such as steam cracking, the material loop can be closed and greenhouse gas emissions that arise from the use of fossil feedstocks and from incineration of collected plastic waste can be reduced. However, using waste-derived feedstocks for steam cracking is still far from industrial application due to the large variety of packaging waste in terms of polymer composition and contamination with residues and metal-containing additives. In fact it has been shown that pyrolysis oils from real post-consumer waste fractions contain more and different contaminants compared to their fossil-based counterparts. The most important contaminants are metals (i.e. aluminum, silicon, sodium and others) and heteroatoms (i.e. chlorine, nitrogen and oxygen) that require special attention in terms of analysis and characterization. Furthermore, a thorough assessment of upgrading techniques is inevitable with these alternative feedstocks to ensure a high-quality petrochemical feedstock while still being economically competitive. In this work, a comprehensive experimental assessment of the pyrolysis oil quality of real post-consumer packaging waste fractions was performed with special emphasis on the impact on crucial steam cracking performance indicators such as coke formation. Based on the pyrolysis oil compositions, potential decontamination techniques are discussed. Our study shows that crude post-consumer plastic waste pyrolysis oils require thorough upgrading before being acceptable for industrial steam crackers without causing disturbances in running processes.

Introduction

The production of plastics exceeded 60 million tons in Europe alone in 2018 and almost 60% of the produced plastics are used for packaging applications [1]. The recycling ratios of collected post-consumer plastic waste are still low with high amounts being landfilled, exported or incinerated for energy recover leading to vast CO_2 emissions. In order to reduce the destruction of ecosystems and environments and to keep the global temperature increase below 1.5 °C, efficient recycling routes are vitally needed. Mechanical recycling, which is considered the most favorable from an energy consumption point-of-view largely depends on high-purity sorted waste fractions such as PET [2, 3]. Next to the high degree of mixing of plastic waste, strict regulations for food-contact applications of recycled plastics exist which poses an additional challenge for mechanical recycling [4]. A promising alternative recycling route is thermochemical recycling, specifically the conversion of plastic waste into an oil-like liquid product via pyrolysis. By subsequently integrating the pyrolysis oils into the petrochemical systems for the production of base chemicals such as ethylene and propylene, the material loop towards new virgin-grade polymers can be closed [5-13]. In this context it has recently been shown that the effect of plastic pyrolysis oil constituents on steam cracking can be beneficial in terms of ethylene yields as reported for pyrolysis oil of virgin polyolefins blended in fossil naphtha [14, 15]. However, several challenges concerning the quality of pyrolysis oils from real waste and their suitability for industrial steam crackers still exist as elaborated in a recent review article [16]. In particular, the large degree of contamination is the main bottleneck for pyrolysis oils as steam cracker feedstocks. Heteroatoms in pyrolysis oils originate from organic or inorganic residues or from wrongly sorted plastics such as PET or PVC. The main source for metal contaminants are additives which are used widely to enhance the polymers' properties [17-19]. The large variety of potential impurities poses a considerable risk for large-scale industrial systems such as steam crackers. It has been recently reported that the typically high concentrations of olefins and aromatics lead to substantially increased formation of heavy products as well as coke formation and fouling. Furthermore, the increased coke formation could be allocated to metal contaminants such as copper, sodium, iron and calcium [20].

It is therefore crucial to assess the quality of pyrolysis oils from a contaminant point-of-view and further to investigate potential upgrading techniques to reach high-quality feedstocks. In this work, we systematically analyzed the pyrolysis products of real post-consumer plastic packaging waste with special emphasis on contaminants such as halogens and metals [21]. This way, we provide comprehensive experimental insight into the relationship between the waste composition and the produced pyrolysis oils. Based on the found compositions, conclusions are drawn in terms of the steam cracking feasibility on the one hand, and necessary upgrading steps on the other hand.

Materials and Methods

Feedstocks: Polyolefin-rich post-consumer plastic packaging waste feedstocks, have been used, namely mixed rigid polyolefins (MPO), rigid PP, and PE films. The waste fractions were collected at curbside and went through typical waste sorting and separation steps such as shredding, cold-washing, float-sink separation, wind sifting and extrusion [20, 21].

Experimental: Pelletized plastic waste samples have been pyrolyzed in a pilot-scale pyrolysis unit as shown in
Fig. **1** [20-22].

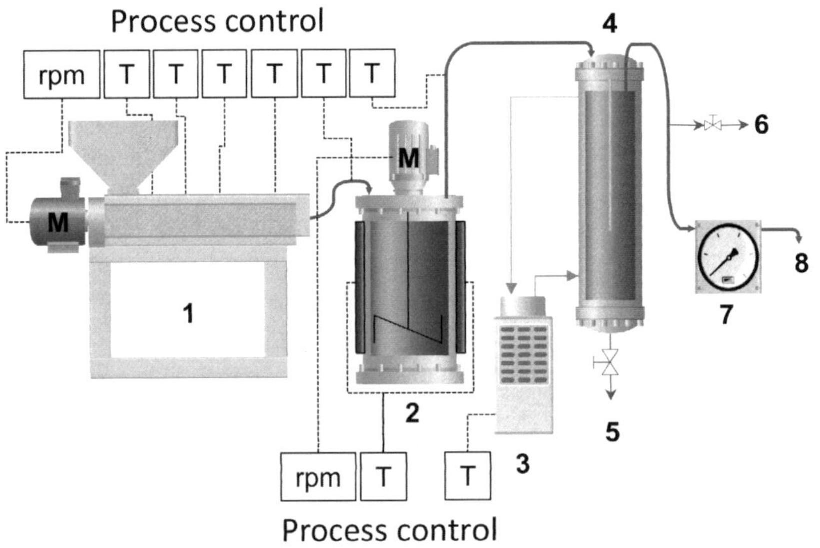

Fig. 1: Pilot-scale pyrolysis unit: (1) extruder, (2) stirred-tank reactor, (3) cooler, (4) condensation section, (5) liquid product sampling, (6) gaseous product sampling, (7) volumetric gas flow-meter, (8) vent, (M) motor [21].

The unit consists of a single-screw extruder which is used as feeding and pre-heating device that is connected to an externally heated stirred-tank reactor (CSTR). The pelletized plastic feedstocks were pyrolyzed at 450 °C and atmospheric pressure. Pyrolysis products leave the reactor in vapor form and are condensed and collected in the downstream section of the reactor. Non-condensable gaseous products are measured using a volumetric gas flow-meter. Solid products remain in the reactor and are removed and weighted separately after completion of each experiment.

Analytics: All pyrolysis oils have been thoroughly characterized using comprehensive two-dimensional gas chromatography (GC × GC) coupled to various detectors such as a flame ionization detector (FID) for the hydrocarbon composition as well as nitrogen and sulfur chemiluminescence detectors (SCD/NCD) for the assessment of trace nitrogen and sulfur. The oxygen concentration was measured using CHNS/O elemental analysis. Furthermore, the halogen concentration was measured using combustion ion chromatography (CIC) and the metal contamination was assessed using inductively coupled plasma – optical emission spectrometry (ICP-OES) [20, 21].

Results and Discussion

Hydrocarbon composition: The hydrocarbon composition of plastic waste pyrolysis oils is substantially different compared to fossil-based feedstocks due to high concentrations of unsaturated hydrocarbons (i.e., olefins/aromatic). In particular polyolefins which are the main constituents of packaging waste yield large amounts of olefins including high degrees of branching [23]. Furthermore, PS which is also a prominent constituent of plastic packaging largely decomposes into its monomer styrene [22]. In terms of steam cracking, olefins (and naphthenes/aromatics) are typically less wanted due to an increased coke formation and fouling of heat exchanger surfaces [24, 25]. The hydrocarbon compositions of the pyrolysis oils analyzed using GC × GC-FID can be seen in Fig. 2.

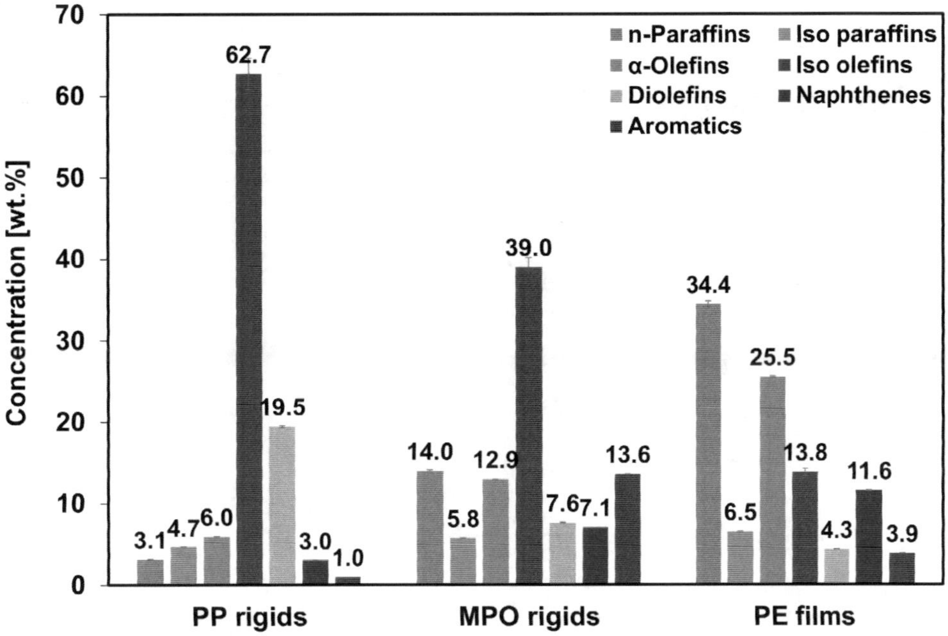

Fig. 2: Hydrocarbon composition of the pyrolysis oils. Data taken from [21].

It can be seen that the pyrolysis oil composition largely depends on the pyrolyzed waste fractions. The total olefin concentration including linear and branched olefins as well as diolefins of PP pyrolysis oil is ~88 %. PE pyrolysis oil contains a total of ~44 % olefins. The mixed polyolefin waste feedstocks which contains roughly equal parts of PE and PP contains ~60 % total olefins. Considering an olefin tolerance of 2 wt.% in commercial steam cracking feedstocks [26], it becomes obvious that the olefin concentrations in all three samples need to be drastically reduced prior to steam cracking either by blending with olefin-free fossil feedstocks or via catalytic hydrogenation. Necessary blending ratios to reduce the olefin concentrations <2 wt.% shown in Table 1.

Table 1: Required dilution ratios with olefin-free fossil feedstocks to reduce the total olefin content (linear, branched and diolefins) below the threshold of 2 wt.% [26]. Data taken from [21].

	Total olefins [wt.%]	Required blending ratio
PP rigids	88.2	1:44
MPO rigids	59.6	1:30
PE films	43.6	1:22

Based on the findings it can be stated that PE-rich pyrolysis oil is more suitable as a steam cracker feedstock due to the lower amount of total olefins and the lower amount of branching [20]. Hence, for polyolefin mixtures it can concluded that the upgrading effort or necessary blending ratios increase with the PP concentration.

Heteroatoms: Next to the degree of unsaturation, branching and aromatics concentration, heteroatoms that are transferred into the pyrolysis oils are important aspects. The most crucial heteroatoms for plastic waste pyrolysis oils are nitrogen, oxygen and chlorine which partly exceed the known threshold for industrial feedstocks substantially [21] (see Table 2).

Table 2: Heteroatom concentrations compared with the industrial threshold values [16, 26].

	PP rigids	MPO rigids	PE film	Industrial threshold [ppm]
Nitrogen [ppm]	29	1143	4	100 (light feedstocks) 2000 (heavy feedstocks)
Sulfur [ppm]	5	46	9	500
Oxygen [ppm]	<1000	<1000	2100	100
Chlorine [ppm]	137	474	143	3
Bromine [ppm]	<6.7	<6.7	22	-

Nitrogen is one of the most important heteroatomic contaminants in plastic waste pyrolysis oils due to the wide range of potential nitrogen sources such as organic and inorganic residues as well as nitrogen containing polymers (i.e. PUR, PA) [16, 21]. In steam cracking, nitrogen contamination might lead to issues such as explosive gum formation in the downstream sections. Potential sources for sulfur in plastic packaging waste are organic residues or certain additives, however, the findings indicate that sulfur is a contaminant of lower concern in thermochemical plastic waste recycling. The threshold for oxygen in steam cracking feedstocks has been reported as 100 ppm which is related to issues such as fouling and corrosion as well as downstream catalyst poisoning issues by formed CO [26, 27]. The main source of oxygen in plastic waste pyrolysis oils is PET that is a large constituent of packaging waste and hence likely to appear in mixed packaging waste fractions [28]. While detailed detection of nitrogen and sulfur compounds is possible via techniques such as nitrogen and sulfur chemiluminescence detectors (NCD/SCD) which can be coupled with GC × GC, oxygen in hydrocarbon samples is difficult to analyze due to the largely overlapping hydrocarbon matrix and the low sensitivity of FID for oxygenates [29]. The concentrations shown in Table 2 indicate that the oxygen concentrations in two of the three samples was below the limit of detection of the used elemental analysis unit (~1000 ppm), which is, however, still substantially higher than the maximum allowable concentration (100 ppm). Hence, alternative techniques are needed such as O-FID, which are, however, typically limited to lighter samples [30]. A possible solution is the atomic emission detector (AED), which, combined with GC × GC is one of the most powerful techniques for the analysis of a wide range of elements and their exact molecular form of appearance [29, 31]. Halogens and especially chlorine belong to the most problematic contaminants in steam cracking due to the formation of corrosive compounds such as HCl leading a maximum allowable concentration of 3 ppm in industrial feedstocks [26]. Chlorine mostly stems from PVC impurities in the sorted waste fraction. Similar to oxygen components, it is still challenging to identify and characterize individual halogenated compounds in complex hydrocarbon matrices. Hence, GC × GC-AED analysis is the most promising analytical technique for halogen detection and quantification which will be subject of future work. No threshold value exists for bromine in steam cracker feedstocks, however, due to its use as flame retardant material it cannot be excluded that traces of bromine occur in pyrolysis oils from mixed plastic waste as verified in the PE film pyrolysis oil (22 ppm). Considering the measured concentrations of chlorine which exceed the maximum allowable concentrations substantially, simple blending with fossil feedstocks prior to steam cracking will not be sufficient to reduce the concentrations sufficiently. In terms of the other heteroatom concentrations, blending with fossil feedstocks in ratios of 5-10 wt.% would level out the most severe impacts.

Metal contaminants: Next to heteroatoms, metal contaminants are important contaminants that stem mostly from additives used in plastic products. Metal contaminants in plastic waste can be transferred from the waste feedstock into the liquid pyrolysis product in form of organometallic complexes in vapor form. Entrainment of solid pyrolysis residue (i.e., ash) with the pyrolysis product vapors is another possibility. All pyrolysis oils contained substantial amounts of metals as shown in Fig. 3. Strict specifications exist for industrial steam cracker feedstocks in terms of metals which relates to issues such as corrosion, fouling, coke formation and especially downstream catalyst poisoning [16, 26, 32]. The maximum allowable concentrations are for instance: 0.5 ppm for calcium, 50 ppm for copper, 1 ppm for iron; 0.005 ppm for mercury; 0.125 ppm for sodium; 100 ppm for nickel; 0.1 ppm for lead and 1 ppm for silicon. In all studied pyrolysis oils, thresholds are exceeded substantially by iron, sodium and lead. The calcium and mercury thresholds are exceeded only by MPO pyrolysis oil, however, the fact that the concentrations were below the limit of detection for PP and PE feedstock does not necessarily imply that the respective metals are indeed absent. It rather implies that additional analyses are needed with more accurate detection techniques such as ICP-MS [16].

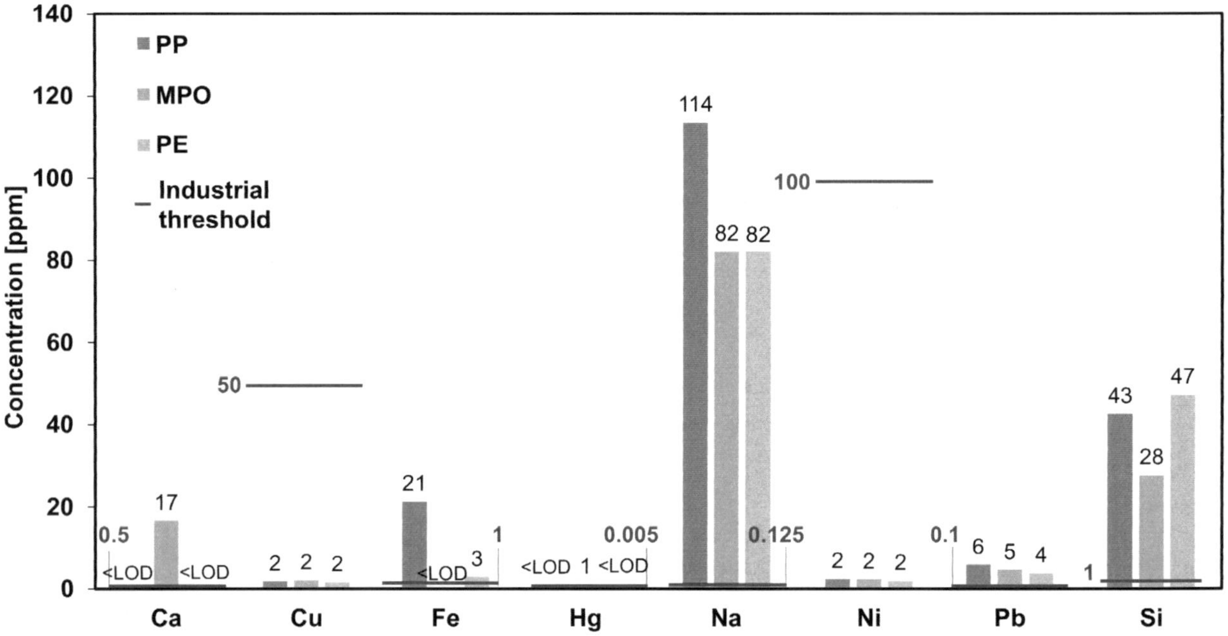

Fig. 3: Metal concentrations in the plastic waste pyrolysis oils compared to the threshold values (in ppm) for industrial steam crackers [21, 26, 32-34].

Furthermore, no industrial limit values exist for metals such as aluminum, antimony or zinc which does not mean that these metals are not causing issues in steam cracking but rather that there is a knowledge gap regarding the impact of these metals in steam cracking. This is mostly because some metals are absent in fossil-based feedstocks. Given that the industrial threshold is partly exceeded by orders of magnitude, simple blending with contaminant-free fossil feedstocks will not be feasible due to the extremely high necessary blending ratios. It is rather implies that thorough upgrading prior to steam cracking is indispensable in terms of the metal contaminants which will be elaborated in a later subsection of this paper.

The general impact of metal components on steam cracking is especially pronounced in terms of coke formation as elaborated in previous work [20]. Two different plastic waste pyrolysis oils were blended with naphtha and steam cracked in a continuous bench-scale steam cracking unit. By virtue of the blending ratio (1:3), the differences in the hydrocarbon composition of the two pyrolysis oils (i.e., olefins, aromatics, etc.) were largely levelled out leaving the metal contaminants as the main difference in the two respective cracked feedstocks. Fig. 4 shows the coke formation compared to the pure fossil naphtha used as an industrial benchmark. It can be seen that steam cracking of pyrolysis oil 1 which contains ~1.7x more total metals compared to pyrolysis oil 2, translated into a substantial coke formation increase compared to fossil naphtha as industrial reference feedstock. The fact that only 25 wt.% of the cracked feedstocks were pyrolysis oil while the other 75 wt.% were fossil naphtha underlines the high impact of the pyrolysis oils in terms of coke formation. Hence it can be stated that the metal contaminants, as the main distinguishing factor between pyrolysis oil 1 and pyrolysis oil 2 are a highly important driver for coke formation during steam cracking [20]. In an industrial context, an increased coke formation by a factor of 2.4, or even 1.4 compared to fossil naphtha would be unacceptable due to a decreased run-lengths and resulting economic losses. Considering the mixing ratios, the pure pyrolysis oils would lead to very high coke formation which would definitely be unfeasible for industrial crackers.

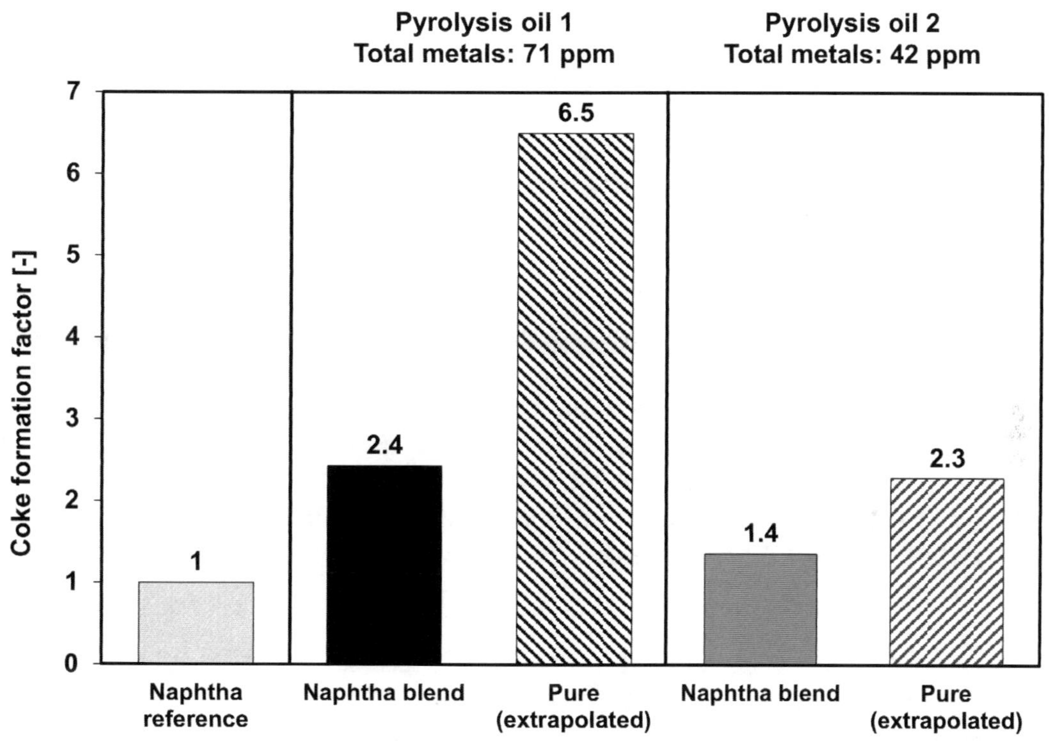

Fig. 4: Radiant coil coking tendency compared to fossil naphtha including extrapolated values for the pure pyrolysis oils (hatched). Data taken from [20].

Upgrading techniques: The findings elaborated above indicate that plastic waste pyrolysis oils are not suitable "as is" for industrial steam crackers. Furthermore, blending with fossil feedstocks as performed in our recent study proved to be insufficient to completely alleviate the impacts of pyrolysis oil-specific contaminants. Hence, thorough upgrading of pyrolysis oils prior to steam cracking is indispensable. Several techniques exist that can be applied throughout the entire thermochemical recycling chain, i.e., pre-treatment of plastic waste for the removal of additives and halogens, use of auxiliary materials during pyrolysis such as catalysts, solvents or supercritical water and the post-treatment of pyrolysis oils to remove metals, heteroatoms and to saturate olefins and aromatics. Given the large variety of upgrading techniques, careful selection of suitable processes is needed in order to find the optimal combination of sufficient final product quality while still being economically competitive.

Improving the quality of pyrolysis oils for subsequent use in steam crackers should be aimed at the most important contaminants nitrogen, oxygen, chlorine and metals. Chlorine removal from PVC-containing waste fractions can be performed by utilizing the lower decomposition temperature of the Cl-bonds compared to the polymer backbone. This way, chlorine is released in form of HCl. Using heated extruders, continuous feeding and chlorine removal can be combined in a single step which has already been applied at larger scale [35]. Removal of nitrogen and oxygen (and other heteroatoms) from pyrolysis oils can be performed via catalytic hydrotreatment [36-39]. However, the most challenging aspect of plastic waste pyrolysis oils is the co-existence of various contaminants that might hamper the catalytic hydrogenation performance. Especially nitrogen compounds are known to inhibit catalytic hydrogenation of sulfur, oxygen and chlorinated compounds [40, 41]. Hence, alternative nitrogen removal techniques are needed such as adsorption via activated carbon [42, 43] or metal organic frameworks [44, 45]. In terms of hydrotreatment it is known that metal contaminants lead to deactivation of catalysts due to deposition on the catalyst surface. Hence, cheap aluminum oxide catalyst guard beds can be installed prior to more precious catalyst beds [46, 47]. Filtration of the crude pyrolysis oil or distillation are promising techniques to remove entrained particles, leaving only the chemically bound metals in the pyrolysis oils [48, 49]. First steam cracking tests using filtered pyrolysis oils show a substantial reduction in coke formation compared to the crude pyrolysis oils which can be directly related to the metal removal. Next to "stand-alone" techniques that are applied to remove individual components from plastic waste samples or pyrolysis oils, integrated processes gain increasing attention. In these processes, plastic waste is directly converted to high-quality petrochemical feedstocks using auxiliary materials such as solvents or continuous on-line upgrading of the product vapors over catalysts [7, 50].

It can be concluded that crude post-consumer plastic waste pyrolysis oils are highly contaminated and cause severe disturbances in steam cracking if not thoroughly upgraded. However, it is worth noticing that fossil steam cracking feedstocks such as naphtha also undergo certain treatment steps from crude oil leading to a stable supply of high-quality feedstocks. Obviously, the comparison between fossil naphtha and crude post-consumer plastic waste pyrolysis oil is not entirely fair. Using a smart combination of waste pre-treatment and pyrolysis oil upgrading would largely alleviate the negative impacts of plastic waste contaminants and hence allow a large-scale application of plastic waste pyrolysis oils as steam cracker feedstocks. In this context, improved waste separation and sorting play a key role. All upgrading efforts could be reduced drastically, if the initial waste feedstock quality improves by, for instance, removal of additives, use of washable inks, or the abandonment of multi-layer foil packaging.

Conclusions and Outlook

Contaminants play a key role in the thermochemical recycling of plastic waste via pyrolysis and steam cracking. It has been shown that pyrolysis oils from real post-consumer plastic waste are heavily contaminated exceeding known industrial feedstock specifications by orders of magnitude. Based on detailed experimental studies, chlorine, nitrogen, oxygen and metals could be identified as the most important contaminants that all have severe negative impacts on steam cracking. Furthermore, detailed steam cracking studies have shown that metal contaminants are strong contributors for coke formation which would lead to a severe run-length reduction in industrial steam crackers and hence to economic penalties. Therefore it can be concluded that crude plastic waste pyrolysis oils are indeed unfeasible as steam cracker feedstocks "as is".

However, application of mild upgrading using dehalogenation as a waste pre-treatment step, filtration and distillation of the pyrolysis oils for the removal of metals as well as hydrotreatment and adsorption for the purification of pyrolysis oils would sufficiently increase the quality of pyrolysis oils towards steam cracking feedstocks. Obviously, careful assessment is needed in order to determine the economic potential of heavily upgraded pyrolysis oils compared to fossil-based feedstocks, however, additional research may shed further light on optimal combinations of processes to reduce the key contaminants. Furthermore, studies elaborating on co-feeding of pyrolysis oils in steam cracking and the individual effects of certain contaminants will provide further insight in terms of acceptable contaminant concentrations.

Acknowledgment

The authors would like to acknowledge financial support by the Catalisti-ICON project (HBC.2018.0262) MATTER (Mechanical and Thermochemical Recycling of mixed plastic waste) funded by Flanders Innovation & Entrepreneurship (VLAIO) and by the Fund for Scientific Research Flanders (FWO) project WASTE.

References

[1] Plastics Europe, "Plastics - The Facts 2020," 2020.
[2] Ragaert, K. et al., *Waste Manage.,* vol. 69, no. 1, pp. 24-58, 2017,
[3] Vilaplana, F. and Karlsson, S., *Macromol. Mater. Eng.,* vol. 293, no. 4, pp. 274-297, 2008,
[4] De Tandt, E. et al., *Waste Manage.,* vol. 119, pp. 315-329, 2021,
[5] Al-Salem, S. M. et al., *Prog. Energy Combust. Sci.,* vol. 36, no. 1, pp. 103-129, 2010,
[6] Al-Salem, S. M. et al., *J Environ Manage,* vol. 197, pp. 177-198, 2017,
[7] Qureshi, M. S. et al., *J. Anal. Appl. Pyrolysis,* vol. 152, p. 104804, 2020,
[8] Coates, G. W. and Getzler, Y. D. Y. L., *Nature Reviews Materials,* vol. 5, no. 7, pp. 501-516, 2020,
[9] Lopez, G. et al., *Renew Sust Energ Rev,* vol. 82, pp. 576-596, 2018,
[10] Lopez, G. et al., *Renew Sust Energ Rev,* vol. 73, pp. 346-368, 2017,
[11] Papari, S. et al., *Materials,* vol. 14, no. 10, 2021,
[12] Vollmer, I. et al., *Angew. Chem. Int. Ed.,* vol. 59, no. 36, pp. 15402-15423, 2020,
[13] Nanda, S. and Berruti, F., *Environ. Chem. Lett.,* vol. 19, no. 1, pp. 123-148, 2021,
[14] Hájeková, E. et al., *J. Anal. Appl. Pyrolysis,* vol. 79, no. 1, pp. 196-204, 2007,
[15] Hájeková, E. and Bajus, M., *J. Anal. Appl. Pyrolysis,* vol. 74, no. 1, pp. 270-281, 2005,
[16] Kusenberg, M. et al., *Waste Manage.,* vol. 138, pp. 83-115, 2022,
[17] Roosen, M. et al., *Environ. Sci. Technol.,* vol. 54, no. 20, pp. 13282-13293, 2020,
[18] Ügdüler, S. et al., *Waste Manage.,* vol. 104, pp. 148-182, 2020,
[19] Ścierski, W., *Applied Sciences,* vol. 11, no. 10, 2021,

[20] Kusenberg, M. et al., *Waste Manage.,* vol. 141, pp. 104-114, 2022,
[21] Kusenberg, M. et al., *Fuel Process. Technol.,* vol. 227, p. 107090, 2022,
[22] Zayoud, A. et al., *Waste Manage.,* vol. 139, pp. 85-95, 2022,
[23] Tsuge, S. et al., *Pyrolysis-gc/ms data book of synthetic polymers pyrograms, thermograms and ms of pyrolyzates*. Elsevier B.V., 2011.
[24] Kopinke, F. D. et al., *Ind Eng Chem Res,* vol. 32, no. 1, pp. 56-61, 1993,
[25] Müller-Steinhagen, H. et al., *Heat Transf. Eng.,* vol. 32, no. 3-4, pp. 189-196, 2011,
[26] Baumgartner, A. J. et al., in *AIChE Spring National Meeting, Ethylene Producers Conference*, New Orleans, Louisiana, 2004.
[27] Ristic, N. D. et al., *Fuel Process. Technol.,* vol. 167, pp. 241-249, 2017,
[28] Anuar Sharuddin, S. D. et al., *Energy Convers. Manage.,* vol. 148, pp. 925-934, 2017,
[29] Tranchida, P. Q. and Mondello, L., *Hyphenations of Capillary Chromatography with Mass Spectrometry*. Elsevier Science, 2019.
[30] Juntarachat, N. et al., *J. Appl. Polym. Sci.,* vol. 130, no. 3, pp. 1863-1872, 2013,
[31] Ross, A. B. et al., *J. Anal. Appl. Pyrolysis,* vol. 58-59, pp. 371-385, 2001,
[32] Sundaram, K. M. and Stancato, B., in *AIChE Spring National Meeting, Ethylene Producers Conference*, Orlando, Florida, 2018, p. 24.
[33] Tatterson, D. F. and Vasalos, I. A., US Patent US4280898A, 1981.
[34] Letzsch, W. S. and Ashton, A. G., in *Stud. Surf. Sci. Catal.*, vol. 76, J. S. Magee and M. M. Mitchell, Eds.: Elsevier, 1993, pp. 441-498.
[35] Okuwaki, A., *Polym. Degrad. Stab.,* vol. 85, no. 3, pp. 981-988, 2004,
[36] Zacher, A. H. et al., *Green Chem.,* 10.1039/C3GC41382A vol. 16, no. 2, pp. 491-515, 2014,
[37] Ware, R. L. et al., *Energy Fuels,* vol. 34, no. 12, pp. 16181-16186, 2020,
[38] Prado, G. H. C. et al., *Energy Fuels,* vol. 31, no. 1, pp. 14-36, 2017,
[39] Murena, F. and Gioia, F., *J. Hazard. Mater.,* vol. 60, no. 3, pp. 271-285, 1998,
[40] Gioia, F. and Murena, F., *J. Hazard. Mater.,* vol. 57, no. 1, pp. 177-192, 1998,
[41] Kerby, M. C. et al., *Catal. Today,* vol. 104, no. 1, pp. 55-63, 2005,
[42] Kim, J. H. et al., *Catal. Today,* vol. 111, no. 1, pp. 74-83, 2006,
[43] Sano, Y. et al., *Energy Fuels,* vol. 18, no. 3, pp. 644-651, 2004,
[44] Wang, Z. et al., *J. Energy Chem.,* vol. 22, no. 6, pp. 869-875, 2013,
[45] Maes, M. et al., *Angew. Chem. Int. Ed.,* vol. 50, no. 18, pp. 4210-4214, 2011,
[46] van Dongen, R. H. et al., *Ind Eng Chem Process Des Dev,* vol. 19, no. 4, pp. 630-635, 1980,
[47] Ali, M. F. and Abbas, S., *Fuel Processing Technology,* vol. 87, no. 7, pp. 573-584, 2006,
[48] Heydariaraghi, M. et al., *J. Anal. Appl. Pyrolysis,* vol. 121, pp. 307-317, 2016,
[49] Wiriyaumpaiwong, S. and Jamradloedluk, J., *Energy Procedia,* vol. 138, pp. 111-115, 2017,
[50] Qureshi, M. S. et al., in *Pyroliq 2019: Pyrolysis and liquefaction of biomass and wastes*, Cork, Ireland, 2019.

639649 Managing Challenges in the Production and Use of Pyrolysis Oil

Abstract:

Pyrolysis oil has become an important part of the circular plastics story, offering a convenient way to convert post-consumer waste and other waste polymer streams into a liquid capable of substituting virgin steam cracker feedstocks for the production of new polymers. However, producing and processing pyrolysis oil is not without its challenges. In this paper we present several of these topics, identifying not only the pitfalls but also offering potential solutions

639497 Bio-Ethylene from Gas Fermentation: Gas Treatment Challenges and Successes

Abstract:

As consumers look for sustainable products, sustainably produced ethylene will become an important intermediate chemical. Ethylene is a building block that can be used to produce consumer products such as sustainable aviation fuel, polyethylene, surfactants, and PET fibers. "Green" ethanol can be dehydrated to ethylene using various commercial technologies. "Green" ethanol can be produced using many sustainable processes. One technology which has been commercialized is gas fermentation of industrial waste gases. These industrial waste gases contain many trace components that can impact fermentation. The unique challenges and novel solutions around mitigation of the fermentation inhibitors present in a variety of industrial waste gas fermentation feedstocks will be discussed.

639369 CS2 Scavenger – New Option to Upgrade Feedstock Quality

Abstract:

Ethylene producers have linked carbon disulfide (CS_2) in petrochemical grade naphtha to corrosion in locations downstream of the cracker and to deterioration of Pygas hydrogenation catalyst activity. CS_2 can poison the selective hydrogenation catalyst that is generally used to remove acetylenes and dienes in olefin plants. CS_2 also contaminates C4 and C5 olefins and diolefins production in polymer plants and can be corrosive when it undergoes transformation to acids. This has led some producers to implement very low specifications (ppm levels) of CS_2 in petrochemical grade naphtha. Other producers have also indicated CS_2 as a primary contaminant particularly in some shale derived feedstocks. Thus CS_2 contamination can effectively limit feedstock flexibility and associated costs.

Reactive removal of CS_2 represents an economically attractive solution. To help meet this growing need, Dorf Ketal is in progress of developing a new chemical for CS_2 removal from Petrochemical grade naphtha/NGL. The new product can remove CS_2 efficiently thereby upgrading contaminated feedstocks. This paper will demonstrate some results of Lab testing, with a variety of commercial feeds, conducted to prove economic viability of an reactive solution.

639552 Ethylene Plant Feed Sulfur Removal

Abstract:

Olefins plant feed flexibility allows operators to maximize profitability. When market conditions shift, it may be possible for traditional ethylene plant feedstocks to have higher value sold as product directly to other users. Plant processes to prepare these feedstocks for the market have occasionally made them unfit for sale to these other users. This has been observed in the case of an ethane/propane import mix. The geography surrounding this importer dictated propane was seasonally more valuable than propylene. The feedstock was separated and propane was to be sold directly to market. However, this propane routinely failed pipeline corrosivity metrics (ASTM 1838 Copper Strip) and had to be fed to the crackers resulting in lower profitability for the site. The contaminant causing these copper strip failures was identified to be sulfur species. Baker Hughes provided a solution to consistently remove this contaminant. In addition to propane, the technology can be applied to C_4 and C_5 streams.

A case history will be presented to demonstrate how this technology was used to facilitate performance improvement with a Baker Hughes proprietary equipment solution.

AIChE Paper 79f

Dihydrodicyclopentadiene – Where does it come from, and can it affect the Ethylene Unit?

Debby Rossana
Sr. Industry Technical Consultant, NALCO Water
An Ecolab Company Downstream Petrochemicals

André Bernard
Process Design Engineer
NOVA Chemicals Corporation

Grace Kim
Research Scientist
Centre for Applied Research, NOVA Chemicals Corporation.

Robert Choi
Senior Technical Engineer
KBR

Prepared for Presentation at the 2022 Spring National Meeting

San Antonio, TX, April 10th – April 14th, 2022

Dihydrodicyclopentadiene – Where does it come from, and can it affect the Ethylene Unit?

Debby Rossana
Sr. Industry Technical Consultant, NALCO Water
An Ecolab Company Downstream Petrochemicals

André Bernard
Process Design Engineer
NOVA Chemicals Corporation

Grace Kim
Research Scientist
Centre for Applied Research, NOVA Chemicals Corporation.

Robert Choi
Senior Technical Engineer
KBR

Abstract

High levels of dihydrodicyclopentadiene (DHDCPD) were found in a stream used as wash oil for the cracked gas compressor. The wash oil stream originates from the plant's downstream aromatic hydrotreating unit. The plant was originally designed as a liquid cracker and had been reconfigured over the years to process a variety of feedstocks. In recent years, it was converted to process 100% ethane feed with limited capabilities for propane and butane cracking. The aromatic unit was kept on stream after the conversion to ethane feed, but operates at reduced rates and processes external streams due to the significant reduction in raw pyrolysis gasoline make from the ethylene cracker. The feed conversion to ethane has significantly affected the quality of the stream used as wash oil. As such, injection of external good quality wash oil was evaluated based on the premise that it could be cycled between the olefins and aromatics units. However, simulation work demonstrated that the external wash oil would blend with the DHDCPD, making wash oil recovery ineffective relative to the targeted quality. DHDCPD is a component for which little if any information exists relative to ethylene plant operation. It is absent from the Ethylene Producer Conference Feedstock Contaminants database. This paper discusses the potential source of DHDCPD, the expected flow path followed by the component through the ethylene unit when the rich stream is used as wash oil to the cracked gas compressor and the potential effect of DHDCPD and its precursors Cyclopentadiene (CPD) and Dicyclopentadiene (DCPD), may have on an ethylene plant.

Introduction

Wash oil is an important element of most compressor antifoulant programs and in fact, 96% of all compressors in North America use a wash oil in some fashion. Wash oil provides solvency for low molecular weight polymers and heavies that can fall out on the internals or interstage coolers of the compressor. Wash oil also provides some scouring effect and is often the carrier for antifoulant programs. The wash oil stream at a North American ethylene plant uses a C9+ overhead stream from the plant's downstream aromatic hydrotreating unit. Use of an internal wash oil stream is fairly common in the industry. Of 80+ wash oil streams analyzed over the last 25 yrs, more than 40% have been internal streams.

The North American ethylene plant in the paper was originally designed as a liquid cracker with NGL feed flexibility introduced within its first decade of operation. Driven by the feedstock competitive advantage of ethane from shale gas in the early 2010's, the plant was reconfigured to process gas only, primarily ethane. The conversion to ethane has significantly affected the quantity and quality of the stream used as wash oil. Due to reduced internal raw pygas feed to the hydrotreater since this conversion, the hydrotreater now processes outside streams as well as the indigenous streams. The importing of raw pygas streams combined with the changes in feed slate has further affected the quality of the wash oil that is obtained from the aromatics unit and has introduced unexpected components.

Recent analysis of this internal wash oil stream found high levels of dihydrodicyclopentadiene (DHDCPD). The presence of DHDCPD has been seen in the past in this wash oil, however, the levels have nearly doubled since this plant had the feed slate conversion, with levels greater than 45% in the wash oil. Figure 1 shows how the aromatics in the wash oil stream dropped as DHDCPD increased.

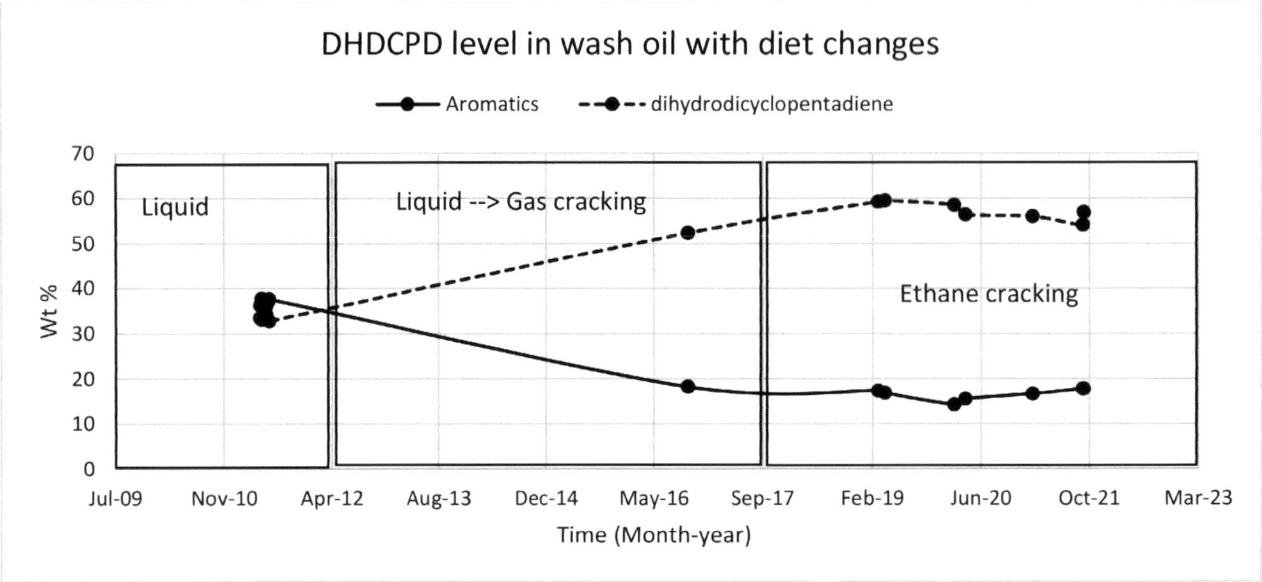

Figure 1: Aromatics and DHDCPD content of wash oil history

Levels this high were a concern being as DHDCPD is a non-aromatic compound and aromaticity is typically an important criteria for a good wash oil. For this reason, the plant considered diluting DHDCPD with a purchased high quality wash oil. There is some merit in mixing a purchased wash oil with an internal stream. The solvency, if low with the internal stream, can be increased and the boiling point range can be smoothed out when diluted with a purchased stream.

Wash Oil Circuit

A typical internal wash oil circulation loop is illustrated on the ethylene process flowsheet extracted from the Ethylene Producer Conference Feedstock Contaminants Database[1] in Figure 2. A C9+ stream from the aromatic hydrotreater (C5+ processing) is recycled and injected in the cracked gas compressor (CGC). Low volatility or high boiling point is one of the characteristics desired for wash oil. As such, wash oil exiting the compressor casings condenses in the inter-coolers and gets collected in the CGC inter-stage knock out drums from where it is recycled to the quench system. The mass of wash oil will mix with raw pyrolysis gasoline (RPG) extracted from the quench water tower settler in a gasoline stripper. The combined wash oil/ RPG stream typically mixes with Debutanizer bottom and is routed to a storage tank, which supplies feed to the aromatic hydrotreating unit.

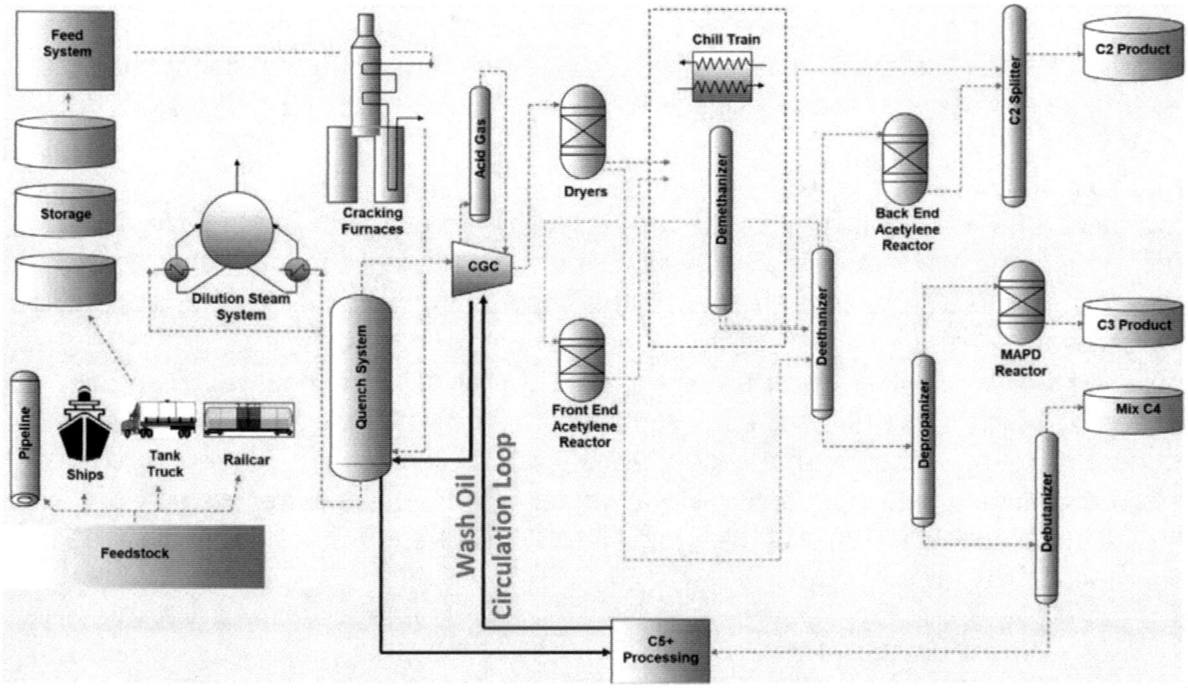

Figure 2: Typical Internal Wash Oil Circulation Loop

Impact of the Wash Oil Loop

A parallel can be made between the wash oil circulation loop illustrated in Figure 2 and Bernard and Fahiminia Debutanizer bottom recycle evaluation[2]. Both recycle loops positively influence the water/ gasoline density differential in the quench water settler. In the Bernard and Fahiminia theoretical evaluation, the hydrocarbon/ water ratio in the quench water settler was estimated at 0.0002 without recycle streams. This study suggested that the hydrocarbon/water ratio could be increased from 0.0002 to 0.0008 by recycling 35% of the Debutanizer bottom stream to the quench water settler. It should be noted that the study did not account for wash oil injection in the CGC. Incremental hydrocarbon/ water density differential was also reported with minimal increase beyond 10% of the Debutanizer bottom stream recycled. The wash oil rate assumed in the wash oil circulation loop illustrated on Figure 2 suggest a potential increase of the hydrocarbon/ water ratio to about 0.01 if the cracked gas compressor (CGC) knock out drums liquids were recycled to the quench water settler instead of a gasoline stripper. Such operation would help the density differential but at values slightly lower then with Debutanizer bottom recycle. The Debutanizer bottom is known to contain fouling precursors such as cyclopentadiene (CPD), dicyclopentadiene (DCPD), isoprene, styrene, indene and other dienes. From a high-level perspective and without rigorous stream analyses, it could be speculated that recycle streams from the aromatic hydrotreater (C5+ processing) would be relatively free of fouling precursors due to hydrogenation of unsaturated hydrocarbons. While the wash oil circulation loop is believed to be beneficial to hydrocarbon/ water separation in the quench water settler; its washing characteristics however deteriorates on ethane cracking primarily due to aromatics deficiencies.

Leveraging on the C5+ processing unit and the current wash oil circuit, significant improvement in wash oil quality could be achievable through minimum external wash oil injection rates. The external wash oil injection would supplement the internal stream aromatic deficiencies. This could be accomplished if the separation train of the C5+ processing unit allows recovery of the external wash oil components. From a boiling point perspective, external wash oil would likely have a boiling curve similar or slightly heavier than the internal C9+ stream, which is currently in use. Based on the typical front end C5+ processing unit illustrated on Figure 3, the internal and external wash oil components would be recovered as overhead product in the Degum tower. However, as described below, recovery of the external wash oil components is affected by blending with DHDCPD and loss of the heavier fraction to the light fuel oil stream.

Wash Oil, Cyclopentadiene (CPD) & Dicyclopentadiene (DCPD) Flow Path

Before further discussion on the potential recovery of external wash oil components, a review of the C5+ processing unit is deemed necessary for understanding the flow path of CPD and DCPD and their relationship with DHDCPD. As shown on Figure 3, the material feeding the C5+ processing unit is the combined stream from the Quench Water Tower RPG and the Debutanizer bottom. The Debutanizer bottom contains significant amounts of CPD and DCPD at compositions that can reach about 10 to 15% for each of those components. RPG from storage, feeds the C5+ processing unit where C5s are first separated in a Depentanizer. The Depentanizer distillate stream feeds the C5 reactor where CPD is converted to DCPD, which is a valued product that is recovered from the DCPD tower. The value of the DCPD contained in the Debutanizer bottom stream is lost, as the boiling point of DCPD causes the component to follow the Depentanizer bottom stream. Note that for maximizing DCPD recovery, there is incentive to minimize the CPD conversion to DCPD in the olefin unit. DCPD and other C6+ components from the bottom of the Depentanizer are then hydrogenated in the 1st stage reactor. The effluent from the 1st stage reactor supplies the BTX (Benzene, Toluene, Xylene) Tower, which separates the C6-C8 from the C9+. The C6-C8 stream undergo a 2nd hydrotreatment (not shown on Figure 3), while the C9+ stream feeds a Degum Tower (sometimes called a Rerun Tower or Denonanizer Tower) for separation of C9/10 components from light fuel oil. Part or the entire C9/10 stream can be use as internal wash oil. However, from a material balance standpoint, a purge in the form of a pygas product must be done to prevent accumulation of the C9/10 material produced from the cracking furnaces.

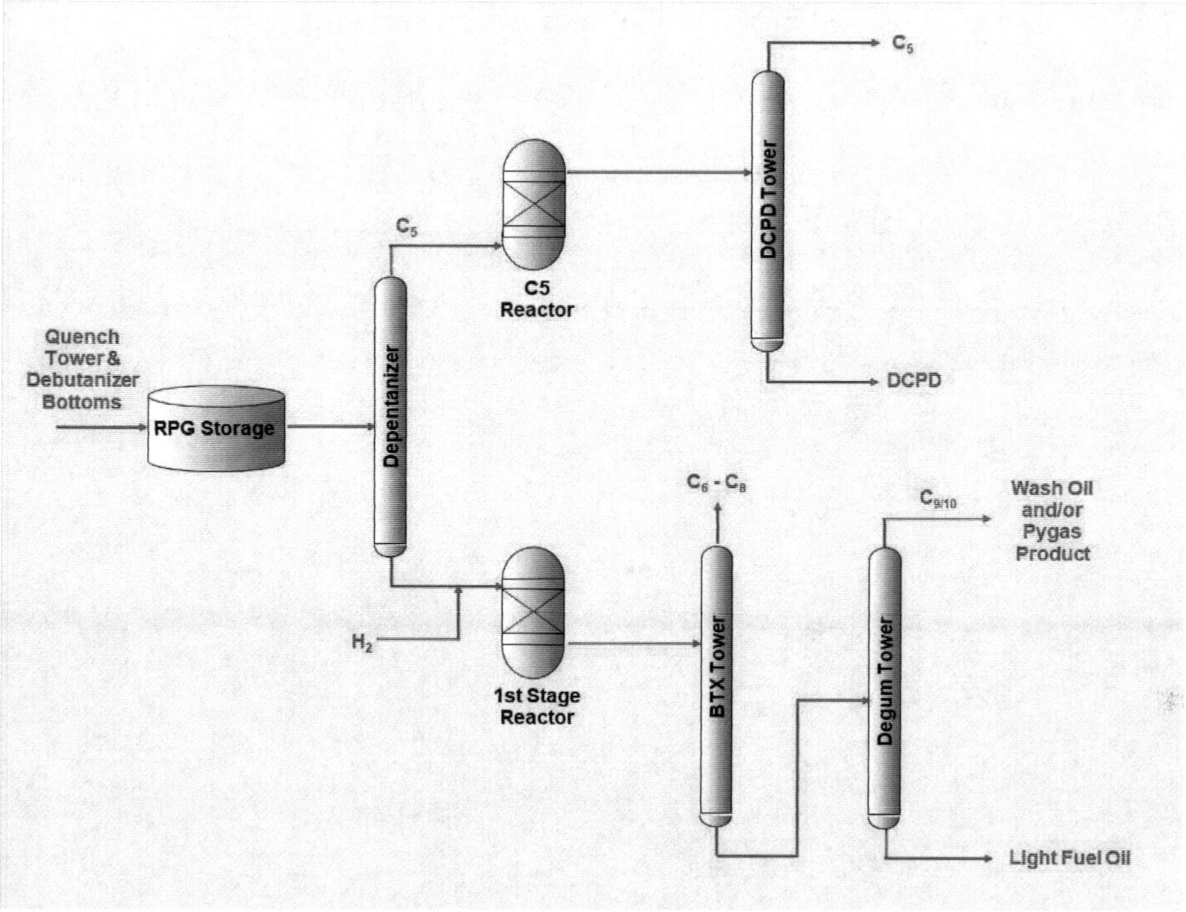

Figure 3: C5 Reactor and 1st Stage Hydrotreater of the C5+ Processing Unit

Simulation Modeling of the Wash Oil Circuit

It is not uncommon to model the wash oil injection stream to the CGC as a model boundary input stream with an assumed composition typically adjusted from laboratory analysis. For convergence and practicality reasons, the list of components included in a process simulation is quite reduced when compared to the large number of chemical compounds produced from cracking. While the light fraction of furnaces' effluent can be analyzed or model accurately, composition resolution becomes less accurate for the heavier fraction due to component lumping and the use of pseudo components. Table 1 below, shows component mapping of a wash oil typical analysis to a simulation component list covering the RPG fraction.

Table 1: Wash Oil Composition Mapping

Carbon Number	Component	Simulation Mapping	Weight Fraction (%)
C5	cyclopentane	1-pentene	0.360
	cyclopentene		
	cyclopentadiene	cyclopentadiene	0.345
C6	2-methylpentane	1-hexene	0.389
	3-methylpentane		
	2-eth-1-butene/n-C6		
	t-2-Me-1,3pentadiene		
	4-Me-1,3-pentadiene		
	methyl cyclopentane		
	other C6's		
	benzene	benzene	0.357
C7	2-methylhexane	1-heptene	0.006
	3-methylhexane		
	n-heptane		
	methyl cyclohexane		
	toluene	toluene	0.041
C8	ethyl benzene	ethyl benzene	0.024
	m,p-xylene	m-xylene	0.149
	o-xylene		
	styrene	styrene	0.350
	other C7 - C8	1-octene	0.247
C9	cumene	Pseudo – C9-325F	4.203
	propyl benzene		
	M/P/O-ethyltoluene		
	mesitylene	Pseudo- 325-350F	12.072
	trimethylbenzene		
	indane		
	indene		
	other C9+		
C10	dicyclopentadiene		
	dihydrodicyclopentadiene	Pseudo – 350-375F	48.9
	tetrahydrodicyclopentadiene	Pseudo – 375-400F	6.6
	naphthalene	Pseudo – 400-425F	2.2
C10-C16	C10-C16	Pseudo – 400-425F	24
Total:			100

The wash oil simulation mapping shown on Table 1 indicates that 98% of the stream is characterized with pseudo components, which provides little insight relative to the wash oil quality. A closer look to the laboratory analysis, however, reveals that the assumption or suspicion previously discussed of a stream containing saturated hydrocarbon downstream the 1st stage reactor is not valid. Almost 50% of the stream is made up of dihydrodicyclopentadiene (DHDCPD), which has one double bond in its structure. The fully hydrogenated analog, tetrahydrodicyclopentadiene (THDCPD), makes up only about 7%.

The DHDCPD molecular formula is $C_{10}H_{14}$, (134.22 molecular weight (MW)), with an approximate boiling point of 180 °C according to available literature. The component is unavailable in commercial process simulator databanks, thus initially, DHDCPD was modelled as a NBP 350-375 F pseudo-component, since the MW and published boiling point indicated this was appropriate.

DHDCPD was also modeled in separate simulations with database components similar in MW and boiling point:

- MDEBZ (m-diethylbenzene), $C_{10}H_{14}$, (134.22 MW), with an approximate boiling point of 182 °C
- DCPD (Dicyclopentadiene), $C_{10}H_{12}$, (132.20 MW), with an approximate boiling point of 170 °C

The results were similar for both compounds and reveals that DHDCPD would follow the flow path of wash oil illustrated on Figure 2.

Simulation Modeling of External Wash Oil Injection & Recovery

Injection of commercially available heavy aromatic wash oil was evaluated by process simulation. The external wash oil stream was characterized from assay published in literature and was heavier than the DHDCPD fraction of the internal wash oil stream described on Table 1. Whereas the DHDCPD portion is in the NBP 350-375 °F range, the heavy aromatic wash oil assay has a substantial portion heavier than NBP 375 °F. Thus, with the C5+ processing unit fractionation configuration, much of the heavy components of the wash oil would no longer be recovered in the C9/10 stream that is part of the wash oil circulation loop but be degraded as light fuel oil. The lighter fraction might cycle up within the wash oil circulation loop, but the DHDCPD levels would remain high.

Where does DHDCPD comes from?

At first glance, it is not clear if DHDCPD originates from the cracking furnaces or if it is formed in a downstream process unit. The high level of DHDCPD was found in the wash oil stream supplied to the CGC as describe in Table 1. The stream being part of semi-closed recycle loop makes it difficult to identify the source. However, pygas analyzes from other ethane crackers using external wash oil do not show any DHDCPD, suggesting that the high DHDCPD level observed are not originating from cracking but likely partially converted from DCPD in the C5+ processing unit 1st stage reactor and due to the recycle.

Cyclopentadiene (CPD) is formed in the cracking furnace. Due to its cyclic structure, CPD is far more likely to undergo a Diels-Alder reaction(3) to form dicyclopentadiene (DCPD) (Figure 4) than participate in free radical polymerization like its C5 linear analogs, isoprene and piperylene.

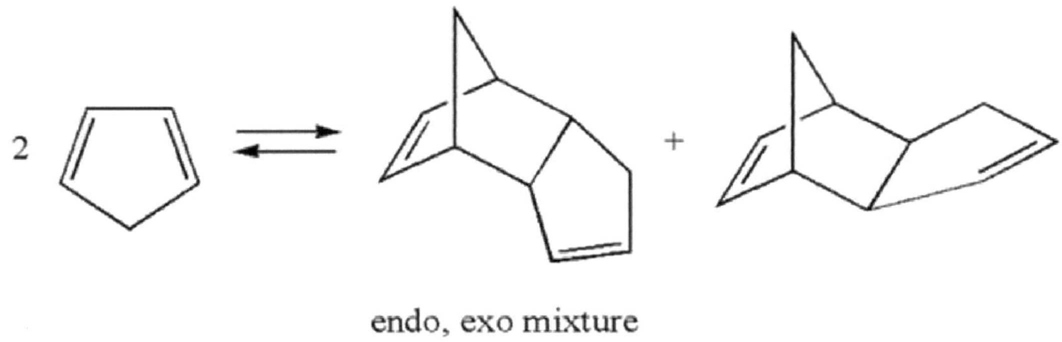

Figure 4: Cyclopentadiene and its Dimers, endo- and exo-Dicyclopentadiene

As the CPD concentrates up in the Debutanizer bottoms, the amount of DCPD will increase and can reach percentage levels as high as 14% or more in the Debutanizer bottoms.

The Debutanizer bottoms are combined with raw pygas streams and are either sold or sent to the Depentanizer associated with the ethylene plant. The bottom of the Depentanizer feeds the aromatics unit. In the aromatics unit of the plant studied in this paper, the DCPD that has formed in the hydrotreater feed is hydrogenated to dihydrodicylcopentadiene (DHDCPD) and tetrahydrodicyclopentadiene (THDCPD), Figures 5a and 5b. Because of the norbornene structure of DCPD, the double bond nearest to the bridged methylene carbon is more strained and will typically get hydrogenated while the other isolated double bond is less likely to be reduced[3] leading predominantly to DHDCPD and not THDCPD.

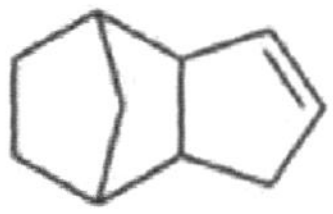

Figure 5a: DHDCPD

Figure 5b: THDCPD (endo structure shown)

DCPD, DHDCPD and THDCPD are all C10s and usually fractionate in the top of the Degum Tower overhead as illustrated in Figure 3. These C10s become part of the wash oil circulation loop that creates a cycle, allowing DHDCPD to build up to some steady state condition. As mentioned earlier, levels as high as 40-60% have been seen in this plant's wash oil stream. Levels this high have been seen in other ethylene plants as well.

CPD/ DCPD Levels in Ethylene Plants

Limited data exists on the typical concentrations of CPD/DCPD in raw pygas streams, but current available data suggests that gas crackers have a higher CPD/DCPD concentration than mixed and liquid crackers. Mixed/liquid crackers presumably see more dilution when heavier materials are cracked. Figure 6 shows the DCPD in a raw pygas stream from a gas cracker and Figure 7 shows the DCPD in the raw pygas stream of a mix cracker.

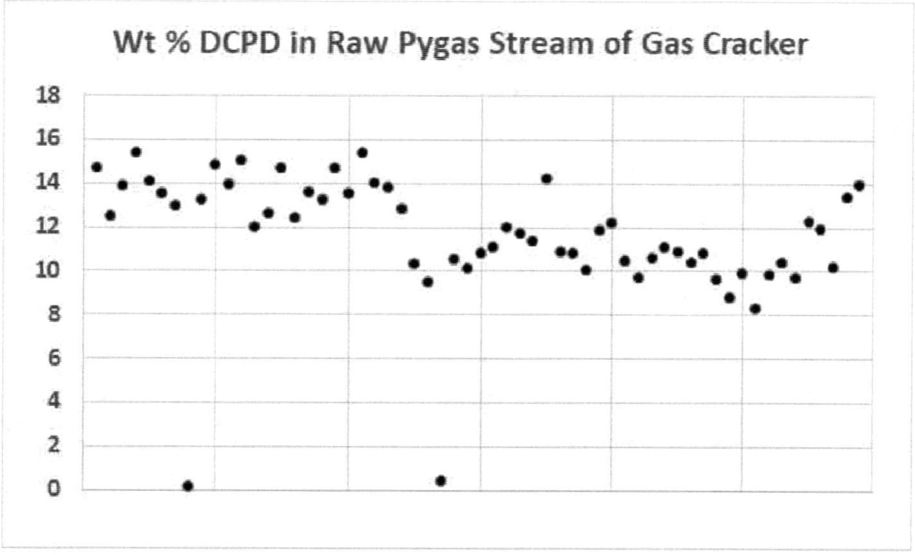

Figure 6: DCPD in a Raw Pygas Stream of a Gas Cracker

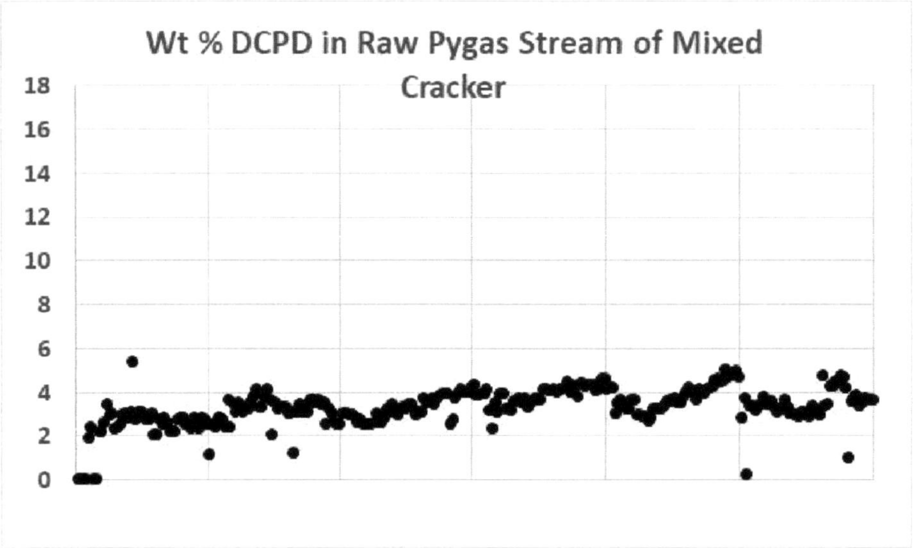

Figure 7: DCPD in a Raw Pygas Stream of a Mixed Cracker

Impacts of CPD, DCPD and DHDCPD on the Ethylene Plant

As mentioned earlier, CPD is a cyclic diene that readily undergoes a Diels-Alder reaction to DCPD. Data is reported that at 50 °C, 100% of CPD is completely dimerized to DCPD in one hr[3]. CPD and DCPD are known to form oligomers and are less likely to form large molecules due to limited solubility. According to the literature[3], oligomers up to about 5 CPDs are typical, **Figure 8**.

Yield of Cyclopentadiene Oligomers at Various Temperatures and Contact Times, wt%

Oligomer	150–160°C 14 h	170–180°C 22 h	200°C 90 h
tricyclopentadiene	40	50	25
tetracyclopentadiene	10	30	45
pentacyclopentadiene	2	5	10
unreacted DCPD	50	10	5

Figure 8: Oligomers of CPD vs time/temperature[3]

Throughout the ethylene plant, evidence of CPD/DCPD effects can be seen. There is no indication, however, that CPD causes issues in quench or process steam systems. However, in the Primary Fractionator of mixed and liquid crackers, DCPD has been measured at fairly high levels and thought to contribute to fouling in some cases.

In the compressor, CPD can form sticky oligomers although CPD concentrations are low, so fouling due to these oligomers is not seen as often in compressors as fouling due to Butadiene (BD) and Styrene-based polymers. However, as this paper mentions, the hydrogenation product of DCPD, DHDPCD, can be found in internal wash oil streams. This will be discussed in a later section of the paper.

In the caustic tower, there are a couple of reactions that could take place with CPD. CPD can undergo Diels Alder cyclization when aldol is dissolved in the hydrocarbon phase, **Figure 9.** Diels Alder reactions with other reactive monomers, such as isoprene, that have condensed in the caustic tower are also possible.

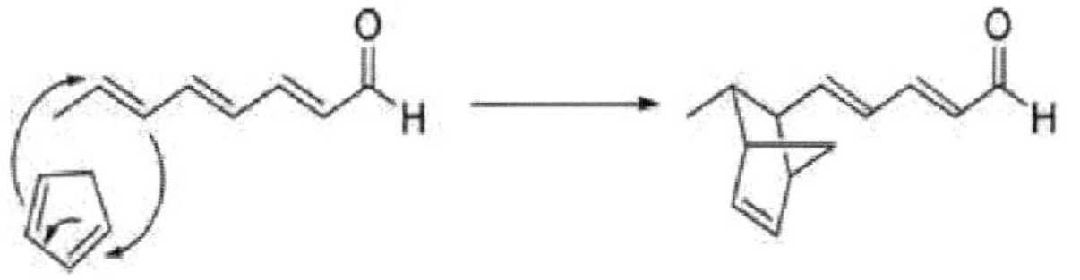

Figure 9: Diels Alder reaction of cyclopentadiene with Aldol polymer

CPD can be deprotanated by caustic forming a cyclic anion that readily forms a fulvene with aldehydes and ketones[3]. Figure 10 shows formation of a fulvene with a growing aldol polymer chain. This reaction can also take place with other aldehydes and ketones in the caustic tower such as acetaldehyde and acetone. The fulvenes are very reactive and can undergo nucleophilic attack at the methylene carbon.

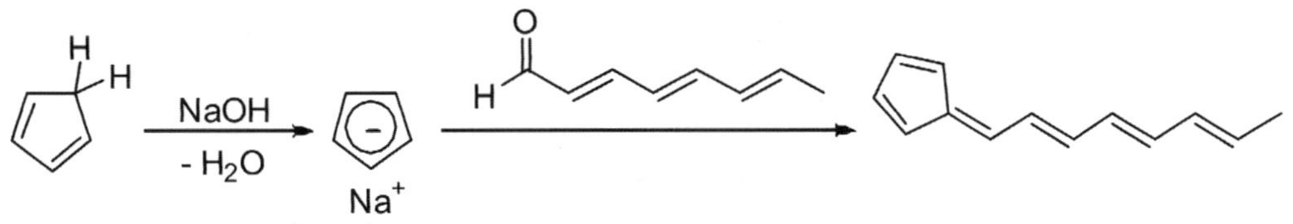

Figure 10: Fulvene formation with cyclopentadiene and Aldol polymer

CPD based polymers are seen occasionally in Depropanizers and more frequently in Debutanizers. The foulant is usually a combination of normal free radical based polymers of butadiene/isoprene/styrene and CPD/DCPD based oligomers. Figure 11 is a picture of a CPD based foulant from a Depropanizer.

Figure 11: Primarily CPD based foulant from a Depropanizer

In addition to the ethylene plant, CPD based fouling can be found in CPD dimerizers to DCPD. Here, the fouling is a mixture of Diels Alder and free radical polymerizations with other C5 reactive dienes.

Effect of DHDCPD on Internal Wash Oils

As discussed earlier, DCPD can get hydrogenated in the 1st stage hydrotreater to dihydrodicylopentadiene, DHDCPD. This C10 molecule will exit the aromatics unit in the C9/C10 overhead stream as shown in Figure 3. This stream is often used as an internal wash oil. A good purchased wash oil typically has a high aromatic content[4, 5] (preferably >80%) and a high initial boiling point (IBP). Consideration to specific gravity differential with water is also taken into account from a separation standpoint. A C9/10 stream typically has a lower than desired IBP but still may have a high aromatics content. However, when DHDCPD levels increase, aromaticity decreases, suggesting it may function poorly as a wash oil (low IBP, low aromatics).

One measurement that is done in the industry to determine the ability of wash oil to dissolve polymers in the compressor is to measure its Kb (Kauri butanol) solubility constant. This test determined the wash oil's solvency. The greater the value, the more powerful the wash oil's solvency. The value is standardized against toluene which has a value of 105 and a mixture of 75% n-heptane and 25% toluene by volume, which has a value of 40. The closer to 100, the better the wash oil. This is a standard ASTM D1133 method although some in the industry use a modified version of this test. Kb measurements for the wash oil stream in this paper are surprisingly decent at (~87). Therefore, while still a light stream, it is not a bad wash oil.

Conclusion

Due to the feedstock changes from liquid to gas, mainly ethane, for ethylene plants over the last decades, the internal wash oil quality has been changing. The pathway of CPD and DCPD has been tracked and it was found that DHDCPD is mainly generated from 1st stage reactor by hydrogenation of DCPD in the C5 + processing unit and recycled back to the cracked gas compressor area as the internal wash oil. Even though it was a concern as the concentration of aromatics decreased with the diet changes, it seems that the solvency of overall internal wash oil was reasonable for the fouling propensity perspective. However, the addition of a good quality, purchased wash oil will provide the benefit for the overall compressor performance.

References

1. Bernard, A., Snider, S., Feedstock Contaminant Database and Interactive Tool, 2008 AIChE Spring Meeting.
2. Bernard, A., Fahiminia, F., Debutanizer Bottoms Recycle for Enhancing Gasoline/ Water Separation in the Quench System, 2019 AIChE Spring Meeting.
3. Cheung, P. T.T., Cyclopentadiene and Dicyclopentadiene, Kirk-Othmer Encyclopedia of Chemical Technology, Vol 8, 2000, pgs 219 – 235.
4. Snider, S., Ethylene Plant Cracker Gas Compressor Fouling, 2006 AICHE Spring Meeting.
5. Dutschmann, D., Megan Harrison, M., Yang, Y., Compressor Wash Oil in a Light Feedslate Environment, 2016 AICHE Spring Meeting.

AIChE Paper Number: 639039
Presented at T4000 Ethylene Plant Technology Fundamentals Session

New Process Technology Development for the Sustainable Production of Olefins

Daniela Ferrari

Associate Director, Packaging & Specialty Plastics and Hydrocarbons R&D

Dow

Georgios Bellos

Sr. Research Scientist, Packaging & Specialty Plastics and Hydrocarbons R&D

Dow

Prepared for Presentation at the 2022 Spring National Meeting

San Antonio, Texas, April 12, 2022

AIChE Paper Number:

New Process Technology Development for the Sustainable Production of Olefins

Daniela Ferrari

Associate Director, Packaging & Specialty Plastics and Hydrocarbons R&D

Dow

Georgios Bellos

Sr. Research Scientist, Packaging & Specialty Plastics and Hydrocarbons R&D

Dow

Abstract:

Industrial decarbonization is one of the major challenges of the XXI century to achieving a sustainable future. As both a recognized leader in the production of olefins and a major user of energy, Dow has a key responsibility. By 2030, Dow has committed to reduce its net annual carbon emissions by 5 million metric tons vs its 2020 baseline (15% reduction). By 2050, Dow intends to be carbon neutral (scopes 1-3 plus product benefits). Dow is conducting research to develop new process technology as well as its integration with existing technology to drive the production of olefins to net-zero emissions.

The presentation will provide an overview of the key projects to produce ethylene and propylene with advantages of reduced capital, improved reliability, simplicity, and versatility along with decreased emissions. Examples of these technologies are Fluidized Catalytic Dehydrogenation (FCDh) of propane and cracking electrification.

Introduction

Industrial decarbonization is one of the major challenges of the XXI century to achieving a sustainable future. The challenge is multifaceted as it poses technical feasibility hurdles along with infrastructure availability and significant capital investments. As one of the leaders in the production of olefins and a major user of energy, Dow is investing in research and implementation of new process technology to produce olefins as well as its integration with existing technology to drive to net-zero emissions. This new olefin production technology will require a significant step change compared to status quo to make an impact in the current footprint. Substantial reduction of CO_2 emissions in an ethylene plant can be achieved in two different ways: by means of replacing heat and energy sources with electricity or by replacing fuel and natural gas with hydrogen. Process electrification as well as hydrogen technology have different advantages and challenges in terms of technology readiness level, process safety and net energy reduction. Similarly, the carbon footprint connected to the propylene production can be significantly reduced by using a completely different approach that integrates catalyst and process design more efficiently.

Dow's proprietary Fluidized Catalytic Dehydrogenation (FCDh) Technology

The shale gas revolution at the beginning of the century led to a dramatic change in the energy, refining and petrochemical sectors and enabled petrochemical companies to produce ethylene from cheap ethane. This created an imbalance in the supply and demand for propylene giving a new impulse to the propylene on purpose industry.

Several fundamental constrains characterize the propane dehydrogenation reaction to propylene. The first is that the reactions are endothermic, hence heat must be added to the process. Secondly, the chemical equilibrium of the dehydrogenation reaction favors olefins at low pressure and high temperature. The third constrain is that the high temperature requirement results in the formation of coke on the catalyst, which depresses catalyst activity. Fourth, the thermal or gas phase reaction has a much lower selectivity to the desired product than the catalytic reaction, requiring the gas residence time at temperature to be minimized.

Dow's proprietary fluidized catalytic dehydrogenation (FCDh) process UNIFINITY™ includes a circulating fluidized bed system (**Error! Reference source not found.**) with an alumina-supported catalyst (1). Propane is introduced into a turbulent or fast-fluidized reactor, followed by a plug-flow riser reactor that transports the catalyst to close-coupled cyclones for rapid product and catalyst separation. The catalyst is then stripped with a stripping agent to recover any product entrained with the catalyst. The spent catalyst is transported to a combustor, where it is reheated in a fluidized bed, using external heating fuel. The catalyst and flue gas then flow through a riser and a termination device. The remaining catalyst is separated from the flue gas in cyclones and returned to a bubbling bed. The catalyst is then regenerated with air to recover the activity in an oxygen soak zone. The proprietary catalyst FLUINITY™ is non-chromium based and contains trace amounts of platinum supported on alumina and it ensures high propane conversion and propylene selectivity under the FCDh process conditions.

The catalyst has optimum fluidization kinetics, high attrition resistance and provides effective combustion of fuel as well as coke in the regenerator.

With this reactor design high conversions at moderate pressures can be achieved, while the robust catalyst formulation ensures performance also in the presence of feed impurities, such as sulfur. High selectivity to propylene is achieved through the low gas residence time at temperature requirements, resulting in lower propane unit ratio than current processes. Furthermore, for the same reason, low levels of coke on the catalyst are obtained during the reaction step.

The proprietary regenerator process in combination with the FLUINITY™ catalyst allow for efficient combustion of fuel, which minimizes any NOx formation relative to competing alternatives.

In addition, by performing the reaction at moderate pressure without H_2 recycle, the process is expected to have 20% lower energy consumption leading to reduced emissions.

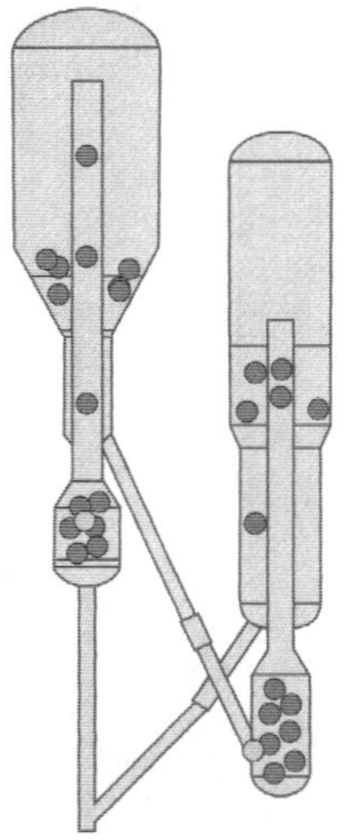

Figure 1 Proprietary catalytic dehydrogenation reactor-regenerator system. The reactor is on the left and the regenerator is on the right.

Cracking Electrification

Ethylene furnaces generate large amounts of high-pressure steam which is used to deliver shaft work to rotating equipment such as cracked gas compressors and ethylene, propylene refrigerators. A part of the steam is released at medium pressure level and used as dilution steam together with make up steam. The ethylene plant is usually integrated with a power plant which consumes natural gas to generate power and steam (*Figure 2*). The overall CO_2 emissions include direct and heat related emissions plus emissions related to power generation.

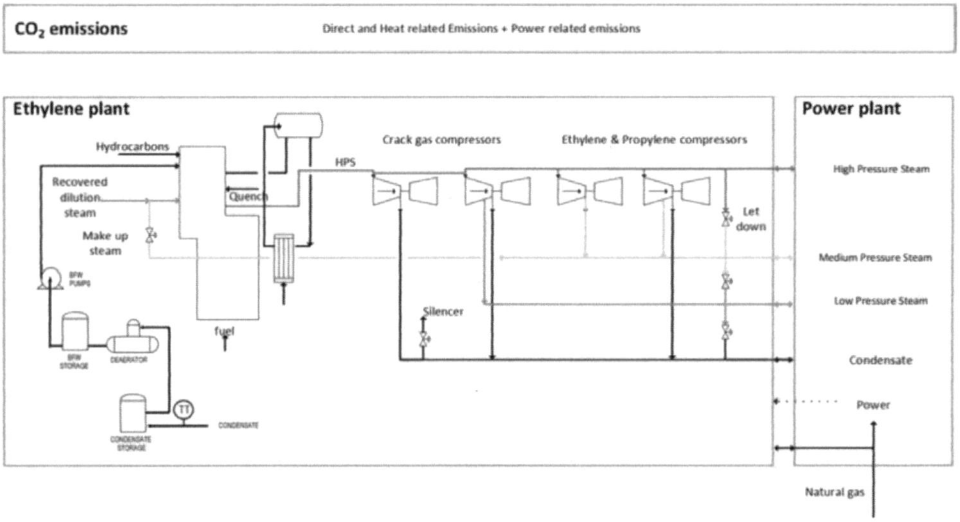

Figure 2 A schematic overview of the integration of an ethylene plant with a power plant (original drawing courtesy of S. Luijten, Dow)

Electrification of the cracking process can reduce energy consumption, but it requires the replacement of the current radiant combustion furnace technology with electrically driven reactors, referred to as electric furnaces (Figure 3). Additionally, the full electrification of the process (Figure 4) requires the replacement of steam turbines with electrical motors (2). This enables the optimization of the advantages of electrification as the efficiency barriers of steam generation and steam turbines are removed.

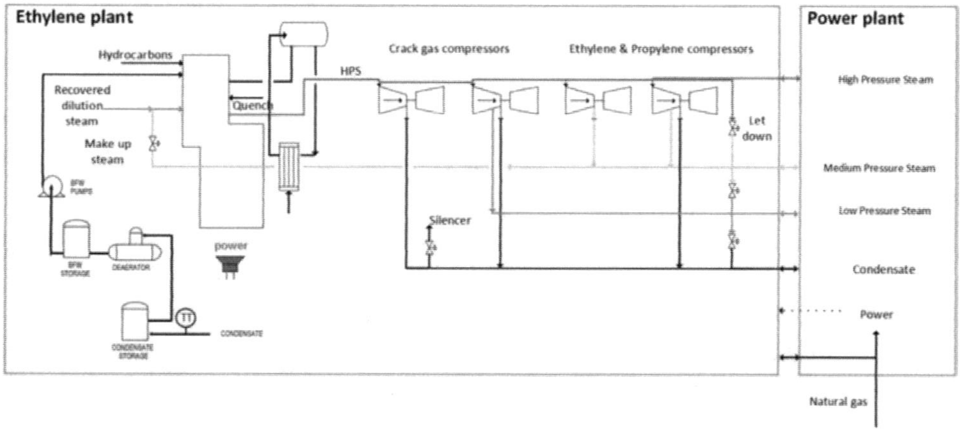

Figure 3 Ethylene plant with electrical furnace

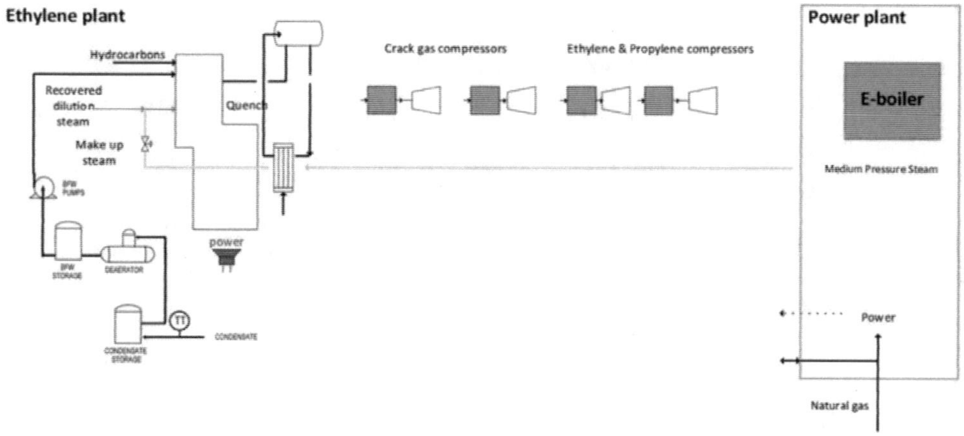

Figure 4 Electrical cracker flowsheet

Process electrification can lead to substantial changes of energy intensity as well as CO2 emissions. The following scenarios were analyzed (Figure 5):

A = Business as usual – this is the conventional best in class ethylene cracker

B = Electrified furnaces – this concept replaces the conventional radiant combustion furnaces with e-furnaces and consumes electricity from conventional power grid (400 kg CO2 / kWh)

C = B+CO2 neutral power –identical with the technology concept B but consumes CO2 neutral power (kg-CO2/kWh = 0)

D = Electrify the entire plant – this concept is based on scenario B but also replaces steam turbines with electrical motors and consumes electricity from conventional power grid

E = D+CO2 neutral power – this concept is based on scenario D but consumes CO2 neutral electricity. This is the ultimate concept

It is noted that CO2 neutral power is generated by wind turbines and solar panels. Electrical efficiency of solar and wind energy is not included in the energy balance of the current article. Figure 5 shows the estimated fuel gas production for a conventional naphtha cracker and its corresponding energy consumption (scenario A). It automatically hints how well a naphtha steam cracker is balanced from energy perspective. Scenario B indicates that electrifying furnaces and maintaining steam consumption at the same level is not going to change the energy intensity. The use of conventional power grid is going to increase energy as the efficiency of power plants is in the range of 40 to 60%. The ultimate scenario E reveals the potential for the deepest energy reduction which can reach the level of 40%. Figure 6 illustrates the impact of each scenario on CO2 emissions. It is clearly shown that the use of CO2 neutral power is a prerequisite to mark a significant reduction of emissions. The presented negative emissions are attributed to a minor export of H2 which is accredited as can replace natural gas.

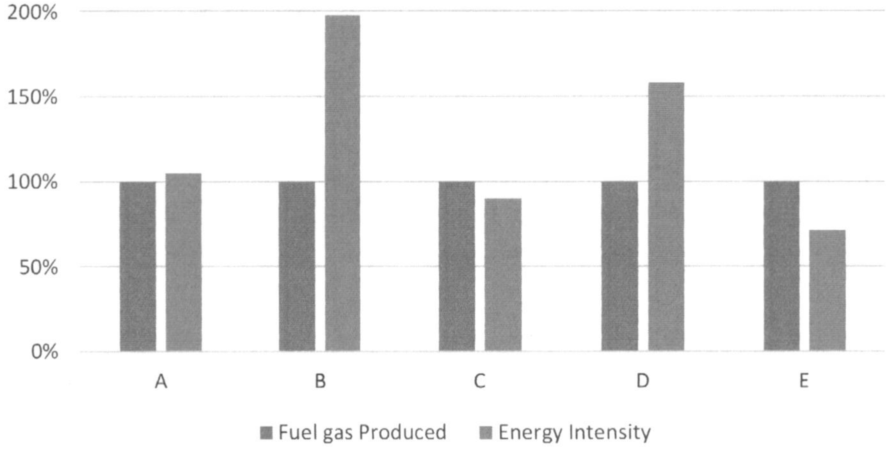

Figure 5 Energy balance for electrified versus conventional ethylene plants (3)

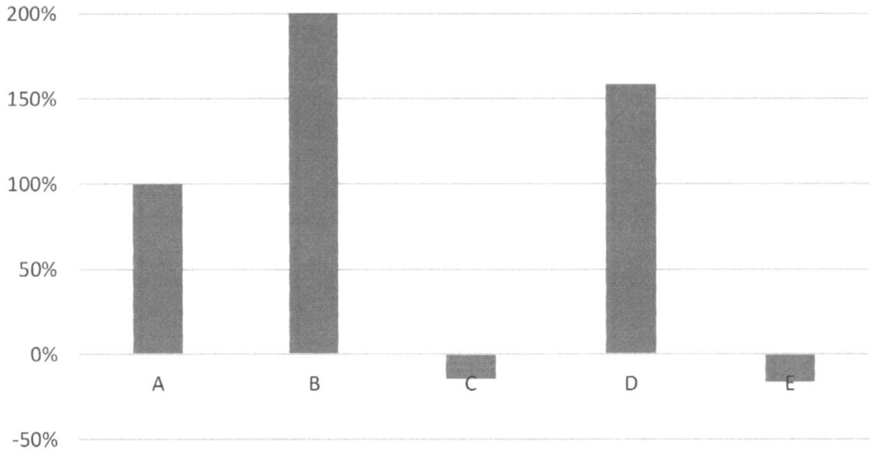

Figure 6 CO2 emissions for electrified versus conventional ethylene plants

The infrastructure requirements to access CO2 neutral power are not to be underestimated. To electrify an ethylene plant of 2000 kTA, 1300 MW are required for the furnaces electrification while full electrification increases the power to 1600 MW.

The next generation wind turbines can generate 15 MW (4) of power (see Figure 7). A conservative farm power density is 3.6 MW/km^2 so that each turbine is not stealing the wind of other's. The power consumption range of 1300-1600 MW implies a wind farm area of 360 to 450 km^2 is required for a 2000 kTA cracker.

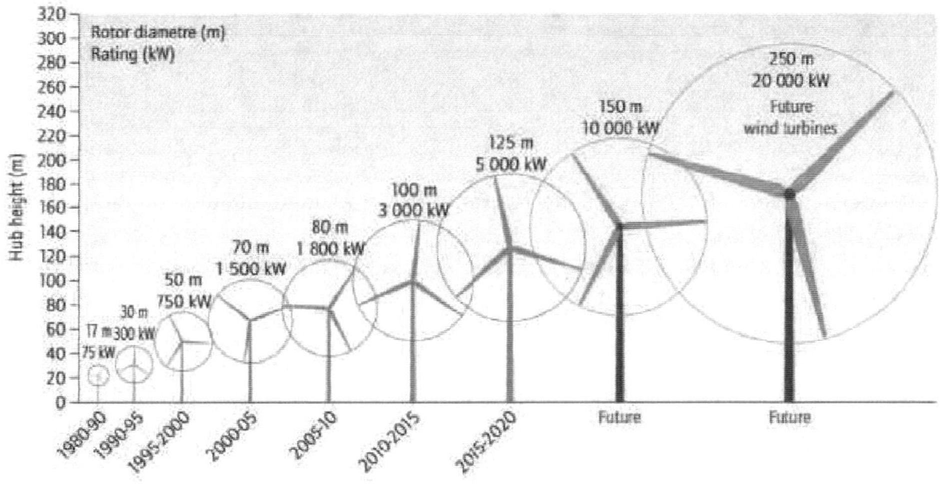

Figure 7 Scaling up journey for wind turbines (5)

The North Sea countries have planned new offshore wind farms (OWF) capacity of 55 GW up to 2030. A study (6) looked plot space which is not used for other purposes and has water depth less than 55 m so that monopiles can be applied. The new plot space can generate 110 GW. This is illustrated with orange polygons in Figure 8 (6). The map includes a scale of 150 km. Therefore it is possible to draw a line and plot a space of 19 x 19 km. In this map, we drew a green/red square to show an area of 360 km^2.

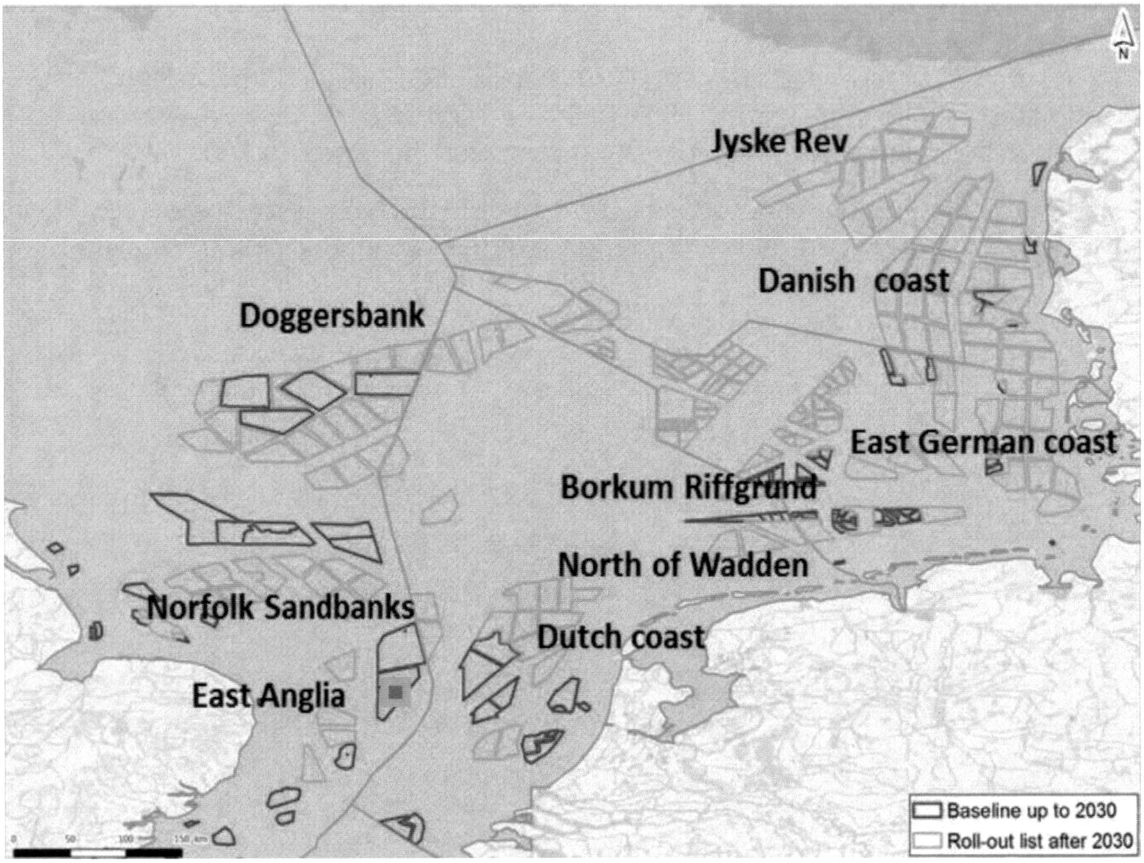

Figure 8 Offshore windmill farms: potential new locations with orange polygons and planned projects by 2030 with red polygons (5). The green/red square shows an area of 360 km² which corresponds to world scale steam crackers power requirements

Conclusion

Industrial decarbonization is one of the major challenges of the XXI century and will require a multiprong solution. Dow is actively working to develop new process technology along with its integration with existing technology to produce olefins at net-zero emissions. This includes optimizing current manufacturing processes, next-generation technologies for emission-free manufacturing, process electrification, and hydrogen production technology coupled with carbon capture, usage, and storage (CCUS) (7, 8).

References

1. M. Pretz, B. Fish, L. Luo, B. Stears, Hydrocarbon Processing, April 2017, pp. 29-36
2. G. Bellos, P. Oud. Low Emission Furnace, IMRPOOF consortium: Dow Open Day, Terneuzen, The Netherlands

3. G. Bellos. Electrification and Thermal Issues for Achieving Zero Net Emissions, Cibo webinar 2020
4. Technology Roadmap – Wind Energy 2013. International Energy Agency 2013.
5. International Energy Agency report, 2013
6. Cost evaluation of North Sea Offshore Wind Post 2030. February 2019, ECN, TNO, Witteveen Bos
7. Dow Benelux outlines roadmap to support Dutch Climate Agreement, achieve CO2 emissions neutrality | Dow Investor Relations, June 2021
8. Dow announces plan to build world's first net-zero carbon emissions ethylene and derivatives complex, October 2021

AIChE Paper Number: 638968

From Steel to Ethylene Super-Dry Reforming of CO_2

Michiel W.F. Van Cauwelaert
Doctoral Researcher
Laboratory for Chemical Technology, Ghent University

Evangelos Delikonstantis
Post-Doctoral Researcher
Laboratory for Chemical Technology, Ghent University

Lukas C. Buelens
Post-Doctoral Researcher
Laboratory for Chemical Technology, Ghent University

Hilde Poelman
Doctoral Assistant
Laboratory for Chemical Technology, Ghent University

Kevin M. Van Geem
Full Professor
Laboratory for Chemical Technology, Ghent University

Vladimir V. Galvita
Associate Professor
Laboratory for Chemical Technology, Ghent University

Prepared for Presentation at the 2022 Spring National Meeting
San Antonio, Texas, April, 12, 2022

From Steel to Ethylene
Super-Dry Reforming of CO_2

Michiel W.F. Van Cauwelaert
Doctoral Researcher
Laboratory for Chemical Technology, Ghent University

Evangelos Delikonstantis
Post-Doctoral Researcher
Laboratory for Chemical Technology, Ghent University

Lukas C. Buelens
Post-Doctoral Researcher
Laboratory for Chemical Technology, Ghent University

Hilde Poelman
Doctoral Assistant
Laboratory for Chemical Technology, Ghent University

Kevin M. Van Geem
Full Professor
Laboratory for Chemical Technology, Ghent University

Vladimir V. Galvita
Associate Professor
Laboratory for Chemical Technology, Ghent University

Abstract:

Closing the carbon cycle in energy-intensive industries is a major objective for the EU to enable a climate-neutral, circular economy by 2050, as exemplified by the Green Deal. A crucial step to address greenhouse gas (GHG) emissions in steel manufacturing is to close the carbon cycle by using the emitted CO_2 to create a value chain and produce carbon-neutral fuels and chemicals. The here discussed value chain firstly converts CO_2 into CO via the innovative super-dry reforming (SDR) technology, with addition of novel elements. Secondly, gas fermentation of CO to ethanol can be performed. Finally, the core technologies of SDR and gas fermentation can be coupled to commercial value chain technologies. The combination of SDR with gas fermentation for ethanol production will be able to achieve a carbon footprint below 20 g CO_2 eq./MJ ethanol (compared to 34 g CO_2 eq./MJ with the current technologies) and hence be a true game-changer in achieving ethanol production with acceptable climate impact. Innovations to the SDR technology will further reduce this impact. Producing isopropanol rather than ethanol, and extending the value chain to produce, in a flexible way, $C_2 - C_3$ olefins such as ethylene and propylene, with current market sizes above 120 Mtonne/y and 300 B€/y, will further consolidate the economics, and expand market opportunities.

Introduction

Nowadays, greenhouse gas (GHG) emissions can be largely divided into four categories, each responsible for around one-fourth of the total emissions, i.e., housing and agriculture, electricity production, transportation and industry [1]. Hence, a large part of the global CO_2 emissions is coming from energy-intensive industries (EIIs) such as iron and steel, refineries, chemical industry and cement industry, where fossil resources are used as a fuel. Moreover, various processes also have inherent CO_2 production, such as e.g. the cement industry where additional CO_2 is emitted in the calcination process for the production of cement itself [2]. Although EIIs have decreased their GHG emissions by almost 40 % between 1990 and 2017, a further decrease is necessary to meet EU targets and EIIs commitments [3]. This is only possible if different actors along the value chain of the EIIs combine forces, resulting in a fadeaway of the traditional borders between these sectors.

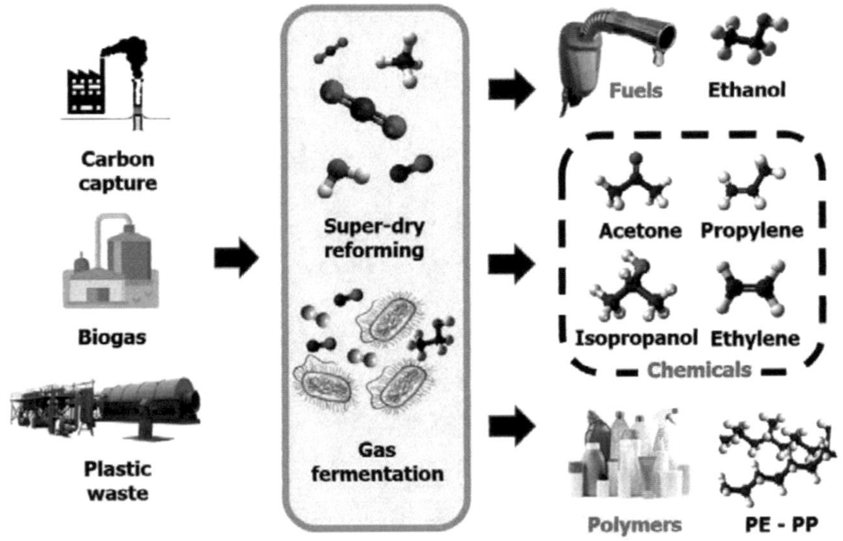

Figure 1. *Value chain for the production of carbon-neutral fuels and chemicals.*

A crucial step to address GHG emissions in steel manufacturing is to close the carbon cycle by using the emitted CO_2 to create a value chain and produce carbon-neutral fuels and chemicals [4]. This approach will amplify the reduction in CO_2 footprint compared to capture/storage-only solutions. In fact, together with a reduction of direct GHG emissions, the substitution of fossil fuels and promotion of direct use of renewable energy results in an enhanced CO_2 equivalent reduction. Moreover, this concept will foster symbiosis and circularity with other industrial sectors (using CO_2 from other EIIs). To achieve a significant GHG and CO_2 equivalent reduction, the following value chain is proposed. Firstly, CO_2 is converted to CO via the super-dry reforming (SDR) technology, to which innovative elements are added, i.e., alternative carbon sources (e.g. non-recyclable plastic waste or biomethane). Secondly, CO is fermented to ethanol. Finally, the core technologies of SDR and gas fermentation are coupled to existing commercial production technologies to complete the value chain, as illustrated in Figure 1.

Super-Dry Reforming

The first step in the value chain for the production of carbon-neutral fuels and chemicals is the conversion of CO_2 to CO. Therefore, CO_2 should first be captured and the corresponding gas stream should be cleaned, as it may contain contaminants such as sulfur or light hydrocarbons [5, 6]. These impurities might otherwise lead to deactivation of the catalysts which are used in subsequent processes. Recently, an intensified CO_2 utilization process, named super-dry reforming (SDR), was developed by Buelens et al. [7]. SDR combines three major processes, i.e., dry reforming of methane, chemical looping redox reactions and calcium looping for CO_2 capture. The global reaction is given in Eq. (1).

$$CH_{4\,(g)} + 3CO_{2\,(g)} \rightarrow 2H_2O_{(g)} + 4CO_{(g)} \qquad \Delta H^{\circ}_{298K} = 110\ \text{kJ mol}^{-1}_{CO_2} \tag{1}$$

$$CH_{4\,(g)} + CO_{2\,(g)} \rightarrow 2H_{2\,(g)} + 2CO_{(g)} \qquad \Delta H^{\circ}_{298K} = 247\ \text{kJ mol}^{-1}_{CO_2} \tag{2}$$

A schematic representation of the process is shown in Figure 2. First, a conventional dry reforming step is performed, Eq. (2). A Ni-based catalyst is used to convert the CH_4 and CO_2 feed into syngas. The latter reduces the iron-based oxygen storage material in the reducer step of the redox chemical looping, according to Eqs. (3) - (4). The iron reduction is enhanced by capturing the formed CO_2 with a calcium oxide sorbent in Eq. (5). The formation of an equilibrium mixture of carbon dioxide, hydrogen, carbon monoxide and water is hence avoided. In conventional dry reforming, this equilibrium mixture limits the conversion through the water-gas shift (WGS) reaction, Eq. (6).

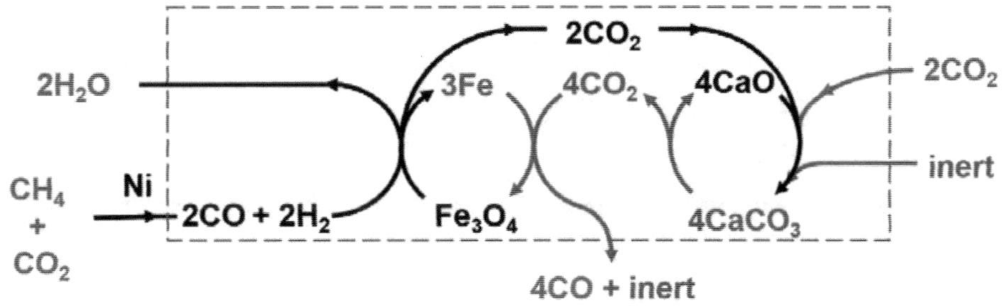

Figure 2. *Schematic representation of the super-dry reforming process. The CO_2 feed on the right represents the carbon dioxide coming from the initial 3:1 CO_2/CH_4 feed which is not converted in the dry reforming step.*

$$Fe_3O_4 + 4CO \leftrightharpoons 3Fe + 4CO_2 \quad (3)$$

$$Fe_3O_4 + 4H_2 \leftrightharpoons 3Fe + 4H_2O \quad (4)$$

$$CaO + CO_2 \leftrightharpoons CaCO_3 \quad (5)$$

$$CO + H_2O \leftrightharpoons CO_2 + H_2 \quad (6)$$

In the oxidizing step, an inert gas is fed. Calcination of $CaCO_3$, as shown in Eq. (7), releases CO_2, which oxidizes iron, producing a high-purity CO stream mixed with the inert, according to Eq. (8).

$$\text{inert} + CaCO_3 \leftrightharpoons CaO + CO_2 + \text{inert} \quad (7)$$

$$\text{inert} + 3Fe + 4CO_2 \leftrightharpoons Fe_3O_4 + 4CO + \text{inert} \quad (8)$$

Compared to conventional dry-reforming no expensive noble metal catalyst is needed to mitigate cokes formation [8]. A Ni-based catalyst can be used, since the operating conditions (1023 K, 1 atm, 3:1 CO_2/CH_4 ratio) are thermodynamically unfavorable for carbon deposition, which results in less deactivation of the used catalyst. Moreover, the heat input required per mole of CO_2 is 55 % lower (110 kJ mol^{-1} compared to 247 kJ mol^{-1}), which makes this process also interesting from an energetic point of view. Per mole of methane three times as much CO_2 can be converted, at the cost of hydrogen which is converted to water in the reducing step. However, when combined with green hydrogen, SDR can be a renewable, versatile and flexible source of syngas.

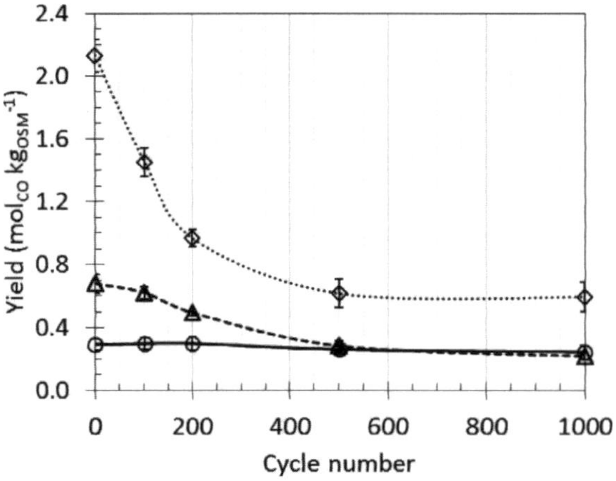

Figure 3. *Specific CO yield (mol$_{CO}$ kg$_{OSM}^{-1}$) during the oxidation step for 1000 redox cycles (p = 1.013 bar, T = 1023 K, F_{tot} 2.4 10^{-4} mol s^{-1}). 10 wt% Fe_2O_3 – $MgAl_2O_4$; 30 wt% Fe_2O_3 – $MgAl_2O_4$; 50 wt% Fe_2O_3 – $MgAl_2O_4$ [9].*

A key challenge is the stability of the oxygen carriers and CO_2 sorbents used in the SDR process, since this determines the maximum allowable operation time. Limited deactivation was observed throughout 1000 H_2/CO_2 redox cycles for an Fe_2O_3 – $MgAl_2O_4$ oxygen storage material (OSM), as shown in Figure 3 [9, 10]. The main deactivation mechanisms are sintering and phase transformation to a $Mg_xFe_{1-x}O$ phase, which causes an oxygen storage capacity loss. The addition of structural promotors will lead to enhanced redox activity as they act as an inert physical barrier against sintering or as a chemically active agent to enhance the reaction rate of iron oxide reduction. After addition of a ZrO_2 promotor for example, almost no deactivation was observed over 100 H_2/CO_2 cycles for a series of hollow structured Fe_2O_3-ZrO_2 oxygen carriers [11]. The increased resistance against sintering is mainly derived from the high melting point and chemical stability of ZrO_2. As Fe_3O_4 and CaO (which is the most common CO_2 sorbent in SDR) are used as raw materials for the iron and steel industry, they can be recycled to the steel production process, once they have reached their end-of-life phase [12].

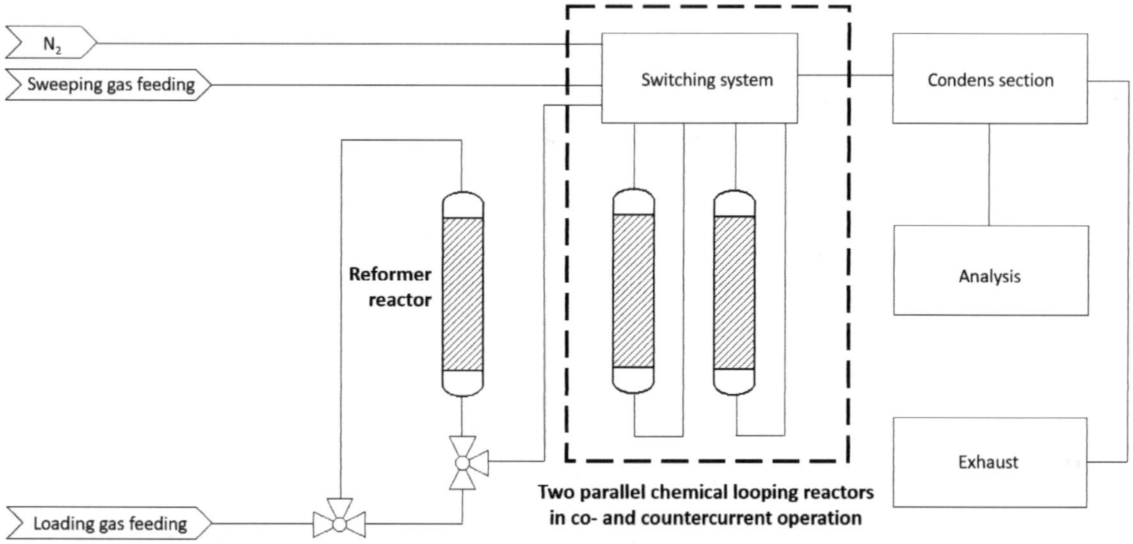

Figure 4. *Illustration of the pilot unit with two dynamically operated fixed-bed reactors.*

A second key challenge is the deactivation of the catalyst by common impurities in CO_2 waste streams. A Ni catalyst supported on $MgAl_2O_4$ showing increased catalytic activity, efficient control of carbon deposition and superior regeneration ability was synthesized by Wang et al. [13]. Compared to an Al_2O_3 support, no inactive $NiAl_2O_4$ phase was formed. The higher activity can be ascribed to the higher support basicity and the stronger interaction between Ni and the support. Resistance of the Ni catalyst against sulfur impurities can be increased by the addition of Rh [14], while the addition of Fe provides redox functionality and reduces carbon deposition [15, 16]. However, Fe – Ni catalysts suffer from deactivation via Fe segregation. This can be prevented by means of Pd addition which forms a thin Fe – Ni – Pd surface layer, acting as a barrier against segregation from the core [17]. Next to sulfur, carbon deposition is a major issue in reforming processes. Catalyst regeneration can however extend the operation time between two catalyst replacements. Depending on the redox functionality of the support, carbon removal either consists of two consecutive contributions, for supports without redox properties such as Al_2O_3 or $MgAl_2O_4$, or two parallel steps, for supports with redox properties such as $MgFe_{0.09}Al_{1.91}O_4$ and $CeZrO_2$ [18]. In the first mechanism, Ni is oxidized to NiO, after which the carbon species located close to the metal are oxidized by the lattice oxygen of NiO. Subsequently, carbon species located further away are oxidized by the lattice oxygen of NiO particles which have become mobile because of the exothermicity of the previous reactions. In the parallel mechanism, the first step remains the same. Carbon deposits located further away are however oxidized through lattice oxygen provided by the support [19].

Currently, Ghent University is building a super-dry reforming unit at pilot level, with a throughput of 1 kg CO_2/h and a materials inventory of approximately 10 kg, illustrated in Figure 4. This unit has two reactors in parallel, which are periodically switched between the reducer and the oxidizer step. This allows to have continuous production of CO and inherent separation of H_2O from CO and CO_2. With more reactors in parallel, a permanent periodic regime, producing a steady stream of CO can be established.

Adding innovative elements to the SDR technology

By combining the exothermic carbonation step in Eq. (5), with the endothermic methane reforming and oxygen storage material reduction, large energy reductions are possible. However, the overall process still remains endothermic. An alternative might be the concept of autothermal chemical looping, which was proposed by Hu et al. [20]. In this process, oxygen is co-fed in the reactor, resulting in exothermic combustion of part of the methane feed. Since this results in a lower amount of CO_2 which can be converted per mole of methane, electrification is a more promising alternative for process heating. Electric heating can be based on resistive (Joule) or inductive heating, both of which are heavily invested in by the chemical process industry [21-25]. Given the large environmental impact of natural gas, alternative carbon sources will be reviewed in this section.

The annual global methane emissions amount to approximately 570 Mtonne. However, there is no consensus on the partitioning of fossil emissions between natural and anthropogenic sources. Recently, Hmiel et al. demonstrated that the former are likely overestimated by one order of magnitude in previous research [26]. This means an additional 50 Mtonne CH_4 per year should be allocated to anthropogenic sources, making society responsible for around 35 % of the annual methane emissions. The largest source of anthropogenic CH_4 emissions is fossil fuel extraction and more specifically leaks from oil and natural gas wells, pipelines and processing plants. Drilling a natural gas well also has a major impact on its surroundings such as, e.g., contamination of water and air pollution, disturbing wildlife, clearing of land, etc. [1]. If natural gas is used as a reductant in the super-dry reforming process, then the environmental impact of the process might significantly increase. Therefore, alternative carbon sources should be considered such as plastic waste or biogas.

Plasma gasification (PG) is a thermochemical process that can convert plastic waste into syngas. However, this technology is not limited to plastic waste, but can also be applied for municipal solid waste (MSW), automotive shredder residue (ASR), tires or mixed waste

[27]. A plasma gasification unit consists of a preprocessing unit, i.e., a shredder to reduce the size of the plastic waste; a furnace with a plasma torch where the gasification takes place; a treatment system for the syngas which is produced; and equipment to process the slag, a by-product from the waste gasification [28]. A plasma torch is used to ionize the molecules, which then decompose into smaller fragments. The organic waste fraction is converted into syngas, the inorganic part leaves the reactor as molten slag. The produced syngas can serve as an alternative for the natural gas, which is used as reductant in super-dry reforming. Via the reverse water-gas shift reaction, Eq. (6), CO_2 will then be converted into CO.

Electricity generation in combined heat and power units is currently the most common utilization of biogas in Europe, despite its potential as a feedstock for the chemical industry [29]. Biogas is a mixture of CH_4 (40 – 70 vol%) and CO_2 (25 – 60 vol%). Because of its delocalized small-scale production, biogas should first be upgraded to biomethane, such that it can be injected into the existing natural gas grid. This renewable methane can then be valorized through super-dry reforming [30]. To meet the standards for injection in the grid, two major steps are required. Trace components such as water vapor, hydrogen sulfide and siloxanes are removed in a cleaning process. Subsequently, CO_2 is separated to raise the calorific value [31]. Typically, the methane fraction in biomethane is higher than 96 vol%.

Gas fermentation of CO to ethanol

During the past decade, several academic and industrial groups developed direct gas fermentation processes to produce fuels and chemicals such as ethanol, 2,3-butanediol, butanol, acetic acid, isopropanol and acetone, using a microbial bioreactor. Feedstocks can originate from industrial (e.g. steel manufacturing, oil refining and chemical production) waste or low-value gases, or from gasification of biomass, MSW, natural gas or coal [32]. The ethanol gas fermentation production process consists of four main steps, i.e., gas acquisition, gas handling, fermentation and (co)-product recovery. Firstly, depending on the feedstock, syngas is produced via gasification of, e.g., biomass or the capturing of waste gases. Secondly, in the gas handling step, the gas stream is deoxygenated and compressed for input in the fermentation reactor. Thirdly, the treated gas is fed into the bioreactor. The microorganisms present in the reactor use CO as an energy and carbon source and secrete ethanol and side products into the surrounding broth. In Eqs. (9) - (12), the overall reaction equations for the production of ethanol are given. When no hydrogen is present in the reactor, CO can still be converted to ethanol, but CO_2 is then also being produced. Reaction (10) will dominate at ratios of H_2 to CO between 0.5 and 1.5. Only at values larger than 1.5, water instead of carbon dioxide becomes the dominant by-product of the reaction.

$$6CO + 3H_2O \leftrightharpoons C_2H_5OH + 4CO_2 \tag{9}$$

$$3CO + 3H_2 \leftrightharpoons C_2H_5OH + CO_2 \tag{10}$$

$$2CO + 4H_2 \leftrightharpoons C_2H_5OH + H_2O \tag{11}$$

$$2CO + 6H_2 \leftrightharpoons C_2H_5OH + 3H_2O \tag{12}$$

Recently, modified pathways have been presented for the production of isopropanol and acetone as well, which are commodity chemicals with annual production volumes above 2 Mtonne for the former and 7 Mtonne for the latter [33, 34]. A syngas fermentation reactor should offer high gas-liquid mass transfer rates and low energy input. This is achieved in a continuous stirred-tank reactor by mixing of the reaction medium. However, considerable energy is required for sufficient mixing. In a bubble column, lower energy inputs are needed, yet mixing is limited. Possible alternatives are biofilm reactors with fixation of the biocatalyst, such as trickle bed or membrane reactors. Continuous operation is preferred over batch operation since it allows for higher volumetric productivities [35].

Coupling of the core technologies of SDR and gas fermentation to commercial value chain technologies

Extending the value chain to produce $C_2 - C_3$ olefins such as ethylene and propylene will further consolidate the economics and expand market opportunities. Therefore, in this section, the dehydration of ethanol and isopropanol to respectively ethylene and propylene are discussed.

Dehydration of isopropanol to propylene

With a global capacity of approximately 140 Mtonne per year, propylene is the second most important feedstock for the production of plastics [36]. Due to the increased exploration of shale gas and the corresponding shift from naphtha to ethane cracking, a shortage of propylene has arisen. This can be attributed to the lower propylene yield when using an ethane feedstock (2 %), compared to a naphtha feedstock (15 %). On purpose propane dehydrogenation and methanol-to-olefins are two technologies that have gained market share by filling this propylene gap [37]. Another pathway could be the selective dehydrogenation of isopropanol to yield propylene, whereby the alcohol can be produced by a microbial fermentation process. This process is still in the research phase and hence not yet employed on an industrial scale.

The selectivity and conversion of dehydration reactions are highly dependent on the reaction conditions and employed catalyst. Recently, Murat et al. developed a Zr-SBA-15 catalyst showing high activity and selectivity for the dehydration of isopropanol [38]. SBA-15 is a mesoporous silica sieve, with a pore diameter that may be adjusted between 5 and 15 nm. SBA-15's key qualities, such as its high surface area, lack of environmental toxicity and high stability under thermal and mechanical influences, make it well-suited for a variety of applications and especially as a support for catalysts [39]. The mesopore volume and median pore diameter can be increased by the addition of zirconium to the support. The best results are obtained when adding 5 wt% zirconium into the structure, with a selectivity to propylene of 100 % at a conversion of 96 % and temperatures between 200 °C and 250 °C [38].

Dehydration of ethanol to ethylene

Ethylene is one of the most important petrochemicals with a production volume of approximately 240 Mtonne per year [40]. Traditionally, ethylene is produced from steam cracking of a naphtha feedstock. Due to the shale gas boom in the United States, ethane has increased its share as a feedstock for steam crackers [37]. A promising route is the dehydration of ethanol, yielding 1 tonne of ethylene from 1.7 tonne of ethanol [41]. Because of the higher feedstock price and the large investments associated with the construction of a new plant, this route is currently not competitive with fossil-derived olefins. Given the drastically increasing production of bioethanol worldwide, economics might improve in the near future.

The dehydration of ethanol to ethylene is a highly endothermic reaction and is typically performed at temperatures between 300 and 500 °C. Ethylene yields of 94 – 99 % have been reached [42]. Currently, the process is mostly run in a fluidized bed reactor with an alumina-based catalyst. Also molecular sieves such as ZSM-5 or SAPO-34 have been subject of research for the ethanol dehydration reaction as they allow to execute the reaction at lower temperatures. Nonetheless, they are currently not applied in industry because of their low stability and high deactivation through carbon deposition. Another challenge for this process is the energy-intensive separation of ethanol and water, which accounts for 60 – 80 % of the total separation cost. Due to the low ethanol concentration (4 – 12 wt%) and the presence of azeotropes, distillation is complex [43].

Carbon footprint of the proposed value chain

Based on material and energy balances, a preliminary carbon footprint of the entire value chain is calculated [44]. LCA modelling is performed in the SimaPro v9.1 software, where the material and energy flows are used as input [45]. The following system elements are considered, production of CO via super-dry reforming, gas fermentation of CO to ethanol and dehydration of ethanol to ethylene. To assess the overall impact, the footprint of additional steps such as CO_2 capture via pressure swing adsorption (PSA), biogas production via anaerobic digestion, natural gas extraction and plasma gasification of plastic waste are also considered. The environmental impact of these steps, together with background processes, e.g., electricity and chemical supply chains, is assessed using inventory data retrieved from the ecoinvent database v3.2 [46]. Data representative for the European market were used for natural gas, water, heat and grid composition.

Unless valorization is possible and economically justifiable, all waste impacts are attributed to the waste producer. In this case, the CO_2 feedstock is assumed burden-free, whereas the environmental impact of additional energy and reduction in CO_2 emissions associated with carbon capture are assigned to the waste producer. For the production of biogas via anaerobic digestion, manure was considered as a feedstock, for which the database allocates impacts to the overall process. If sewage sludge would be used as a feedstock, ecoinvent does not allocate impacts to biogas production [47]. Four different scenarios have been examined, i.e., SDR with natural gas as a feedstock and energy carrier, SDR with biogas and gasified plastic waste as a feedstock and heated via the plasma reactor, SDR with biogas as a feedstock and heated with natural gas and finally, SDR with biogas as a feedstock and heat supplied via electricity. The carbon footprint of fossil-based heat and grid electricity (*left*) is compared to renewable heat and electricity (*right*) in Figure 5.

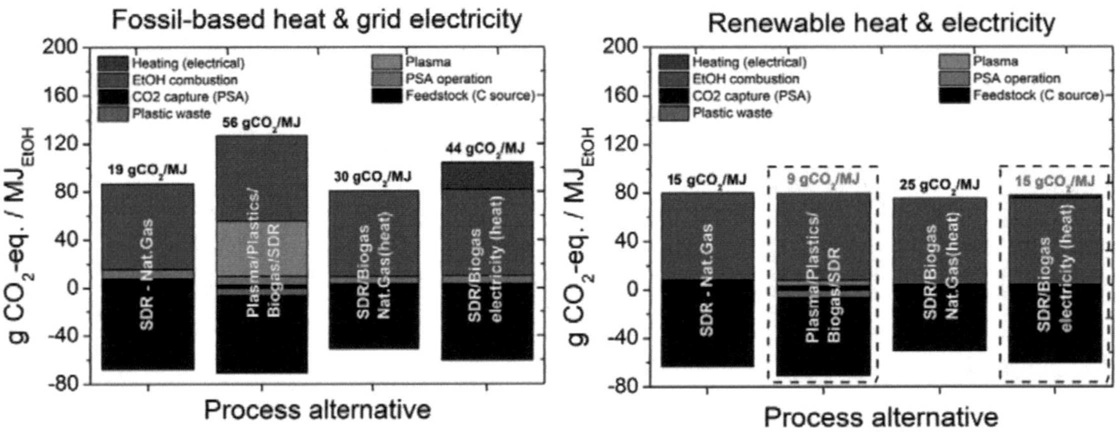

***Figure 5**. Preliminary carbon footprint calculation of the four different scenarios. Energy provided by fossil-based heat and grid electricity **(left)**; Energy provided by renewable heat and electricity **(right)**. The overall footprint value is indicated on top of the bars. As valorization of plastic waste avoids the incineration of non-recyclable plastics, this is considered to reduce the carbon footprint.*

The current calculations show that the carbon footprint of ethanol production can be reduced from 34 g CO_2 eq./MJ ethanol to 19 g CO_2 eq./MJ ethanol under current market conditions. Because of the composition of the (grey) grid electricity and the associated carbon footprint, plastic waste gasification is currently not competitive with other technologies. If renewable heat and electricity are available, the carbon footprint can be further reduced to 15 g CO_2 eq./MJ ethanol. Under these conditions, electrification of the SDR and switching from natural gas to biogas results in the same carbon footprint. The largest reduction can be achieved by SDR with a combined feed of biogas and plasma-assisted plastics pyrolysis, resulting in a footprint of 9 g CO_2 eq./MJ ethanol.

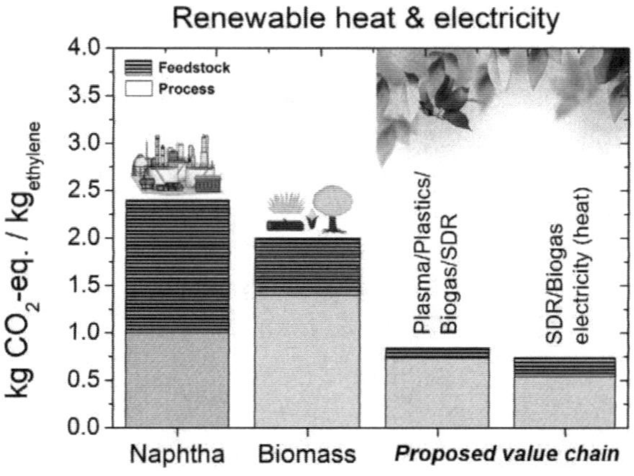

Figure 6. *Preliminary carbon footprint calculation for the production of ethylene.*

In Figure 6, the carbon footprint of the conventional ethylene production process (naphtha steam cracking), is compared to the proposed value chain and emerging technologies such as ethylene from biomass via sugar hydrolysis. The largest emissions arise from the conventional technology (2.4 kg CO_2 eq./kg ethylene). Even though the bio-based ethylene has a significantly lower footprint (2.0 kg CO_2 eq./kg ethylene), the proposed value chain results in the most sustainable ethylene production at only 0.8 – 1.0 kg CO_2 eq./kg ethylene. Alternative technologies, such as CO_2 to methanol followed by methanol to olefins, have carbon footprints almost five times higher than the proposed value chain.

Conclusion

A new value chain for the production of carbon-neutral fuels and chemicals has been proposed. The first step in this value chain is the conversion of CO_2 to CO via the innovative super-dry reforming technology. This is a strongly intensified CO_2 conversion process as it converts up to three moles of CO_2 molecules per mole of CH_4 into a pure CO stream (compared to one in conventional dry reforming). As the overall process is endothermic, electrification can reduce its environmental impact. Alternative carbon sources can further lower the carbon footprint of SDR, two promising options in this regard are plasma gasification of plastic waste and biogas. Subsequently, an ethanol production technology is applied for gas fermentation of CO to ethanol. Modified pathways have been proposed for the production of chemicals such as acetone and isopropanol. Finally, the large volume base chemicals ethylene and propylene can be produced from dehydration of the corresponding alcohols. A carbon footprint analysis demonstrates that the impact of ethanol production under current market conditions can be reduced from 34 g CO_2 eq./MJ ethanol to 19 g CO_2 eq./MJ ethanol. When electricity and heat are sourced from renewable energy, further reductions up to 9 g CO_2 eq./MJ ethanol can be achieved. Extending the value chain to produce, in a flexible way, C_2 – C_3 olefins such as ethylene and propylene, will further consolidate the economics, and expand market opportunities.

Acknowledgements

Michiel Van Cauwelaert and Vladimir Galvita acknowledge a personal grant from the Research Fund of Ghent University (BOF; 01D08521; 01N16319). The research leading to these results has also received funding from the European Research Council under the European Union's Horizon 2020 research and innovation program / ERC grant agreement n° 818607.

References

[1] U.S. Environmental Protection Agency, "Inventory of U.S. Greenhouse Gas Emissions and Sinks: 1990-2019," 2021.

[2] M. B. Ali, R. Saidur, and M. S. Hossain, "A review on emission analysis in cement industries," *Renewable and Sustainable Energy Reviews,* vol. 15, no. 5, pp. 2252-2261, 2011, doi: 10.1016/j.rser.2011.02.014.

[3] S. de Bruyn, C. Jongsma, B. Kampman, B. Görlach, and J.-E. Thie, "Energy-intensive industries - Challenges and opportunities in energy transition," *Study for the committee on Industry, Research and Energy (ITRE), Policy Department for Economic, Scientific and Quality of Life Policies, European Parliament, Luxemburg,* 2020.

[4] K. De Ras, R. Van de Vijver, V. V. Galvita, G. B. Marin, and K. M. Van Geem, "Carbon capture and utilization in the steel industry: challenges and opportunities for chemical engineering," *Current Opinion in Chemical Engineering,* vol. 26, pp. 81-87, 2019, doi: 10.1016/j.coche.2019.09.001.

[5] F. García-Labiano, L. F. de Diego, P. Gayán, J. Adánez, A. Abad, and C. Dueso, "Effect of Fuel Gas Composition in Chemical-Looping Combustion with Ni-Based Oxygen Carriers. 1. Fate of Sulfur," *Industrial & Engineering Chemistry Research,* vol. 48, no. 5, pp. 2499-2508, 2009, doi: 10.1021/ie801332z.

[6] J. Adánez, C. Dueso, L. F. de Diego, F. García-Labiano, P. Gayán, and A. Abad, "Effect of Fuel Gas Composition in Chemical-Looping Combustion with Ni-Based Oxygen Carriers. 2. Fate of Light Hydrocarbons," *Industrial & Engineering Chemistry Research,* vol. 48, no. 5, pp. 2509-2518, 2009, doi: 10.1021/ie8013346.

[7] L. C. Buelens, V. V. Galvita, H. Poelman, C. Detavernier, and G. B. Marin, "Super-dry reforming of methane intensifies CO2 utilization via Le Chatelier's principle," *Science,* vol. 354, no. 6311, pp. 449-452, 2016, doi: 10.1126/science.aah7161.

[8] D. Pakhare and J. Spivey, "A review of dry (CO2) reforming of methane over noble metal catalysts," *Chemical Society Reviews,* vol. 43, no. 22, pp. 7813-7837, 2014, doi: 10.1039/C3CS60395D.

[9] L. C. Buelens, A. N. V. R. Dharanipragada, H. Poelman, Z. Zhou, G. B. Marin, and V. V. Galvita, "Exploring the stability of Fe2O3-MgAl2O4 oxygen storage materials for CO production from CO2," *Journal of CO2 Utilization,* vol. 29, pp. 36-45, 2019, doi: 10.1016/j.jcou.2018.11.008.

[10] L. C. Buelens, A. Van Alboom, H. Poelman, C. Detavernier, G. B. Marin, and V. V. Galvita, "Fe2O3–MgAl2O4 for CO Production from CO2: Mössbauer Spectroscopy and in Situ X-ray Diffraction," *ACS Sustainable Chemistry & Engineering,* vol. 7, no. 10, pp. 9553-9565, 2019, doi: 10.1021/acssuschemeng.9b01036.

[11] B. Jin, H. Poelman, C. Detavernier, Z. Liang, G. B. Marin, and V. V. Galvita, "Microstructured ZrO2 coating of iron oxide for enhanced CO2 conversion," *Applied Catalysis B: Environmental,* vol. 292, p. 120194, 2021, doi: 10.1016/j.apcatb.2021.120194.

[12] V. Singh, L. C. Buelens, H. Poelman, M. Saeys, G. B. Marin, and V. V. Galvita, "Carbon monoxide production using a steel mill gas in a combined chemical looping process," *Journal of Energy Chemistry,* 2021, doi: 10.1016/j.jechem.2021.12.042.

[13] H. Wang *et al.*, "Hierarchical Fe-modified MgAl2O4 as a Ni-catalyst support for methane dry reforming," *Catalysis Science & Technology,* vol. 10, no. 20, pp. 6987-7001, 2020, doi: 10.1039/D0CY01119C.

[14] S. A. Theofanidis *et al.*, "Effect of Rh in Ni-based catalysts on sulfur impurities during methane reforming," *Applied Catalysis B: Environmental,* vol. 267, p. 118691, 2020, doi: 10.1016/j.apcatb.2020.118691.

[15] S. A. Theofanidis, V. V. Galvita, C. Konstantopoulos, H. Poelman, and G. B. Marin, "Fe-Based Nano-Materials in Catalysis," *Materials,* vol. 11, no. 5, 2018, doi: 10.3390/ma11050831.

[16] S. A. Theofanidis *et al.*, "Fe-Containing Magnesium Aluminate Support for Stability and Carbon Control during Methane Reforming," *ACS Catalysis,* vol. 8, no. 7, pp. 5983-5995, 2018, doi: 10.1021/acscatal.8b01039.

[17] S. A. Theofanidis, V. V. Galvita, M. Sabbe, H. Poelman, C. Detavernier, and G. B. Marin, "Controlling the stability of a Fe–Ni reforming catalyst: Structural organization of the active components," *Applied Catalysis B: Environmental,* vol. 209, pp. 405-416, 2017/07/15/ 2017, doi: DOI: 10.1016/j.apcatb.2017.03.025.

[18] S. A. Theofanidis *et al.*, "Mechanism of carbon deposits removal from supported Ni catalysts," *Applied Catalysis B: Environmental,* vol. 239, pp. 502-512, 2018, doi: 10.1016/j.apcatb.2018.08.042.

[19] S. A. Theofanidis, R. Batchu, V. V. Galvita, H. Poelman, and G. B. Marin, "Carbon gasification from Fe–Ni catalysts after methane dry reforming," *Applied Catalysis B: Environmental,* vol. 185, pp. 42-55, 2016, doi: 10.1016/j.apcatb.2015.12.006.

[20] J. Hu, L. Buelens, S.-A. Theofanidis, V. V. Galvita, H. Poelman, and G. B. Marin, "CO2 conversion to CO by auto-thermal catalyst-assisted chemical looping," *Journal of CO2 Utilization,* vol. 16, pp. 8-16, 2016, doi: 10.1016/j.jcou.2016.05.006.

[21] S. T. Wismann *et al.*, "Electrified methane reforming: A compact approach to greener industrial hydrogen production," *Science,* vol. 364, no. 6442, pp. 756-759, 2019, doi: 10.1126/science.aaw8775.

[22] M. G. Vinum *et al.*, "Dual-Function Cobalt–Nickel Nanoparticles Tailored for High-Temperature Induction-Heated Steam Methane Reforming," *Angewandte Chemie International Edition,* vol. 57, no. 33, pp. 10569-10573, 2018, doi: 10.1002/anie.201804832.

[23] "Accelerating electrification with the "Cracker of the Future" Consortium." https://www.borealisgroup.com/news/accelerating-electrification-with-the-cracker-of-the-future-consortium (accessed 07/01, 2022).

[24] "BASF, SABIC and Linde join forces to realize the world's first electrically heated steam cracker furnace." https://www.basf.com/global/en/who-we-are/sustainability/whats-new/sustainability-news/2021/basf-sabic-and-linde-join-forces-to-realize-wolds-first-electrically-heated-steam-cracker-furnace.html (accessed 07/01, 2022).

[25] M. Bonheure, L. Vandewalle, G. B. Marin, and K. M. Van Geem, "Dream or reality? Electrification of the chemical process industries," *Chemical Engineering Progress,* vol. 117, no. 3, pp. 37-42, 2021.

[26] B. Hmiel *et al.*, "Preindustrial 14CH4 indicates greater anthropogenic fossil CH4 emissions," *Nature,* vol. 578, no. 7795, pp. 409-412, 2020, doi: 10.1038/s41586-020-1991-8.

[27] A. An'shakov, V. Faleev, A. Danilenko, E. Urbakh, and A. Urbakh, "Investigation of plasma gasification of carbonaceous technogeneous wastes," *Thermophysics and Aeromechanics,* vol. 14, pp. 607-616, 2007, doi: 10.1134/S0869864307040105.

[28] S. Achinas, "An Overview of the Technological Applicability of Plasma Gasification Process," in *Contemporary Environmental Issues and Challenges in Era of Climate Change*. Singapore: Springer, 2020, ch. 15, p. 293.

[29] M. Pöschl, S. Ward, and P. Owende, "Evaluation of energy efficiency of various biogas production and utilization pathways," *Applied Energy,* vol. 87, no. 11, pp. 3305-3321, 2010, doi: 10.1016/j.apenergy.2010.05.011.

[30] K. Verbeeck, L. C. Buelens, V. V. Galvita, G. B. Marin, K. M. Van Geem, and K. Rabaey, "Upgrading the value of anaerobic digestion via chemical production from grid injected biomethane," *Energy & Environmental Science,* vol. 11, no. 7, pp. 1788-1802, 2018, doi: 10.1039/C8EE01059E.

[31] E. Ryckebosch, M. Drouillon, and H. Vervaeren, "Techniques for transformation of biogas to biomethane," *Biomass and Bioenergy,* vol. 35, no. 5, pp. 1633-1645, 2011, doi: 10.1016/j.biombioe.2011.02.033.

[32] R. M. Handler, D. R. Shonnard, E. M. Griffing, A. Lai, and I. Palou-Rivera, "Life Cycle Assessments of Ethanol Production via Gas Fermentation: Anticipated Greenhouse Gas Emissions for Cellulosic and Waste Gas Feedstocks," *Industrial & Engineering Chemistry Research,* vol. 55, no. 12, pp. 3253-3261, 2016, doi: 10.1021/acs.iecr.5b03215.

[33] T. Walther and J. M. François, "Microbial production of propanol," *Biotechnology Advances,* vol. 34, no. 5, pp. 984-996, 2016, doi: 10.1016/j.biotechadv.2016.05.011.

[34] B. Schiel-Bengelsdorf and P. Dürre, "Pathway engineering and synthetic biology using acetogens," *FEBS Letters,* vol. 586, no. 15, pp. 2191-2198, 2012, doi: 10.1016/j.febslet.2012.04.043.

[35] I. K. Stoll, N. Boukis, and J. Sauer, "Syngas Fermentation to Alcohols: Reactor Technology and Application Perspective," *Chemie Ingenieur Technik,* vol. 92, no. 1-2, pp. 125-136, 2020, doi: 10.1002/cite.201900118.

[36] L. Fernández, "Global production capacity of propylene 2018-2030," 2021. [Online]. Available: https://www.statista.com/statistics/1065879/global-propylene-production-capacity/

[37] A. Al-Douri, D. Sengupta, and M. M. El-Halwagi, "Shale gas monetization – A review of downstream processing to chemicals and fuels," *Journal of Natural Gas Science and Engineering,* vol. 45, pp. 436-455, 2017, doi: 10.1016/j.jngse.2017.05.016.

[38] M. Murat, Z. Tišler, J. Šimek, and J. M. Hidalgo-Herrador, "Highly Active Catalysts for the Dehydration of Isopropanol," *Catalysts,* vol. 10, no. 6, 2020, doi: 10.3390/catal10060719.

[39] Q. N. Khanh Nguyen, N. T. Yen, N. D. Hau, and H. L. Tran, "Synthesis and Characterization of Mesoporous Silica SBA-15 and ZnO/SBA-15 Photocatalytic Materials from the Ash of Brickyards," *Journal of Chemistry,* vol. 2020, p. 8456194, 2020, doi: 10.1155/2020/8456194.

[40] L. Fernández, "Production capacity of ethylene worldwide from 2018 to 2025," 2021. [Online]. Available: https://www.statista.com/statistics/1067372/global-ethylene-production-capacity/

[41] A. Mohsenzadeh, A. Zamani, and M. J. Taherzadeh, "Bioethylene Production from Ethanol: A Review and Techno-economical Evaluation," *ChemBioEng Reviews,* vol. 4, no. 2, pp. 75-91, 2017, doi: 10.1002/cben.201600025.

[42] M. Zhang and Y. Yu, "Dehydration of Ethanol to Ethylene," *Industrial & Engineering Chemistry Research,* vol. 52, no. 28, pp. 9505-9514, 2013, doi: 10.1021/ie401157c.

[43] A. Singh and G. P. Rangaiah, "Review of Technological Advances in Bioethanol Recovery and Dehydration," *Industrial & Engineering Chemistry Research,* vol. 56, no. 18, pp. 5147-5163, 2017, doi: 10.1021/acs.iecr.7b00273.

[44] E. Delikonstantis, M. Scapinello, and G. D. Stefanidis, "Process Modeling and Evaluation of Plasma-Assisted Ethylene Production from Methane," *Processes,* vol. 7, no. 2, 2019, doi: 10.3390/pr7020068.

[45] E. Delikonstantis *et al.*, "An assessment of electrified methanol production from an environmental perspective," *Green Chemistry,* 2021, doi: 10.1039/d1gc01730f.

[46] G. Wernet, C. Bauer, B. Steubing, J. Reinhard, E. Moreno-Ruiz, and B. Weidema, "The ecoinvent database version 3 (part I): overview and methodology," *The International Journal of Life Cycle Assessment,* vol. 21, no. 9, pp. 1218-1230, 2016, doi: 10.1007/s11367-016-1087-8.

[47] E. Delikonstantis, E. Igos, M. Augustinus, E. Benetto, and G. D. Stefanidis, "Life cycle assessment of plasma-assisted ethylene production from rich-in-methane gas streams," *Sustainable Energy & Fuels,* vol. 4, no. 3, pp. 1351-1362, 2020, doi: 10.1039/c9se00736a.

AIChE Paper Number: 6388489

NOVEL LOW-EMISSION ETHYLENE PLANT

Author: ***Peter Oud***
Product Development Manager Ethylene
Technip Energies, Zoetermeer, The Netherlands

Prepared for Presentation at the 2022 Spring National Meeting
San Antonio, Texas, USA, April 11-14, 2022

NOVEL LOW-EMISSION ETHYLENE PLANT

Author:

Peter Oud
Product Development Manager Ethylene
Technip Energies, Zoetermeer, The Netherlands

Abstract:

Global warming and CO_2 reduction are taking center-stage and this is driving the ethylene industry to look for economically viable ways to reduce its emissions. Last year, Technip Energies presented the low-emission cracking furnace paper and also introduced the first stage of the low-emission ethylene plant demonstrating how this proprietary furnace technology can be implemented.

This paper extends the low-emission ethylene plant concept presented last year with two additional stages for reducing CO_2 emissions, to allow the ethylene plant to cost-effectively run on green hydrogen as fuel. This is an alternative for running the plant on blue hydrogen, as will be presented in a separate paper by Nick Rogers.

The green hydrogen-based ethylene plant concept presented in this paper offers CO_2 emission reduction by more than 50 percent without the need for carbon capture, which would certainly be of interest for those ethylene producers that have no access to a carbon capture and storage infrastructure. CO_2 emission reductions come from improved energy efficiency and phasing out fossil fuels firing.

The first stage of the concept utilizes the methane rich excess fuel gas made available by applying low-emission cracking furnaces instead of conventional ones to replace major steam turbines by electric motors. Firing this fuel gas in a combined cycle power plant (CCGT) instead of the cracking furnaces increases net power production due to the substantially higher heat-to-power conversion level in a CCGT compared to that of a conventional ethylene plant. In addition to maintaining a secured plant power supply, this also enables the export of generated excess power. Alternatively, excess fuel can be exported to displace fossil fuels and power with a low carbon foot print can be imported.

The second stage introduces renewable energy to further reduce the plant's emissions. This is used to significantly extend the power export to the external grid by up to 50 percent. One of the main issues to be dealt with is that it is difficult to handle the intermittent nature of renewable power and it storage. This issue is solved by temporarily storing the ethylene plant's excess fuel production and converting it to power at high efficiency in the CCGT during periods when there is a renewable power deficit. By storing the fuel surplus, within the premises or externally by the fuel grid manager, the ethylene plant can be used to support the introduction of renewables to the power grid by providing an efficient power-to-gas solution.

The final stage uses green hydrogen, produced from water by electrolysis, to fire as fuel in the cracking furnaces, thus substantially increasing the fuel gas surplus. For the generation of hydrogen, renewable power is used. Compared to the flow sheet of Stage 2, a 100 percent increase of exported power is feasible on a continuous basis. The fluctuations in the renewables are handled by the CCGT and the fuel storage. The CCGT can be run at its optimal design point on a continuous basis. This operation allows the CCGT to operate essentially full-time at an optimum heat-to-power efficiency, allowing the CO_2 emissions to be minimized.

An economic evaluation shows that this green hydrogen-based ethylene plant concept presented herein demonstrates how low-carbon ethylene production can be achieved without major impact on the operating margins. In a favorable economic environment, with relatively cheap renewable power and medium to high carbon credits, these can even improve substantially, providing enough margin to support the necessary investments. This concept can be used to make renewable power suitable for export to the grid and to phase out fossil fuels. It is particularly well-suited for regions with abundant, available renewable power sources to provide a mix of these with a high daily availability. Using a minimal amount of green hydrogen, a more than 50 percent reduction of the CO_2 emission level is feasible with the fully implemented concept, covering both the ethylene plant direct emissions and the emissions related to firing the excess fuel gas in the power plant.

Introduction

Presented last year

Last year Technip Energies presented the low-emission cracking furnace paper and also introduced the first stage of the low-emission ethylene plant using this proprietary furnace technology[1].

The first stage was aimed at meeting the 2020 emission targets of the European Union and was developed as part of the IMPROOF project that was sponsored by the Horizon 2020 program of the European Union. A new cracking furnace design, the low-emission cracking furnace, was presented by Technip Energies that can meet said target. By modifying the heat recovery scheme, fuel consumption and the associated CO_2 emissions can be reduced by 30 percent. In addition, a flow sheet was presented demonstrating how this furnace design could be integrated in the ethylene plant, whereby the overall CO_2 emissions are significantly reduced, taking into account not only the emissions from the ethylene plant itself, but also the emissions related to the power generation.

This year's contribution

This year's paper introduces two additional stages to the low-emission ethylene plant concept for reducing CO_2 emissions, to allow the ethylene plant to run on green hydrogen as fuel. This is an alternative for running the plant on blue hydrogen as will be presented in a separate paper by Nick Rogers.

As already indicated in last year's paper, Stage 1 of the low-emission ethylene plant concept can serve as the foundation for additional modifications of the ethylene plant to include renewable power sources in the power grid to reach the European Union's target of 55 percent CO_2 reduction by 2030.

An overview of the green hydrogen-based ethylene plant concept is shown in Figure 1 below.

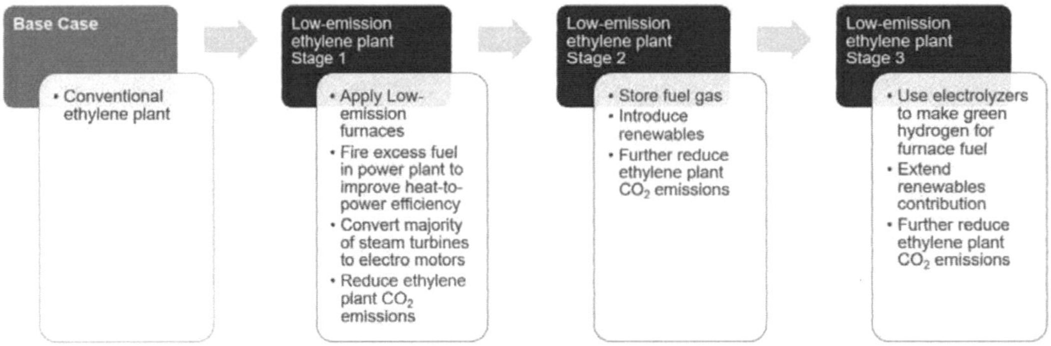

Figure 1: Green hydrogen-based ethylene plant concept

The overview shows how a conventional plant is converted to a green hydrogen firing plant in three stages:

- Stage 1 uses low-emission cracking furnaces. It is aimed at improving fuel efficiency by firing the excess fuel gas in a power plant, located inside or outside the premises, at a higher heat-to-power efficiency. With the power produced, most of the condensing steam turbines can be replaced by electro motors. The power surplus can be exported, reducing the specific emissions of the ethylene plant.
- In Stage 2, the excess fuel gas is stored to be able to import renewable power, when this is available and economically attractive. The stored fuel gas is used when the renewable power is not available or too expensive. This allows an increase of the power export and a further reduction of the emissions.
- In Stage 3, hydrogen produced by electrolyzers running on renewable power, so-called "green hydrogen", is used to run the cracking furnaces. This displaces part of the traditional fuel gas. This displaced fuel gas can be stored and used to generate even more export power, further reducing the specific emissions.

This concept can be used to make renewable power suitable for export to the grid and can be used to phase out fossil fuels. With an availability factor of the renewable power of 33 percent, a more than 50 percent reduction of the specific emissions is feasible, while green hydrogen consumption is minimal, as will be demonstrated below.

Low-emission Ethylene Plant Concept Development

The low-emission ethylene plant flow sheet can be developed in various stages, starting from a conventional ethylene plant flow sheet. Hence, the issues arising from implementation of this technology in an actual ethylene plant are not addressed here.

The design is meant to give direction to the ethylene plant decarbonization discussion, rather than addressing the issues related to implementation of this technology in each and every ethylene plant, each having its own complexities.

The approach is very conceptual and the examples are simplifications of an actual ethylene plant design. To demonstrate the concept in a nutshell, and make it accessible for a large audience, a more generally applicable and very basic representation of the ethylene plant has been selected. To be more specific, a single naphtha furnace will be used to represent the furnace section, fired by the plant's internally produced fuel gas. The shaft power for all machines driven by the VHP steam generated in the cracking furnace are represented by a single steam turbine and a compressor. These machines can also be located outside of the ethylene plant battery limit if VHP steam is exported. In this way, the most important parts of the ethylene plant effected by the introduction of the low-emission furnace are highlighted. The power plant, methane storage, renewable power supply and electrolyzers are treated in a similar way. Each of these systems is complex and many different options are available for the same purpose. The associated complexities are discussed after the detailed presentation of the concept.

As indicated by Van Geem et al.[2] in their life cycle assessment of the IMPROOF project, ethylene plant decarbonization has a significant effect on not only the plant emissions themselves, but also on the emissions from the power generation. Low-emission cracking furnaces are producing significantly less VHP steam, which is used for shaft power generation. The reduced VHP steam production results in a power deficit that has to be compensated with power from the grid. The overall CO_2 emission level was demonstrated to be highly dependent on the carbon foot print of the local grid. In a predominantly grey power scenario, overall CO_2 emissions would increase, rather than decrease, when connected to the grid. The low-emission cracking furnace was found to be less dependent on the carbon footprint of the grid, and was considered to be a relatively mature design, suitable for both grey and green power scenario.

In addition to the reduction of VHP steam, the fuel gas reduction realized by the low-emission cracking furnace results in an excess of fuel gas. For the part of the fuel gas that is generated by the ethylene plant, a low carbon destination needs to found. One such destination could be the production of blue hydrogen. The blue hydrogen route offers a very high level of CO_2 emission reduction, using hydrogen generated from reforming of the excess unused methane produced by the cracking process, and effectively capturing the CO_2 in the hydrogen plant at high pressure using T.EN's proprietary BlueH_2 by T.EN^{TM} process. This blue hydrogen route is presented in a separate paper by Nick Rogers.

Another destination for the ethylene plant's excess fuel gas is a combined cycle power plant (CCGT). The substantially higher heat-to-power conversion level of more than 50 percent increases the overall power output of the fuel gas. Part of the power generated can be used to secure the power supply to the low-emission ethylene plant, while the excess power generated can be exported to the local grid. This concept is particularly attractive in combination with green hydrogen firing as will be demonstrated below. Full hydrogen firing by the cracking furnaces would reduce all CO_2 emissions from the cracking furnace, without the need for carbon capturing. Green hydrogen, generated from water by electrolysis using renewable power, is still expensive in terms of both CAPEX and OPEX, but is expected to rapidly become available at competitive cost. The high fuel efficiency of the low-emission furnace significantly reduces these associated costs significantly. Power plant CO_2 emissions can be captured in due time, provided there is an infrastructure for the sequestration of the CO_2.

In the flow sheets presented below, the power plant is treated as a dedicated power plant for the ethylene plant, but it can also be part of a larger nearby power facility.

Conventional ethylene plant flow sheet

The conventional ethylene plant flow sheet is shown in Figure 2 below.

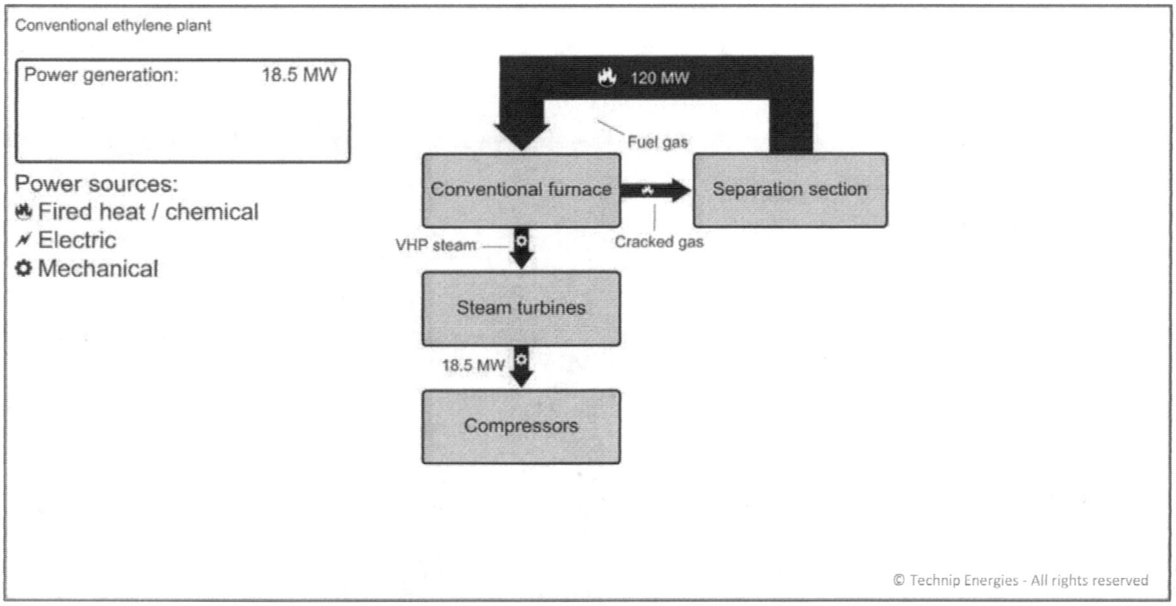

Figure 2: Conventional ethylene plant flow sheet

The conventional naphtha furnace, having a production capacity of approximately 140 kilo tonnes of ethylene per year, uses 120 MW of fired heat combusting fuel gas, mainly methane and some hydrogen, separated from the cracked gas in the separation section of the plant. In the furnace, this fired heat is used to produce the products, including 17.4 metric tonnes of ethylene per hour from the hydrocarbon feedstock, which in this case is naphtha. Part of the fired heat is used to generate VHP steam. Approximately 18.5 MW of shaft power is produced by the steam turbines to drive the ethylene plant machines, mainly compressors and a few pumps. The operation is fully autonomous, as far as the main power consumers are concerned.

Low-emission cracking furnace

For a low-emission ethylene plant, the low-emission cracking furnace developed under the European Union funded IMPROOF project can be used. This T.EN proprietary cracking furnace technology is a key technology in the T.EN handbook and is presented in Figure 3 below.

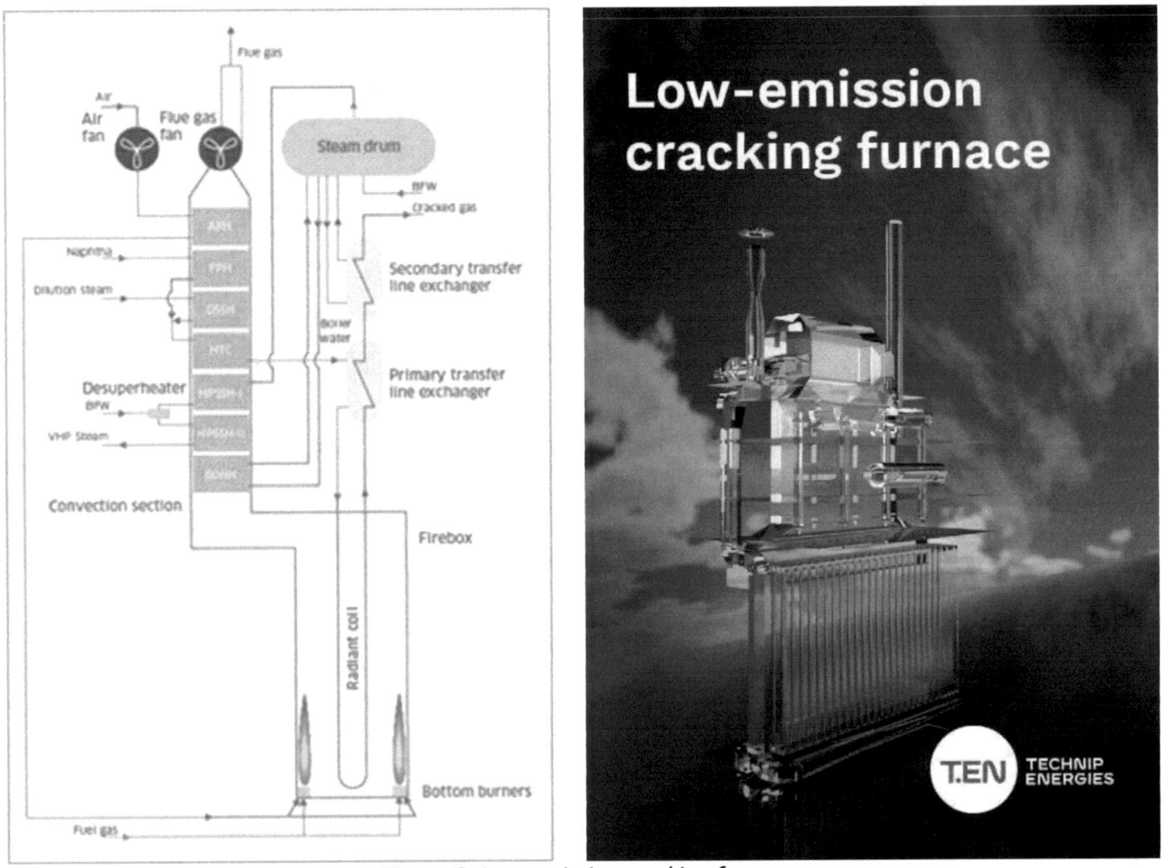

Figure 3: Low-emission cracking furnace

This type of furnace was presented in detail in last year's presentation[1] during the EPC 2021.

Air preheat and firing in the lower part of the firebox are used to raise firebox efficiency and reduce fuel consumption by more than 30 percent.

In the context of this paper, the low-emission cracking furnace is used to fire as much green hydrogen as possible with the renewable power available. With full hydrogen firing, the ethylene plant direct emissions are fully eliminated. As the generation of hydrogen is relatively expensive, high fuel efficiency is of paramount importance.

The revised heat integration scheme, featuring a feed-effluent type primary transferline exchanger, enables the radiant coil inlet temperature to be maintained at the same level as in conventional applications. This is necessary to limit fuel consumption, but is at the expense of a lower VHP steam production. The impact of the latter on the low-emission plant flow scheme is explained below.

This type of furnace was demonstrated to be cost-effective. The additional cost for a low-emission cracking furnace, relative to that of a conventional cracking furnace, can be paid back within one year based on the improved operating margin.

Low-emission ethylene plant flow sheet Stage 1

Stage 1 of the ethylene plant flow sheet is shown in Figure 4 below. A patent has been filed by Technip Energies for this flow sheet.

The first stage of the concept presented last year utilized the methane rich excess fuel gas from T.EN's proprietary low-emission cracking furnaces design to upgrade the power efficiency of the plant. By firing this excess fuel gas at a substantially higher heat-to-power conversion level in a high efficiency power plant, the power supply to the ethylene plant is secured and the excess power generated can be exported.

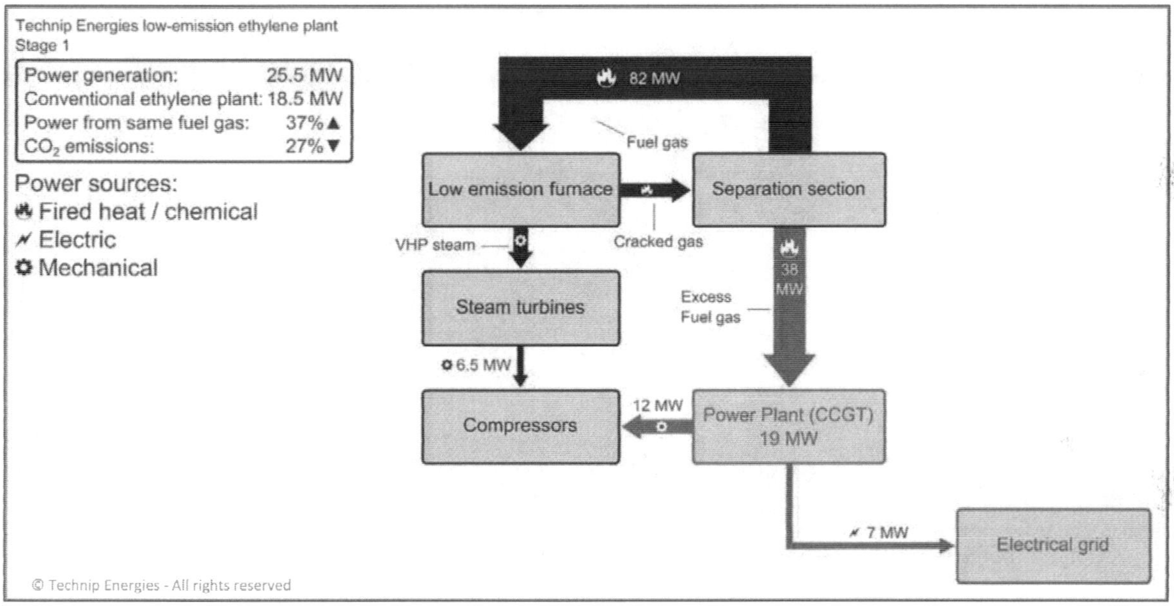

Figure 4: Low-emission ethylene plant flow sheet Stage 1

When applying low-emission cracking furnaces, the fuel consumption of the furnace reduces from 120 MW to 82 MW, a more than 30 percent reduction. However, the steam production also declines, reducing the generated shaft power from 18.5 MW to 6.5 MW. This roughly two-thirds reduction leaves a power deficit of 12 MW that needs to be compensated for by replacing the majority of the steam turbines by electro motors. The power for this is supplied by a 19 MW combined cycle power plant, indicated as CCGT. This power plant converts the excess fuel gas to power with a heat-to-power efficiency of more than 50 percent. This results in a net power export of 7 MW, allowing the ethylene plant to remain autonomous as far as the main power consumers are concerned. As the power generated from the same fuel gas increases

by 37 percent, from 18.5 to 25 MW, the specific CO_2 emission* reduces by 27 percent overall.
* CO_2 emissions per MWh of power produced

Low-emission ethylene plant flow sheet Stage 2

Stage 2 of the ethylene plant flow sheet is shown in Figure 5 below, showing daily average generation and consumption figures of fuel and power. A patent has been filed by Technip Energies for this flow sheet as well.

With the plant's power supply ensured, the second stage presented herein introduces renewable energy as an alternative to further reduce the plant's emissions and extend the power export to the external grid by 50 percent. By storing the excess fuel, the ethylene plant can be used to help the introduction of renewables to the power grid.

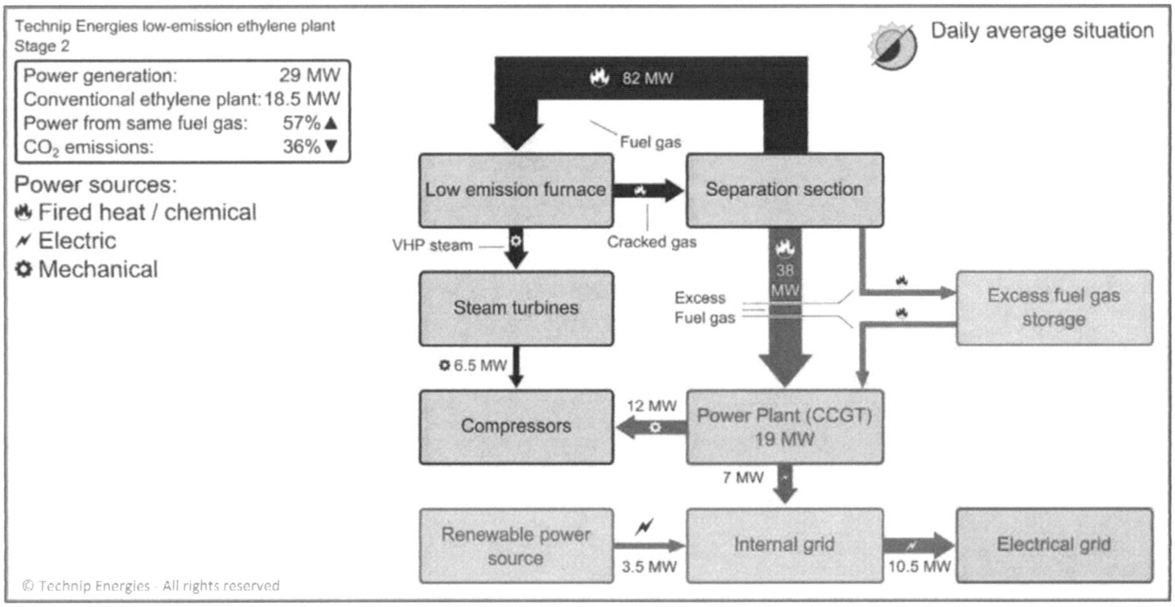

Figure 5: Low-emission ethylene plant flow sheet Stage 2 - daily average situation
(based on 33% renewable power availability)

The production of excess fuel gas offers the opportunity to temporarily store this fuel gas. This allows the plant to run on renewable power, when this is available and economically attractive, and run on stored excess fuel when this power is absent. The picture shows the daily average situation, based on a renewable power availability factor of 33 percent. With this flow scheme, the power export can be increased from 7 MW to 10.5 MW, a 50 percent increase. The total power generated from the same fuel gas increases by 57 percent from 18.5 MW for the base Case to 29 MW for Stage 2. This reduces the specific CO_2 emission by 36 percent overall.

Low-emission ethylene plant flow sheet Stage 3

The final step, Stage 3, uses green hydrogen, produced from water by electrolysis, to fire as fuel in the cracking furnaces, thus substantially increasing the excess methane-rich fuel gas production. For the generation of hydrogen, renewable power is used. Compared to the flow sheet of Stage 2, a 100 percent increase of exported power is feasible on a continuous basis. The fluctuations in the renewables are handled by the CCGT and the fuel storage. The CCGT can be run at its optimal design point on a continuous basis. This operation allows the CCGT to operate essentially full-time at an optimum heat-to-power efficiency, allowing the CO_2 emissions to be minimized.

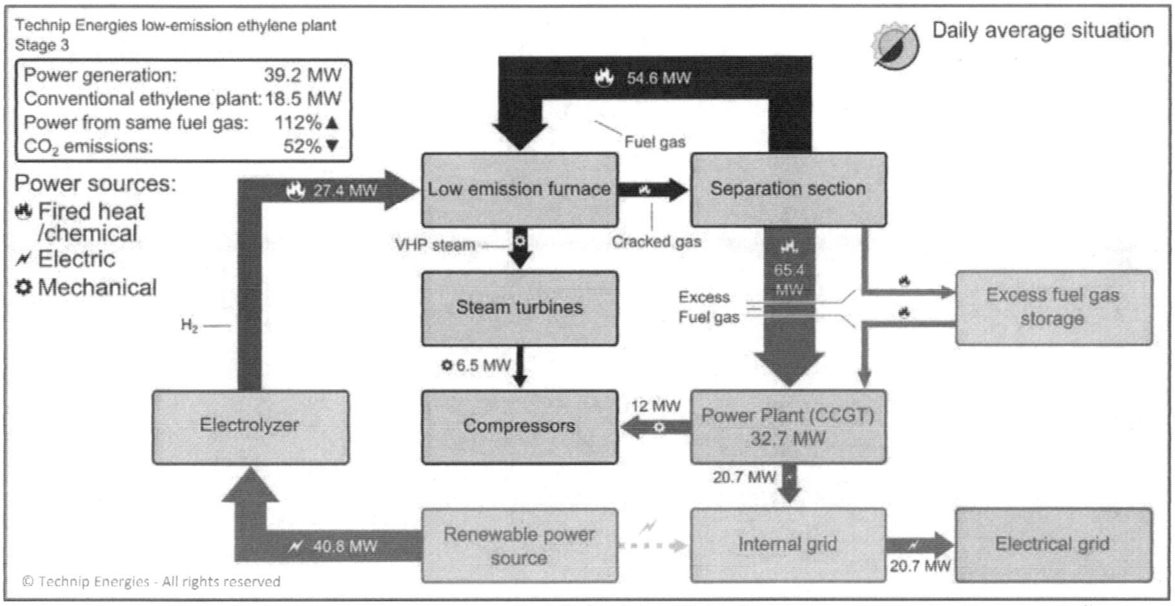

Figure 6: Low-emission ethylene plant flow sheet Stage 3 – daily average situation (based on 33% renewable power availability)

As we have seen in Stage 1, it is more efficient to fire fuel gas in the power plant than in the ethylene plant, as the power output increases by roughly 60 percent. In Stage 3, we take advantage of this by replacing the fuel gas by green hydrogen, using renewable power and an electrolyzer. This allows us to export more fuel gas to the power plant. By replacing, on a daily average basis, one-third of the 82 MW fired heat required for the low-emission furnaces by green hydrogen, excess fuel gas to the power plant can be raised by 27.4 MW of fired heat, from 38 MW to 65.4 MW. This 65.4 MW can be converted in the power plant to 32.7 MW, of which 20.7 MW can be exported to the grid, raising the export by 10.2 MW, which is nearly double that of Stage 2. Compared to the Base Case, the power generated from the same fuel gas increases by 112 percent, from 18.5 MW for the Base Case to 39.2 MW for Stage 3. This reduces the specific CO_2 emission by 52 percent overall.

Overview green hydrogen-based ethylene plant concept

Table 1below shows an overview of the green hydrogen-based ethylene plant concept for the various stages.

Table 1: Green hydrogen-based ethylene plant concept overview

Stage	Hardware changes	Resulting configuration	CO2 impact
Base case	• **Conventional furnace** with VHP steam production • **Condensing steam turbine drives** for compressors	• Fuel gas from the separation section drives furnace • Fully autonomous operation	Base

Stage 1	• **Low emission furnace** with reduced VHP steam production and fuel gas consumption
	• **Electric motors** for majority compressors
	• **Power plant (CCGT)** to improve heat-to-power efficiency (50%)
		Ability to maintain autonomous and export power
		• Excess fuel gas firing in furnace and CCGT
		• 19 MW CCGT power for:
		  • Electrified compressors drives: 12 MW
		  • Export: 7MW
			Including power plant
			• -27%

Stage 2	Same as Stage 1 + additionally • **Methane storage** to allow for renewable power variability • 3.5 MW **renewable power import**	Same as Stage 1 + additionally **Ability to import renewable power and convert to constant power** • 3.5 MW extra power export due to renewable power	Including power plant • -36%
Stage 3	Same as Stage 2 + additionally • **Electrolyzer** to make green hydrogen for furnace fuel • 37.3 MW **extra renewable power import** • **Larger methane storage and CCGT**	Same as Stage 2 + additionally **Ability to convert green hydrogen to methane and store it** • 10.2 MW extra power export	Including power plant • -52%

Renewable power

In the conceptual approach above, the renewable power daily average availability has been taken as 33 percent. This is an over-simplification of the actual situation. In reality, wind and solar power, complemented with hydropower, are seasonal and vary over the day depending on the weather conditions. Often, a considerably higher peak capacity needs to be installed to realize the required daily average capacity. Temporary excess power needs to be stored and power shortages need to be compensated by, for instance, increased gas firing or producing hydropower. The power grid architecture and handling is well beyond the scope of this paper, but this concept could be one of the means for the grid manager to balance the grid with fluctuating renewable power sources. This needs to be further carefully studied.

The best mix of renewable power sources depends highly on the local situation. A mix of various renewable power sources needs to be selected, such that the daily availability factor is as high as possible, and suitable means for storage of the energy need to be available. The green-hydrogen based ethylene plant can be one of a number of provisions in the energy grid for temporary power storage.

Higher hydrogen production

To increase the daily load factor of electrolyzers, power from a renewable power grid can be used to increase the daily hydrogen production, provided that this power is relatively cheap. The additional excess fuel gas can be exported. Additional investment in a local power station is not required, as the ethylene plant is already fully autonomous.

Carbon capturing

Carbon capturing of the flue gas either from the power station, or the ethylene furnaces, or both, is a way to further reduce the CO_2 emissions. This requires a carbon capturing infrastructure and means to store the CO_2. Alternatively, carbon capture can be done on reformate in a reforming scheme. These can be implemented in any stage.

Excess fuel storage

The power required for fuel storage is not shown, so as to not complicate the matter.

For liquified methane storage, it is 1MW on a daily basis, and comes fully from renewable sources, as storage is only required when this renewable power is available. The stored excess fuel gas reduces the power requirements of the ethylene plant by reducing the load on the refrigeration machines when reintroduced into the cold box. This reduces the storage power to 0.7 MW.

Alternatively, the fuel can be exported and fuel storage can be left to the fuel grid manager. In this case, for instance, pressurized storage in caverns is one of the options.

Fuel gas export/import

In reality, the low-emission furnace produces more fuel than required and a fuel export line should have been shown. It was decided to simplify and not explicitly show it. By fixing the fuel gas production to that fired in the conventional ethylene plant, the base line CO_2 emissions are fixed, and one can focus on the overall deviations.

As will be explained below, instead of supplying the power from a power plant on the premises of the ethylene plant, the power can also be supplied from a nearby power station, provided the carbon footprint is below (or close to) that of the firing of excess fuel gas. In this case, 0.38 metric tonnes per MWh electric power produced. The power plant is shown here as it is relevant to the evaluation of the total CO_2 emissions from the exported excess fuel gas. In this case, the fuel can be exported and imported when required and storage can be handled by the fuel grid manager.

Power generated by the VHP steam

The power generated by the VHP steam is shown to be driving compressors in the back-end of the plant and providing this power as mechanical power (shaft power). The VHP steam in a real plant is not restricted to the ethylene plant, but is also used to drive other equipment outside the ethylene plant scope or used is for direct heating. Within the ethylene plant, some of the steam is condensed in the dilution steam generators after power recovery. In other words, the VHP steam system is complex, and extends past the boundary conditions, and is different for each and every complex. For the examples, the majority of the VHP steam is used for power generation and a minor part for dilution steam generators after power recovery to set a base

line. Whether the power is used within the boundaries of the ethylene plant, or not, is considered to be irrelevant. Even if it is used outside of the ethylene plant, the loss of power needs to be compensated by electrification of the relevant driver. This approach allows for very fast benchmarking and makes the effect of various different conceptual design changes to the energy infrastructure relatively easy to evaluate.
In this case, the power requirement of the backend of the plant is approximately 16.5 MW and the remaining 2 MW is outside of the ethylene plant BL.

Cooling water reduction due to VHP steam reduction

The reduction of the VHP steam will reduce the duty of the surface condensers of the condensing turbines and the corresponding cooling water demand. Part of the cooling water demand will be shifted to the condensing turbines of the power plant, but as the efficiency of heat-to-power conversion of the system improves, there will be a net reduction of the cooling water consumption. This was demonstrated by Nick Rogers[3]. His paper demonstrated the validity of the Stage 1 flowsheet concept for actual ethylene plants handling both gas feed and liquid feed.

Recycle furnace

A real naphtha cracker would have a recycle furnace. This was felt to unnecessarily complicate the evaluation of the various design concepts and is kept out of the evaluation. Note that for gas cracking, the low-emission furnace shows even higher reductions in fuel gas. The effect of the recycle furnace can be evaluated more closely during a real plant feasibility study. At this point, the effect of the modifications on the actual VHP steam system can be studied more closely.

Emissions comparison

Ethylene plant direct specific emission, excluding the power plant

The ethylene plant specific emissions from the furnace stack, expressed as emissions per metric ton of ethylene produced, are shown in Figure 7, for all cases in absolute and relative terms. The furnace stack is the only emission source within the ethylene plant itself, and can be seen as Scope 1 of the greenhouse gas protocol, if the power plant is considered to be outside of the scope.

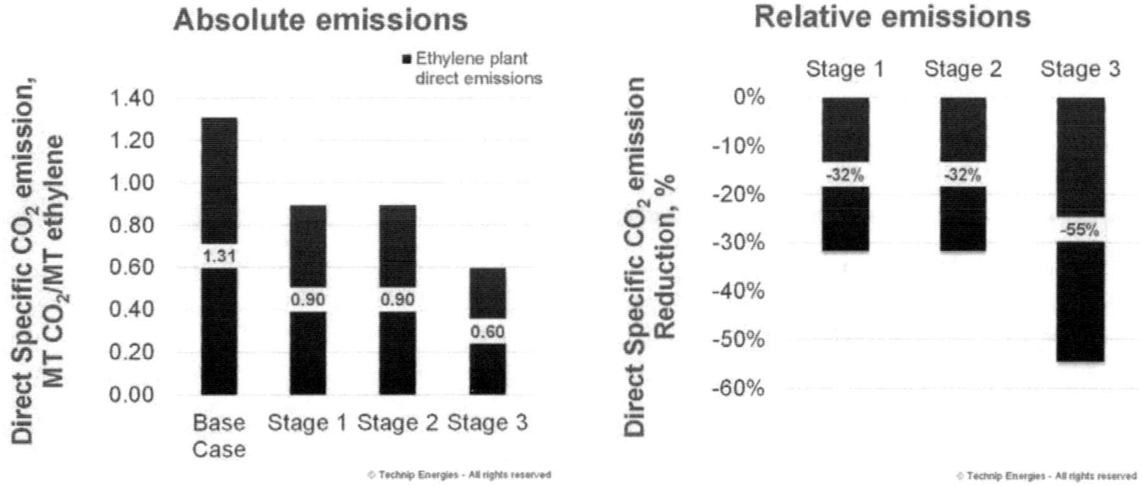

Figure 7: Ethylene plant direct specific CO_2 emissions

The direct emissions are determined by the fuel consumption of the cracking furnaces. Due to the reduction of the fired heat, from 120 MW to 82 MW in Stages 1 and 2, the CO_2 emissions per ton of ethylene produced reduce from 1.31 MT CO_2 per MT to 0.90 MT, a 32 percent reduction. In Stage 3, the fuel consumption is further reduced to 54.6 MW on a daily basis, reducing the direct emissions to 0.60 MT CO_2 per MT of ethylene produced. This is a reduction of 55 percent.

Ethylene plant total specific emission, excluding the power plant

The ethylene plant total specific emissions, including the emissions from the furnace stack, as well as those related to the power supply, expressed as emissions per metric ton of ethylene produced, are shown in Figure 8, for all cases in absolute and relative terms.

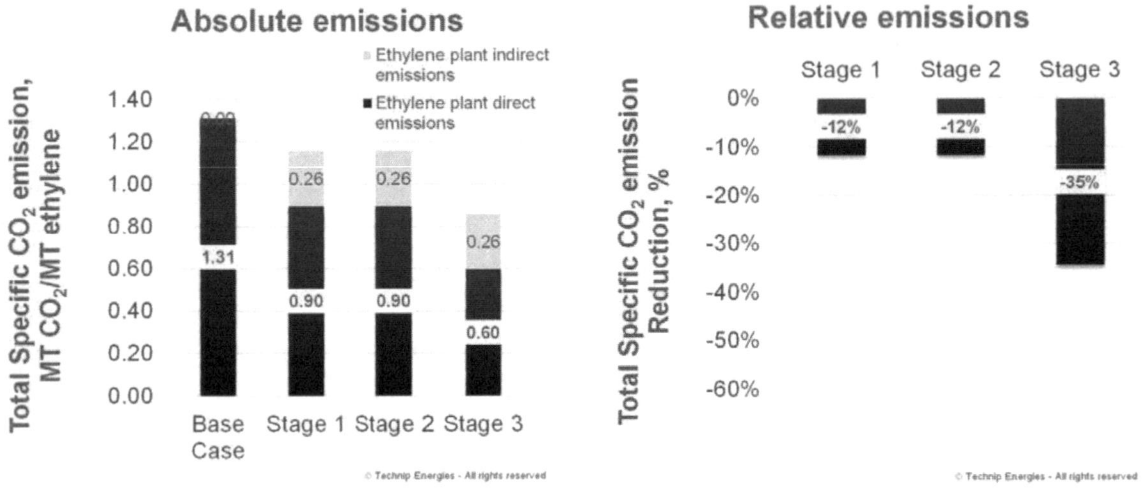

Figure 8: Ethylene plant total specific CO_2 emissions

If the 12 MW of imported power, to drive the compressors to compensate for the reduction of VHP steam, would be from a renewable power source, for instance by applying carbon capturing on the power plant, the indirect emissions would be 0 MT CO_2 per MT of ethylene produced, and the total emissions would be the same as the direct emissions as shown in Figure 7 above.

In the flow sheets presented, the plant's own excess fuel gas is used to fire the power plant. The carbon footprint of this fuel gas is 0.38 metric tonnes per MWh electric power produced. For a power requirement of 12 MWh electric power per 17.43 metric tonnes of ethylene produced, this corresponds to an indirect specific emission of 0.26 metric tonnes of CO_2 per metric tonnes of ethylene produced for all three stages. This has to be added to the direct emissions as presented in Figure 7, as has been done in Figure 8.

The absolute total emissions of the ethylene plant in this case are 1.31 for the Base Case, 1.16 for Stages 1 and 2, and 0.86 for Stage 3, all expressed in MT CO_2 per MT of ethylene produced.

The relative total emission reduction of the ethylene plant in this case is 12 percent for Stages 1 and 2, and 35 percent for Stage 3.

To put this in perspective, the carbon footprint of conventional coal-fired plants is 1.2 metric tonnes per MWh electric power produced, more than three times higher than that of the excess fuel gas, while that of natural gas is 0.4 metric tonnes per MWh electric power produced. Importing power from such a coal-fired plant would raise the indirect emissions by 0.56 (0.82-0.26) MT CO_2 per MT of ethylene produced, while importing power from a natural gas-fired power plant would only raise the emissions marginally by 0.014 MT. This means that as long as the power grid has considerable contributions from coal firing, importing power could increase the total emissions, even over that of the Base Case.

As long as the carbon foot print of the local power grid is below (or close to) that of the firing of excess fuel gas, in this case 0.38 metric tonnes per MWh electric power produced as the hydrogen content is relatively low, the net production of excess fuel gas can be exported, and power from the grid can be imported, to further reduce the ethylene plant total specific emissions level. In this case, the power generation and methane storage can be handled by the energy provider, provided that the energy generation is reliable enough for uninterrupted ethylene plant operations.

Low-emission ethylene plant total emissions, including the power generation

In the flow sheets discussed above, the fuel gas consumption has been kept at the same level for all flow sheets. In the Base Case, all this fuel gas is consumed by the cracking furnaces, while in the other flow sheets the fuel gas is diverted partly or fully to the power plant. This means that the absolute emissions do not reduce, but that the specific emissions per tonne of ethylene produced do reduce, due to the increased power generation for the same amount of fuel as is shown in Table 1, where the power exported to the grid is indicated for each case.

Table 2: Power export per case in MWh

Case	Exported power
Base Case	0
Stage 1	7
Stage 2	10.5
Stage 3	20.7

The exported power in all cases is generated without raising the fuel consumption. This allows the phasing out of both coal and gas firing.

The exported power displaces power from another source, and as long as the other power source has a higher carbon foot print, the net effect is a reduction of the specific CO_2 emission. This is shown in Figure 9 and Figure 10 below for two cases respectively:

- Phasing out coal firing
- Phasing out fossil gas fuel firing

For these two cases, the ethylene plant total specific emissions, including the emissions from the furnace stack and those related to the power supply, expressed as emissions per metric ton of ethylene produced, are shown for all stages, in absolute and relative terms.

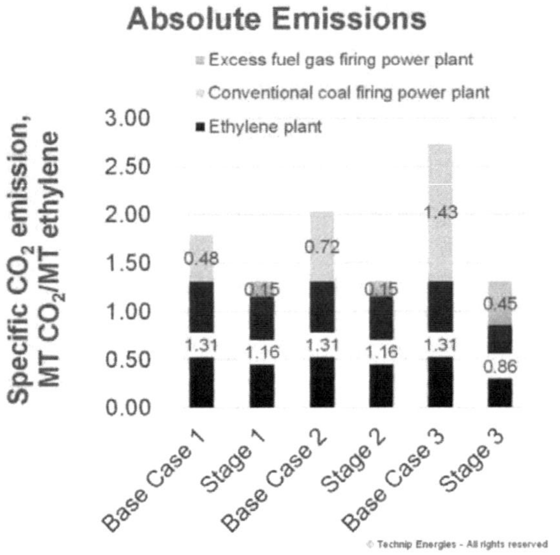

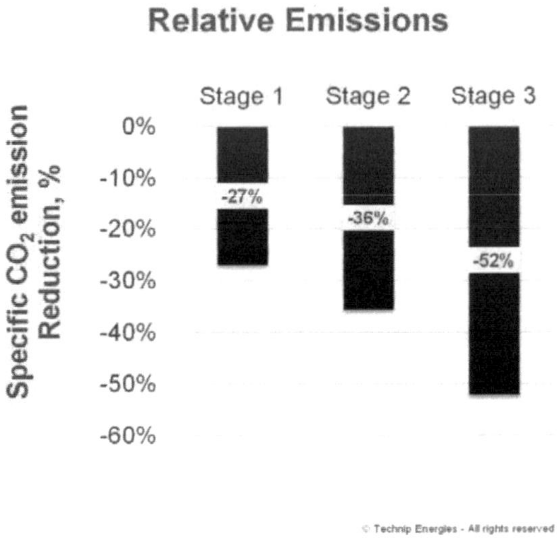

Figure 9: Low-emission ethylene plant total specific emissions, including the power generation – phasing out coal firing case

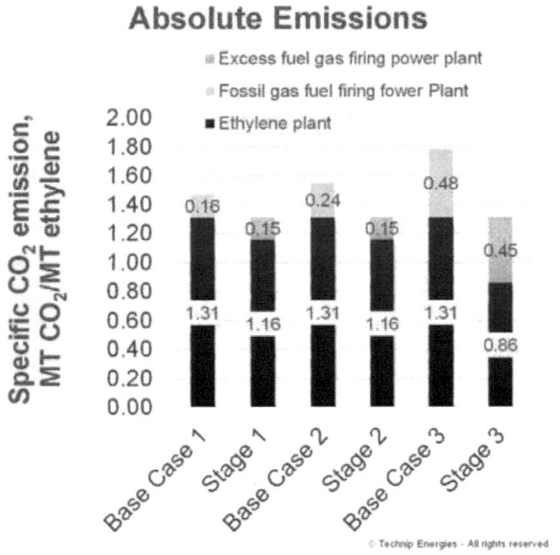

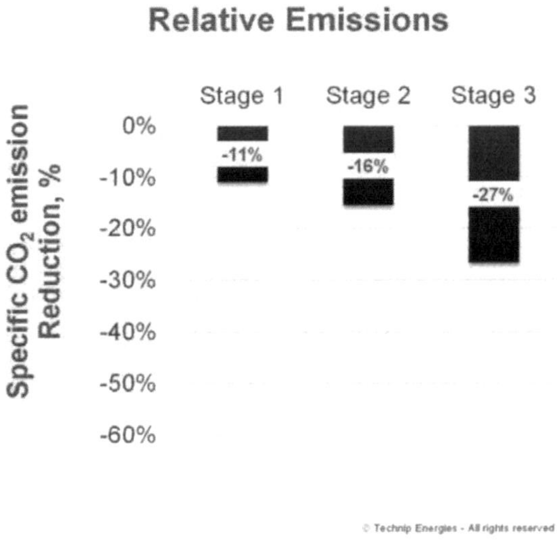

Figure 10: Low-emission ethylene plant total specific emissions, including the power generation – phasing out fossil gas fuel firing case

Both graphs show the total absolute specific emissions of the ethylene plant, in line with those presented in Figure 8. In addition, the total absolute specific emissions of the power generation are also included.

The Base Case of each stage includes the specific emission related to the phased out power supply as reported in Table 2. For Figure 9, this is based on a carbon footprint of conventional coal fired power plants of 1.2 metric tonnes per MWh electric power produced. For Figure 10, this is based on a fossil gas fuel-fired power plants of 0.4 metric tonnes per MWh electric power produced. These are the avoided emissions.

For each of the low-emission plant stages, the specific emission of the exported power is represented based on the a carbon footprint of the excess fuel gas firing power plant (CCGT) of 0.38 metric tonnes per MWh electric power produced. For Stage 1 and 3, this is the same as those in Table 1, but for Stage 2, this is slightly lower because 3.5 MW out of 10.5 MW of the export is from renewable power with no carbon footprint.

The total emission for all the low-emission plant stages is the same as the Base Case ethylene plant total specific emission of 1.31 MT CO_2 per MT of ethylene produced, as all of these cases are based on the same fuel gas being consumed. It is the Base Case total emissions that are different for each stage, due to the different amount of phased-out emissions for each stage from the different amount of power exported and the different power source being phased out. These avoided emissions are summarized in Table 2 below, and form the basis for the CO_2 credits calculation, also reported in Table 2.

For the gas firing cases, the power plant emissions are very similar to those from the excess fuel gas firing, as both have similar carbon footprints. For the coal firing cases, these contributions are 3 times higher. In Base Case Stage 3, the total emission from the coal firing power plant is even higher than that of the ethylene plant. In that case, the low-emission plant total emission can be reduced by 52 percent relative to the Base Case, which is a considerable achievement without applying any carbon capturing. Even when all coal firing is removed from the grid, there is still potential left for phasing out fossil gas fuel firing. For Stage 3, this is 27 percent.

CO_2 credits from avoided emissions

The avoided emissions mentioned above serve as basis for potential carbon credits during the period that fossil fuels are being phased out. Table 2 shows the situation for the two steps in the energy transition:

- Phasing out of conventional coal firing power plants
- Phasing out of fossil fuel gas firing power plants

The table is based on a 2021 price level of 55 €/MT CO_2 for the carbon credits. See Table 3 below for a complete overview of the price set used.

Table 3: Avoided specific emissions and associated credit levels

Avoided specific emission, MT CO_2/MT ethylene	Stage 1	Stage 2	Stage 3
Phasing out coal	0.48	0.72	1.43
Phasing out fossil gas fuel	0.16	0.24	0.48
Corresponding CO_2 credits, €/MT ethylene	**Stage 1**	**Stage 2**	**Stage 3**
Phasing out coal	26.4	39.6	78.6
Phasing out fossil gas fuel	8.8	13.2	26.4

Whether these carbon credits will be recognized and awarded will be subject to the emissions trading system in place.

Economic evaluation

Low-emission ethylene plant economics

An economic evaluation has been performed to evaluate overall economics of the various stages of the low-emission ethylene plant relative to the conventional ethylene plant.

The price-set, as shown in Table 4 below, has been used for this purpose. This table shows the 2021 price level of the CO_2 credits. In addition, it shows the power and natural gas prices on European level (Eur28) for large industrial consumers.

Table 4: Price-set utilities and CO2 credits

Price-set	Units	2021
CO_2 Emissions *	€/MT CO_2	55
Power **	€/MWh	58.3
Natural Gas **	€/MWh	20.3

* Source : https ://sandbag.be/index.php/carbon-price-viewer/ (Value at June 24 2021)
**Data source: Eurostat – Data Explorer.
Electricity: Band IG Eur28 2021S1
Natural gas: Band I6 Eur28 2021S1

Figure 11 and Figure 12 present the results of this economic evaluation of the low-emission ethylene plant for the various stages as compared to a conventional ethylene plant. This evaluation is done for two cases respectively and includes the power plant:

- Phasing out coal firing
- Phasing out fossil gas fuel firing

For each of these two cases, the contribution to the operating margins of the avoided emissions level is different, as already mentioned above in Table 2. The contributions in the second phase are three times lower than in the first phase.

Another aspect effecting the operating margins is the ratio between the price of the renewable power and that of the price received for the power exported to the grid. For the latter, the export grid price, the power price reported in Table 3 has been used. The renewable price has been taken as a percentage of this. On this basis, four cases have been evaluated:

- Same price:
  Renewable/grid power price =100 percent
- Renewable power price 20 percent lower than export grid price:
  Renewable/grid power price = 80 percent
- Renewable power price 40 percent lower than export grid price :
  Renewable/grid power price = 60 percent
- Renewable power price 60 percent lower than export grid price :
  Renewable/grid power price = 40 percent

For each of these four cases, the effect on the operating margins has been shown in both figures.

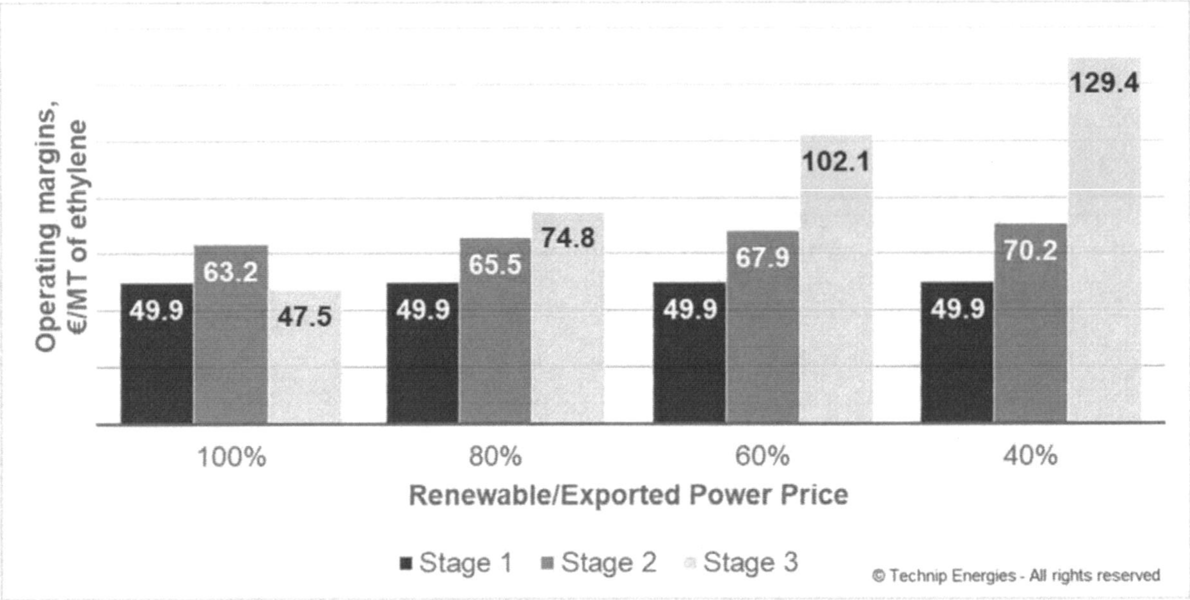

Figure 11: Low-emission ethylene plant operating margins for various ratios between renewable power price and export grid power price – phasing out conventional coal firing phase

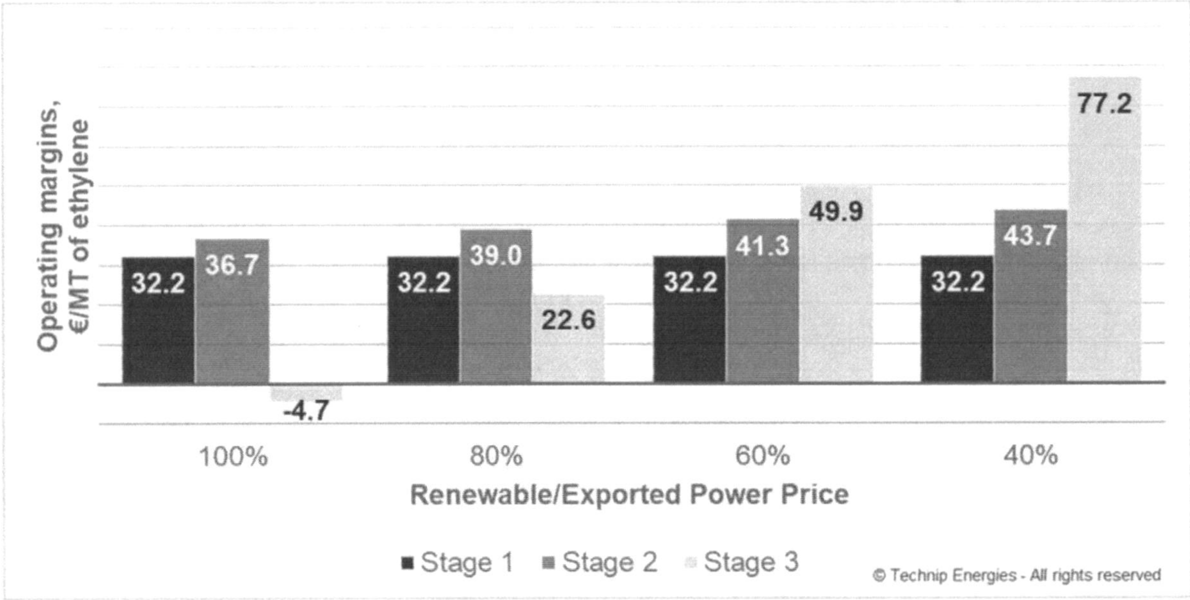

Figure 12: Low-emission ethylene plant operating margins for various ratios between renewable power price and export grid power price – phasing out fossil gas fuel firing phase

The difference between the operating margins in the two figures is the difference in the contributions of the avoided emissions, which are 17.6, 26.4 and 52.3 €/MT ethylene produced for Stages 1, 2 and 3 respectively. For instance, all the Stage 3 figures in Figure 12 are 52.3 €/MT ethylene produced lower than those in Figure 14 for this reason.

In the first phase of the energy transition, as depicted in Figure 11, the phasing out of coal, all stages have positive operating margins, even if the renewable power is the same price as that of the export grid. This is the stage where investments would have the most impact.

In the second phase of the energy transition, as depicted in Figure 12, the phasing out of fossil gas fuel firing, most of the stages have a positive operating margin, except for Stage 3 when the renewable power is at the same price level as that of the export grid. This is due to the contribution of the green hydrogen. When the renewable energy is expensive, this has a prominent effect on the operating margin. This is no reason not to build the Stage 3 flow sheet. It simply means that the amount of green hydrogen firing has to be reduced, until the renewable power price reduces by approximately 20 percent, to compensate for the CO_2 credits reduction. Alternatively, a mild fee can be charged for upgrading the renewable power to grid power.

Overview

Low-emission ethylene plant benefits

The low-emission ethylene plant concept has a considerable number of benefits:

- It provides a method to considerably reduce CO_2 emissions without requiring carbon capturing. The ethylene plant direct emission level can be reduced by 55 percent and the concept is also effective in phasing out coal and fossil gas-fired power plants.
- The required power infrastructure for low-emission operation is considerable and will take time to build. This infrastructure can be implemented in stages with minimum impact on operating margins. The intermediate stages show sufficiently high positive operating margins to recover these investments.
- As the concept is mostly based on available technology, the technology readiness level of the major equipment is high. This allows the concept to be fully implemented within a reasonable time span.
- Ethylene plant availability is of paramount importance. Investing in a nearby power plant, whether on the premises or not, and firing it with the excess fuel gas produced by the ethylene plant itself, would provide the plant with a reliable power source and maintain the plant's autonomy.
- One of the most interesting aspects of this concept is the ability of the ethylene plant to handle and store renewable power. This ability allows a stable, continuous output to the power grid with a high renewable contribution. In Stage 3, the renewable contribution is two-thirds. The ethylene plant handles all renewable power fluctuations and produces a considerable continuous power to the grid. This type of power handling and storing features are advantageous in the energy transition, and ethylene plants can aid to fill this niche.
- Another interesting aspect of this concept is that cracking furnaces can accept green hydrogen at a relatively low pressure, and that it doesn't need to be stored at very high pressure, such as is the case for use in mobility sector. This avoids the CAPEX, OPEX and CO_2 emissions related to high pressure hydrogen storage. Basically the ethylene plant converts the major part of the hydrogen to methane, which can be stored much easier with existing infrastructure (gas grids, gas storage facilities).
- The ethylene plant produces various fuel gas streams. Methane-rich fuel gas, rather than hydrogen-rich fuel gas, can be sent to storage. This stored methane-rich gas is low in hydrogen and very suitable for use in conventional gas turbines.
- To reduce the CO_2 emissions from the gas turbine, it is important to run the gas turbine as close as possible to its design point. This minimizes fuel consumption and allows for a high heat-to-power conversion efficiency. The Stage 3 flow sheet is able to run at design capacity.
- Due to the improved heat-to-power conversion, overall less cooling water is required for the surface condensers of the condensing turbines.

Example of a 1400 kta low-emission ethylene plant

Using the numbers of the Stage 3 example above for a 1,400 kta low-emission ethylene plant, as is presented in Figure 13, the operating margins are in the range from 50 – 102 €/MT of ethylene produced, if the daily average renewable power price is 40 percent cheaper than the grid. This means that the operating margins range from 70 – 142 Million € per year from exporting 207 MWh of electric power to the grid.

Figure 13: 1400 kta low-emission ethylene plant Stage 3 overview

The corresponding annual CO_2 reduction is as follows:

- During the phasing out conventional coal firing: ~2000 kta
- During the phasing out fossil gas fuels firing: ~670 kta

Any future price developments of natural gas, electricity and carbon credits are not taken into account in these numbers, as they are highly dependent on the local economic environment. They are merely shown to indicate that with this flow scheme, it is possible to reduce CO_2 emissions, while at the same time, raise the plant's overall operating margins. With the example that is being presented in this paper, individual operating companies can assess the economic viability of the low-emission ethylene plant flow sheet for their own specific economic environment.

It is worth specifically noting here, that the OPEX evaluation of the low-emission ethylene plant as being depicted above does not include an evaluation of the CAPEX. It should be fully understood that an investment in a power plant is not a short-term investment, contrary to the payback time of the low-emission cracking furnace itself as presented above. The payback time still needs to be determined, but is considered to be outside of the scope of the original IMPROOF project, which was aimed exclusively at cracking furnace emission reduction, and not at ethylene plant reduction. Yet, the article is aimed at presenting a practically feasible roadmap to reduce the CO_2 level for the ethylene plant, as well as its power supply, as these are being affected by the low-emission cracking furnace concept.

Global market outlook

It is beyond the scope of this paper to present a full market outlook. The situation can be different for each and every region in the world, and for each and every nation, depending on its own national resources. For a global market outlook, reference is made to the IEA World Energy Outlook 2020[4] and the BP Energy Outlook 2020[5]. Both outlooks present similar trends.

In its World Energy Outlook 2020, the IEA sketches the development for the global energy market over the coming 20 years for a number of scenarios, taking into account the impact of the COVID-19 pandemic.

Based on this report, and in particular the outlook for the fuel supply section, coal is the first fossil fuel to be phased out gradually over the coming 30 years, followed by oil[6]. The natural gas market is the least affected, as this fuel source is expected to play a dominant role in the transition phase, during the phasing out of coal and oil, as it is seen as the lesser evil in terms of CO_2 emissions. Especially in the emerging economies, the energy demand is expected to grow rapidly over the coming 10 years[7] . In Asia, which is expected to account for half of the world's energy consumption by 2050[8], electricity demand is expected to double by 2050. Natural gas and renewables will be the main drivers to meet this strong increase in demand. The natural gas price is expected to increase in these regions, as is illustrated in Figure 14 below[8]. This figure shows the expected evolution of the natural gas price in the region for the period 2019 through 2050. It anticipates a natural gas price annual growth rate of 3.9 through 8.1 percent in China, and slightly lower growth rates in the neighboring countries identified as the Other Non-OECD Asia countries[9].

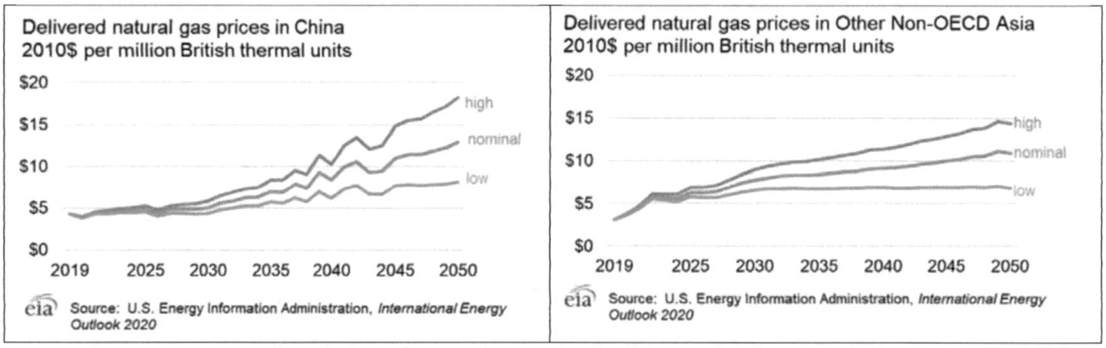

Figure 14: Expected evolution of Natural gas price in China and Other Non-OECD Asia

The low-emission plant concept can play a major role in this energy transition by supplying not only the fuel gas, but also generating the electricity to meet this increasing energy demand. This would replace part of the demand for LNG. At the same time, the ethylene plant operator would have a reliable power source, independent from the external grid.

Conclusions

The novel low-emission plant concept:

- Offers an economically interesting, partial decarbonization solution, not requiring a carbon capture infrastructure
- Support the energy transition by enabling large-scale storage of energy, in the form of excess methane, and coping with renewable power supply intermittency.
- Globally, it can be used to meet the rising energy demand, especially in the emerging economies, reducing the dependence on fossil fuels.
- Is particularly attractive for solar power as a basis for renewable energy, and other renewable power sources, such as wind and hydropower.

Acknowledgment

The work leading to this invention has received funding from the European Union Horizon H2020 Programme (H2020-SPIRE-04-2016) under grant agreement n°723706.

References

1. Oud, P; Van Boekel, T; "LOW-EMISSION CRACKING FURNACE", Paper 8b, AIChE Spring National Meeting, 2021.
2. Van Geem, K.M; Mynko, O; Marin, G.B; Brown, D.J; Amghizar, I; "How to Reduce CO2 Emissions of Steam Cracking Furnaces: The Million Dollar Question", Paper 8a, AIChE Spring National Meeting, 2021.
3. Rogers, N; "THE END OF THE STEAM AGE?", Paper 114c, AIChE Spring National Meeting, 2021.
4. IEA (2020), "World Energy Outlook 2020", IEA, Paris https://www.iea.org/reports/world-energy-outlook-2020
5. BP (2020), Energy Outlook 2020 edition, https://www.bp.com/en/global/corporate/energy-economics/energy-outlook.html
6. IEA, Global fuel supply by scenario, 2010-2040, IEA, Paris https://www.iea.org/data-and-statistics/charts/global-fuel-supply-by-scenario-2010-2040
7. IEA, Electricity demand outlook in selected regions/countries in the Stated Policies Scenario, 2019-2030, IEA, Paris https://www.iea.org/data-and-statistics/charts/electricity-demand-outlook-in-selected-regions-countries-in-the-stated-policies-scenario-2019-2030
8. U.S. Energy Information Administration (2020), Effects of Changes in Natural Gas Prices and Renewable Capital Costs on the Electric Sector in Asia https://www.eia.gov/outlooks/ieo/section_issue_Asia.php
9. See https://www.eia.gov/outlooks/ieo/pdf/appm.pdf for the region definitions.

AIChE Paper Number: 118d

Novel Catalyst for Single Step Conversion of Plastic Waste with 85% Olefin Yield

Andreas Eschenbacher
Post-Doctoral Researcher
Laboratory for Chemical Technology, Ghent University

Robin John Varghese
Post-Doctoral Researcher
Laboratory for Chemical Technology, Ghent University

Evangelos Delikonstantis
Post-Doctoral Researcher
Laboratory for Chemical Technology, Ghent University

Oleksii Mynko
Doctoral Researcher
Laboratory for Chemical Technology, Ghent University

Farnoosh Goodarzi
Post-Doctoral Researcher
DTU Chemistry, Technical University of Denmark

Kasper Enemark-Rasmussen
Post-Doctoral Researcher
DTU Chemistry, Technical University of Denmark

Jogchum Oenema
Post-Doctoral Researcher
Laboratory for Chemical Technology, Ghent University

Mehrdad SeifaliAbbas-Abadi
Post-Doctoral Researcher
Laboratory for Chemical Technology, Ghent University

Georgios Stefanidis
Professor
Laboratory for Chemical Technology, Ghent University

Kevin M. Van Geem
Full Professor
Laboratory for Chemical Technology, Ghent University

Prepared for Presentation at the 2022 Spring National Meeting
San Antonio, Texas, April, 12, 2022

Novel Catalyst for Single Step Conversion of Plastic Waste with 85% Olefin Yield

Andreas Eschenbacher
Post-Doctoral Researcher
Laboratory for Chemical Technology, Ghent University

Robin John Varghese
Post-Doctoral Researcher
Laboratory for Chemical Technology, Ghent University

Evangelos Delikonstantis
Post-Doctoral Researcher
Laboratory for Chemical Technology, Ghent University

Oleksii Mynko
Doctoral Researcher
Laboratory for Chemical Technology, Ghent University

Farnoosh Goodarzi
Post-Doctoral Researcher
DTU Chemistry, Technical University of Denmark

Kasper Enemark-Rasmussen
Post-Doctoral Researcher
DTU Chemistry, Technical University of Denmark

Jogchum Oenema
Post-Doctoral Researcher
Laboratory for Chemical Technology, Ghent University

Mehrdad SeifaliAbbas-Abadi
Post-Doctoral Researcher
Laboratory for Chemical Technology, Ghent University

Georgios Stefanidis
Professor
Laboratory for Chemical Technology, Ghent University

Kevin M. Van Geem

Full Professor
Laboratory for Chemical Technology, Ghent University

Abstract:

The interest in the chemical recycling of plastic waste via thermochemical processes is on the rise since it bears the potential to promote a carbo-circular economy. Within this study, the catalytic fast pyrolysis of polyolefinic waste streams was investigated to recover valuable base chemicals at high selectivity. We compare the energy demand and carbon footprint of this direct route with other state-of-the-art processes. Different feeds, (virgin and post-consumer polyolefins), catalyst/feed ratios, and reaction temperatures were tested in a micropyrolysis reactor coupled to two-dimensional gas chromatography. HZSM-5 zeolite was used as catalyst for upgrading the pyrolysis vapors. After a steam-treatment at 800 °C, a phosphorus-stabilized microporous HZSM-5 and its desilicated version showed high selectivity for production of valuable base chemicals when upgrading pyrolysis vapors at 600 °C, with ~80% C2-C4 olefins and ~5% aromatics produced from virgin PE. Increasing the reaction temperature from 600 to 700 °C, increased the aromatics yield to ~10 wt. % while producing similarly high overall yields of C2-C4 olefins. Processing post-consumer mixed polyolefin resulted in 5-10 wt. % lower olefins yield than using virgin PE but produced ~5 wt. % more aromatics due to its contamination with PS and PET. While parent HZSM-5 rapidly deactivated for repeated runs owing to its high coking propensity, phosphorus-modified and steamtreated HZSM-5 showed almost no deactivation during 130 runs with stable conversion towards C5+ aliphatics and high C2-C4 olefins selectivity with a 70% reduction in coke deposition. The performance of this direct olefins production route was further evaluated in a plantwide context. It was found that it requires ~38% lower energy input than the step-wise counterpart (plastics pyrolysis followed by pyrolytic oil steam cracking), while it results to at least an order of magnitude lower carbon footprint as compared to peer state-of-the-art processes.

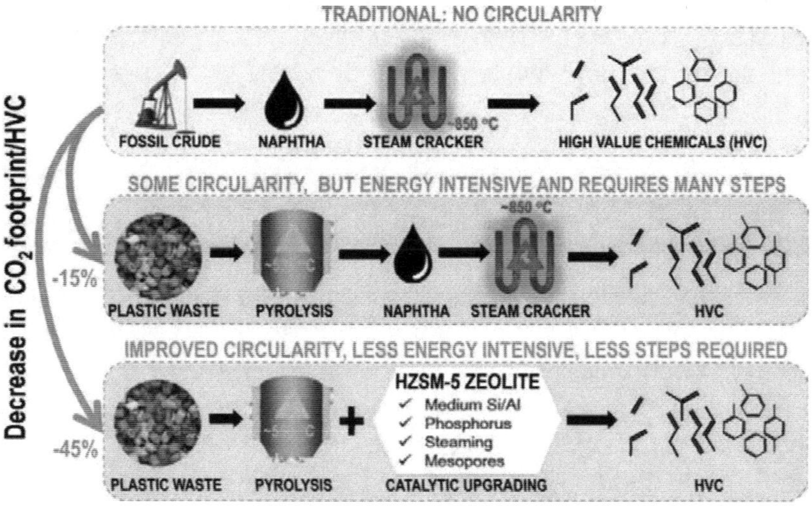

Introduction

Olefins (i.e., ethylene, propylene, and butenes) and aromatics (i.e. benzene, toluene, and xylenes) constitute the initial building blocks in the production processes of many materials used in our daily lives. For the production of thermoplastics (PE, PP, PET, PS, etc.), the International Energy Agency forecasted an increased production of roughly 600 Mt of thermoplastics in 2050, double the production level of 2010 [1]. Accordingly, the demand for the above-mentioned base chemicals continues to increase, and to date, these are primarily produced from fossil resources via steam cracking and to a lesser extent as a by-product from fluid catalytic cracking [2–5]. Producing plastics from fossil recourses has a high carbon footprint due to the energy-intensive production processes. In addition, polyolefins are challenging to recycle mechanically since the end-of-life mixed polyolefins waste is often contaminated with other plastics (PET, PS, PU, PA) and inorganics such as metals in pigments, fillers, or dirt [6–9]. Besides being "downcycled" by mechanical recycling to lower value products, their energy content is often valorized by incineration, leading to further fossil-based CO_2 emissions [10].

Chemical recycling is seen as a promising technology for the recycling of polyolefins to produce new plastics with identical properties to virgin, fossil-derived plastics. Via thermal pyrolysis, the polyolefins are cracked to mostly liquid product containing a broad distribution of carbon number ranging from C_1 to heavy waxes having different structures [11,12]. After subsequent treatments removing contaminants, the pyrolysis liquid can be used to produce monomers with ~65 wt. % yield by further processing in a steam cracker [10]. However, at current crude oil prices (app. $70/bbl, 07-2021) this approach is more expensive than directly processing fossil feed and requires intensive processing to remove contaminants from the pyrolysis liquid to very low levels [13,14]. To date, with less than <1% being recycled back to monomers [15], there is a clear demand for more efficient monomer recovery development.

Within this work, the direct upgrading of polyolefin pyrolysis vapors over a catalyst in the vapor phase was studied using modified H-ZSM-5 zeolites. This integrated process comprises fewer steps compared to the incumbent approach of pyrolysis, oil de-contamination, and steam cracking, while the use of a catalyst allows to operate at lower temperatures further benefitting process efficiency [16–20]. By physically separating the pyrolysis and the close-coupled catalytic upgrading stage in an ex-situ mode, the direct contact of non-volatile inorganic contaminants with the zeolite is avoided. This can prevent the poisoning of acid sites, as was reported when processing ash-containing biomass in direct catalyst contact [20–23].

In the present work, HZSM-5 with high to medium Al density was studied (Si/Al ~ 23) and mesopores were incorporated via a desilication procedure [24–26], thereby creating hierarchical zeolites with both micro-and mesoporosity. Since the mesopores are connected to the micropores and the external surface of the crystallites, the micropore diffusion pathway of reactants/products is shortened, thereby limiting undesired secondary reactions [27–30].

In this study, for the first time, we attempted to harness the benefits of mesoporosity, phosphorus modification, and steam treatment of HZSM-5 zeolite to maximize the recovery of C_2-C_4 olefins from polyolefins via catalytic upgrading of pyrolysis vapors. The use of two-dimensional gas chromatography provided detailed quantitative product distribution using different virgin polyolefin types and post-consumer mixed polyolefin (MPO), different catalysts, and different operating conditions (temperature and catalyst/feed ratio, i.e. contact times). Lastly, also the catalytic activity of a selection of the best performing catalyst during repeated reaction/regeneration cycles, and the change in product distribution, was studied. Furthermore, the catalyst performance was evaluated in a plantwide context, focusing on the energy and carbon footprints of such a catalytic process. The results were eventually benchmarked against other peer processes to stress and highlight the significance of the current catalyst impact. The combination of these different aspects makes these investigations increasingly relevant for industrial applications.

Results

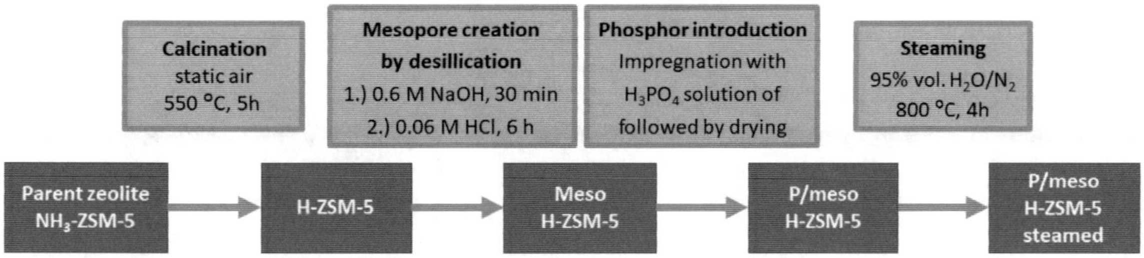

Figure 1. Brief summary of the catalyst synthesis.

In Figure 1 the synthesis procedure of the P/mesoZ55-ss catalyst is schematically depecited. Table 1 summarizes the physicochemical properties of the studied catalysts, whereas in Figure 2 the properties of P/MesoZ55 are highlighted.

Table 1. Physicochemical properties of phosphorus modified and steam-treated ("ms or ss" in suffix) catalysts in comparison unmodified parent versions

	Z55	*P/Z55-ss*	*P/mesoZ55-ss*
V_{micro}^{a} [cc/g]	0.22	0.06	0.17
V_{micro}^{b} [cc/g]	0.13	0.07	0.07
V_{meso}^{c} [cc/g]	0.09	0.06	0.11
V_{total}^{d} at p/p_0 =0.99	0.31	0.30	0.43
BET area^{d} [m^2/g]	423	401	349
Acidity^{e} [mmol NH_3/g]	0.65	0.10	0.09
Si/Al^{f}	28.5	28.8	19.2
Framework Si/Al^{e}	55.6	84	85
P/Al	-	1.2	0.6
P loading (wt. %)	-	1.9	1.4
Brønsted acid sites ($mmol \cdot g^{-1}$)^{g}	0.26	0.05	0.02
Lewis acid sites ($mmol \cdot g^{-1}$)^{h}	0.04	0.02	0.01
Brønsted/Lewis acidity	6.5	4.7	1.3

[a]determined using argon physisorption and NL-DFT method [26]; [b]obtained from N_2 physisorption and *t-plot* method; [c]obtained from N_2 physisorption and BJH analysis of adsorption branch; [d]obtained from N_2 physisorption; [e]determined from NH_3-TPD in desorption range 100-600 °C; [f]Determined with XRF; [e]determined with ^{29}Si MAS NMR peak simulations; [f]from Pyridine-IR, baseline-to-baseline integration of 1545 cm^{-1}; [g]from Pyridine-IR, baseline-to-baseline integration of 1455 cm^{-1};

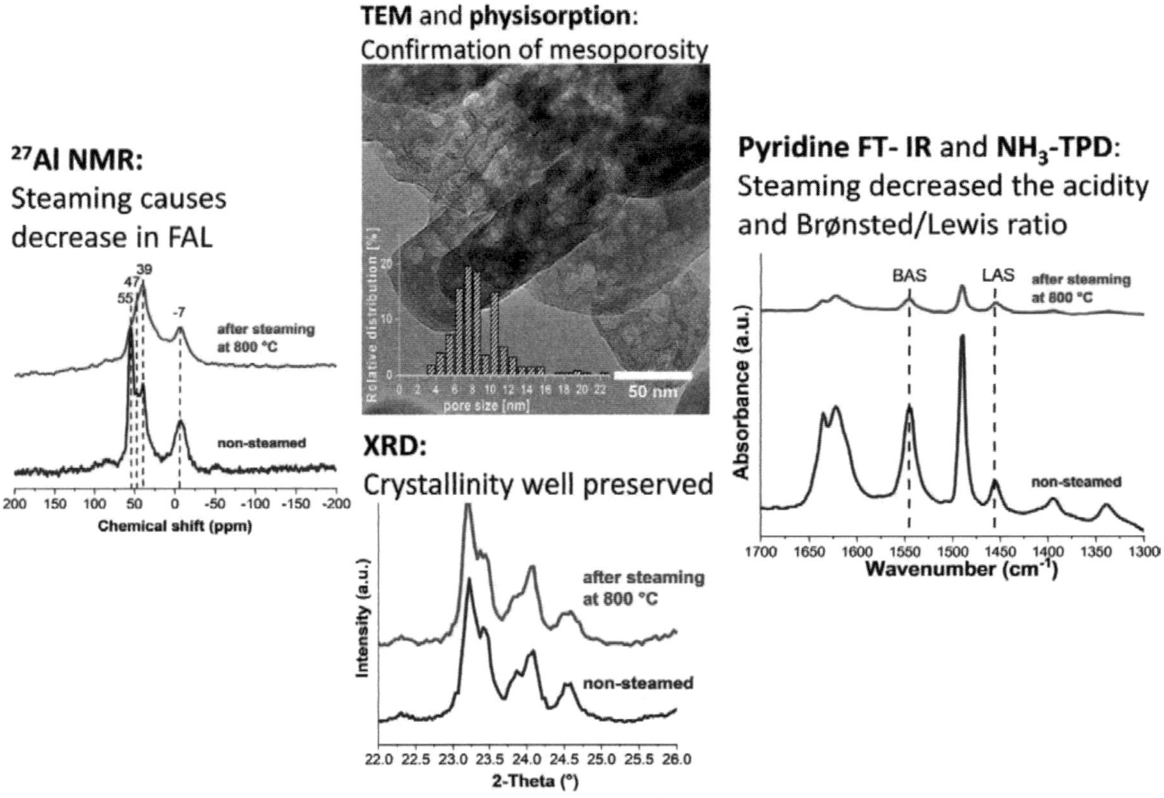

Figure 2. Main catalyst characteristics summarized for P/mesoZ55

Vapor upgrading with catalyst

Yields obtained with mesoporous HZSM-5 derived from desilication of parent HZSM-5 with SiO_2/Al_2O_3 = 55 that was also modified with a similar P/Al ratio of ~0.6 and subjected to severe steaming at 800 °C showed a favorable yield structure with low yields of aromatics and naphtha-range aliphatics while producing high yield of propylene and C_4 olefins (Table 2). Therefore, the following investigations focused on the *P/mesoZ55-ss* and investigated in more details the effect of catalyst temperature and using different feeds on the production of monomers.

Table 2. Product distribution using LDPE as feed and steamed phosphorous-modified mesoporous HZSM-5 obtained from parent materials with different SiO_2/Al_2O_3 ratios. Catalyst temperature = 600 °C and catalyst/feed ratio = 80

catalyst	**CH_4**	**C_2-C_4 olefins**	**Ethylene**	**Propylene**	**1,3-Butadiene**	**Other C_4=**	**C_4 alkanes**	**Aromatics**	**Aromatic-free C_5-C_{11}**	**C_{12}-C_{20}**	**C_{21}-C_{35}**
P/mesoZ55-ss	0.4	73.7	8.5	39.1	1.6	24.5	1.4	5.5	11.3	5.5	2.1

With increasing reaction temperature, the yield of CH_4, ethylene, 1,3-butadiene, and aromatics increased (Figure 3), while there was a decreasing trend for C_{5+} aliphatics. This was particularly pronounced between 600 and 700 °C, since the yield of benzene increased slightly from 0.9 wt. % at 500 °C to 1.4 wt. % at 600 °C but then more than doubled to 3.8 wt. % at 700 °C. Similar trends are observed for ethylene and 1,3-butadiene (Figure 3), and also the yield of H_2 first increased from 0.05 wt. % (500 °C) to 0.11 wt. % (600 °C), and then jumped to 0.30 wt. % at 700 °C. H_2 yields often correlate with the yields of coke and aromatics (=coke precursors) due to their lower H/C ratio compared to the feed. Interestingly, propylene yield seemed less affected and likely even higher catalyst temperatures (or higher activity/acidity) would have been needed to further crack propylene.

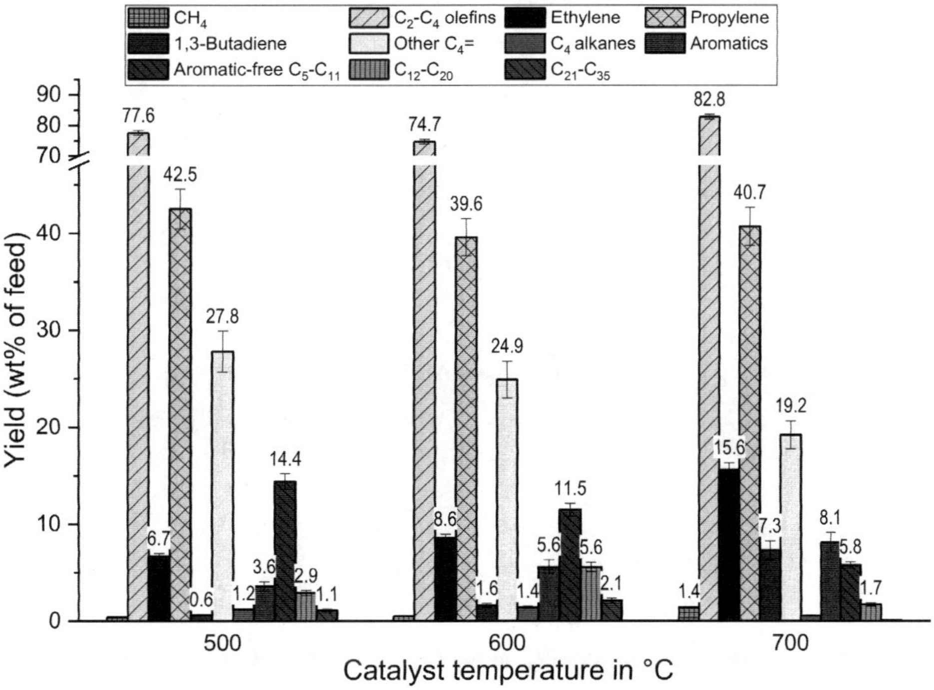

Figure 3. Yields obtained for a single reaction using LDPE as feed and *P/mesoZ55-ss* as catalyst at different catalyst temperatures. Catalyst loading was 32 mg and catalyst/feed ratio = 80.

Product distribution using P/mesoZ55 as catalyst and different feedstock

As shown in

Table **3**, the product yield distribution using different PE-types was comparable, while there were also clear differences. In particular, from LLDPE a ~10 wt. % more propylene was produced compared to LDPE. On the other hand, upgrading of HDPE pyrolysis vapors produced the highest yield of C_4 olefins. Using PP as feed resulted in a lower yield of C_4 olefins while C_5-C_{20} aliphatics were more prevalent. In contrast to PE, these aliphatics comprised more branched isomers as a consequence of the branched structure of PP and in agreement with findings from literature [78–83]. With MPO as feed, the monomer production was slightly lower with 72 wt. % C_2-C_4 olefins and ~11 wt. % aromatics, primarily attributed to the higher aromatics content in the feed (from PET/PS contamination) which are more difficult to crack.

Table 3. Yields obtained for a single reaction with different feeds using *P/mesoZ55-ss* as catalyst at 600 °C and a catalyst/feed ratio of 80.

feed	CH_4	C_2-C_4 olefins	Ethylene	Propylene	1,3-Butadiene	Other C_4=	C_4 alkanes	Aromatics	Aromatic-free C_5-C_{11}	C_{12}-C_{20}	C_{21}-C_{35}
LDPE	0.4	73.7	8.5	39.1	1.6	24.5	1.4	5.5	11.3	5.5	2.1
LLDPE	0.4	83.6	5.6	51.0	2.2	24.8	1.1	4.2	7.7	2.7	0.3
HDPE	0.3	82.8	7.1	37.2	2.6	36.0	0.9	4.1	9.9	1.6	0.4
PP	0.7	70.0	11.2	37.9	1.4	19.5	0.6	6.8	13.3	8.0	0.5
MPO[c]	0.7	73.2	12.1	38.5	1.6	21.0	1.0	9.0	10.3	4.5	1.3

[a]composition: 23 wt. % LLDPE, 7.5 wt. % LDPE, 29.5 wt. % HDPE, 40 wt. % PP;
[b]post-consumer mixed polyolefins after waste sorting, cold-washing, and pelletization

It should be stressed that even a microporous *Z55* that was impregnated with 2 wt. % P and steam-treated at 800 °C (*P/Z55-ss*) showed a very promising yield distribution (Table 4) and there was no apparent benefit (in terms of improved monomer yields) of the added mesoporosity when comparing the initial product yield structure. If the time-on-stream of the catalyst would be very short followed by catalyst regeneration—as is the case in fluid catalytic cracking—the more economic catalyst formulation would be *P/Z55-ss* as it does not require the additional treatment steps of mesopore addition. However, for a longer time on stream—as would be the case in a fluid or fixed bed—the risk that large fractions of active sites inside the zeolite particle centers become inaccessible upon coke deposits blocking poremouths may be reduced in a hierarchical zeolite. This was investigated in the next section.

Table 4. Yields obtained using LDPE and MPO as feed and microporous *P/Z55-ss* or hierarchical *P/mesoZ55-ss* as catalyst for ex-situ vapor upgrading. Catalyst temperature was 600 °C and catalyst/feed ratio = 80.

Feed + catalyst	CH_4	C_2-C_4 olefins	Ethylene	Propylene	1,3-Butadiene	Other C_4=	C_4 alkanes	Aromatics	Aromatic-free C_5-C_{11}	C_{12}-C_{20}	C_{21}-C_{35}
LDPE+ *P/Z55-ss*	0.5	81.6	9.9	40.4	1.9	29.4	1.3	4.4	6.2	4.4	1.4
LDPE + *P/mesoZ55-ss*	0.4	77.5	6.2	41.9	1.9	27.6	1.6	4.5	8.6	5.6	1.9
MPO + *P/Z55-ss*	0.7	73.5	12.3	39.4	1.7	20.2	1.1	7.5	8.0	7.2	1.5
MPO + *P/mesoZ55-ss*	0.7	73.2	12.1	38.5	1.6	21.0	1.0	9.0	10.3	4.5	1.3

Catalyst stability against deactivation by coke

Initially, the deactivation was studied with HDPE feed and *P/Z55-ss* catalyst at a high catalyst temperature of 700 °C and a catalyst/feed ratio of 80 since at 700 °C the highest C_2-C_4 olefin yield was observed with LDPE and a higher coking propensity was expected at the increased cracking severity due to the increased yield of aromatics (Figure 3). However, no deactivation of *P/Z55-ss* was observed after passing 16 HDPE pyrolysis vapor pulses over the catalyst, amounting to a cumulative feed/catalyst ratio of 0.2 (Table 5). The accumulated coke yield was 1.0 wt. %, corresponding to a coke/catalyst loading of 0.2 wt. %, or 5.1 µg coke/m^2 catalyst. Advantageously, the limited number of products contained mainly C_2-C_4 olefins, monoaromatics, and some C_5-C_{20} aliphatics; these were primarily alkanes, in contrast to thermal conditions where the products contain a high concentration of α-olefins and di-olefins with unsaturation at both ends of the linear chain, termed α,ω-olefins. Favorably, despite the high catalyst temperature, the formed aromatics mainly comprise monoaromatics (primarily benzene and toluene, and few alkylated monoaromatics and do not contain higher two and three-ring aromatics. The absence of higher aromatics indicates a low coking propensity, which could be attributed to the high shape-selectivity of conventional HZSM-5.

Even at an 8 times lower catalyst loading (catalyst/feed ratio = 10), upgrading 21 pyrolysis vapor pulses, i.e., reaching a cumulative feed/catalyst ratio of 2.1, no change in product distribution due to deactivation by coking was observed (

Table **6**). Under these conditions, due to the higher cumulative feed/catalyst ratio, more coke had accumulated on the catalyst, amounting to 0.62 wt. % of the catalyst weight and 15.6 µg coke/m^2 catalyst. The coke yield, however, was only 0.30 wt. % (instead of 1 wt. % described above for catalyst/feed ratio = 80 and cumulative feed/catalyst ratio = 0.2) due to the lower conversion at the lowered catalyst loading.

Comparing the product distribution obtained with catalyst/feed ratio = 80 (Table 5) and catalyst/feed ratio = 10 (

Table **6**) shows that at a lower catalyst/feed ratio (corresponding to a shorter contact time), the yield of C_{12+} products was higher, while the higher catalyst/feed ratio (= higher contact time) allowed to crack these. This resulted in ~10 wt. % higher ethylene yields, while also a small improvement in propylene yield was observed (Table 5). The yield of C_4 olefins (excl. 1,3-butadiene), however, was lower at the higher catalyst/feed ratio, suggesting that these may have cracked to ethylene or reacted further, e.g., with ethylene and propylene forming aromatics during Diels-Alder reaction. The higher yield of aromatics observed at the higher catalyst/feed ratio agrees with the higher coke yield since aromatics are coke precursors.

Table 5. Yields obtained from repeatedly passing HDPE pyrolysis vapors through a fixed bed of 32 mg *P/Z55-ss* at 700 °C. Catalyst/feed ratio = 80 per injection.

Reaction # (cumulative feed/catalyst ratio	CH_4	C_2-C_4 olefins	Ethylene	Propylene	1,3-Butadiene	Other C_4=	C_4 alkanes	Aromatics	Aromatic-free C_5-C_{11}	C_{12}-C_{20}	C_{21}-C_{35}
1 (0.01)	2.1	82.3	30.0	37.6	4.4	10.3	0.8	10.4	4.0	0.3	0.0
5 (0.06)	2.6	80.3	28.5	36.0	4.1	11.6	0.7	10.5	5.7	0.3	0.0
10 (0.12)	2.8	77.3	29.0	34.3	4.0	10.0	0.7	10.5	8.5	0.3	0.0
16 (0.20)	2.9	78.8	30.1	34.8	4.0	10.0	0.5	11.5	5.6	0.6	0.0

Table 6. Yields obtained from repeatedly passing HDPE pyrolysis vapors pyrolysis reactor (550 °C) and through a fixed bed of 4 mg *P/Z55-ss* at 700 °C. Catalyst/feed ratio = 10 per injection.

Reaction and cum. Feed/catalyst ratio	CH_4	C_2-C_4 olefins	Ethylene	Propylene	1,3-Butadiene	Other C_4=	C_4 alkanes	Aromatics	Aromatic-free C_5-C_{11}	C_{12}-C_{20}	C_{21}-C_{35}
1 (0.1)	1.9	71.7	20.1	30.2	5.2	16.2	1.1	4.9	6.6	14.1	1.0
11 (1.1)	1.7	71.4	19.9	29.4	3.9	18.2	1.0	5.5	5.8	17.7	0.8
21 (2.1)	1.9	72.7	19.3	31.7	4.6	17.2	1.2	4.3	3.6	16.6	1.2

The high catalyst temperature of 700 °C may not be ideal for the intended industrial process due to the increased energy demand. In addition, at a high temperature such as 700 °C the extent of thermal cracking contribution increases, which could assist in cracking the waxes and therefore effectively delay the catalyst deactivation. A change in product distribution due to catalyst deactivation by coke may therefore more easily be studied at a moderate catalyst temperature and using a feed with a high coking propensity. Therefore, in the next approach, the more industrially relevant MPO was used as a feed since its contamination with other plastics such as PET, PS, and PA was expected to form more coke on the catalyst compared to using virgin PE/PP as feed [84,85]. Product yields as function of increasing run number were compared at a catalyst temperature of 600 °C using 8 mg unmodified parent *Z55* and 32 mg *P/Z55-ss* as catalyst. The difference in loading was required to obtain a similar level of conversion over the fresh catalyst, since *P/Z55-ss* had a lower acidity of 0.10 mmol NH_3/g compared to *Z55* (0.65 mmol NH_3/g). It is noted that the number of acid sites loaded to the reactor was still 58% higher using 8 mg *Z55* compared to using 32 mg *P/Z55-ss*. Despite this difference, the yields were remarkably constant during 130 runs when using *P/Z55-ss* while *Z55* rapidly lost its activity for converting C_{5+} aliphatics (**Figure 4**). Likewise, there was a steeper decline in the yield of C_2-C_4 olefins, particularly propylene, compared to the stable yield profiles obtained with the *P/Z55-ss*. For both catalysts, it is interesting to note an initial increase in propylene and C_4 olefins, while ethylene and aromatics are highest for the first run and then stabilize or decrease at a low rate with increasing run numbers. This may be attributed to higher coking rates and/or overcracking of the C_3/C_4 olefins over a fresh catalyst.

The yields of CO_2, CO, and H_2 during the deactivation studies shown in Figure 4 were ~2 wt. %, 0.6 wt. %, and 0.1 wt. %, respectively, approximately double the values observed with α-Al_2O_3. Considering the low coke yields (vide infra), this shows that the maximum error made by normalizing the results without taking into account carbon oxides, H_2, and coke is ~3%.

Comparing the deactivation behavior of *P/mesoZ55-ss* with *P/Z55-ss* shows that initially there was a slight decline in C_2-C_4 olefins using *P/mesoZ55-ss,* and after that, both catalysts displayed rather stable operation. This indicates that only the initialization behavior of these catalyst is different. To draw conclusions on stability, future studies should monitor the product yields over time for continuous and longer operation until there is a steep decline in activity observed.

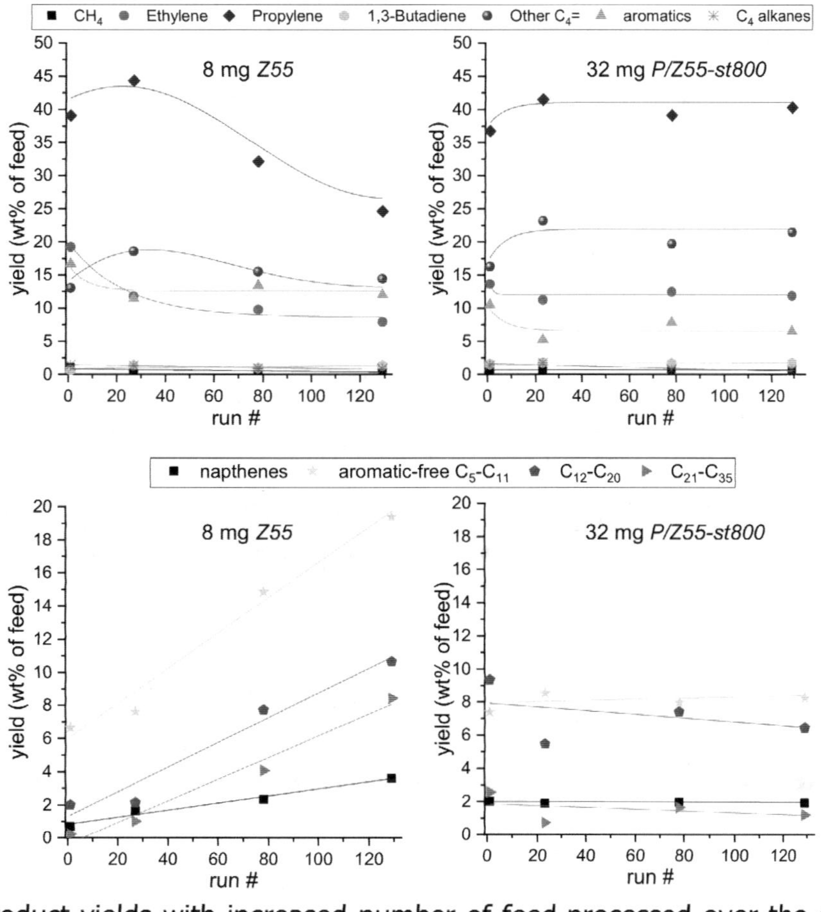

Figure 4. Product yields with increased number of feed processed over the catalyst using (a) 8 mg of the parent unmodified *Z55* and (b) 32 mg of *P/Z55-ss*. Yields are shown based on ash-free feed. The fitted lines are guides to the eye.

The catalytic upgrading significantly decreased the yield of higher α-olefins and diolefins with the unsaturation in the alpha and end position compared to the thermal reference case. This can be explained by the higher reactivity of α-olefins and dienes since the end-chain unsaturation provides high electron density and little steric limitations, thereby facilitating reaction with acid sites. While for *Z55* the selectivity to *n*-paraffins rapidly declined with increased run number at the expense of α-olefins and dienes, the selectivity to *n*-paraffins remained high and only at run number 130 some dienes were observed in the product slate (Figure 5).

Alkanes may potentially also result from hydrogen transfer reactions, and there was a decreasing trend in the C_4 hydrogen transfer index (HTI)—defined as the combined yields of iso-butane and *n*-butane divided by the total yield of C_4 alkanes and alkenes [86,87], as shown in the insert of Figure 5. Generally, however, the extent of hydrogen transfer is quite low in the *ex-situ* vapor phase upgrading since hydrogen transfer as an exothermic reaction with a slow reaction rate is not favoured by the high reaction temperature and short reaction time [88]. For *in-situ* catalytic upgrading with the polymer melt in direct contact with the catalyst, higher HTI values of ~0.2-0.4 were reported when using HZSM-5 in fluid bed reactors [89,90].

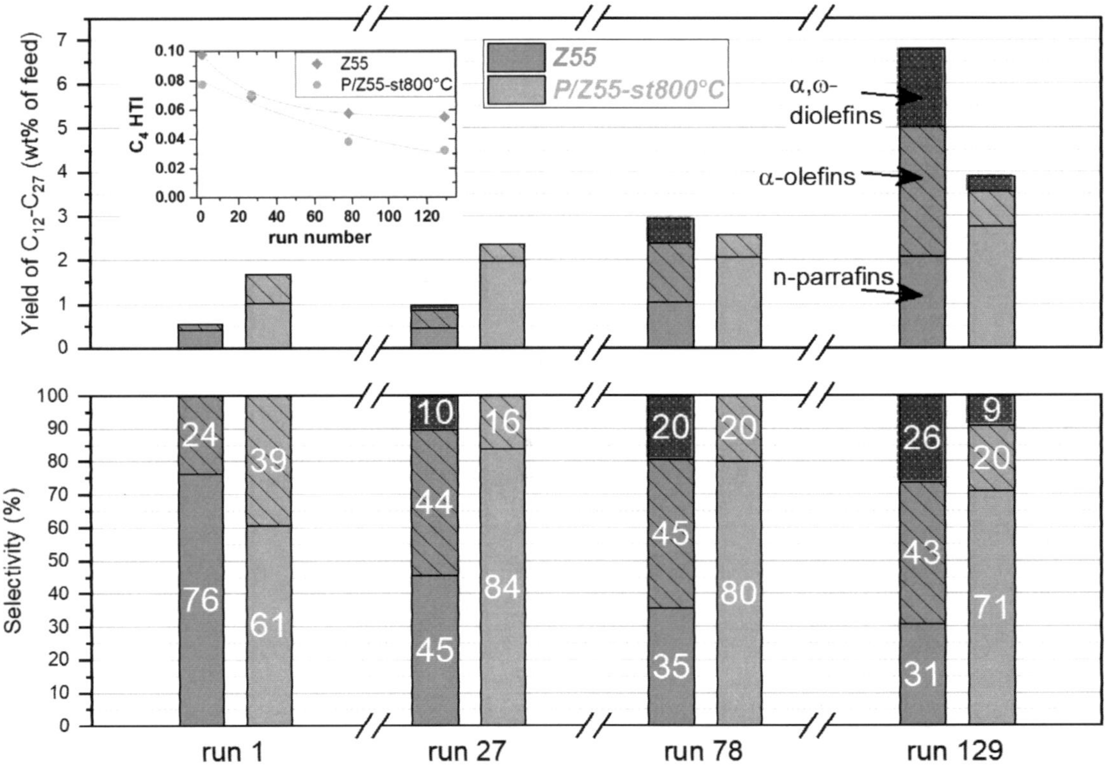

Figure 5. Yields of C_{12}-C_{27} *n*-paraffins, α-olefins, and *α,ω*-diolefins. Minor yields of beta-olefins, iso-olefins/paraffins at much lower selectivites compared to the three main product groups not shown. Insert in top figure shows C_4 hydrogen transfer index.

Coke quantification

Even though a lower amount (8 mg) of *Z55* was loaded to obtain a similar initial conversion of C_{12+} products as with 32 mg of *P/Z55-ss*, after 130 runs the coke yields using 8 mg *Z55* was very similar with 0.17 wt. % compared to 0.20 wt. % using 32 mg *P/Z55-ss*. This clearly demonstrates the higher coking propensity of the unmodified parent *Z55*, particularly apparent when comparing the coke deposited per surface area, which was 8 μg/m^2 using *P/Z55-ss* and 27 μg/m^2 for the parent *Z55*. This can explain the faster loss in activity observed for the parent *Z55* (Figure 4).

2.3.3. Catalyst stability during reaction/regeneration cycles

Since the catalyst was already steam-treated under high severity (800 °C), there was no loss in activity apparent for repeated cycles of catalyst regeneration at 700 °C, and reaction at 600 °C (
Table **7**).

Table 7. Yields obtained from six reaction-regeneration cycles using HDPE as feed. Conditions: catalyst = *P/mesoZ55-ss*, catalyst/feed ratio per reaction = 80, pyrolysis and catalyst temperature were 550 and 600 °C, respectively. Catalyst regeneration was achieved by increasing the catalyst temperature to 700 °C and dosing pulses of oxygen until no more production of CO_2 and CO was observed.

Reaction cycle	H_2	CH_4	C_2-C_4 olefins	Ethylene	Propylene	1,3-Butadiene	Other C_4=	C_4 alkanes	Aromatics	Aromatic-free C_5-C_{11}	C_{12}-C_{20}	C_{21}-C_{35}
1	0.1	0.3	80.0	5.9	36.8	2.2	34.6	0.9	4.2	11.7	2.4	0.4
2	0.1	0.3	81.8	5.5	43.4	2.5	29.8	1.0	3.6	11.4	1.6	0.3
3	0.1	0.4	81.1	6.3	43.6	3.5	26.3	0.9	5.4	9.9	2.0	0.3
4	0.1	0.4	80.1	5.4	40.3	2.9	30.9	1.0	3.7	12.4	2.1	0.3
5	0.1	0.3	79.8	5.5	40.0	2.7	26.4	1.0	3.7	12.5	2.2	0.4
6	0.1	0.4	80.8	5.6	41.7	2.3	30.7	0.9	3.5	12.0	2.1	0.3

Plantwide process evaluation

The performance of the studied catalytic pyrolysis of plastic waste streams to valuable base chemicals was further evaluated in a plantwide context. Particularly, the energy and carbon footprint of the direct catalytic process studied in the present work at a catalyst temperature of 600 °C (=Case 1) were estimated and compared against the conventional two-step counterpart (Case 2) comprising pyrolysis of the plastic waste streams followed by pyrolytic oil steam cracking. To this end, detailed process flow diagrams (PFDs) of the reaction section were developed for both cases and optimized with respect to heat minimization. Beside fossil-based heating (provided by fired heaters), electrically conductive heating (applied by electrified heaters) was also considered in the analysis. The latter constitutes a modern and efficient alternative that can (potentially) fully run on renewable electricity, thereby, preventing direct CO_2 emissions sourced by fuel combustion [91].

The following analysis was performed on the basis of i) the feed: 1 kg of plastic waste; polyethylene was considered as a model compound and ii) the final product: 1 kg of high value chemicals (HVC) comprising a mixture of ethylene (C_2H_4), propylene (C_3H_6), benzene (C_6H_6), hydrogen (H_2) and butadiene (C_4H_6) to facilitate a fair benchmarking over other peer processes.

Energy footprint

A heat integration targeting to the reactors energy duty minimization (i.e. catalytic reactor, pyrolyzer and steam cracker) was initially performed. The energy footprints by means of total energy demand of the studied (Case 1) and conventional (Case 2) processes are presented in Figure 6. Both fossil-based (Figure 6a) and electrically conductive heating (Figure 6b) alternatives are included. The former heating alternative always resulted in a higher total energy demand than the latter in both process cases. The reasoning lies in the higher heat losses occurring in the heating elements which subsequently drive the heater efficiency: 65.6% [92] and 95% [93] are representative efficiency values for the fired and electrified heaters, respectively. Although multiple heat exchangers are placed in series to further minimize the heat losses in the flue gas when fired heaters are used, the high operating temperatures (usually higher than that of the flue gas) do not always allow for high heat integration potential. Notably, no flue gas is present when electrified heaters are used.

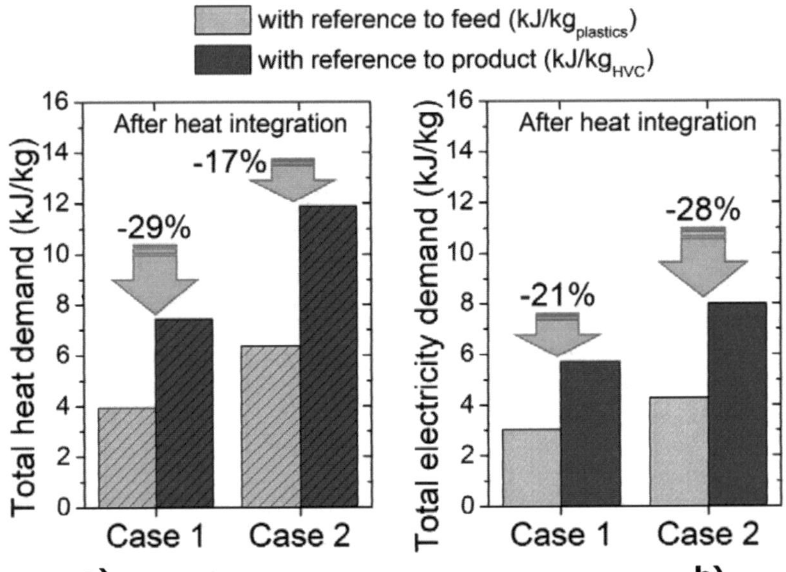

a) **b)**

Figure 6. Energy footprint by means of total energy demand of the studied (Case 1) and conventional (Case 2) processes after heat integration: a) use of fired heaters running on natural gas and b) use of electrified heaters. The downwards pointing arrows indicate the reducing in heat/electricity demand after heat integration, in comparison to the demand before heat integration.

Collectively, 3.9 $kJ_{heat}/kg_{plastics}$ (7.4 kJ_{heat}/kg_{HVC}) and 6.4 $kJ_{heat}/kg_{plastics}$ (11.9 kJ_{heat}/kg_{HVC}) are required for Case 1 and Case 2, respectively, when fossil-based heating is applied, whereas 3.0 $kJ_{el.}/kg_{plastics}$ (5.7 $kJ_{el.}/kg_{HVC}$) and 4.3 $kJ_{el.}/kg_{plastics}$ (8.0 $kJ_{el.}/kg_{HVC}$) are required for Case 1 and Case 2, respectively, when electrically conductive heating is applied. Overall, in the studied process (Case 1), the energy duty is about 38% lower than in the conventional counterpart (Case 2) when fired heaters are employed and about 30% lower when electrified heaters are employed, showing that plastics cracking and catalytic upgrading integrated in one unit can lead to significantly reduced energy inputs.

Carbon footprint

Herein, a preliminary ex-ante life cycle assessment focusing on the process carbon footprint (expressed in kg CO_2-eq./kg) was performed on the basis of the energy demand that was reported in the previous section. The foreground data (i.e. energy demand) was linked to background data from the ecoinvent v3.6 database [94]. Background processes (i.e. electricity and heat supply) were mostly representative of European market (heat and electricity supply). The cut-off system model was applied [95]: the waste stream impacts were fully ascribed to the waste producer but the treatment activity impacts were attributed to the party that is in charge of the waste stream valorization since there is an economic interest (chemicals production). Following this reasoning, the plastic waste stream was considered burden-free while the energy demand associated to the treatment activity was charged on the foreground processes (Case 1 and Case 2). The catalyst synthesis carbon footprint was neglected [96].

The carbon footprint of both the studied (Case 1) and conventional (Case 2) processes considering fired heaters running on natural gas and electrified heaters running on grid, solar and wind harvested electricity is shown in Figure 7 (Figure 7a: with reference to feed; Figure 7b:with reference to product:). Both processes attain the highest carbon footprint (even higher than that of fossil-based heat as energy source) when grid electricity is utilized as energy source; specifically, 0.36 kg CO_2-eq./$kg_{plastics}$ (0.68 kg CO_2-eq./kg_{HVC}) and 0.51 kg CO_2-eq./$kg_{plastics}$ (0.95 kg CO_2-eq./kg_{HVC}) for Case 1 and Case 2, respectively. Although a mix of fossil fuels and renewable sources is considered in grid electricity generation, the unitary impact of grid electricity gets higher than that of fossil-based heat after applying the conversion efficiency (0.42 kg CO_2-eq./$kWh_{grid\ el.}$ versus 0.27 kg CO_2-eq./kWh_{heat}). Yet, both processes attain significantly low carbon footprint when renewable electricity solely drives the heaters: about 80% lower environmental burden is attained when solar harvested electricity is utilized instead of grid electricity. In case of wind harvested electricity utilization, the lowest carbon footprint is attained: 0.03 kg CO_2-eq./$kg_{plastics}$ (0.05 kg CO_2-eq./kg_{HVC}) and 0.04 kg CO_2-eq./$kg_{plastics}$ (0.07 kg CO_2-eq./kg_{HVC}) for Case 1 and Case 2, respectively. The unitary impact of solar harvested electricity is about two times the wind harvested one since higher environmental burden is imposed along the life cycle of a photovoltaic panel as compared to that of a wind turbine.

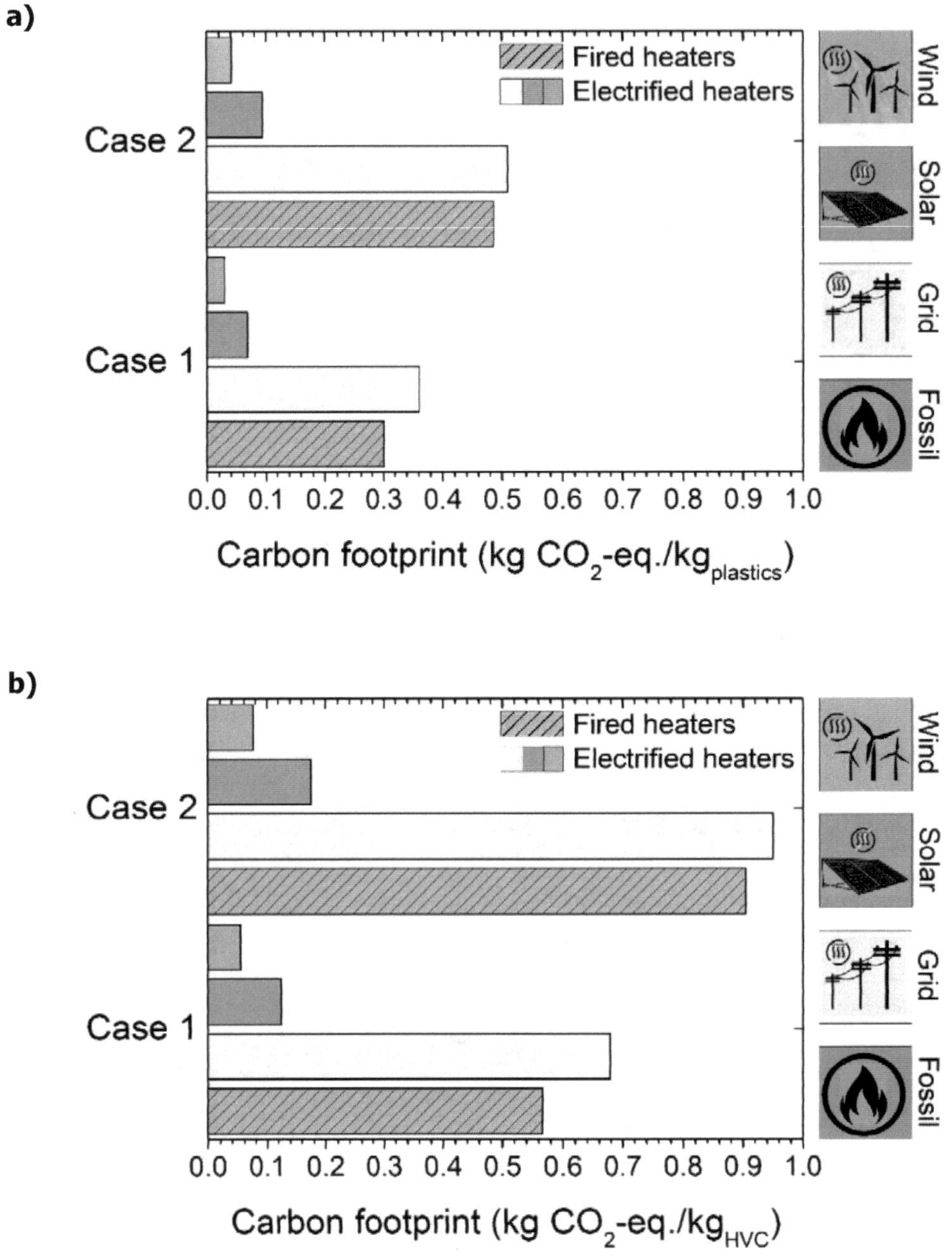

Figure 7. Carbon footprint of both the studied (Case 1) and conventional (Case 2) processes after heat integration for fired heaters running on natural gas and electrified heaters running on grid, solar and wind electricity: a) with reference to feed (expressed in kg CO_2-eq./$kg_{plastics}$) and b) with reference to product (expressed in kg CO_2-eq./kg_{HVC}).

Peer processes benchmarking

a)

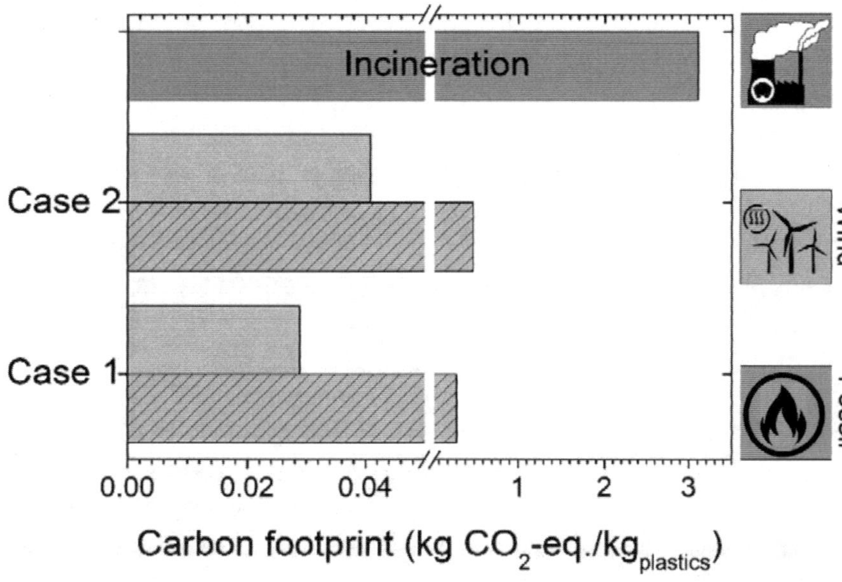

b)

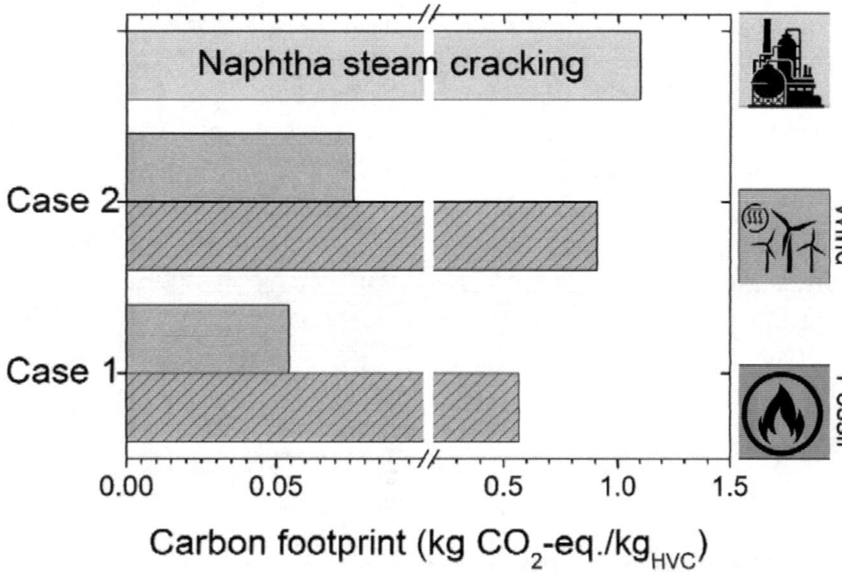

Figure 8. Carbon footprint benchmarking of both the studied (Case 1) and conventional (Case 2) processes for fired heaters (running on natural gas) and electrified heaters (running on wind electricity) against traditional peer processes: a) plastic waste streams incineration and b) high value chemicals synthesis via naphtha steam cracking.

A process benchmarking from an environmental perspective is presented in Figure 8. The traditional plastic waste incineration (Figure 8a) and naphtha steam cracking (Figure 8b) were considered as benchmarks. Figure 7a shows the environmental impacts of Case 1 and Case 2, as potential plastic waste valorization processes, in comparison to that of incineration. The functional unit for this comparison was 1 kg of plastic waste valorization, while a gate-to-gate approach was applied. Figure 8b shows the environmental impacts of Case 1 and Case 2, as alternative processes for HVC synthesis, in comparison to naphtha steam cracking. The functional unit for this comparison was 1 kg of HVC synthesis, while a cradle-to-gate approach was applied. The carbon footprint value of the incineration process was calculated based on the plastic waste combustion reaction, assuming polyethylene as the main plastic waste stream and the carbon footprint value of the naphtha steam cracking was obtained from literature data [97] after considering an intermediate value (1.1 kg CO_2-eq./kg_{HVC}) between world-average and state-of-the-art processes.

Both Case 1 and Case 2 constitute far more sustainable (plastic waste) valorization process alternatives than the traditional incineration, regardless the primary energy source utilized (Figure 8a). At least an order of magnitude lower carbon footprint can be attained when fired heaters running on natural gas are employed whereas, notably, two orders of magnitude lower carbon footprint can be attained if the fired heaters are replaced to electrified heaters running on wind harvested electricity. In all scenarios, the studied catalytic process (Case 1) results in the lowest possible environmental burden (Figure 8a).
Figure 8b shows the same trend when Case 1 and Case 2 are proposed for HVC synthesis. At least 45% and 15% reduction in the carbon footprint of Case 1 and Case 2, respectively, is attained when HVC are produced by plastic waste in fired heaters than oil-based naphtha in thermally driven steam crackers. Although a higher amount of energy is required to melt and crack the long carbon chains of the plastic waste as compared to naphtha steam cracking, a lower environmental impact is attained eventually. This is ascribed to the fact that plastic waste that serves as carbon source is burden-free, unlike oil-based naphtha which accounts for almost 30% of the total carbon footprint [97]. Remarkably, the environmental burden drops to an order of magnitude lower values upon switching to electrified heaters. Likewise the previous comparison, the proposed catalytic process (Case 1) results in the lowest possible environmental burden.

Conclusion

Phosphorous-modified conventional and mesoporous HZSM-5 zeolites were demonstrated to be suitable catalysts for producing valuable base chemicals such as C_2-C_4 olefins and aromatics from different virgin and post-consumer waste polyolefins. After steam treatment at 800 °C, the acidity of all catalyst was reduced to ~0.10-0.15 mmol NH_3/g, resulting in a lower activity. However, these catalysts can be considered stable under upgrading and regeneration conditions which are not expected to exceed 700 °C in the process. Varying the catalyst/feed ratio between 10 and 80 and the catalyst temperature between 500 and 700 °C affected the proportion of different monomers. This provides great flexibility for plant operators to steer the product slate towards the highest market demand and maximize C_4 olefins, ethylene, propylene, or aromatics production, all while keeping carbon losses to CH_4 and coke reasonably low, below 1.5% and 0.5 %, respectively.

The yield structure obtained from a synthetic polyolefins mixture of LLDPE, LDPE, HDPE and PP matched the relative contributions of the single polymers, and a monomers yields of 84 wt. %, i.e., 79 wt. % C_2-C_4 olefins and 5 wt. % aromatics was obtained. When using post-consumer MPO feed, the monomer yield was slightly lower with 74 wt. % C_2-C_4 olefins and 9 wt. % aromatics, and this is attributed to the higher aromatics content in the feed, which is more difficult to crack.

While parent HZSM-5 rapidly deactivated and showed a high coking propensity (27 µg of coke /m^2 catalyst), HZSM-5 that was modified with 2 wt. % P and steam treated at 800 °C showed almost no deactivation during 130 runs with stable conversion of C_{5+} aliphatics and high C_2-C_4 olefins selectivity with a much lower coke deposition of 8 µg/m^2.

Finally, it was found that catalytic pyrolysis requires ~38% lower energy input than plastics pyrolysis followed by pyrolytic oil steam cracking. The studied catalytic process constitutes a sustainable alternative process for plastic waste valorization since it leads to (at least) an order of magnitude (0.3 versus 3.1 kg CO_2-eq./$kg_{plastics}$) lower environmental burden than the current practice (incineration). In addition, it constitutes an excellent paradigm of circular economy by closing the carbon balance upcycle: plastic wastes are transformed to fresh building blocks (i.e. HVC) which can further be used for new polymers production, while this cycle can (theoretically) be repeated numerous times.

Future investigations should be directed to study the techno-economic analysis of the direct pyrolysis vapor upgrading process compared to steam cracking of pyrolysis-derived liquid.

Acknowledgements

This work was performed in the framework of the Catalisti clusterSBO project WATCH (HBC.2019.0001 "Plastic waste to chemicals"), with the financial support of VLAIO (Flemish Agency for Innovation and Entrepreneurship). The research leading to these results has also received funding from the European Research Council (ERC) under the European Union's Horizon 2020 Programme (P8/2007-2013)/ERC grant agreement no. 818607 (OPTIMA). In addition, the authors would like to acknowledge the financial support from the Fund for Scientific Research Flanders (FWO) for the project WASTE. We are indebted to Prof. Dr. Kim Ragaert for providing the cold-washed mixed polyolefin pellets used as feed in present work and we thank Ph.D. student Oğuzhan Akin for assisting with the preparation of micropyrolysis tests.

References

[1] M. Worsdorfer, The future of petrochemicals, in: Futur. Petrochemicals, IEA, 2018. https://doi.org/10.1787/9789264307414-en.

[2] O. Awayssa, N. Al-Yassir, A. Aitani, S. Al-Khattaf, Modified HZSM-5 as FCC additive for enhancing light olefins yield from catalytic cracking of VGO, Appl. Catal. A Gen. 477 (2014) 172–183. https://doi.org/10.1016/j.apcata.2014.03.021.

[3] A. Akah, M. Al-Ghrami, Maximizing propylene production via FCC technology, Appl. Petrochemical Res. 5 (2015) 377–392. https://doi.org/10.1007/s13203-015-0104-3.

[4] X. Meng, C. Xu, J. Gao, L. Li, Studies on catalytic pyrolysis of heavy oils: Reaction behaviors and mechanistic pathways, Appl. Catal. A Gen. 294 (2005) 168–176. https://doi.org/10.1016/j.apcata.2005.07.033.
[5] A.I. Hussain, A.M. Aitani, M. Kubů, Jıϵ. Čejka, S. Al-Khattaf, Catalytic cracking of Arabian Light VGO over novel zeolites as FCC catalyst additives for maximizing propylene yield, Fuel. 167 (2016) 226–239. https://doi.org/10.1016/j.fuel.2015.11.065.
[6] M. Roosen, N. Mys, M. Kusenberg, P. Billen, A. Dumoulin, J. Dewulf, K.M. Van Geem, K. Ragaert, S. De Meester, Detailed Analysis of the Composition of Selected Plastic Packaging Waste Products and Its Implications for Mechanical and Thermochemical Recycling, Environ. Sci. Technol. 54 (2020) 13282–13293. https://doi.org/10.1021/acs.est.0c03371.
[7] K. Ragaert, L. Delva, K. Van Geem, Mechanical and chemical recycling of solid plastic waste, Waste Manag. 69 (2017) 24–58. https://doi.org/10.1016/j.wasman.2017.07.044.
[8] J.-P. Lange, Towards circular carbo-chemicals – the metamorphosis of petrochemicals, Energy Environ. Sci. (2021). https://doi.org/10.1039/d1ee00532d.
[9] K. Kleinhans, M. Hallemans, S. Huysveld, G. Thomassen, K. Ragaert, K.M. Van Geem, M. Roosen, N. Mys, J. Dewulf, S. De Meester, Development and application of a predictive modelling approach for household packaging waste flows in sorting facilities, Waste Manag. 120 (2021) 290–302. https://doi.org/10.1016/j.wasman.2020.11.056.
[10] J.-P. Lange, Towards circular carbo-chemicals – the metamorphosis of petrochemicals, Energy Environ. Sci. (2021). https://doi.org/DOI: 10.1039/D1EE00532D.
[11] M. Arabiourrutia, G. Elordi, G. Lopez, E. Borsella, J. Bilbao, M. Olazar, Characterization of the waxes obtained by the pyrolysis of polyolefin plastics in a conical spouted bed reactor, J. Anal. Appl. Pyrolysis. 94 (2012) 230–237. https://doi.org/10.1016/j.jaap.2011.12.012.
[12] W. Kaminsky, Chemical recycling of plastics by fluidized bed pyrolysis, Fuel Commun. 8 (2021) 100023. https://doi.org/10.1016/J.JFUECO.2021.100023.
[13] M. Kusenberg, A. Zayoud, M. Roosen, H.D. Thi, M.S. Abbas-Abadi, U. Eschenbacher, Andreas Kresovic, S. De Meester, K.M. Van Geem, A comprehensive experimental investigation of plastic waste pyrolysis oil quality and its dependence on the plastic waste composition, Fuel Process. Technol. (2021). https://doi.org/10.1016/j.fuproc.2021.107090.
[14] M. Kusenberg, K.M. Van Geem, Opportunities and challenges for the application of post-consumer plastic waste pyrolysis oils as steam cracker feedstocks: to decontaminate or not to decontaminate, Waste Manag. accepted (2021).
[15] Recycling and the future of the plastics industry | McKinsey, (n.d.). https://www.mckinsey.com/industries/chemicals/our-insights/how-plastics-waste-recycling-could-transform-the-chemical-industry (accessed June 22, 2021).
[16] M. Solis, S. Silveira, Technologies for chemical recycling of household plastics – A technical review and TRL assessment, Waste Manag. 105 (2020) 128–138. https://doi.org/10.1016/j.wasman.2020.01.038.
[17] R. Palos, A. Gutiérrez, F.J. Vela, M. Olazar, J.M. Arandes, J. Bilbao, Waste Refinery: The Valorization of Waste Plastics and End-of-Life Tires in Refinery Units. A Review,

Energy and Fuels. 35 (2021) 3529–3557. https://doi.org/10.1021/acs.energyfuels.0c03918.
[18] L.O. Mark, M.C. Cendejas, I. Hermans, The Use of Heterogeneous Catalysis in the Chemical Valorization of Plastic Waste, ChemSusChem. 13 (2020) 5808–5836. https://doi.org/10.1002/cssc.202001905.
[19] S.D. Anuar Sharuddin, F. Abnisa, W.M.A. Wan Daud, M.K. Aroua, A review on pyrolysis of plastic wastes, Energy Convers. Manag. 115 (2016) 308–326. https://doi.org/10.1016/j.enconman.2016.02.037.
[20] D.P. Serrano, J. Aguado, J.M. Escola, Developing advanced catalysts for the conversion of polyolefinic waste plastics into fuels and chemicals, ACS Catal. 2 (2012) 1924–1941. https://doi.org/10.1021/cs3003403.
[21] A. Eschenbacher, P.A. Jensen, U.B. Henriksen, J. Ahrenfeldt, C. Li, J.Ø. Duus, U.V. Mentzel, A.D. Jensen, Impact of ZSM-5 deactivation on bio-oil quality during upgrading of straw derived pyrolysis vapors, Energy & Fuels. 33 (2019) 397–412. https://doi.org/10.1021/acs.energyfuels.8b03691.
[22] A. Eschenbacher, P.A. Jensen, U.B. Henriksen, J. Ahrenfeldt, C. Li, J.Ø. Duus, U.V. Mentzel, A.D. Jensen, Deoxygenation of wheat straw fast pyrolysis vapors using HZSM-5, Al2O3 , HZSM-5/Al2O3 extrudates, and desilicated HZSM-5/Al2O3 extrudates, Energy & Fuels. 33 (2019) 6405–6420. https://doi.org/10.1021/acs.energyfuels.9b00906.
[23] K.G. Kalogiannis, S.D. Stefanidis, A.A. Lappas, Catalyst deactivation, ash accumulation and bio-oil deoxygenation during ex situ catalytic fast pyrolysis of biomass in a cascade thermal-catalytic reactor system, Fuel Process. Technol. 186 (2019) 99–109. https://doi.org/10.1016/j.fuproc.2018.12.008.
[24] J.C. Groen, L.A.A. Peffer, J.A. Moulijn, J. Pérez-Ramírez, Mesoporosity development in ZSM-5 zeolite upon optimized desilication conditions in alkaline medium, Colloids Surfaces A Physicochem. Eng. Asp. 241 (2004) 53–58. https://doi.org/10.1016/j.colsurfa.2004.04.012.
[25] J.C. Groen, J. a. Moulijn, J. Peréz-Ramírez, Desilication: on the controlled generation of mesoporosity in MFI zeolites, J. Mater. Chem. 16 (2006) 2121. https://doi.org/10.1039/b517510k.
[26] A. Eschenbacher, P.A. Jensen, U.B. Henriksen, J. Ahrenfeldt, S. Ndoni, C. Li, J.Ø. Duus, U.V. Mentzel, A.D. Jensen, Catalytic deoxygenation of vapors obtained from ablative fast pyrolysis of wheat straw using mesoporous HZSM-5, Fuel Process. Technol. 194 (2019) 106119. https://doi.org/10.1016/J.FUPROC.2019.106119.
[27] M. Holm, E. Taarning, K. Egeblad, C.H. Christensen, Catalysis with hierarchical zeolites, Catal. Today. 168 (2011) 3–16. https://doi.org/10.1016/j.cattod.2011.01.007.
[28] D.P. Serrano, R.A. García, G. Vicente, M. Linares, D. Procházková, J. Čejka, Acidic and catalytic properties of hierarchical zeolites and hybrid ordered mesoporous materials assembled from MFI protozeolitic units, J. Catal. 279 (2011) 366–380. https://doi.org/10.1016/j.jcat.2011.02.007.
[29] D. Verboekend, M. Milina, S. Mitchell, J. Pérez-Ramírez, Hierarchical zeolites by desilication: Occurrence and catalytic impact of recrystallization and restructuring, Cryst. Growth Des. 13 (2013) 5025–5035. https://doi.org/10.1021/cg4010483.

[30] J. Garcia-Martinez, K. Li, Mesoporous Zeolites : Preparation , Characterization and Applications, Wiley-VCH Verlag GmbH & Co KGaA, 2015.
[31] M.A. Bari Siddiqui, A.M. Aitani, M.R. Saeed, S. Al-Khattaf, Enhancing the production of light olefins by catalytic cracking of FCC naphtha over mesoporous ZSM-5 catalyst, Top. Catal. 53 (2010) 1387–1393. https://doi.org/10.1007/s11244-010-9598-1.
[32] A. Eschenbacher, F. Goodarzi, A. Saraeian, S. Kegnæs, B.H.B.H. Shanks, A.D.A.D. Jensen, Performance of Mesoporous HZSM-5 and Silicalite-1 Coated Mesoporous HZSM-5 Catalysts for Deoxygenation of Straw Fast Pyrolysis Vapors, J. Anal. Appl. Pyrolysis. 145 (2019) 104712. https://doi.org/10.1016/J.JAAP.2019.104712.
[33] A. Eschenbacher, J.A. Andersen, A.D. Jensen, Catalytic conversion of acetol over HZSM-5 catalysts – influence of Si/Al ratio and introduction of mesoporosity, Catal. Today. (2020). https://doi.org/10.1016/j.cattod.2020.03.041.
[34] L.Y. Jia, M. Raad, S. Hamieh, J. Toufaily, T. Hamieh, M.M. Bettahar, G. Mauviel, M. Tarrighi, L. Pinard, A. Dufour, Catalytic fast pyrolysis of biomass: Superior selectivity of hierarchical zeolites to aromatics, Green Chem. 19 (2017) 5442–5459. https://doi.org/10.1039/c7gc02309j.
[35] T. Blasco, A. Corma, J. Martínez-Triguero, Hydrothermal stabilization of ZSM-5 catalytic-cracking additives by phosphorus addition, J. Catal. 237 (2006) 267–277. https://doi.org/10.1016/j.jcat.2005.11.011.
[36] L. Huang, Q. Li, Enhanced Acidity and Thermal Stability of Mesoporous Materials with Post-treatment with Phosphoric Acid, Chem. Lett. (1999) 829–830.
[37] J.N. Louwen, L. van Eijck, C. Vogt, E.T.C. Vogt, Understanding the Activation of ZSM-5 by Phosphorus: Localizing Phosphate Groups in the Pores of Phosphate-Stabilized ZSM-5, Chem. Mater. (2020). https://doi.org/10.1021/ACS.CHEMMATER.0C03411.
[38] H.E. Van Der Bij, F. Meirer, S. Kalirai, J. Wang, B.M. Weckhuysen, Hexane cracking over steamed phosphated zeolite H-ZSM-5: Promotional effect on catalyst performance and stability, Chem. - A Eur. J. 20 (2014) 16922–16932. https://doi.org/10.1002/chem.201404924.
[39] G. Caeiro, P. Magnoux, J.M. Lopes, F.R. Ribeiro, S.M.C. Menezes, A.F. Costa, H.S. Cerqueira, Stabilization effect of phosphorus on steamed H-MFI zeolites, Appl. Catal. A Gen. 314 (2006) 160–171. https://doi.org/10.1016/j.apcata.2006.08.016.
[40] A. Corma, J. Mengual, P.J. Miguel, Stabilization of ZSM-5 zeolite catalysts for steam catalytic cracking of naphtha for production of propene and ethene, "Applied Catal. A, Gen. 421–422 (2012) 121–134. https://doi.org/10.1016/j.apcata.2012.02.008.
[41] O.D. Mante, F.A. Agblevor, S.T. Oyama, R. McClung, The effect of hydrothermal treatment of FCC catalysts and ZSM-5 additives in catalytic conversion of biomass, Appl. Catal. A Gen. 445–446 (2012) 312–320. https://doi.org/10.1016/j.apcata.2012.08.039.
[42] W. Yao, J. Li, Y. Feng, W. Wang, X. Zhang, Q. Chen, S. Komarneni, Y. Wang, Thermally stable phosphorus and nickel modified ZSM-5 zeolites for catalytic co-pyrolysis of biomass and plastics, RSC Adv. 5 (2015) 1–4. https://doi.org/10.1039/C5RA02947C.
[43] I. Torre, J.M. Arandes, P. Castano, M. Azkoiti, J. Bilbao, H.I. de Lasa, Catalytic cracking of plastic pyrolysis waxes with vacuum gasoil: effect of HZSM-5 zeolite in the FCC catalyst, Int. J. Chem. React. Eng. 4 (2006).

[44] R.H. Harding, A.W. Peters, J.R.D. Nee, New developments in FCC catalyst technology, Appl. Catal. A Gen. 221 (2001) 389–396. https://doi.org/10.1016/S0926-860X(01)00814-6.

[45] H.E. van der Bij, B.M. Weckhuysen, Phosphorus promotion and poisoning in zeolite-based materials: synthesis, characterisation and catalysis, Chem. Soc. Rev. 44 (2015) 7406–7428. https://doi.org/10.1039/C5CS00109A.

[46] H.E. van der Bij, B.M. Weckhuysen, Local silico-aluminophosphate interfaces within phosphated H-ZSM-5 zeolites, Phys. Chem. Chem. Phys. 16 (2014) 9892. https://doi.org/10.1039/c3cp54791d.

[47] T.F. Degnan, G.K. Chitnis, P.H. Schipper, History of ZSM-5 fluid catalytic cracking additive development at Mobil, Microporous Mesoporous Mater. 35–36 (2000) 245–252. https://doi.org/10.1016/S1387-1811(99)00225-5.

[48] Y.J. Lee, J.M. Kim, J.W. Bae, C.H. Shin, K.W. Jun, Phosphorus induced hydrothermal stability and enhanced catalytic activity of ZSM-5 in methanol to DME conversion, Fuel. 88 (2009) 1915–1921. https://doi.org/10.1016/j.fuel.2009.04.007.

[49] N. Xue, X. Chen, L. Nie, X. Guo, W. Ding, Y. Chen, M. Gu, Z. Xie, Understanding the enhancement of catalytic performance for olefin cracking: Hydrothermally stable acids in P/HZSM-5, J. Catal. 248 (2007) 20–28. https://doi.org/10.1016/j.jcat.2007.02.022.

[50] G. Zhao, J. Teng, Z. Xie, W. Jin, W. Yang, Q. Chen, Y. Tang, Effect of phosphorus on HZSM-5 catalyst for C4-olefin cracking reactions to produce propylene, J. Catal. 248 (2007) 29–37. https://doi.org/10.1016/j.jcat.2007.02.027.

[51] N. Xue, L. Nie, D. Fang, X. Guo, J. Shen, W. Ding, Y. Chen, Synergistic effects of tungsten and phosphorus on catalytic cracking of butene to propene over HZSM-5, Appl. Catal. A Gen. 352 (2009) 87–94. https://doi.org/10.1016/j.apcata.2008.09.029.

[52] A. Corma, J. Mengual, P.J. Miguel, Stabilization of ZSM-5 zeolite catalysts for steam catalytic cracking of naphtha for production of propene and ethene, Appl. Catal. A Gen. 421–422 (2012) 121–134. https://doi.org/10.1016/j.apcata.2012.02.008.

[53] Y. Xue, Thermochemical conversion of organic and plastic waste materials through pyrolysis, (2017).

[54] A. Eschenbacher, R.J. Varghese, M. SeifaliAbbas-Abadi, K.M. Van Geem, Maximizing light olefins and aromatics as high value base chemicals via single step catalytic conversion of plastic waste, Chem. Eng. J. (2021) 132087. https://doi.org/10.1016/J.CEJ.2021.132087.

[55] A. Eschenbacher, F. Goodarzi, R.J. Varghese, K. Enemark-Rasmussen, S. Kegnæs, M.S. Abbas-Abadi, K.M. Van Geem, Boron-Modified Mesoporous ZSM-5 for the Conversion of Pyrolysis Vapors from LDPE and Mixed Polyolefins: Maximizing the C 2 –C 4 Olefin Yield with Minimal Carbon Footprint, ACS Sustain. Chem. Eng. 9 (2021) 14618–14630. https://doi.org/10.1021/acssuschemeng.1c06098.

[56] C.A. Emeis, Determination of integrated molar extinction coefficients for infrared absorption bands of pyridine adsorbed on solid acid catalysts, J. Catal. 141 (1993) 347–354. https://doi.org/10.1006/jcat.1993.1145.

[57] K. Schofield, The enigmatic mechanism of the flame ionization detector: Its overlooked implications for fossil fuel combustion modeling, Prog. Energy Combust. Sci. 34 (2008) 330–350. https://doi.org/10.1016/j.pecs.2007.08.001.

[58] Y. Xue, P. Johnston, X. Bai, Effect of catalyst contact mode and gas atmosphere during catalytic pyrolysis of waste plastics, Energy Convers. Manag. 142 (2017) 441–451.
[59] Y. Shen, Z. Qin, S. Asahina, N. Asano, G. Zhang, S. Qian, Y. Ma, Z. Yan, X. Liu, S. Mintova, The inner heterogeneity of ZSM-5 zeolite crystals, J. Mater. Chem. A. 9 (2021) 4203–4212. https://doi.org/10.1039/d0ta11023j.
[60] R.M. Dessau, E.W. Valyocsik, N.H. Goeke, Aluminum zoning in ZSM-5 as revealed by selective silica removal, Zeolites. 12 (1992) 776–779. https://doi.org/10.1016/0144-2449(92)90049-U.
[61] D. Fodor, F. Krumeich, R. Hauert, J.A. Van Bokhoven, Differences between individual ZSM-5 crystals in forming hollow single crystals and mesopores during base leaching, Chem. - A Eur. J. 21 (2015) 6272–6277. https://doi.org/10.1002/chem.201406182.
[62] D. Verboekend, S. Mitchell, M. Milina, J.C. Groen, J. Pérez-Ramírez, J. Peréz-Ramírez, J. Pérez-Ramírez, Full compositional flexibility in the preparation of mesoporous MFI zeolites by desilication, J. Phys. Chem. C. 115 (2011) 14193–14203. https://doi.org/10.1021/jp201671s.
[63] H. Chen, H. Cheng, F. Zhou, K. Chen, K. Qiao, X. Lu, P. Ouyang, J. Fu, Catalytic fast pyrolysis of rice straw to aromatic compounds over hierarchical HZSM-5 produced by alkali treatment and metal-modification, J. Anal. Appl. Pyrolysis. 131 (2018) 76–84. https://doi.org/10.1016/j.jaap.2018.02.009.
[64] A. Yamaguchi, D. Jin, T. Ikeda, K. Sato, N. Hiyoshi, T. Hanaoka, F. Mizukami, M. Shirai, Deactivation of ZSM-5 zeolite during catalytic steam cracking of n-hexane, Fuel Process. Technol. 126 (2014) 343–349. https://doi.org/10.1016/J.FUPROC.2014.05.013.
[65] L.H. Ong, M.M. Dömök, R. Olindo, A.C. Van Veen, J.A. Lercher, A.C. van Veen, J.A. Lercher, Dealumination of HZSM-5 via steam-treatment, Microporous Mesoporous Mater. 164 (2012) 9–20. https://doi.org/10.1016/j.micromeso.2012.07.033.
[66] A.P. Hawkins, A. Zachariou, S.F. Parker, P. Collier, N. Barrow, I.P. Silverwood, R.F. Howe, D. Lennon, Effect of steam de-alumination on the interactions of propene with H-ZSM-5 zeolites, RSC Adv. 10 (2020) 23136–23147. https://doi.org/10.1039/d0ra03871g.
[67] X. Gao, Z. Tang, H. Zhang, C. Liu, Z. Zhang, G. Lu, D. Ji, High performance phosphorus-modified ZSM-5 zeolite for butene catalytic cracking, Korean J. Chem. Eng. 27 (2010) 812–815. https://doi.org/10.1007/s11814-010-0134-6.
[68] H.E. Van Der Bij, B.M. Weckhuysen, Local silico-aluminophosphate interfaces within phosphated H-ZSM-5 zeolites, Phys. Chem. Chem. Phys. 16 (2014) 9892–9903. https://doi.org/10.1039/c3cp54791d.
[69] G.L. Woolery, G.H. Kuehl, On the nature of framework Brønsted and Lewis acid sites in ZSM-5, Zeolites. 19 (1997) 288–296. https://doi.org/10.1016/S0144-2449(97)00086-9.
[70] J. Zhuang, D. Ma, G. Yang, Z. Yan, X. Liu, X. Liu, X. Han, X. Bao, P. Xie, Z. Liu, Solid-state MAS NMR studies on the hydrothermal stability of the zeolite catalysts for residual oil selective catalytic cracking, J. Catal. 228 (2004) 234–242. https://doi.org/10.1016/j.jcat.2004.08.034.
[71] K. Damodaran, J.W. Wiench, S.M. Cabral de Menezes, Y.L. Lam, J. Trebosc, J.P. Amoureux, M. Pruski, Modification of H-ZSM-5 zeolites with phosphorus. 2. Interaction

between phosphorus and aluminum studied by solid-state NMR spectroscopy, Microporous Mesoporous Mater. 95 (2006) 296–305. https://doi.org/10.1016/j.micromeso.2006.05.034.
[72] H.E. Van Der Bij, L.R. Aramburo, B. Arstad, J.J. Dynes, J. Wang, B.M. Weckhuysen, Phosphatation of zeolite H-ZSM-5: A combined microscopy and spectroscopy study, ChemPhysChem. 15 (2014) 283–292. https://doi.org/10.1002/cphc.201300910.
[73] G. Lischke, R. Eckelt, H.G. Jerschkewitz, B. Parlitz, E. Schreier, W. Storek, B. Zibrowius, G. Öhlmann, Spectroscopic and physicochemical characterization of P-Modified H-ZSM-5, J. Catal. 132 (1991) 229–243. https://doi.org/10.1016/0021-9517(91)90259-7.
[74] G. Seo, R. Ryoo, 31P, 27Al, and 129Xe NMR study of phosphorus-impregnated HZSM-5 zeolite catalysts, J. Catal. 124 (1990) 224–230. https://doi.org/10.1016/0021-9517(90)90117-3.
[75] T.M. Duncan, D. DOUGLASS, ON THE P-31 CHEMICAL-SHIFT ANISOTROPY IN CONDENSED PHOSPHATES, Chem. Phys. 87 (1984) 339–349.
[76] D.B. Parekh, Y.C. Rotliwala, P.A. Parikh, Synergetic pyrolysis of high density polyethylene and Jatropha and Karanj cakes: A thermogravimetric study, J. Renew. Sustain. Energy. 1 (2009) 033107. https://doi.org/10.1063/1.3153904.
[77] M.P. Kothandaraman, M. Somasundaram, Non-isothermal kinetic study on copyrolysis of Juliflora and low-density polyethylene, Biomass Convers. Biorefinery. (2019). https://doi.org/10.1007/s13399-019-00559-2.
[78] S. Tsuge, Y. Sugimura, T. Nagaya, Structural characterization of polyolefins by pyrolysis-hydrogenation glass capillary gas chromatography, J. Anal. Appl. Pyrolysis. 1 (1980) 221–229. https://doi.org/10.1016/0165-2370(80)80005-2.
[79] L. Sojáka, R. Kubineca, H. Jurdáková, E. Hájeková, M. Bajus, Gc-Ms of Polyethylene and Polypropylene Thermal Cracking Products, Pet. Coal. 48 (2006) 1–14.
[80] I. Çit, A. Sinağ, T. Yumak, S. Uçar, Z. Misirlioğlu, M. Canel, Comparative pyrolysis of polyolefins (PP and LDPE)and PET, Polym. Bull. 64 (2010) 817–834. https://doi.org/10.1007/s00289-009-0225-x.
[81] M.T.S.P. De Amorim, C. Comel, P. Vermande, Pyrolysis of polypropylene, J. Anal. Appl. Pyrolysis. 4 (1982) 73–81. https://doi.org/10.1016/0165-2370(82)80028-4.
[82] P. Kusch, Application of Pyrolysis-Gas Chromatography/Mass Spectrometry (Py-GC/MS), Elsevier Ltd, 2017. https://doi.org/10.1016/bs.coac.2016.10.003.
[83] L. Ballice, R. Reimert, Classification of volatile products from the temperature-programmed pyrolysis of polypropylene (PP), atactic-polypropylene (APP) and thermogravimetrically derived kinetics of pyrolysis, Chem. Eng. Process. 41 (2002) 289–296. https://doi.org/10.1016/S0255-2701(01)00144-1.
[84] J. Scheirs, Overview of Commercial Pyrolysis Processes for Waste Plastics, 2006. https://doi.org/10.1002/0470021543.ch15.
[85] S. Hafeez, E. Pallari, G. Manos, A. Constantinou, Catalytic conversion and chemical recovery, Elsevier Inc., 2018. https://doi.org/10.1016/B978-0-12-813140-4.00006-6.
[86] S. Teketel, W. Skistad, S. Benard, U. Olsbye, K.P. Lillerud, P. Beato, S. Svelle, Shape selectivity in the conversion of methanol to hydrocarbons: The catalytic performance of one-dimensional 10-ring zeolites: ZSM-22, ZSM-23, ZSM-48, and EU-1, ACS Catal. 2 (2012) 26–37. https://doi.org/10.1021/cs200517u.

[87] J. Scheirs, W. Kaminsky, Feedstock recycling and pyrolysis of waste plastics: Converting Waste Plastics into Diesel and Other Fuels, 2006. https://doi.org/10.1016/s1351-4180(06)71853-0.
[88] A. Akah, M. Al-Ghrami, M. Saeed, M.A.B. Siddiqui, Reactivity of naphtha fractions for light olefins production, Int. J. Ind. Chem. 8 (2017) 221–233. https://doi.org/10.1007/s40090-016-0106-8.
[89] M. del R. Hernández, A. Gómez, Á.N. García, J. Agulló, A. Marcilla, Effect of the temperature in the nature and extension of the primary and secondary reactions in the thermal and HZSM-5 catalytic pyrolysis of HDPE, Appl. Catal. A Gen. 317 (2007) 183–194. https://doi.org/10.1016/j.apcata.2006.10.017.
[90] P.N. Sharratt, Y.H. Lin, A.A. Garforth, J. Dwyer, Investigation of the Catalytic Pyrolysis of High-Density Polyethylene over a HZSM-5 Catalyst in a Laboratory Fluidized-Bed Reactor, Ind. Eng. Chem. Res. 36 (1997) 5118–5124. https://doi.org/10.1021/ie970348b.
[91] K.M. van Geem, V. V. Galvita, G.B. Marin, Making chemicals with electricity, Science (80-.). 364 (2019) 734–735. https://doi.org/10.1126/science.aax5179.
[92] M. Shekarchian, F. Zarifi, M. Moghavvemi, F. Motasemi, T.M.I. Mahlia, Energy, exergy, environmental and economic analysis of industrial fired heaters based on heat recovery and preheating techniques, Energy Convers. Manag. 71 (2013) 51–61. https://doi.org/10.1016/j.enconman.2013.03.008.
[93] A.M. Bassily, G.M. Colver, Modelling and performance analysis of an electric heater, Int. J. Energy Res. 28 (2004) 1269–1291. https://doi.org/10.1002/er.1029.
[94] G. Wernet, C. Bauer, B. Steubing, J. Reinhard, E. Moreno-Ruiz, B. Weidema, The ecoinvent database version 3 (part I): overview and methodology, Int. J. Life Cycle Assess. 21 (2016) 1218–1230. https://doi.org/10.1007/s11367-016-1087-8.
[95] E. Delikonstantis, E. Igos, S.A. Theofanidis, E. Benetto, G.B. Marin, K. Van Geem, G.D. Stefanidis, An assessment of electrified methanol production from an environmental perspective, Green Chem. 23 (2021) 7243–7258. https://doi.org/10.1039/d1gc01730f.
[96] E. Delikonstantis, E. Igos, M. Augustinus, E. Benetto, G.D. Stefanidis, Life cycle assessment of plasma-assisted ethylene production from rich-in-methane gas streams, Sustain. Energy Fuels. 4 (2020) 1351–1362. https://doi.org/10.1039/c9se00736a.
[97] T. Ren, M.K. Patel, K. Blok, Steam cracking and methane to olefins: Energy use, CO2 emissions and production costs, Energy. 33 (2008) 817–833. https://doi.org/10.1016/j.energy.2008.01.002.
[98] P.J. Donaj, W. Kaminsky, F. Buzeto, W. Yang, Pyrolysis of polyolefins for increasing the yield of monomers' recovery, Waste Manag. 32 (2012) 840–846. https://doi.org/10.1016/j.wasman.2011.10.009.
[99] M. Artetxe, G. Lopez, G. Elordi, M. Amutio, J. Bilbao, M. Olazar, Production of light olefins from polyethylene in a two-step process: Pyrolysis in a conical spouted bed and downstream high-temperature thermal cracking, Ind. Eng. Chem. Res. 51 (2012) 13915–13923. https://doi.org/10.1021/ie300178e.
[100] J.A. Conesa, R. Font, A. Marcialla, A.N. Garcia, Pyrolysis of polyethylene in a fluidised bed reactor, CHISA 2008 - 18th Int. Congr. Chem. Process Eng. (2008) 1238–1246.

[101] M. Della Zassa, M. Favero, P. Canu, Two-steps selective thermal depolymerization of polyethylene. 1: Feasibility and effect of devolatilization heating policy, J. Anal. Appl. Pyrolysis. 87 (2010) 248–255. https://doi.org/10.1016/j.jaap.2010.01.003.
[102] M. Del Remedio Hernández, Á.N. García, A. Marcilla, Study of the gases obtained in thermal and catalytic flash pyrolysis of HDPE in a fluidized bed reactor, J. Anal. Appl. Pyrolysis. 73 (2005) 314–322. https://doi.org/10.1016/j.jaap.2005.03.001.
[103] B.J. Milne, L.A. Behie, F. Berruti, Recycling of waste plastics by ultrapyrolysis using an internally circulating fluidized bed reactor, J. Anal. Appl. Pyrolysis. 51 (1999) 157–166. https://doi.org/10.1016/S0165-2370(99)00014-5.
[104] R.W.J. Westerhout, J. Waanders, J.A.M. Kuipers, W.P.M. Van Swaaij, Development of a continuous rotating cone reactor pilot plant for the pyrolysis of polyethene and polypropene, Ind. Eng. Chem. Res. 37 (1998) 2316–2322. https://doi.org/10.1021/ie970703y.
[105] C. Dean, Naphtha catalytic cracking for propylene production, Digit. Refin. 18 (2013) 1–5. https://www.digitalrefining.com/article/1000787.
[106] A. Corma, E. Corresa, Y. Mathieu, L. Sauvanaud, S. Al-Bogami, M.S. Al-Ghrami, A. Bourane, Crude oil to chemicals: Light olefins from crude oil, Catal. Sci. Technol. 7 (2017) 12–46. https://doi.org/10.1039/c6cy01886f.
[107] N. Nikolopoulos, R.G. Geitenbeek, G.T. Whiting, B.M. Weckhuysen, Unravelling the effect of impurities on the methanol-to-olefins process in waste-derived zeolites ZSM-5, J. Catal. 396 (2021) 136–147. https://doi.org/10.1016/j.jcat.2021.02.015.

AIChE Paper Number: 118e

PROCESSING OF RENEWABLE GASOIL IN A STEAMCRACKER

presented at AICHE conference
at Ethylene Plant Technology Fundamentals session 118E

Thomas Hora
Furnace Process Engineer
Monomer & Utilities Technology
LyondellBasell

Jens Becker
Technology Manager Monomer & Utilities
LyondellBasell

Prepared for Presentation at the Year (2022) Spring National Meeting
San Antonio, TX., April 10-15, 2022

PROCESSING OF RENEWABLE GASOIL IN A STEAMCRACKER

Thomas Hora
Furnace Process Engineer
Monomer & Utilities Technology
LyondellBasell
Jens Becker
Technology Manager Monomer & Utilities
LyondellBasell

Abstract

A key option for steam crackers to reducing the carbon footprint of olefins and associated final products is the substitution of fossil oil based feedstocks by sustainably-produced raw materials. The paper and associated presentation provides an actual overview on recent processing of renewable diesel feedstock (Hydrogenated Vegetable Gas Oil, HVGO) in a commercial steam cracker of an European Petrochemical Site with HVGO as the major component in blended feed to multiple furnaces.

It includes a discussion of pre-processing assessments and tasks based on the available feedstock information. As part of a mandatory management of change process in a current production infrastructure typical preparation tasks are recommended in order to evaluate operating windows and to anticipate regulatory constraints. Furthermore the paper provides a summary of the analysis method, which was applied to establish traceability of processed renewable fractions in the product streams. This is added by an overview on ICCS PLUS based certification and the associated mass balance method.

Finally, the paper compares the outcomes of trials against anticipated aspects of the pre-assessment and proposes according targets for future production trials.

Introduction

Climate change is one of the most important global challenges both now and for future generations. As such, the Paris Agreement was signed in December 2015, where it was agreed to limit global warming to well below 2 °C, preferably 1.5 °C, by reducing Green House Gas (GHG) emissions, which include CO2.

As a significant contributor to CO2 emissions the petrochemical industry is also under significant pressure to improve their processes and target for carbon neutrality. The climate impact from steam crackers is related to two sources. One is the high energy needed to crack feedstock, which is currently delivered by combustion of fuel gas, mainly methane. The other source is the fossil feedstock that is used to produce light olefins as intermediate products which are further converted, mainly to plastics. In the following sections the replacement of fossil feedstocks by renewables will be discussed.

Fossil feedstocks consist of biomass that has been converted over millions of years and under very specific conditions to gas, oil and coal. When hydrocarbon byproducts produced from fossil fuels are combusted any uncaptured CO2 released is out of balance with natural biomass growth, and the CO2 concentration in the atmosphere increases. The basis for a sustainable approach considers the CO2 balance of the feedstock.

In this view renewable feeds are one building block in reducing CO2 emissions of the petrochemical industry. Renewable feeds, also called Hydrogenated Vegetable (Gas) Oil (HVO / HVGO), can be produced from all types of vegetable and waste oil. In comparison to fossil feedstocks that result in CO2 accumulation in the atmosphere, renewable feeds are more sustainable as the raising of the crop is happening in short cycles and absorbs CO2 from the atmosphere again and again.

Production of Hydrogenated Vegetable Gas Oil (HVGO)

Bio based oils have first been used as partial substitute to fossil fuels, especially in the automotive industry. This first generation of renewable fuels is produced by esterification of vegetable oils to Fatty Acid Methyl Esters (FAME), that contain a high amount of organic acids. These components are highly corrosive and can only be blended into conventional fuels in limited concentration.

To compensate for this a second generation of renewable fuels has been developed resulting in a highly paraffinic oil that has very similar specifications to fossil based diesel fuel or gas oil. The basic processes are shown in the figure below.

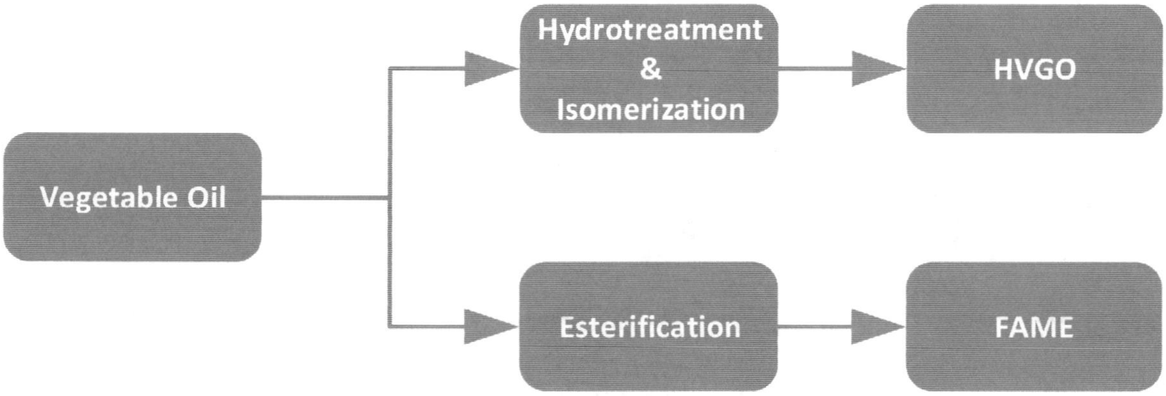

Fig. 1: Simplified flow diagram for production of FAME and HVGO

This new development has made it also possible to use these oils as alternative feed for steam crackers. Although the oil contains mainly iso-paraffins its general paraffinic character, with > 95 wt-% iso-paraffins, makes it a potential and favorable feedstock for the petrochemical industry. The production process from vegetable oil to HVGO is very similar to that for fossil fuels.

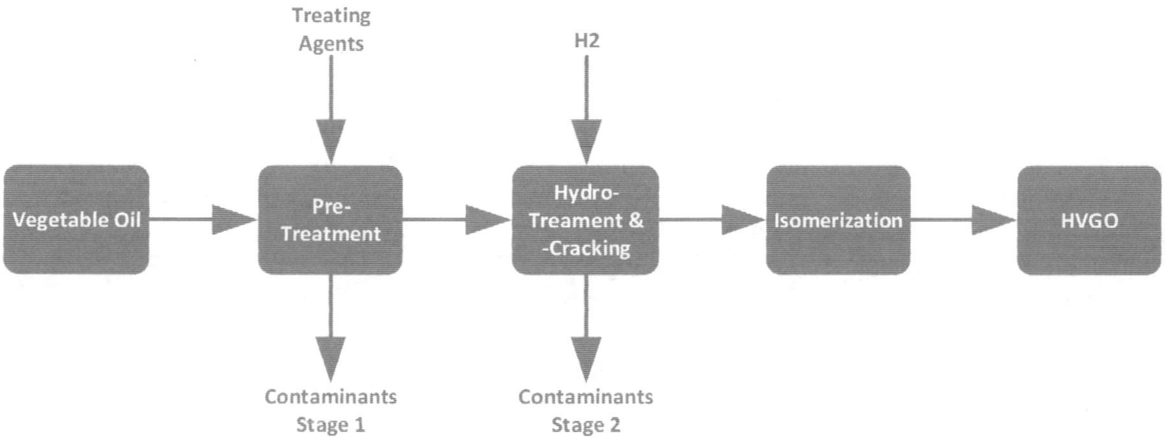

Fig. 2: Principal HVGO production process steps

The first pre-treatment step, which is specific to the nature of the vegetable oil feed, is needed to protect the downstream processes from early aging due to catalyst poisoning and coking. The pretreated oil then enters a hydrotreatment and hydrocracking process step, which are catalytic processes distributed over several beds.

The catalytic hydrocracking is necessary to control the hydrocarbon chain length and subsequently the boiling curve. This is needed to maintain consistent thermodynamic and physical properties of the oil, so that the properties are compatible with those of fossil feedstock.

The hydrotreatment step has two purposes. It firstly saturates unsaturated hydrocarbons and radicals. This also helps control the chain length, by limiting further reaction to long chained oil. The second purpose is the further removal of contaminants in the form of heteroatoms. This involves primarily oxygen-, nitrogen- and sulfur-containing molecules, that are undesired in the final product.

The last step is an isomerization reaction which converts normal- into iso-paraffins. For automotive fuels it is favorable to have predominantly iso-paraffins to control the cloud point of the fuel. Normal paraffins tend to have a lower cloud point, which can be a problem for diesel fuel especially in cold regions. For the petrochemical industry this process step is not be favored, but as the current HVGO production facilities are designed to produce automotive fuel, this fact has to be accepted for the steam cracker process.

Due to the described production process of HVGO, it has many advantages as a feed to the steam cracking process. Besides being a potential pathway to advance sustainability, its paraffinic nature promises good yields, especially for ethylene and propylene. At the same time it reduces the yields for heavy oils and fuel gas, which are undesired products from steam cracking.

In addition to these yield benefits HVGO also contains very low levels of contaminants, as the production process includes two steps of contaminant removal. This makes it a very clean feedstock for a steam cracker. This aspect will be further discussed in a later section of the paper.

A downside to HVGO is that the process is developed to produce diesel like fuels. Translated to steam cracker feed, this is comparable to a typical gas oil fraction in terms of physical and thermodynamic properties. In the world of petrochemicals gas oil is normally a very expensive feedstock and for that reason most cracking furnaces are not designed for such feed. This requires a detailed assessment of the cracking furnaces to avoid the risk of damage to the equipment.

Certification and tracing of sustainability

Renewable feeds are one method to increase sustainability and lower the carbon footprint of plastic material. At the same time, they are currently very expensive feeds, which also impacts the price of the resulting polymer. To maintain transparency to the final customer and the society in general, it is therefore necessary to have a process in place that is able to document the amount of sustainable feed contained in the final product. This can be achieved in two ways: one is to utilize a mass balance approach to document the distribution of renewable feed to final products, and the other is to directly analyze for renewables in the final products.

For the mass balance approach the renewable content of products is documented with a certification system like the International Sustainability & Carbon Certification (ISCC PLUS). The certificates provided through this system are controlled by independent companies that are conducting regular audits. The certificates are related to the volume of sustainable feed processed and are assigned to final products, without requiring product analysis. This a reasonable approach which limits the complexity of sustainable production, especially on the logistic side. The physical separation of sustainable feed and product is for most production sites very challenging or even impossible. The mass balance approach allows for advancing sustainable plastics production in these situations.

The second option to certify sustainable products is to directly analyze for renewable content. This can be achieved through the method of radiocarbon- or C14-dating.

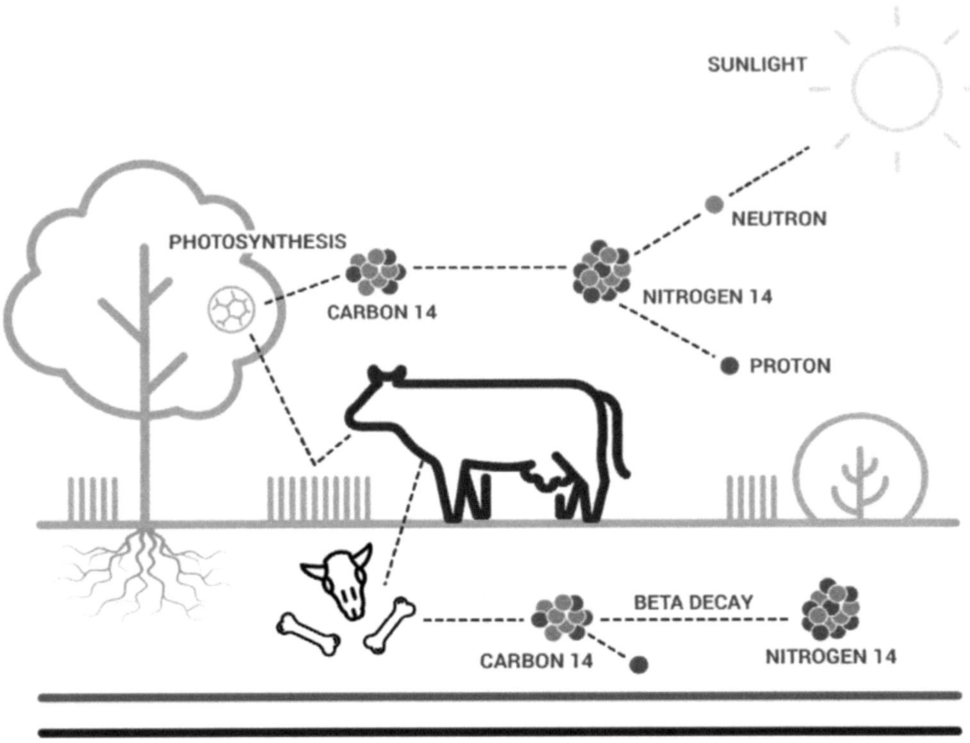

Fig. 3: C-14 origin and route

C14-dating is a globally accepted method for determining the age of organic material. C14 is a radioactive carbon isotope that is produced by photosynthesis and is thus contained in all plants. Due to its radioactive character C14 decreases once the plant has died, so this method can be used to determine the time of death of a plant making use of the half-value period of C14 isotope.

In the case of renewable feed the death of the plant is very shortly before its processing in a steam cracker. As a result the C14-content is still at initial level. For fossil feeds, which have been stored in the ground for many millions of years, the C14-content cannot be determined anymore, and is essentially zero. Radiocarbon testing of final products that are a mixture of fossil and renewable carbon sources will yield an average C14 content that can be used to determine the contribution of the fossil and renewable carbon. This is a reasonable approach to analyze renewable content of products at any stage of the process. This can also help increase transparency that the utilization of renewable feed is traceable through the whole process chain until final product. It should be noted that radiocarbon testing is only applicable for the tracing of renewable feeds, and does not work for all sustainable feeds, such as pyrolysis oil generated by chemical recycling of waste polymers.

This paper is focused on campaigns where the C14-method has been traced from imported HVGO feed through several process stages until final plastic pellets or powder. The renewable feedstock portion was in a significant range and kept through the whole production chain in a reasonable constant concentration. It was processed on liquid furnace types capable to run on typical refinery feedstocks ranging from Naphtha to hydrocracker bottoms.

The work was performed with HVGO from a single supplier, which is not exclusive. Also any other manufacturer is possible and the results of the trials are widely applicable.

Pre-Processing assessments

Preprocessing assessments are applied as indispensable part of a Management of Change Procedure to anticipate important aspects with regard to plant safety or performance.
Since HVGO exhibits a hydrocarbon composition which deviates from fossil feedstocks processed in the cracker several aspects are to be considered. This evaluation comprises not only the furnaces but also downstream units in the cracker as well as pipelines and storage facilities. Therefore the following topics are identified for further assessment or monitoring:

Furnaces

- Evaporation behavior in the convection sections
- Margins to mechanical design
- Fouling / premature coking
- Operating window (coil outlet temperature, steam:oil ratio, capacities)
- Runtime, yields

Other Cracker Units

- Accumulation of contaminants
- Material degradation risks
- Impairment of catalyst performance and process bottlenecks

Logistics / Regulatory

- Suitability of tanks (materials, type, mixabilities)
- Conflicts with existing operating permits (e.g. permitted feedstocks)

To cover the different aspects with an appropriate number of investigations the following evaluation route is followed:

- Collection of available feedstock information and analysis data from feedstock manufacturer (e.g. MSDS, boiling curve and raw hydrocarbon composition data)
- Laboratory Analysis (hydrocarbon composition, boiling curves,, density, contaminants)
- Full Furnace Simulations (expected yields/runtime/fouling risks/operating window)

Pre-processing results

Hydrocarbon components analysis

The hydrocarbon compositions of the processed HVGO qualities showed a high content of paraffinic and isoparaffinic components whereas Olefins, Naphthenic or Aromatic components were in a low percentage range (< 5%). The density of the feedstock did not vary in the different campaigns.

In comparison to the standard fossil feedstocks processed in the cracker this translates into the following expected performance-predictions (compared to a Standard Naphtha):

- Higher ethylene and propylene yields
- Lower oil and fuel gas production
- Comparable runtimes

ASTM D86 distillation results show significantly different curve shapes in comparison to conventional fossil feedstocks and places the feedstock between Naphtha and Straight Run Gasoil. The boiling curve differences have to be considered especially with respect to the given furnace design. A relatively "flat" inclination of the curve between 10% and 95% cuts indicates early and high evaporation rates within a small temperature range.

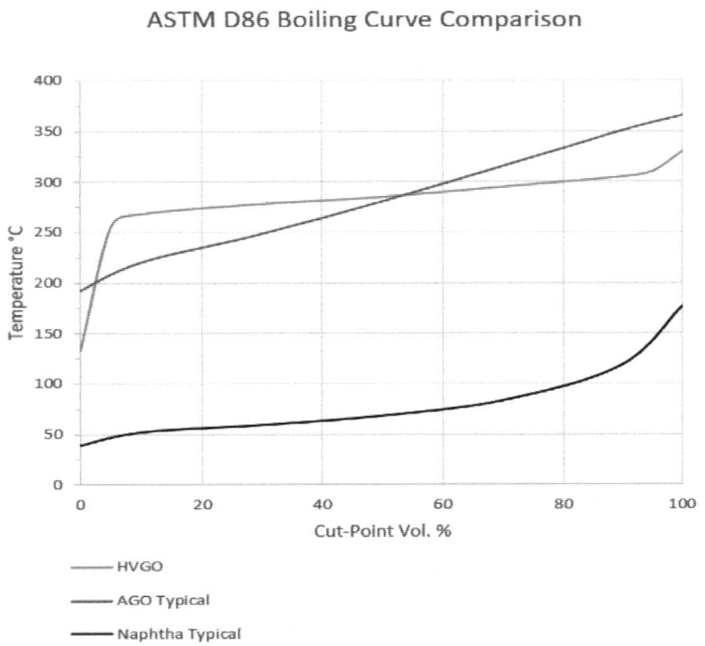

Figure 4: Schematic boiling curves of different feedstock types

Density analysis results showed very little variation throughout the different campaigns. Also residuals content and color is comparable in all samples. Therefore one feedstock composition is seen to be sufficient for pre-assessment furnace simulation runs.

Contaminant and Laboratory Analysis

Analysis parameters are selected according to the following table and the laboratory results are rated based on long term operational experiences with standard feedstocks.

Table 1: Contaminants risk rating

Parameter	***Typical Associated Risks***	***Approximate Measured Range***	***Risk Level***
Cloud Point (°C)	Settling/Plugging/Fouling Risks in Tanks, Equipment and Piping	-34°C to -24°C	Low
Pour Point (°C)	Settling/Plugging/Fouling Risks in Tanks, Equipment and Piping	-63°C	Low
Water Content (ppm)	Discharge of Free Water Phase to Furnaces (Tank Bottom)	30 ppm – 200 ppm	Medium
Halides (Cl, F...)	Wet/Hot Corrosion of Convection Section / Coils, Stainless Steel Stress Cracking, Acid/CO/CO2 Formation	0.1 ppm – 0.2 ppm	Low
Oxygenates (ppm)	PH-Value Shifts (Emulsions), Corrosion, Fouling, Coking, CO Formation	2 ppm – 10 ppm	Medium
Nitrogen-Compounds (ppm)	PH-Value Shifts (Emulsions), Cold Box Risks (NH4 Salts, NOX Gums)	0.3 ppm – 1.7 ppm	Medium
Sulfur Compounds (ppm)	Poisoning of Catalysts, Fouling in Caustic System, Corrosion, Pyrophoric Iron Formation, Waste Water Treatment Issues	0.1 ppm – 3 ppm	Low
Sodium (ppm)	Corrosion (Na-S) Convection Section / Coils	20 ppb	High
Mercury (ppm)	Catalyst Poison / Cold Box Risks (Embrittlement of Al-Alloys); Environmental, Occupational Safety	1 ppb – 50 ppb	Low to High
Arsine	Catalyst Poison, Environmental, Occupational Safety	1 ppb – 100 ppb	Low to Medium
Lead	Catalyst Poison, Environmental, Occupational Safety	10 ppb – 100 ppb	Low to Medium
Other Metals (Al, Fe, Mn, V, Zn)	Coke Promoters, Catalyst Poisons, Corrosion	Total ~ 1 ppm	Low

Furnace simulations

Furnace simulations are performed with one feedstock composition but with varying furnace operating parameters, which are well inside the standard operating window for Naphtha to gasoil-like feedstocks.

Applied input variations are mainly related to:

- Capacity
- Dilution steam superheating
- OSBL feed-preheating temperatures
- Feed-routing (furnace feed-inlet)

The furnace simulation outcomes indicate the following aspects, that are considered to have relevant risk potential and have to be controlled accordingly:

- Full feed evaporation in feed-preheaters
- Reduction of material design margins in convection section exchangers
- Lower high pressure steam outlet temperature
- Lower coil inlet temperature
- Lower flue gas outlet temperature

Assumed associated risks:

- Elevated fouling risk at hot tube surfaces (e.g. feed preheater outlet)
- Higher stress on mechanical components (potential lifetime reduction)
- Incomplete evaporation of heavies (premature coking risk)
- Unfavorable duty shifts (low high pressure steam temperature)
- Runtime impact / firing constraints
- Capacity reduction
- Energy efficiency reduction

Mitigation of anticipated risks

The processing of the renewable feedstock is performed under different starting conditions.

The steam oil ratio is adjusted in same order of magnitude as for gas oil cracking and a straight coil outlet temperature is established.

For all runs with the feedstock a conservative risk prevention strategy for the cracker and especially the furnaces is applied.

An individual set of actions is developed to control all relevant risks, which are identified in the pre-assessment phase.

This can include general precautions, such as:

- *Not all furnaces running on renewable feed*
- *Feed-transition "Furnace by Furnace"*
- *Preferred risk based selection of furnaces*
- *No max. capacity or max runtime runs during first trials*
- *Contaminant analysis of supplies before inventorying to tank farm*
- *Tight control of tank switch-over-operations*
- *Adjustment of OSBL feedstock supply temperature level*

More specific mitigation actions can be grouped into adjustments of single operating parameters, analysis of selected feedstock quality parameters and an intense online observation of units especially the furnaces, quenchwater-separation and convertor performance or known plant bottlenecks.

Table 2: Relevant risks and selected control-measures

Risks	**Assumed Impacted Units**	**Associated Cause**	**Risk Control**	**Side Effects of Controls**
Fouling / Coking	Furnace Convection Section	Heavies/Boiling Behavior	Reduction of OSBL Preheat	DENOX Efficiency Reduction
Fouling / Emulsification	Quench Water Separation / CGC	Oxygenates / Nitrogen Compounds	Quality Testing	
Corrosion (Hot/Wet)	Furnace Convection Section	Metals (Na), Halides, Oxygenates	Quality Testing	
Catalyst Deactivation	Reactors	Total Metals, Hg	Unit Performance Monitoring	
Hazardous Compound Formation	Cold Box (NOX fouling)	Nitrogen Compounds	Quality testing	
Equipment Damage	Furnaces (Coils)	Free Water (Flashing/Hammering)	Dewatering of Tanks / Quality Testing	
Equipment Damage	Aluminum Equipment	Amalgamation (Hg)	Quality Testing	
Exceedance Of Material Design T	Furnace Convection Section	Duty shifts / Early Evaporation	Furnace Load Reduction	Lower Plant Utilization
Operability Issues	Backend Turbines	Low Temperature of High Pressure Steam	Reduce Boiler Feed Water Injection	
Operability Issues	Primary Fractionator, CGC	Effluent Composition Change (Lights/Oils)	Adjustment of Cracking Parameters	Flexibility Losses / Yield Changes
Unexpected Shutdowns	Furnaces	Rapid Feed Transition (Abrupt Firing Changes)	Optimization of Tank Levels / Slow Transition / Manual Firing	
Environmental / Occupational Risk	Tanks, Cold Box, Adsorbers/ Catalysts, Waste Water Treatment	Accumulation of Contaminants (Hg, As, Pb)	Quality Testing	

Run results

Overall HVC yields and a lower production of fuel/oils are in the expected range. Post run mass balance data show a relatively high mismatch of propylene content compared to simulation results during the pre-assessment phase. This can become a limitation for the precision of early economical evaluations.

Table 3: Comparison of simulated and mass balanced yields

	Mass Balance vs. Simulated Yields Deviations
H2	<< ± 1%
CH4	< ± 1 %
Ethylene	***<< ± 1 %***
Propylene	***< ± 2 %***
CC4	<< ± 1%
Pygas	< ± 1%
Cracker Oil	< ± 1%

C-14 tracing has shown a very good consistency of detected concentrations in ISBL streams with the results gained from final PE and PP product testing. Depending on the existing plant conditions the maximum number and capacity of furnaces processing renewable feed might be limited. This determines the maximum achievable concentration of renewable phase in the final products. In this respect the application of mass balance based certificates can be the better alternative.

The described risk control measures were performed with appropriate efforts and were proven to be effective to prevent anticipated impact on furnaces and downstream units of the cracker.

Good OSBL/ISBL coordination was a key to manage rapid changes during transition and the processing of this kind of renewable feedstock became a viable task.

Summary and outlook

The cracking of renewable feed in a significant amount has been proven to be possible and even traceable until the final polymer product. For that purpose such material has been processed with HVGO as the major component in blended feed to multiple furnaces in a commercial steam cracker at an European petrochemical site.

However there are various challenges that have to be managed.

First of all a detailed analysis and characterization of the feed is necessary as well as a detailed assessment of impacts on the steam cracker unit. Especially the furnace section needs a lot of attention, as the majority of furnaces are not designed for that specific type of feedstock. It is recommended to review the related equipment in detail, including the use of rigorous simulations.

In the second step before starting to process HVGO, a detailed and well-designed procedure has to be developed, that maintains a sustainable production chain from import of the feed via storing, feed transition until the final product. This is especially complex when using direct analysis of C14 content in the polymer.

Besides that it is necessary to describe the process in depth to the cracker operators, starting with the transition phase, which needs special attention. The cracking conditions have to be fully specified and to avoid potential damage to furnaces supervised by experienced personnel. This should also involve a well-designed monitoring program.

As the performed campaigns have been limited in time, trials have not lasted a full runtime cycle yet (~ 50% of expected runtime: either by starting with decoked furnace or in the middle of a run cycle). Therefore it is one of the next targets to perform maximum runtime trials. It is mandatory to understand long term impact especially on the furnace operation to determine coking and fouling tendency. Besides that it is also necessary to further understand and optimize cracking conditions. Also potential minor modifications at the furnaces might be necessary to further establish the cracking of HVGO, which is part of our product portfolio going forward.

Disclaimer

All information ("Information") contained herein is provided without compensation and is intended to be general in nature. You should not rely on it in making any decision. LyondellBasell accepts no responsibility for results obtained by the application of this Information, and disclaims liability for all damages, including without limitation, direct, indirect, incidental, consequential, special, exemplary or punitive damages, alleged to have been caused by or in connection with the use of this Information. LyondellBasell disclaims all warranties, including, but not limited to, the implied warranties of merchantability and fitness for a particular purpose, that might arise in connection with this information.

AIChE Paper Number: 639099
Session T4000 Ethylene Plant Fundamentals

118F. REFORMING OF CRACKER FUEL GAS

Author & Presenter: ***Nick Rogers***
Technology Manager
Technip Energies, Milton Keynes, U.K.

Prepared for Presentation at the 2022 Spring National Meeting
April 2022.

Abstract:

The high CO_2 emissions from ethylene plants are well documented, which stems from the use of internally produced fuel gas, which is combusted in the furnaces and boilers to provide reaction energy and to raise steam for internal compressor drivers and process heating duties.
Gas crackers are inherently lower CO_2 emitters since they have both lower furnace heat duties, plus the internally produced fuel gas is higher in hydrogen content. Therefore, maximum CO_2 reduction will occur when the hydrogen content of the fuel gas is maximized.
In the decarbonized cracker concept, combining a cracker with a hydrogen reformer, the cracker combustion energy demand is supplied purely from the firing of high purity hydrogen, with the principal source of this hydrogen being from the reforming of the entire cracker fuel gas product.
Technip Energies has recently announced a suite of blue hydrogen technologies under the "BlueH$_2$ by T.EN^{TM}" banner. In the blue hydrogen process, the CO_2 produced from reforming process is captured. The high purity hydrogen product is then utilized to provide the fuel for both the Blue Hydrogen unit and for the cracker, as furnace and boiler fuel gas.
The decarbonized cracker will produce significantly lower direct (Scope 1) CO_2 emissions, with the majority of equivalent CO_2 emissions coming from the CO_2 burden associated from electrical power generation (Scope 2).
This paper will address whether the cracker and reformer combination can effectively be applied to liquid and gas crackers and provide an indication of the overall cost of CO_2 emission reduction. It will also provide an overview of the different types of reforming available and their merits when compared to the green hydrogen production route via the electrolysis of water.

1. Introduction

For all cracker types, the furnaces are fired from light hydrocarbons, which when burnt lead to large emissions of CO_2 from the furnace stacks. Gas crackers have fuel gas that is high in hydrogen content and can achieve the lowest CO_2 emissions for olefin production, without consideration for post-combustion carbon capture. The CO_2 emissions per tonne of high value chemical (HVC) products from liquids crackers are typically twice those of a comparable gas cracker, as the high methane fuel gas produces more direct CO_2 from combustion.

If CO_2 emissions from the olefins industry are to be reduced, whilst ethylene production is still forecast to increase, then new approaches will be needed in order to achieve these aims.

Concepts for CO_2 reduction will need to be applied both to existing ethylene plants, in order that they remain able to operate, and for new ethylene plants, such that they are sanctioned for investment and contribute to the CO_2 reduction targets set by either operating companies themselves or from national and international regulatory measures introduced to enforce the targets set.

In this paper, the key factor under consideration is whether it is practical and economic for the internally produced fuel gas stream to be utilized to produce a very high hydrogen stream to meet the energy demands of the cracker combustion processes. By burning hydrogen rich fuel gas, CO_2 emissions will be drastically reduced in the stack emissions from the furnaces and boilers.

Reforming of methane to produce hydrogen is a mature set of technologies. Recent developments are the emergence of blue hydrogen units that can capture the CO_2 produced either directly from the synthesis gas and/or from combustion products in the reformer stack.

The alternative production route for green hydrogen from electrolysis of water is compared to the use of blue hydrogen.

The potential savings in CO_2 emissions are readily calculated from residual methane content of the hydrogen fuel and the CO_2 emission rate for generation of electricity in the particular region, or as published for each country.

Definitions:

- Grey hydrogen is the most common form and is generated from natural gas, or methane, through the process of steam reforming.

- Blue hydrogen is also mainly produced from natural gas, again using steam reforming, but the generated CO_2 is captured with a view to storage or use.

- Green hydrogen is made by using clean electricity from renewable energy sources, such as solar or wind power, to electrolyze water.

2. The Need for Change

The global ethylene capacity in 2020 was around 195 MMt/a, producing approximately 250 MMt/a of CO_2 as direct emissions. In most crackers, 80-95 percent of all Scope 1 CO_2 emissions are from firing of fuel gas in the cracking furnaces, the residual is from boilers for steam generation and occasionally small amounts for process heating.

It is recognized in Europe, the USA, and beyond, that action to reduce greenhouse gas emissions is essential and legislation is in place, or being discussed, that will incentivize companies to meet the targets that are being set.

If we take Europe as an example, in June 2021, the EU adopted a European Climate Law, establishing the aim of reaching net zero greenhouse gas emissions (GHG) in the EU by 2050. The law sets an intermediate target of reducing GHG by at least 55 percent by 2030 compared to 1990 levels[1]. In addition, there are insufficient emission rights to cover for all emissions in the Emissions Trading Scheme (ETS). Crackers will get a part of their CO_2 allocations for free based on the cracker benchmark (the current European Union ETS benchmark value is 0.681 tonnes of CO_2 per tonne of HVC for the 4th period, which covers the timeframe of 2021-2030).

The data below in Figure 1 shows that most European crackers are above the CO_2 benchmark – for most plants a 20-30 percent reduction in CO_2 emissions will bring them below the benchmark.

Operators which make significant reductions in their emissions due to technology change, may therefore end up having more allowances than they are emitting and can sell these extra allowances (or buy fewer allowances).

Figure 1 - CO_2 Emissions from EU Crackers 2016-7 (Data from CEFIC/EEPC)

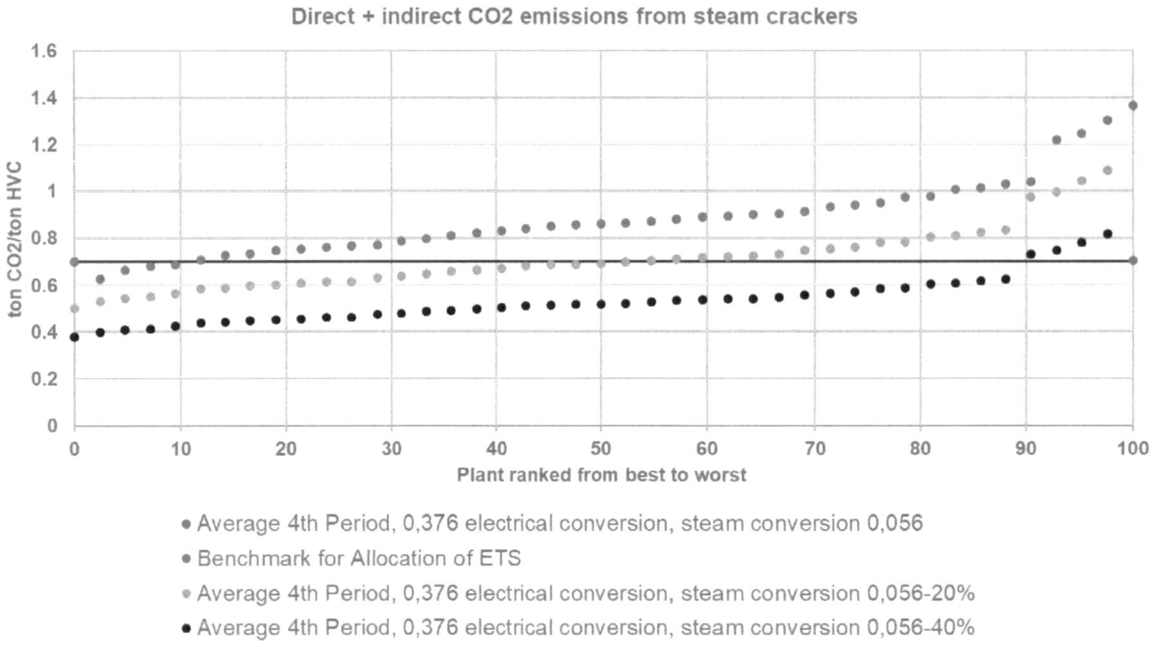

The price of CO_2 in the ETS has risen considerably, the current price (at the start of 2022) is ~80 Euro/Tonne) [2]

Figure 2 – The Price of EU Carbon Permits

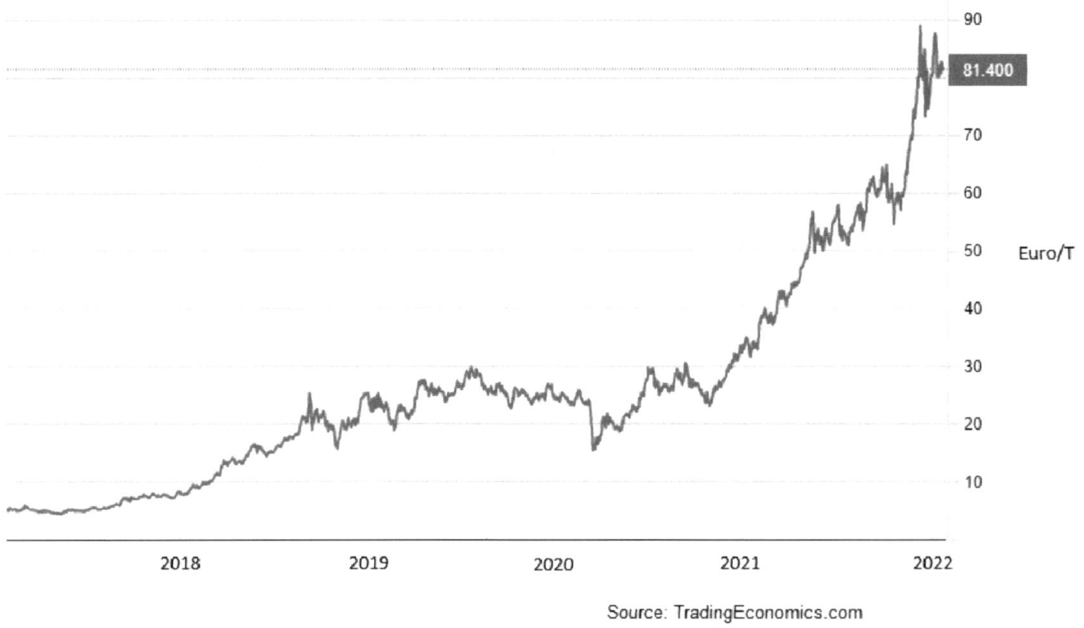

3. Current Emissions from Steam Crackers

The starting point for any evaluation is the current CO_2 emissions from a cracker with fuel gas fired in the furnaces and auxiliary steam boilers.

The CO_2 burden, best expressed as tonnes of CO_2 per tonne of high value chemicals, is provided in Table 1, where the typical high value chemicals are considered to be ethylene, propylene, butadiene and the aromatic components in the pyrolysis gasoline.

Table 1 – Typical Cracker CO_2 Emissions

Feed Type	**t CO_2 / t Ethylene**	**t CO_2 / t HVC**
Ethane	0.35 - 0.45	0.35 - 0.45
Mixed Feed (Ethane, LPG and Naphtha)	0.6 – 0.8	0.45 – 0.55
Naphtha	1.0 - 1.2	0.55 - 0.65

Depending on the feedstock we can expect Scope 1+2 CO2 emissions to be 0.4 to 0.7 t CO2/ t HVC for a modern cracker. There are many older crackers where this value is considerably higher and by definition, 90 percent of the crackers in Europe are above the Emission Trading Scheme benchmark value of 0.681 which is valid until 2030.

In order to improve these numbers, the current commercially proven options exist:

- Modifications to furnaces to improve radiant efficiency, such as changes to coils and the addition of pre-heating of combustion air
- Improvements in major machinery efficiency
- Reduce firing demand by raising less steam / electrification of main compressors
- Post combustion carbon capture from furnace flue gas
- Increase the hydrogen content of the furnace fuel gas

The aim is to eliminate the need for post combustion carbon capture by dramatically increasing the hydrogen content of the furnace fuel. To do this we can source hydrogen directly from electrolysis or make our on-demand hydrogen from the conversion of the cracker fuel gas in a blue hydrogen unit. This could be based around any of the current reforming technologies.

4. Current Emissions from (Grey) Hydrogen Plants

Reviewing the baseline numbers for a hydrogen plant is also useful.

CO_2 is produced in the Water-Gas Shift stage of the process from the conversion of CO to CO_2, which boosts hydrogen production. The endothermic process requires fuel gas combustion for energy, so the second contributor for CO_2 emissions for a conventional hydrogen plant will be the reformer or preheater flue gas.

The Grey Hydrogen process emits 9-12 kg of CO_2 per kg of hydrogen product. This number seems alarmingly high, but is due to low weight of the main hydrogen product. As always, the final number is dependent on a number of factors including the flow scheme and the CO_2 burden from the imported utilities such as the production of BFW, steam and electricity.

5. Decarbonized Cracker Concept

Liquid crackers typically have an excess of internally produced fuel gas.

We know that the higher the hydrogen content of the fuel gas, the lower the CO_2 emissions, so to maximise hydrogen production we will determine the impact if we send all of the cracker fuel gas to a blue hydrogen plant.

CO_2 capture from syngas is easier than from flue gas – it's at pressure and is contained in a smaller stream and at higher concentration when compared with flue gas capture.

We can also apply this concept to a gas cracker, where we normally need to add external natural gas to satisfy the fuel balance demand. A gas cracker will already have the benefit that internally produced fuel gas is rich in hydrogen, but there will still be improvements by going purer. We do however need to ensure that there is always enough hydrogen available.

Figure 3 – Decarbonized Cracker Arrangement

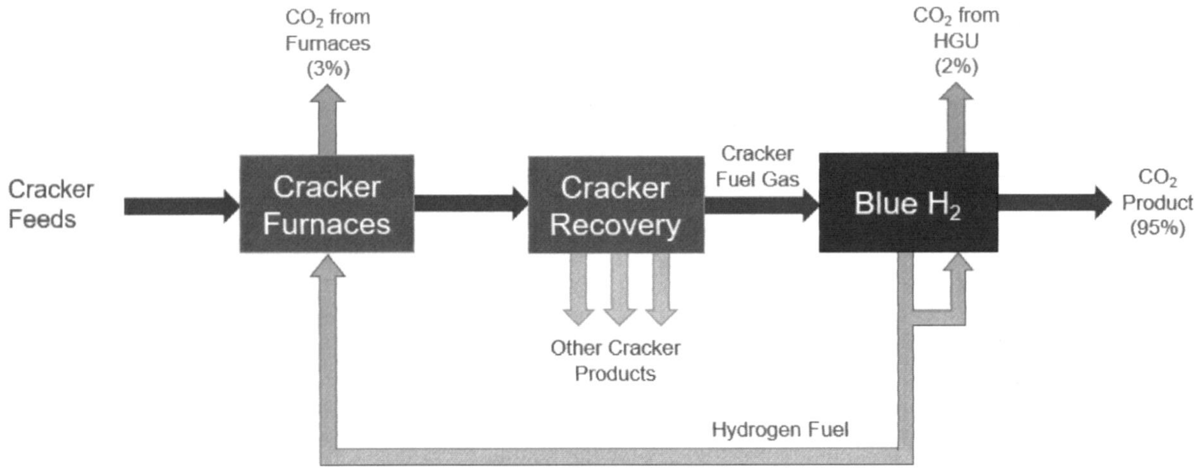

The selection of hydrogen plant technology can be reviewed on a case by case basis, but for this study the blue hydrogen from steam methane reforming, based on BlueH2 by T.EN™, was selected based on the relative merits of the key performance indicators for the available technologies.

By comparison, the alternatives of Autothermal Reforming (ATR) and Partial Oxidation (POx) are both processes for producing syngas, composed of hydrogen and carbon monoxide, by partially oxidizing a hydrocarbon feed with oxygen and steam and subsequent catalytic reforming. The cost and energy for producing the oxygen in an air-separation plant is included in Table 2.

Table 2

	Grey H_2	BlueH$_2$ by TEN		Electrolysis
		SMR	ATR/POx	
Carbon Capture?	no	yes	yes	n/a
HC Consumption	low	lowest	high	n/a
Power Demand	lowest	low	high	highest
Steam Balanced?	tailored	tailored	long	n/a
ISBL C Footprint	highest	low	high	lowest
OSBL C Footprint	lowest	low	high	highest
CAPEX	lowest	low	high	highest
LCOH (Cost of H_2)	lowest	low	high	highest

For lowest CAPEX and lowest cost of hydrogen, then Grey H_2 is a clear winner, but this does nothing to improve the carbon footprint. Electrolysis has the lowest ISBL carbon impact since no hydrocarbons are needed for the process, but its large OSBL power demand gives it a higher overall CO_2 impact. Electrolysis is also currently the most expensive technology.
The benefits of carbon removal are best achieved from a hydrogen plant with Carbon Capture. The ATR/POX technologies are higher in CAPEX and cost of hydrogen due to the need to have an additional air separation unit in order to make oxygen. In the example below the oxygen demand would be over 100 t/h if an ATR process was chosen.

6. Decarbonized Cracker Concept Applied to New Liquids Cracker

For this study, the target ethylene production capacity is taken as 1500kta – a large modern megacracker size. It is assumed that a future cracker would utilize a reduced CO_2 furnace design and partial electrification of the main train compressors, where we can already readily achieve a 10-25 percent reduction in CO_2 emissions, the amount being dependent on the CO_2 burden from the imported electricity. The higher the renewable content, the larger the overall CO_2 reduction.

The produced fuel gas is sent the hydrogen plant to produce a hydrogen product with around 98 mol percent purity. This is a value that gives a balance between recovery percentage and purity against minimizing the associated utilities in order to achieve the product. As demonstrated in the table below, the hydrogen plant produces enough nett hydrogen to satisfy the cracker fired heater energy demand, bearing in mind that it will also have produced enough hydrogen to also satisfy the fired duty of the hydrogen plant's reformer.

The cracker can also be used to satisfy any steam demand deficit in the hydrogen plant.

Table 3 – Hydrogen / Energy Balance (Liquids Cracker)

	Units	Value	Notes
Fuel Gas from Cracker to Hydrogen Generation (HGU)	t/h	91.2	LHV=54,019 kJ/kg
Hydrogen from HGU to Cracker	t/h	38.0	LHV=104,971 kJ/kg
Hydrogen Purity	mol%	98.2	
Energy in Hydrogen to Cracker	MW	1107	
Cracker Furnace Energy Demand	MW	995	Low Steam Furnace Design

For each cracker permutation studied, the total CO_2 emissions are calculated from fuel gas combustion, steam imports/exports and from electricity supplied with the EU ETS base value of 376 t of CO_2 / GWh.

Table 4 – CO_2 Emission Reduction (Liquids Cracker)

	CO_2 Emissions, MMt/a	t of CO_2 / t of HVC
Base Cracker CO_2 Emissions	1.83 (100%)	0.631 t/t
Electrified Cracker CO_2 Emissions	1.60 (88%)	0.554 t/t
Elect. Cracker + HGU CO_2 Emissions	0.68 (35%)	0.235 t/t

With a high renewable content in the electricity supply, the impact will be even greater, with up to an 80 percent overall reduction in CO_2 emissions.
Investment decisions for such a project will be heavily linked on the cost to pollute. A saving of over a million tonnes of CO_2 is worth 90 million Euros/per year when the CO_2 trading value is 80 Euro per tonne. To compare with other means of carbon capture, the levelized cost of CO_2 capture, established over a 25 year investment period is around 50-60 Euros per tonne.

7. Decarbonized Cracker Concept applied to New Gas Cracker

For a target ethylene production capacity of 1500 kta – a large modern megacracker size; it is again assumed that a future cracker would utilize a reduced CO_2 furnace design and have partial electrification of the main train compressors, where we can already readily achieve a 5-10 percent reduction in CO_2 emissions, the amount being dependent on the CO_2 burden from the imported electricity.

The produced fuel gas is sent the hydrogen plant to produce a hydrogen product with around 99 mol percent purity. This is achieved with a combination of hydrogen pre-separation combined with the hydrogen plant.

Figure 4

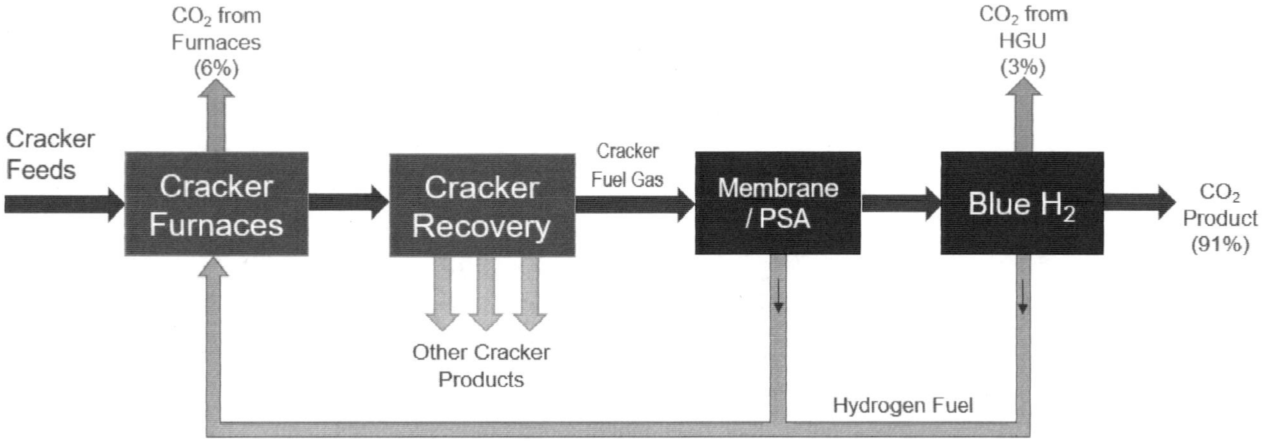

As demonstrated in the table below, the hydrogen plant produces enough net hydrogen to satisfy the cracker fired heater energy demand and potentially having a small excess that could be used for low CO_2 electricity production in a gas turbine. As in the liquid cracker example above, the cracker can also be used to satisfy any steam demand deficit in the hydrogen plant.

Table 5 – Hydrogen / Energy Balance (Gas Cracker)

	Units	**Value**	**Notes**
Fuel Gas from Cracker to Hydrogen Generation (HGU)	t/h	31.9	LHV=79,228 kJ/kg
Hydrogen from HGU to Cracker	t/h	20.9	LHV=110,825 kJ/kg
Hydrogen Purity	mol%	98.8	
Energy in Hydrogen to Cracker	MW	645	
Cracker Furnace Energy Demand	MW	562	Low Steam Furnace Design

For each cracker permutation studied, the total CO_2 emissions are calculated from fuel gas combustion, steam imports/exports and from electricity supplied with the EU ETS base value of 376 t of CO_2 / GWh.

Table 6 – CO_2 Emission Reduction (Gas Cracker)

	CO2 Emissions, MMt/a	**t of CO_2 / t of HVC**
Base Cracker CO_2 Emissions	0.55 (100%)	0.366 t/t
Electrified Cracker CO_2 Emissions	0.56 (103%)	0.376 t/t
Elect. Cracker + HGU CO_2 Emissions	0.34 (62%)	0.228 t/t

With a high renewable content in the electricity supply, the impact will be even greater, with up to an 80 percent overall reduction in CO_2 emissions. The CO_2 reduction is worth 17 million Euros/per year when the CO_2 trading value is 80 Euro per tonne. To compare with other means of carbon capture, the levelized cost of CO_2 capture, established over a 25 year investment period is around 70-140 Euros per tonne.

8. Comparison with Hydrogen from Electrolysis

The examples given above in Sections 6 and 7 have applied the cracker + blue hydrogen concept, so why have we not considered using hydrogen from the green hydrogen route?

The most commercially advanced technology for electrolysis is Alkaline Electrolysis (AE). It has been applied in areas such as ammonia production in conjunction with hydro-electric power. This technology is therefore investigated for suitability to supply the cracker.

In order to produce the same energy stream as the liquids cracker example, an electrolysis unit would need to produce around 30 t/h of pure hydrogen. Using data published by DNV in their Technology Progress Report 2021[3], the energy needed to produce this quantity of hydrogen is 1.5 GWh. The CAPEX for such a unit would also be large at over 1.5 billion dollars or roughly three times the cost of the equivalent Blue H_2 Unit.

With EU ETS standard CO_2 burden electricity, the CO_2 emissions from purchased electricity would be equivalent to 2.5 standard liquid crackers.

With renewable electricity, the Scope 1 and 2 CO_2 emissions are much reduced for hydrogen production from electrolysis, but still an order of magnitude higher than those from the Blue H_2 unit. The average wind turbine produces 2 MWh at a capacity factor of 33 percent, so over 2400 wind turbines would be needed – and are not included in the CAPEX cost analysis.

Based on OPEX, CAPEX, CO_2 emissions and practicality, electrolysis is not considered to be at the stage where it can supply the cracker fuel gas demand.

9. Conclusions

The examples above demonstrate that when Cracker fuel gas is fed to a hydrogen plant it can potentially produce enough hydrogen rich fuel gas to satisfy the cracker furnace and boiler demands.

For both cracker types studied, no natural gas import is needed to the hydrogen unit or to the cracker but is an option if further hydrogen product is beneficial – the CO_2 would still be captured in a blue hydrogen unit. An outlet for the captured CO_2 is necessary, so may not be suited to every site until further maturity in the CO_2 to chemicals routes to compliment the use of CO_2 for enhanced oil recovery or straight deep hole sequestration.

Any excess hydrogen produced could be utilized for low carbon power generation or other benefit.

The study shows that the Decarbonized Cracker Concept can readily achieve a CO_2 reduction of up to 80 percent.

The cost of CO_2 capture/reduction is equivalent to other carbon capture technologies, but each opportunity should be assessed on its merits to establish whether post combustion capture is more suited.

All technology utilized is fully commercialized and can be implemented into projects now.

Finally, hydrogen from electrolysis is not yet competitive at this scale of hydrogen production and would need to be reliable. Using green hydrogen would also leave a large cracker fuel gas stream looking for another low carbon destination.

10. References

1. "European Commission launches proposals to reach 55% emissions reduction by 2030". https://www.unsdsn.org/european-commission-launches-proposals-to-reach-55-emissions-reduction-by-2030#:~:text=In%20June%202021%2C%20the%20EU,2030%20compared%20to%201990%20levels.
2. "EU Carbon Permits" https://tradingeconomics.com/commodity/carbon
3. DNV Technology Progress Report 2021. Downloadable from https://eto.dnv.com/technology-progress-report-2021

ETHYLENE RISK AND TECHNOLOGY REVIEW
WHAT'S THE OUTLOOK?

Steve Smith
Manager, Environmental Issues
Lyondell Chemical Company

Prepared for Presentation at the 2022 Spring National Meeting
San Antonio, April 13, 2022

ETHYLENE RISK AND TECHNOLOGY REVIEW WHAT'S THE PATH FORWARD?

Steve Smith
Manager, Environmental Issues
Lyondell Chemical Company

Abstract:

EPA promulgated the Ethylene Risk and Technology Review (RTR) National Emission Standard for Hazardous Air Pollutants (NESHAP, 40 CFR, Part 63, Subpart XX and Subpart YY) on July 6, 2020. It is applicable to new and existing ethylene production units. The final rule included requirements for furnace and decoke operations, flares, process vents, heat exchange systems and pressure relief devices. The final rule eliminated the exemption for startup, shutdown and malfunction operations, and requires the emission limits must be met at all times.

This paper will discuss the rules and the potential impact on the ethylene production units.

Ethylene production units are regulated, at the Federal level, by 40 CFR, Part 63, Subpart XX and Subpart YY, commonly referred to as EMACT (Ethylene Maximum Available Control Technology). Every eight years MACT standards go through a risk and technology review. The Ethylene Risk and Technology Review (RTR) final rule was promulgated July 6, 2020 and the effective date is July 6, 2023.

Affected sources

- All storage vessels that store liquids containing organic HAP (Hazardous Air Pollutant)[1]
  Storage vessel or tank means a stationary unit that is constructed primarily of non-earthen materials (such as wood, concrete, steel, fiberglass, or plastic) that provides structural support and is designed to hold an accumulation of liquids or other materials. Storage vessel includes surge control vessels and bottoms receiver vessels. For the purposes of regulation under the storage vessel provisions of this subpart, storage vessel does not include vessels permanently attached to motor vehicles such as trucks, railcars, barges, or ships; pressure vessels designed to operate in excess of 204.9 kilopascals (29.7 psi) and without emissions to the atmosphere; or wastewater storage vessels. Wastewater storage vessels are covered under the wastewater provisions of § 63.1106.[2]
- All ethylene process vents from continuous unit operations.
  Ethylene process vent means a gas stream with a flow rate greater than 0.005 standard cubic meters per minute (0.176 cubic feet per minute) containing greater than 20 parts per million by volume HAP that is continuously discharged during operation of an ethylene production unit. On and after July 6, 2023, ethylene process vent means a gas stream with a flow rate greater than 0.005 standard cubic meters per minute (0.176 cubic feet per minute) containing greater than 20 parts per million by volume HAP that is continuously or periodically discharged during operation of an ethylene production unit. Ethylene process vents are gas streams that are discharged to the atmosphere (or the point of entry into a control device, if any) either directly or after passing through one or more recovery devices. Ethylene process vents do not include:
    (A) Pressure relief device discharges;
    (B) Gaseous streams routed to a fuel gas system, including any flares using fuel gas, of which less than 50 percent of the fuel gas is derived from an ethylene production unit;
    (C) Gaseous streams routed to a fuel gas system whereby any flares using fuel gas, of which 50 percent or more of the fuel gas is derived from an ethylene production unit, comply with § 63.1103(e)(4) beginning no later than the compliance dates specified in § 63.1102(c);
    (D) Leaks from equipment regulated under this subpart;
    (E) Episodic or nonroutine releases such as those associated with startup, shutdown, and malfunction until July 6, 2023;
    (F) In situ sampling systems (online analyzers) until July 6, 2023; and
    (G) Coke combustion emissions from decoking operations beginning no later than the compliance dates specified in § 63.1102(c).[3]

---

[1] 40 CFR part 63, Subpart XX, Hazardous Air Pollutants – Benzene, 1,3 Butadiene, Cumene, Ethylbenzene, Hexane, Naphthalene, Styrene, Toluene, o-Xylene, m-Xylene, p-Xylene

[2] 40 CFR Part 63, Subpart YY, §63.1101, Storage vessel or tank definition

[3] 40 CFR Part 63, Subpart YY, §63.1103(e)(2), Ethylene process vent definition

- Transfer racks that load HAP-containing material

  Transfer rack means the collection of loading arms and loading hoses at a single loading rack that is used to fill tank trucks and/or railcars with organic HAP. Transfer rack includes the associated pumps, meters, shutoff valves, relief valves, and other piping and valves.[4]
- Equipment that contains or contacts organic HAP

  Equipment means each of the following that is subject to control under this subpart: pump, compressor, agitator, pressure relief device, sampling collection system, open-ended valve or line, valve, connector, instrumentation system in organic hazardous air pollutant service as defined in § 63.1103 for the applicable process unit, whose primary product is a product produced by a source category subject to this subpart.[5]
- Waste streams associated with an ethylene production unit.

  Waste stream means the waste generated by a particular process unit, product tank, or waste management unit. Examples of a waste stream include process wastewater, product tank drawdown, sludge and slop oil removed from waste management units, and landfill leachate.[6]
- All heat exchange systems associated with an ethylene production unit.

  Heat exchange system means any cooling tower system or once-through cooling water system (e.g., river or pond water). A heat exchange system can include more than one heat exchanger and can include an entire recirculating or once-through cooling system.[7]
- All ethylene cracking furnaces and associated decoking operations.

Potential Impacts

There are some differences in some requirements based on whether it is an existing unit, that is, commenced construction or reconstruction on or before October 9, 2019 or a new unit, that commenced construction or reconstruction after October 9, 2019. The compliance date for existing units is upon initial startup or July 6, 2023, whichever is later. The compliance date for new units is upon initial startup, or July 6, 2020, whichever is later. These compliance dates are applicable to the following requirements: storage vessel; ethylene process vents; transfer rack; equipment; bypass line; decoking; flare; maintenance vents; operation and maintenance monitoring and testing; recordkeeping; and reporting.

Storage Vessel

The control requirements are dependent of the size of the vessel and the maximum true vapor pressure of the total organic HAP.

[4] 40 CFR Part 63, Subpart YY, §63.1103(e)(2), Transfer rack definition

[5] 40 CFR Part 63, Subpart YY, §63.1101, Equipment definition

[6] 40 CFR Part 63, Subpart YY, §63.1103(e)(2), Waste stream definition

[7] 40 CFR Part 63, Subpart YY, §63.1082(b), Heat exchange system definition

Table 1 – Storage Vessel Requirements[8]

Total Organic HAP True Vapor Pressure	Vessel Size	Control Requirements
≥0.49 psi but <11.1 psi	>1,056 gallons but <25,096 gallons	Fill through a submerged fill pipe or Comply with the requirements for vessels >25,056 gallons
	>25,096 gallons	Comply with Subpart WW National Emission Standards for Storage Vessels (Tanks) – Control Level 2, or Reduce emissions of total organic HAP by 98 weight-percent by venting emissions through a closed vent system to a flare, or Reduce emissions of total organic HAP by 98 weight-percent by venting emissions through a closed vent system to any combination of non-flare control devices, or Reduce emissions of total organic HAP by 98 weight-percent by routing emissions to a fuel gas system or process
>11.1 psi	No size threshold	Reduce emissions of total organic HAP by 98 weight-percent by venting emissions through a closed vent system to a flare, or Reduce emissions of total organic HAP by 98 weight-percent by venting emissions through a closed vent system to any combination of non-flare control devices, or Reduce emissions of total organic HAP by 98 weight-percent by routing emissions to a fuel gas system or process

Ethylene Process Vents

The previous definition of ethylene process vents applied to vents from continuous unit operations. As outlined above the ethylene process vent definition now includes intermittent vents. The impact is nonroutine or intermittent vents that previously were not covered are now covered as ethylene process vents. As stated below the threshold for control is dependent on whether the unit is an existing unit or a new unit.

If the process vent is at an existing unit and the vent stream has a flow rate ≥0.388 scfm and a total organic HAP concentration ≥50 parts per million by volume on a dry basis; or the process

[8] 40 CFR Part 63, Subpart YY, §63.1103(e) Table 7 – What Are My Requirements If I Own or Operate an Ethylene Production Existing or New Affected Source

vent is at a new source and the vent stream has a flow rate ≥0.282 scfm and a total organic HAP concentration ≥30 parts per million by volume on a dry basis, then reduce emissions of total organic HAP by 98 weight-percent; or reduce organic HAP or TOC to a concentration of 20 parts per million by volume on a dry basis corrected to 3-percent oxygen; whichever is less stringent, by venting emissions through a closed vent system to a flare; or to any combination of non-flare control devices.[9]

Transfer Racks

If materials loaded have a true vapor pressure of total organic HAP ≥0.49 psi and ≥20,077 gallons per day (averaged over any consecutive 30-day period) of HAP-containing material is loaded, then reduce emissions of organic HAP by 98 weight-percent; or reduce organic HAP or TOC to a concentration of 20 parts per million by volume on a dry basis corrected to 3-percent oxygen; whichever is less stringent, by venting emissions through a closed vent system to any combination of control devices or install process piping designed to collect the HAP-containing vapors displaced from tank trucks or railcars during loading and to route it to a process, a fuel gas system, or a vapor balance system.[10]

Equipment

If the equipment contains or contacts ≥5 weight-percent organic HAP; and the equipment is not in vacuum service then comply with the requirements of 40 CFR Subpart UU, National Emission Standards for Equipment Leaks – Control Level 2 Standards. Subpart UU applies to pumps, compressors, agitators, pressure relief devices, sampling connection systems, open-ended valves or lines, valves, connectors, instrumentation systems, and closed vent systems and control devices used to meet the requirements of subpart UU. Lines and equipment not containing process fluids are not subject to the provisions of subpart UU. Subpart UU sets the requirements for leak detection and repair. There are two specific exclusions: instead of complying with the pressure relief device requirements of subpart UU, meet the requirements of §63.1107(h), and in lieu of the flare requirement of subpart UU, comply with the requirements specified in §63.1103(e)(4). These two Subpart YY sections will be discussed further below.[11]

Waste Streams

A waste stream that contains any of the following: benzene; cumene; ethylbenzene; hexane; naphthalene; styrene; toluene; o-xylene; m-xylene; p-xylene; or 1,3-butadiene must comply with the waste requirements of Part 63, subpart XX.[12]

Process wastewater is water which comes in contact with benzene or butadiene during manufacturing or processing operations conducted within an ethylene production unit. Process wastewater is not organic wastes, process fluids, product tank drawdown, cooling water blowdown, steam trap condensate, or landfill leachate. Process wastewater includes direct-contact cooling water.[13]

[9] Ibid.
[10] Ibid.
[11] Ibid.
[12] Ibid.
[13] 40 CFR, Part 63, Subpart XX, §63.1082, Process wastewater definition

A continuous butadiene waste stream is the continuously flowing process wastewater from the following equipment: the aqueous drain from the debutanizer reflux drum, water separators on the C4 crude butadiene transfer piping, and the C4 butadiene storage equipment; and spent wash water from the C4 crude butadiene carbonyl wash system. The continuous butadiene waste stream does not include butadiene streams generated from sampling, maintenance activities, or shutdown purges. A continuous butadiene waste stream contains less than 10 parts per million by weight (ppmw) of benzene.[14]

Treat continuous butadiene waste streams that contain greater than or equal to 10 ppmw 1,3-butadiene and have a flow rate greater than or equal to 0.005 gallons per minute, by routing the stream to a treatment process or wastewater treatment system used to treat benzene waste streams or comply with the process wastewater provisions of 40 CFR Part 63, Subpart G.[15]

If the continuous butadiene waste stream is routed to a treatment process or wastewater treatment system used to treat benzene waste streams then the following requirements apply:

- Determine the butadiene concentration of the waste stream, unless you designate that the stream must be controlled;[16]
- For each waste management unit that receives or manages the waste stream must comply with the 40 CFR Part 61, Subpart FF benzene waste operations requirements for tanks, surface impoundments, containers, individual drains systems, and oil water separators;[17]
- Comply with the 40 CFR Part 61, Subpart FF benzene waste operations recordkeeping requirements, except substitute "1,3-butadiene" for each occurrence of "benzene" and "continuous butadiene waste stream" for each occurrence of "waste stream";[18]
- Comply with the 40 CFR Part 61, Subpart FF benzene waste operations reporting requirements, except substitute "1,3-butadiene" for each occurrence of "benzene" and "continuous butadiene waste stream" for each occurrence of "waste stream";[19] and
- If a steam-assisted, air-assisted, non-assisted, or pressure-assisted multi-point flare is used to comply with 40 CFR Part 61, Subpart FF, then you must comply with the requirements § 40 CFR Part 63, Subpart YY, 63.1103(e)(4) in lieu of 40 CFR Part 60, Subpart A §60.18
- Comply with the process wastewater requirements of 40 CFR Part 63, Subpart G (NESHAP for SOCMI Process Vents, Storage Vessels, Transfer Operations and Wastewater). In the Notification of Compliance Status and in the Periodic Reports submit the information detailed in Subpart G.[20]
- If the total annual benzene quantity from waste at your facility is less than 11 tons/year, then the following requirements also apply.[21]
    - Before July 6, 2023, comply with the requirements of this section at all times except during periods of startup, shutdown, and malfunction, if the startup, shutdown, or malfunction precludes the ability of the affected source to comply with the

[14] 40 CFR Part 63, Subpart XX, §63.1082(b) Continuous butadiene waste stream definition
[15] 40 CFR Part 63, Subpart XX, §63.1095(a)
[16] 40 CFR Part 63, Subpart XX, §63.1095(a)(1)(i)
[17] 40 CFR Part 63, Subpart XX, §63.1095(a)(1)(ii)
[18] 40 CFR Part 63, Subpart XX, §63.1095(a)(1)(iii)
[19] 40 CFR Part 63, Subpart XX, §63.1095(a)(1)(iv)
[20] 40 CFR Part 63, Subpart XX, §63.1095(a)(2)
[21] 40 CFR Part 63, Subpart XX, §63.1095(a)

requirements of this section and the owner or operator follows the provisions for periods of startup, shutdown, and malfunction, as specified in § 63.1111.[22]
- Beginning on July 6, 2023 you must comply with the requirements of this section at all times.[23]

Dilution steam blowdown waste stream is any continuously flowing process wastewater stream resulting from the quench and compression of cracked gas at an ethylene production unit and is discharged from the unit. This stream typically includes the aqueous or oily-water stream that results from condensation of dilution steam, blowdown from dilution steam generation systems, and aqueous streams separated from the process between the cracking furnace and the cracked gas dehydrators. The dilution steam blowdown waste stream does not include blowdown that has not contacted HAP-containing process materials. Beginning July 6, 2023, the dilution steam blowdown streams generated from sampling, maintenance activities, or shutdown purges are included in the definition of dilution steam blowdown waste stream.[24]

Spent caustic waste stream is the continuously flowing process wastewater stream that results from the use of a caustic wash system in an ethylene production unit. Beginning July 6, 2023, the spent caustic streams generated from sampling, maintenance activities, or shutdown purges are included in the definition of spent caustic waste stream.[25]

For waste streams that contain benzene, you must comply with the requirements of 40 CFR Part 61, Subpart FF, except as specified in Table 2[26] to this subpart. You must manage and treat waste streams that contain benzene as specified in either of the following two paragraphs .[27]

If the total annual benzene quantity from waste at your facility is less than 11 tons/year, then you must manage and treat spent caustic waste streams and dilution steam blowdown waste streams according to 40 CFR Part 61 Subpart FF §61.342(c)(1) through (c)(3)(i). Before July 6, 2023, the requirements of this paragraph shall apply at all times except during periods of startup, shutdown, and malfunction, if the startup, shutdown, or malfunction precludes the ability of the affected source to comply with the requirements of this section and the owner or operator follows the provisions for periods of startup, shutdown, and malfunction, as specified in 40 CFR Part 63 Subpart YY, §63.1111 Startup, shutdown, and malfunction. Beginning on July 6, 2023, the requirements of this paragraph shall apply at all times.[28]

If the total annual benzene quantity from waste at your facility is greater than or equal to 11 tons/year, then you must manage and treat waste streams according to any of the options in 40 CFR Part 61 Subpart FF §61.342(c)(1) through (e) or transfer waste off-site. If you elect

---

[22] 40 CFR Part 63, Subpart XX, §63.1095(a)(3)

[23] 40 CFR Part 63, Subpart XX, §63.1095(a)(3)

[24] 40 CFR Part 63, Subpart XX, §63.1082(b) Dilution steam blowdown waste stream definition

[25] 40 CFR Part 63, Subpart XX, §63.1082(b) Spent caustic waste stream definition

[26] 40 CFR Part 63, Subpart XX, Table 2 to Subpart XX of Part 63 - Requirements of 40 CFR Part 61, Subpart FF, Not Included in the Requirements for This Subpart and Alternate Requirements

[27] 40 CFR Part 63, Subpart XX, §63.1095(b)

[28] 40 CFR Part 63, Subpart XX, §63.1095(b)(1)

to transfer waste off-site, then you must comply with the requirements of 40 CFR Part 63 Subpart XX, §63.1096.[29]

Beginning no later than initial startup (if construction or reconstruction began on or before October 9, 2019) or July 6, 2023, whichever is later, if you use a steam-assisted, air-assisted, non-assisted, or pressure-assisted multi-point flare to comply with 40 CFR Part 61, Subpart FF, then you must comply with the requirements § 40 CFR Part 63, Subpart YY, 63.1103(e)(4).[30]

Heat Exchange Systems

Comply with the heat exchange system requirements of 40 CFR Part 63 Subpart XX, §63.1086.[31] Monitor cooling water in each heat exchange system for the following Hazardous Air Pollutants (HAPs): benzene; 1,3-butadiene; cumene; ethylbenzene; hexane; naphthalene; styrene; toluene; o-xylene; m-xylene; and p-xylene (either total or speciated) or other representative substances (e.g., total organic carbon or volatile organic compounds (VOC)) that indicate the presence of a leak according to the requirements in the five following paragraphs.[32]

You define the equipment that comprises each heat exchange system. A heat exchange system may consist of an entire heat exchange system or any combinations of heat exchangers such that, based on the rate of cooling water at the entrance and exit to each heat exchange system and the sensitivity of the test method being used, a leak of 6.75 lbs./hr. or greater of the HAPs listed above would be detected. For example, if the test you decide to use has a sensitivity of 1 ppmv for total HAP, you must define the heat exchange system so that the cooling water flow rate is 1,802 cubic foot per minute or less so that a leak of 6.75 lbs./hr. can be detected.[33]

For existing sources, monitor cooling water monthly for 6 months, both initially and following completion of a leak repair. If no leaks are detected by monitoring monthly for a 6-month period, monitor quarterly thereafter until a leak is detected. If a leak is detected, monitor monthly until the leak has been repaired and then monitor monthly for 6 months. At new sources monitor weekly for 6 months, both initially and following completion of a leak repair. If no leaks are detected by monitoring weekly for a 6-month period, monitor monthly thereafter until a leak is detected. If a leak is detected, monitor weekly until the leak has been repaired. Upon completion of the repair, monitor weekly for 6 months.[34]

Determine the concentration of the monitored substance in the heat exchange system cooling water using any method listed in 40 CFR Part 136 – Guidelines Establishing Test Procedures for the Analysis of Pollutants. Use the same method for both entrance and exit samples. Alternative methods may be used upon approval by the Administrator.[35]

Take a minimum of three sets of samples at each entrance and exit.[36]

---

[29] 40 CFR Part 63, Subpart XX, §63.1095(b)(2)
[30] 40 CFR Part 63, Subpart XX, §63.1095(b)(3)
[31] 40 CFR Part 63 subpart YY, Table 7 to §63.1103(e)(h)
[32] 40 CFR Part 63 Subpart XX, §63.1086(a)
[33] 40 CFR Part 63 Subpart XX, §63.1086(a)(1)
[34] 40 CFR Part 63 Subpart XX, §63.1086(a)(2)
[35] 40 CFR Part 63 Subpart XX, §63.1086(a)(3)
[36] 40 CFR Part 63 Subpart XX, §63.1086(a)(4)

Calculate the average entrance and exit concentrations, correcting for the addition of make-up water and evaporative losses, if applicable. Using a one-sided statistical procedure at the 0.05 level of significance, if the exit mean concentration is at least 10 percent greater than the entrance mean of the HAP (total or speciated) listed above or other representative substance, and the leak is at least 6.75 lbs./hr. you have detected a leak.[37]

Beginning no later than initial startup (if construction or reconstruction began on or before October 9, 2019) or July 6, 2023, whichever is later, perform monitoring to identify leaks of total strippable hydrocarbons from each heat exchange system subject to the requirements of this subpart according to the procedures in in the five following paragraphs.[38]

For each closed loop recirculating heat exchange system, you must collect and analyze a sample from either of the locations described below:

- each cooling tower return line or any representative riser within the cooling tower prior to exposure to air for each heat exchange system; or
- selected heat exchanger exit line(s), so that each heat exchanger or group of heat exchangers within a heat exchange system is covered by the selected monitoring location(s). [39]

For each once-through heat exchange system, you must collect and analyze a sample from the following location(s): selected heat exchanger exit line(s), so that each heat exchanger or group of heat exchangers within a heat exchange system is covered by the selected monitoring location(s). The selected monitoring location may be at a point where discharges from multiple heat exchange systems are combined provided that the combined cooling water flow rate at the monitoring location does not exceed 165,000 gallons per minute You may also elect to collect and analyze an additional sample from the following location(s): The inlet water feed line for a once-through heat exchange system prior to any heat exchanger. If multiple heat exchange systems use the same water feed (*i.e.,* inlet water from the same primary water source), you may monitor at one representative location and use the monitoring results for that sampling location for all heat exchange systems that use that same water feed.[40]

If you comply with the total strippable hydrocarbon concentration leak action level as specified in in the following paragraph, you must determine the total strippable hydrocarbon concentration (in parts per million by volume (ppmv) as methane) at each monitoring location using the “Air Stripping Method (Modified El Paso Method) for Determination of Volatile Organic Compound Emissions from Water Sources” using a flame ionization detector analyzer for on-site determination as described in the Modified El Paso Method. If you comply with the total hydrocarbon mass emissions rate leak action level as specified in the following paragraph: you must determine the total strippable hydrocarbon concentration (in parts per million by volume (ppmv) as methane) at each monitoring location using the “Air Stripping Method (Modified El Paso Method) for Determination of Volatile Organic Compound Emissions from Water Sources” using a flame ionization detector analyzer for on-site determination as described in the Modified El Paso Method; and convert the total strippable hydrocarbon

[37] 40 CFR Part 63 Subpart XX, §63.1086(a)(5)
[38] 40 CFR Part 63 Subpart XX, §63.1086(e)
[39] 40 CFR Part 63 Subpart XX, §63.1086(e)(1)
[40] 40 CFR Part 63 Subpart XX, §63.1086(e)(2)

concentration (in ppmv as methane) to a total hydrocarbon mass emissions rate (as methane) using the calculations in "Air Stripping Method (Modified El Paso Method) for Determination of Volatile Organic Compound Emissions from Water Sources" using a flame ionization detector analyzer for on-site determination as described in the Modified El Paso Method.[41]

For each heat exchange system, you must comply with the applicable monitoring frequency and leak action level, as defined below:

- For each heat exchange system that is part of an ethylene production affected source that commenced construction or reconstruction on or before December 6, 2000, you must monitor quarterly using a leak action level defined as a total strippable hydrocarbon concentration (as methane) in the stripping gas of 6.2 ppmv or, for heat exchange systems with a recirculation rate of 10,000 gallons per minute or less, you may monitor quarterly using a leak action level defined as a total hydrocarbon mass emissions rate from the heat exchange system (as methane) of 0.39 lb./hr. If a leak is detected as specified in the leak definition below, then you must monitor monthly until the leak has been repaired. Once the leak has been repaired quarterly monitoring for the heat exchange system may resume.
- For each heat exchange system that is part of an ethylene production affected source that commences construction or reconstruction after December 6, 2000 and on or before October 9, 2019, you must monitor weekly for 6-months and monitor monthly thereafter using a leak action level defined as a total strippable hydrocarbon concentration (as methane) in the stripping gas of 6.2 ppmv or, for heat exchange systems with a recirculation rate of 10,000 gallons per minute or less, you may monitor using a leak action level defined as a total hydrocarbon mass emissions rate from the heat exchange system (as methane) of 0.39 lb./hr. If a leak is detected, then you must monitor weekly until the leak has been repaired. Once the leak has been repaired, monthly monitoring for the heat exchange system may resume.
- For each heat exchange system that is part of an ethylene production affected source that commences construction or reconstruction after October 9, 2019, you must initially monitor weekly for 6-months beginning upon startup and monitor monthly thereafter using a leak action level defined as a total strippable hydrocarbon concentration (as methane) in the stripping gas of 6.2 ppmv or, for heat exchange systems with a recirculation rate of 10,000 gallons per minute or less, you may use a leak action level defined as a total hydrocarbon mass emissions rate from the heat exchange system (as methane) of 0.39 lbs./hr. if the heat exchange system has a recirculation rate of 10,000 gallons per minute or less. If a leak is detected, then you must monitor weekly until the leak has been repaired. Once the leak has been repaired, monthly monitoring for the heat exchange system may resume.[42]

For once-through heat exchange systems for which the inlet water feed is monitored, a leak is detected if the difference in the measurement value of the sample taken from the heat exchanger exit line(s) and the measurement value of the corresponding sample taken from the inlet water feed equals or exceeds the leak action level. For all other heat exchange systems

[41] 40 CFR Part 63 Subpart XX, §63.1086(e)(3)
[42] 40 CFR Part 63 Subpart XX, §63.1086(e)(4)

a leak is detected if a measurement value of the sample equals or exceeds the leak action level[43]

Cracking Furnaces and Associated Decoke Processes

Beginning no later than initial startup (if construction or reconstruction began on or before October 9, 2019) or July 6, 2023, whichever is later, the owner or operator must comply with the following paragraphs.

During normal operations, conduct daily inspections of the firebox burners and repair all burners that are impinging on the radiant tube(s) as soon as practical, but not later than 1 calendar day after the flame impingement is found. The owner or operator may delay burner repair beyond 1 calendar day using the following procedures provided the repair cannot be completed during normal operations, the burner cannot be shut down without significantly impacting the furnace heat distribution and firing rate, and action is taken to reduce flame impingement as much as possible during continued operation.

- If a shutdown for repair would cause greater emissions than the potential emissions from delaying repair, repair must be completed following the next planned decoking operation (and before returning the ethylene cracking furnace back to normal operations) or during the next ethylene cracking furnace complete shutdown (when the ethylene cracking furnace firebox is taken completely off-line), whichever is earlier.
- If a shutdown for repair would cause lower emissions than the potential emissions from delaying repair, then shutdown of the ethylene cracking furnace must immediately commence and the repair must be completed before returning the ethylene cracking furnace back to normal operations

An inspection may include, but is not limited to: visual inspection of the radiant tube(s) for localized bright spots (this may be confirmed with a temperature gun), use of luminescent powders injected into the burner to illuminate the flame pattern, or identifying continued localized coke build-up that causes short runtimes between decoking cycles. A repair may include, but is not limited to: Taking the burner out of service, replacing the burner, adjusting the alignment of the burner, adjusting burner configuration, making burner air corrections, repairing a malfunction of the fuel liquid removal equipment, or adding insulation around the radiant tube(s).

At least two of the control measures specified in the following bullets must be utilized to minimize coke combustion emissions from the decoking of the radiant tube(s) in each ethylene cracking furnace.

- During decoking operations, beginning before the expected end of the air-in decoke time, continuously monitor (or use a gas detection tube or equivalent sample technique every three hours to monitor) the CO_2 concentration in the combined decoke effluent downstream of the last component being decoked for an indication that the coke combustion in the ethylene cracking furnace radiant tube(s) is complete. The owner or operator must immediately initiate procedures to stop the coke combustion once the CO_2 concentration at the outlet consistently reaches a level that indicates combustion of coke is complete and site decoke completion assurance procedures have been concluded.
- During decoking operations, continuously monitor the temperature at the radiant tube(s) outlet when air is being introduced to ensure the coke combustion occurring inside the radiant tube(s) is not so aggressive (*i.e.,* too hot) that it damages either the radiant tube(s)

[43] 40 CFR Part 63 Subpart XX, §63.1086(e)(5)

or ethylene cracking furnace isolation valve(s). The owner or operator must immediately initiate procedures to reduce the temperature at the radiant tube(s) outlet once the temperature reaches a level that indicates combustion of coke inside the radiant tube(s) is too aggressive.

- After decoking, but before returning the ethylene cracking furnace back to normal operations, verify that decoke air is no longer being added.
- After decoking, but before returning the ethylene cracking furnace back to normal operations and/or during normal operations, inject materials into the steam or feed to reduce coke formation inside the radiant tube(s) during normal operation.[44]

Beginning no later than initial startup (if construction or reconstruction began on or before October 9, 2019) or July 6, 2023, whichever is later, the owner or operator must conduct cracking furnace isolation valve inspections as specified in the following paragraphs.

Prior to decoking operation, inspect the applicable ethylene cracking furnace isolation valve(s) to confirm that the radiant tube(s) being decoked is completely isolated from the ethylene production process so that no emissions generated from decoking operations are sent to the ethylene production process. If poor isolation is identified, then the owner or operator must rectify the isolation issue prior to continuing decoking operations to prevent leaks into the ethylene production process.

Prior to returning the ethylene cracking furnace to normal operations after a decoking operation, inspect the applicable ethylene cracking furnace isolation valve(s) to confirm that the radiant tube(s) that was decoked is completely isolated from the decoking pot or furnace firebox such that no emissions are sent from the radiant tube(s) to the decoking pot or furnace firebox once the ethylene cracking furnace returns to normal operation. If poor isolation is identified, then the owner or operator must rectify the isolation issue prior to continuing normal operations to prevent product from escaping to the atmosphere through the decoking pot or furnace firebox.[45]

Flares

Beginning no later than initial startup (if construction or reconstruction began on or before October 9, 2019) or July 6, 2023, whichever is later if a steam-assisted, air-assisted, non-assisted, or pressure-assisted multi-point flare is used as a control device for an emission point subject to the requirements applicable to ethylene production source (Table 7 to §63.1103(e))[46], then the owner or operator must meet the applicable requirements for flares as specified in 40 CFR Part 63 Subpart CC §63.670 Requirements for flare control devices and 63.671 Requirements for flare monitoring systems, including the provisions in 40 CFR Part 63 Subpart CC Table 12 Individual Component Properties and Table 13 Calibration and Quality Control Requirements for CPMS, except as specified in 40 CFR Part 63 Subpart YY, §63.1103 (e)(4)(i) through (xiv) as summarized below. This requirement also applies to any flare using fuel gas from a fuel gas system, of which 50 percent or more of the fuel gas is derived from an ethylene production unit, being used to control an emission point subject to the requirements in Table 7 of this section.[47]

---

[44] 40 CFR Part 63 Subpart YY, §63.1103(e)(7)
[45] 40 CFR Part 63 Subpart YY, §63.1103(e)(8)
[46] 40 CFR Part 63 Subpart YY, §63.1103(e)
[47] 40 CFR Part 63 Subpart YY, §63.1103(e)(4)

- The owner or operator may elect to comply with the alternative means of emissions limitation requirements, and submit the alternative means of emissions limitation request following the requirements in § 63.1113.[48]
- Develop and implement the flare management plan no later than July 6, 2023. [49]
- If required to develop a flare management plan and submit it to the Administrator, then the owner or operator must also submit all versions of the plan in portable document format (PDF) to the EPA via CEDRI.[50]
- Comply with the maximum flare tip velocity operating limit at all times. [51]
- Substitute "ethylene production unit" for each occurrence of "petroleum refinery."[52]
- Each occurrence of "refinery" does not apply. [53]
- If a pressure-assisted multi-point flare is used as a control device for an emission point subject to the requirements in Table 7, then comply with the following requirements:[54]
    - The owner or operator is not required to comply with the flare tip velocity requirements;
    - Substitute "800" for each occurrence of "270" in § 63.670(e) of subpart CC;
    - Determine the 15-minute block average NHVvg using only the direct calculation method;
    - If a pressure-assisted multi-point flare uses cross-lighting on a stage of burners rather than having an individual pilot flame on each burner, operate each stage of the pressure-assisted multi-point flare with a flame present at all times when regulated material is routed to that stage of burners. Each stage of burners that cross-lights in the pressure-assisted multi-point flare must have at least two pilots with at least one continuously lit and capable of igniting all regulated material that is routed to that stage of burners. The pilot flame(s) on each stage of burners that use cross-lighting must be continuously monitored by a thermocouple or any other equivalent device used to detect the presence of a flame;
    - If a stage of burners on the flare uses cross-lighting, the distance between any two burners in series on that stage can be no more than 6 feet when measured from the center of one burner to the next burner. A distance greater than 6 feet between any two burners in series may be used provided a performance demonstration confirms the pressure-assisted multi-point flare will cross-light a minimum of three burners using the spacing between the burners and location of the pilot flame as represented in the projected installation.
    - Install and operate pressure monitor(s) on the main flare header, as well as a valve position indicator monitoring system for each staging valve
    - If a pressure-assisted multi-point flare is operating under the requirements of an approved Alternative Means of Emission Limitations (AMEL), then

[48] 40 CFR Part 63 Subpart YY, §63.1103(e)(4)(i)
[49] 40 CFR Part 63 Subpart YY, §63.1103(e)(4)(ii)
[50] 40 CFR Part 63 Subpart YY, §63.1103(e)(4)(iii)
[51] 40 CFR Part 63 Subpart YY, §63.1103(e)(4)(iv)
[52] 40 CFR Part 63 Subpart YY, §63.1103(e)(4)(v)
[53] 40 CFR Part 63 Subpart YY, §63.1103(e)(4)(vi)
[54] 40 CFR Part 63 Subpart YY, §63.1103(e)(4)(vii)

continue to comply with the terms of the AMEL or comply with the provisions listed above
    - Net Heating Value (NHV) may be determined with a continuous process mass spectrometer[55]
- An owner or operator using a gas chromatograph or mass spectrometer for compositional analysis for net heating value may choose to use the Calibration Error (CE) of $NHV_{measured}$ versus the cylinder tag value NHV as the measure of agreement for daily calibration and quarterly audits in lieu of determining the compound-specific CE. The CE for NHV at any calibration level must not differ by more than 10 percent from the certified cylinder gas value. The CE for must be calculated using the following equation:[56]
- Keep the flare monitoring records specified in § 63.1109(e).[57]
- Comply with the reporting requirements specified in § 63.1110(d) and (e)(4).[58]
- When determining compliance with the pilot flame requirements substitute "pilot flame or flare flame" for each occurrence of "pilot flame."[59]
- When determining compliance with the flare tip velocity and combustion zone operating limits specified in § 63.670(d) and (e), the requirement effectively applies starting with the 15-minute block that includes a full 15 minutes of the flaring event.[60]
- An owner or operator may submit a request for approval of an alternative test method. The alternative test method must be able to demonstrate on an ongoing basis at least once every 15-minutes that the flare meets 96.5% combustion efficiency and provide a description of the alternative recordkeeping and reporting that would be associated with the alternative test method. The alternative test method request may also include a request to use the alternative test method in lieu of the pilot or flare flame monitoring requirements.[61]

Atmospheric Pressure Relief Devices

Except during a pressure release, operate each PRD in organic HAP service with an instrument reading of less than 500 ppm as detected by Method 21.[62]

For pressure relief devices in organic HAP gas or vapor service, the owner or operator must comply with either Subpart YY, §63.1107, (e)(2)(i) or (ii), summarized below, following a pressure release, as applicable

- If the PRD doesn't include a rupture disk, then within five days of returning to organic HAP service verify, using Method 21, the PRD is operating with an instrument reading of less than 500 ppm[63]
- If the PRD consists of or includes a rupture disk, install a replacement disk as soon as practicable after a pressure release, but no later than five calendar days after the pressure release[64]

[55] 40 CFR Part 63 Subpart YY, §63.1103(e)(4)(viii)
[56] 40 CFR Part 63 Subpart YY, §63.1103(e)(4)(ix)
[57] 40 CFR Part 63 Subpart YY, §63.1103(e)(4)(x)
[58] 40 CFR Part 63 Subpart YY, §63.1103(e)(4)(xi)
[59] 40 CFR Part 63 Subpart YY, §63.1103(e)(4)(xii)
[60] 40 CFR Part 63 Subpart YY, §63.1103(e)(4)(xiii)
[61] 40 CFR Part 63 Subpart YY, §63.1103(e)(4)(xiv)
[62] 40 CFR Part 63 Subpart YY, §63.1107(e)(1)
[63] 40 CFR Part 63 Subpart YY, §63.1107(e)(2)(i)
[64] 40 CFR Part 63 Subpart YY, §63.1107(e)(2)(ii)

Except as specified in the following paragraph, emissions of organic HAP to the atmosphere from pressure relief devices in organic HAP service are prohibited, and the owner or operator must comply with the requirements specified in Subpart YY, §63.1107(e)(3)(i) and (ii), summarized below, for all pressure relief devices in organic HAP service.[65]

- Equip each PRD in organic HAP service with a device(s) or parameter monitoring system capable of:[66]
    - identifying a the pressure release;
    - recording the time and duration of the pressure release; and
    - notifying operators immediately that a pressure release is occurring. The device or monitoring system may be either specific to the pressure relief device itself or may be associated with the process system or piping, sufficient to indicate a pressure release to the atmosphere.[67]
- If any pressure relief device in organic HAP service releases to atmosphere as a result of a pressure release event, the owner or operator must calculate the quantity of organic HAP released during each pressure release event and report this quantity. Calculations may be based on data from the pressure relief device monitoring alone or in combination with process parameter monitoring data and process knowledge.[68]
    - Pressure relief management plan – includes at least three documented redundant prevention measures for each affected PRD
    - Examples of prevention measures include:
        - flow, temperature, level and pressure indicators with monitors, or automatic actuators;
        - documented routine inspection and programs and/or operator training;
        - deluge systems;
        - staged relief systems where initial PRD (with lower set pressure) discharges to a flare or other closed vent system and control device
- If an affected PRD releases to atmosphere, then a root cause analysis and corrective analysis must be completed
- PRD releases will be tracked on an annual basis and releases associated with *force majeure* events will be noted
- Certain PRD releases will be considered violations of the pressure release management work practice standard
        - a PRD release when the root cause is operator error or poor maintenance
        - a second PRD release from a single PRD in a three year period for the same root cause for the same equipment
        - a third PRD release from a single PRD in a three year period for any reason
        - the exception is PRD releases due to a *force majeure* event

If a pressure relief device in organic HAP service is designed and operated to route all HAP emissions from pressure releases through a closed vent system to a control device or to a process, fuel gas system, or drain system, the owner or operator is not required to comply with the paragraphs above for that pressure relief device.[69]

---

[65] 40 CFR Part 63 Subpart YY, §63.1107(e)(3)

[66] 40 CFR Part 63 Subpart YY, §63.1107(e)(3)(i)

[67] 40 CFR Part 63 Subpart YY, §63.1107(e)(3)(i)(C)

[68] 40 CFR Part 63 Subpart YY, §63.1107(e)(3)(ii)

[69] 40 CFR Part 63 Subpart YY, §63.1107(e)(4)

For pressure relief devices in organic HAP service, keep records of the information summarized below, as applicable.[70]

- A list of identification numbers for pressure relief devices that vent to a fuel gas system, process, drain system, or closed-vent system and control device.[71]
- A list of identification numbers for pressure relief devices that vent to atmosphere.[72]
- A list of identification numbers for pressure relief devices equipped with rupture disks, that vent to atmosphere.[73]
- The dates and results of the monitoring following a pressure release for each pressure relief device that vents to atmosphere. The results shall include:[74]
  - The background level measured during each compliance test.
  - The maximum instrument reading measured at each piece of equipment during each compliance test.
- For pressure relief devices in organic HAP service subject to pressure relief management requirements noted above, keep records of each pressure release to the atmosphere, including the following information:[75]
  - The source, nature, and cause of the pressure release.
  - The date, time, and duration of the pressure release.
  - An estimate of the quantity of total HAP emitted during the pressure release and the calculations used for determining this quantity.
  - The actions taken to prevent this pressure release.
  - The measures adopted to prevent future such pressure releases.

Operation and Maintenance Requirements

The requirements of this paragraph apply beginning no later than July 6, 2023. The emission limitations and established parameter ranges shall apply at all times except during periods of startup, shutdown, malfunction, or non-operation of the affected source resulting in cessation of the emissions to which Part 63 Subpart YY applies.[76] Equipment leak requirements apply at all times except during periods of startup, shutdown, or malfunction, process unit shutdown, or non-operation of the affected source in which the lines are drained and depressurized resulting in cessation of the emissions to which the equipment leak requirements apply.[77] During startups, shutdowns, and malfunctions when the emission standards do not apply as outlined above, the owner or operator shall implement, to the extent reasonably available, measures to prevent or minimize excess emissions. The measures to be taken shall be identified in the startup, shutdown, and malfunction plan (if applicable), and may include, but are not limited to, air pollution control technologies, recovery technologies, work practices, pollution prevention, monitoring, and/or changes in the manner of operation of the affected source. Back-up control devices are not required, but may be used if available.[78] For batch unit operations, shutdown does not include the normal periods between batch cycles; and startup does not include the recharging of batch unit operations, or the transitional conditions due to

---

[70] 40 CFR Part 63 Subpart YY, §63.1107(f)
[71] 40 CFR Part 63 Subpart YY, §63.1107(f)(1)
[72] 40 CFR Part 63 Subpart YY, §63.1107(f)(2)
[73] 40 CFR Part 63 Subpart YY, §63.1107(f)(3)
[74] 40 CFR Part 63 Subpart YY, §63.1107(f)(4)
[75] 40 CFR Part 63 Subpart YY, §63.1107(f)(5)
[76] 40 CFR Part 63 Subpart YY, §63.1108(a)(1)
[77] 40 CFR Part 63 Subpart YY, §63.1108(a)(2)
[78] 40 CFR Part 63 Subpart YY, §63.1108(a)(5)

changes in product.[79] Malfunctions shall be corrected as soon as practical after their occurrence.[80] Operation and maintenance requirements established pursuant to section 112 of the Clean Air Act are enforceable, independent of emissions limitations or other requirements in relevant standards.[81]

The requirements of this paragraph apply beginning no later than July 6, 2023. The emission limitations and established parameter ranges of Subpart YY shall apply at all times except during periods of non-operation of the affected source (or specific portion thereof) resulting in cessation of the emissions to which this subpart applies. Equipment leak requirements shall apply at all times except during periods of non-operation of the affected source (or specific portion thereof) in which the lines are drained and depressurized resulting in cessation of the emissions to which the equipment leak requirements apply.[82] At all times, the owner or operator must operate and maintain any affected source, including associated air pollution control equipment and monitoring equipment, in a manner consistent with safety and good air pollution control practices for minimizing emissions.[83]

Start-up, Shutdown and Malfunction

A 2015 court decision set aside the start-up, shutdown and malfunction (SSM) language that allowed for the affirmative defense for excess emissions during SSM activities. The effect of the decision is there is no exemption from emission limits during SSM activities and SSM language is being removed from each affected rule. Therefore, because of the court decision the SSM language and exceptions embedded in individual definitions/standards for periods of startup, shutdown, episodic, maintenance or other transient periods have been removed.

[79] 40 CFR Part 63 Subpart YY, §63.1108(a)(3)

[80] 40 CFR Part 63 Subpart YY, §63.1108(a)(6)

[81] 40 CFR Part 63 Subpart YY, §63.1108(a)(7)

[82] 40 CFR Part 63 Subpart YY, §63.1108(a)(4)(i)

[83] 40 CFR Part 63 Subpart YY, §63.1108(a)(4)(ii)

640078 Cracking of Renewable Feedstock.

Abstract:

Borealis and our commitment with sustainability, is working in ways to offer viable alternatives to conventional feed stocks, which will not only reduce the Borealis' plant's overall CO_2 footprint, but also help our customers maintain high product quality while meeting their own sustainability goals. Increasing renewable and bio-based feed stocks in our steam cracking facilities in Porvoo, Finland and Stenungsund Sweden as well as our Propane Dehydrogenation Plant in Kallo, Belgium, is the foundation for achieving this goal, both through base chemicals as well as the Bornewables^{TM}, the Borealis portfolio of premium circular polyolefins.

On this journey, Borealis has processed second generation renewable feeds at its petrochemical facilities. In this paper, we will describe the way Borealis has set up the process and the results of the tests that have been performed. This paper will focus on the preparations for the tests, including a detailed analysis of the feed, a risk assessment on furnace and recovery section, simulations performed to select the most suitable furnace and specific parameters that were monitored.

During the tests, follow up was done by monitoring these identified meters and analysers. Finally, the tests results were compared to the simulated values.

AICHE PAPER NUMBER 154C

SHELL AND THE PLASTIC CIRCULAR ECONOMY: INNOVATION, COLLABORATION AND INVESTMENT

Luis Grau Lisnier
Senior Technologist Base Chemicals
Shell Global Solutions International B.V., Amsterdam The Netherlands

Nicolas Genty
Principal technologist Base Chemicals
Shell Global Solutions International B.V., Amsterdam The Netherlands

Prepared for Presentation at the 2022 Spring National Meeting
San Antonio, TX, (April 13th, 2022)

SHELL AND THE PLASTIC CIRCULAR ECONOMY: INNOVATION, COLLABORATION AND INVESTMENT

Luis Grau Lisnier
Senior Technologist Base Chemicals
Shell Global Solutions International B.V., Amsterdam The Netherlands

Nicolas Genty
Principal technologist Base Chemicals
Shell Global Solutions International B.V., Amsterdam The Netherlands

ACKNOWLEDGEMENTS

Many thanks to Chris van der Ree, Chief Technologist & Development at BlueAlp, Ian Budd and Guus van Rossum of Shell, for their invaluable contributions in writing this paper.

ABSTRACT

Plastic products bring important benefits to society - helping to improve living standards, and deliver health, safety, and sustainability benefits. Customers use it to make products such as medical equipment, furniture and home appliances. Plastic is also integral to the energy transition, for example, in components of solar panels and wind turbines; and building insulation and light-weighting transportation; helping society to lower carbon emissions.

Plastic, however, is often associated with disposable and throwaway packaging; and in places where waste management infrastructure and traditional recycling don't exist, or plastic waste is not managed appropriately, it can end up as litter. Like our customers, we are concerned about the global plastic waste issue. Plastics don't belong in our oceans, rivers or landscapes. To solve this societal challenge industry, governments and customers need to work together.

Annual global plastics production is approximately 367 million metric tons/yr. It is estimated that of the plastic waste generated, 260 million metric tons, only 16% is recycled, and the remainder goes to landfill, incineration or is unmanaged after use. Plastic recycling rates vary widely by region and are far below the global recycling rates for products like paper (60%) and iron and steel (~80%)[1].

[1] McKinsey: Plastics recycling: Using an economic-feasibility lens to select the next moves

Unlocking the potential of the plastic circular economy

Shell supports the need for improved circularity of global plastics markets - encouraging reduction, reuse and recycling of plastics to mitigate the release of plastic into the environment.

We are playing an active and prominent role through innovation, collaboration and investment to find lasting solutions.

Shell is a leading member of the Alliance to End Plastic Waste. This is an alliance of global companies including chemicals and plastic manufacturers, consumer goods companies and waste management companies, along with the World Business Council for Sustainable Development. It has committed $1.5 billion over the next five years to projects across four strategic pillars:

- Infrastructure: Catalyze investment in enhanced waste infrastructure to prevent plastic from reaching the environment.
- Innovation: Innovate new sustainable materials, designs, technologies and business models based on sound science and economics.
- Education: Engage and collaborate with governments, industry and consumers to drive effective solutions.
- Clean-up: Develop solutions to address and clean up areas of existing plastic waste in the environment.

Shell is responding to customer demands for more chemical products with less virgin-hydrocarbons. We are developing and investing, with industry partners, in advanced chemical recycling. Using a special heating process called pyrolysis, it transforms hard-to-recycle plastic waste, which would otherwise be incinerated or go to landfill, into a liquid feedstock that can be used to produce chemicals. This liquid, referred to as pyrolysis oil, is used in a cracker at a Shell chemicals plant to produce a range of chemicals, including plastics.

Delivering on Shell ambition

Shell has an ambition to recycle one million metric tons of plastic waste in our global chemicals plants by 2025; and we are working with partners at different stages of the value chain to grow the market for pyrolysis oil to industrial and profitable quantities and deliver on our ambition.

In 2019, collaborating with Nexus Fuels, Shell successfully started using pyrolysis oil in our Norco petrochemicals plant in the U.S.; and over the next three years Shell will use 60,000 metric tons of pyrolysis oil from Nexus Fuels at Norco. We are exploring ways to increase these volumes. In Europe, we have also completed successful pyrolysis oil pilots at our facilities and are about to do the same in Asia; and we are investing in pyrolysis technology and securing more supply volumes.

In 2021, Shell invested in BlueAlp[2] and together we are working to develop, scale and deploy BlueAlp's plastic waste to chemicals feedstock technology. Already at a commercial scale, Shell's technology experience and size will help BlueAlp to further improve and scale-up the capacity of BlueAlp's technology to recycle larger volumes of plastic waste. BlueAlp's technology will be licensed to others to support the exposition of pyrolysis oil production globally.

In the Netherlands, Shell and BlueAlp (through a joint-venture company) will build two conversion units with the combined capacity to process approximately 35,000 metric tons of tough-to-recycle plastic waste, resulting in 25,000 of pyrolysis oil. The two units are expected to be operational in 2023/24 and Shell will use 100 percent of the pyrolysis oil at its Moerdijk (The Netherlands) and Rheinland (Germany) plants. Shell is also exploring licensing two additional units for deployment within Asia to supply the Shell Energy and Chemicals Park Singapore.

Additionally, Shell has also secured supply for its European crackers from Pryme. Currently under construction, Pryme's Rotterdam plant is scheduled to be operational in 2022 and is forecast to convert 60,000 metric tons of plastic waste into pyrolysis oil annually.

Inconsistent purity of pyrolysis oil is an industry-wide challenge and one that impedes the growth of the market. Chemical crackers generally have a limited tolerance for lower-grade pyrolysis oil and this restricts the volumes that are useable. Shell is tackling this with investment in proprietary upgrader technology, which removes contaminants enabling Shell to process much larger volumes of pyrolysis oil. Shell's first pyrolysis oil upgrader unit will be at the Shell Energy and Chemicals Park Singapore, the unit will have a capacity of 50,000 tonnes per year and start-up is due in 2023.

Our U.S. sites (Norco and Geismar) and plants in Singapore, Germany and The Netherlands have been certified by the ISCC [International Sustainability and Carbon Certification], an independent third party, to produce circular chemical products from pyrolysis oil. While volumes are currently limited, we are looking to grow supply quickly.

After a brief introduction describing the issue at large and Shell's response and ambition, this paper will focus on the collaboration between Shell and BlueAlp to produce pyrolysis oil, the deployment of Shell's propriety technology to upgrade the purity of that pyrolysis oil at Shell assets to make it suitable for processing in our steam crackers and the cracking of pyrolysis oil in Shell steam crackers in Moerdijk, Rhineland and the Shell Energy and Chemicals Park Singapore. The paper will conclude on the progress made to date and the future developments required to achieve Shell's ambition.

---

[2] Shell has 21.25 equity stake in BlueAlp

1. Unlocking the potential of the plastic circular economy

Shell is a founding member of the Alliance to End Plastic Waste (AEPW[3]). This not-for-profit organization is bringing together top minds from across the plastics value chain (chemical and plastic manufacturers, consumer goods companies, retailers, converters and waste management companies) and partnering with the financial community, governments and civil society. The AEPW has committed $1.5 billion over the next five years to help end plastic waste in the environment.

In 2020, the alliance published its first progress report setting out projects it supports, including community projects in Africa, India and South-east Asia that aim to prevent waste plastic from entering nature, rivers and the sea. The report also outlines technology projects that aim to unlock economic value from post-consumer waste. The alliance contributes funds to projects as well as the expertise of its members, for example, engineers, safety professionals and scientists.

Shell is also active in other parts of the world, including the USA. Shell is part of an industry-wide effort to ensure that plastics converted into recycled chemical feedstocks via pyrolysis are appropriately classified and recognized by state and local governments. In the USA, we are at the beginning of this journey with positive results in several key states.

Shell is taking tangible, positive steps to provide our customers with sustainable, circular chemicals; and we will continue to work with multiple players across the value chain in Asia, Europe and the U.S. to enable its growth and be a leader in this market.

2. Delivering on Shell ambition

Shell Ventures BV and BlueAlp Holding BV announced a strategic partnership[4] to develop, scale and deploy BlueAlp's plastic waste to chemical feedstock technology. The technology transforms plastic waste which is tough to recycle into a recycled feedstock (i.e. pyrolysis oil) that can be used to make sustainable chemicals. Shell has taken a 21.25% equity stake in BlueAlp as part of the agreement.

With BlueAlp's innovative technology and Shell's size and experience we can advance the plastic waste recycling technology needed to meet growing customer demand for sustainable chemicals. This partnership is one of the important steps Shell is taking to reach our ambition of recycling one million metric tons of plastics waste a year in our global chemicals plants by 2025.

Shell is also working with other pyrolysis oil suppliers (Nexus, Pryme and Environmental Solutions Asia) and across the whole value chain to provide customers with a secure supply of

---

[3] https://endplasticwaste.org/en

[4] https://www.bluealp.nl/news

high-quality circular products including collaborating with industry partners to drive the development of the infrastructure needed to collect and sort plastic waste.

Under the agreement, Shell and BlueAlp are forming a joint-venture company to build two new conversion units in The Netherlands, which are forecast to convert more than 30,000 metric tons of plastic waste per year. The units are planned to be operational in 2023 and will supply 100% of their pyrolysis oil as feedstock to Shell's Moerdijk and Rhineland crackers. Shell is exploring licensing a further two units for deployment within Asia to supply the Shell Energy and Chemicals Park Singapore.

Shell is providing its expertise and support to scale up the BlueAlp™ technology to commercial size. Our immediate focus is to increase the technology's current processing capacity. BlueAlp is licensing out the technology also to others, this will help communities worldwide put hard to recycle plastic waste to better use.

Shell's technology team, will work with BlueAlp to improve and scale up the technology's capacity to recycle larger volumes of plastic that cannot be mechanically recycled.

The first step has been to process this crude pyrolysis oil (untreated) directly in our crackers in order to get knowledge and experience processing this material and at the same time helping the pyrolysis oil companies increase their production and improve their technology. This step has been carried out in several Shell crackers at different facilities.

Shell has developed and is currently building its own technology to upgrade the purity of the pyrolysis oil at its assets, in order to convert it into suitable Steam Cracker feed(s). The first upgrader will be built in Singapore – Bukom, with a capacity of 50,000 metric tons per annum[5] and it will be an intermediary step towards the 2025 ambition of processing 1 million metric tons per annum of waste plastic worldwide.

The Shell upgrader and steam crackers would be able to process the full boiling range of the pyrolysis oil. This is possible by optimizing the type of furnace, furnace operating parameters, feed boiling curve range fractionation and co-craking.

One of the biggest challenges is the pyrolysis oil quality, the contaminants level and the variability of these contaminants. This would hamper the Steam Cracker performance, increase plant corrosion, decrease catalyst life, amongst other concerns.

Deep knowledge in the pyrolysis oil properties, behavior and detailed characterization are key to understanding plant performance, feed flexibility/opportunities and risks. This could only be obtained by analyzing and processing a wide range of pyrolysis oils and having years of

---

[5] https://www.reuters.com/markets/europe/shell-builds-singapore-pyrolysis-oil-unit-convert-plastic-waste-chemicals-2021-11-23/

experience in cracking "more difficult" feedstocks (wider boiling range and high content of unsaturated components).

3. BlueAlp™ Process

An overview of the BlueAlp™ process in given in Figure 1 and a picture of a BlueAlp™ plant in Oostende, Belgium in Figure 2. The plastic feed consists mainly of long-chained hydrocarbons, like polyethlene (PE), most commonly used plastic in the world, and polypropylene (PP), the second most commonly used plastic in the world.

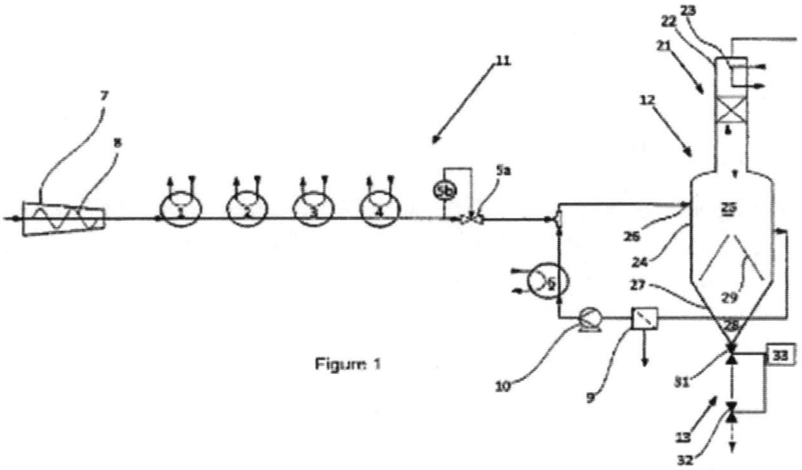

Figure 1: Overview of the BlueAlp™ process

The plastics feed is admitted or introduced to an electrically heated extruder (at low throughput and during startup) to densify and heat the feed, extract air and remove water from it. In normal operation, the friction of the extruder is sufficient. The waste plastic feed slowly gets melted into a thick, highly viscous liquid, with most hydrocarbons in the liquid phase, through a series of four shell and tube heat exchangers placed in series. The exit temperature of the melted process fluid exiting the extruder is about 200-340°C.

Figure 2: Impression of the BlueAlp™ plant located in Oostende (Belgium)

The pressure required for melting the plastic depends on the type of feed/plastics is typically 20 bar, to limit the gas content, at a temperature of 450°C.

After the last heat exchanger, a pressure-reduction valve controls the pressure to limit the gas bell size, to guarantee the heat transfer. The plastic melt is then flashed and admitted to the pyrolysis reactor, which is operated close to atmospheric pressure to reduce un-desired reactions in the gas phase. The reactor operates at a temperature of up to 450 °C.

The reactor contains three outlets:

- Top outlet: This allows the exit of the gaseous product containing condensable gases (i.e. products) and non-condensable gases
- Side Outlet: This outlet is used for recirculating the reactor liquid. The liquid is recirculated and sent through a heat exchanger. The heated liquid is mixed with fresh feed in order to heat up the fresh feed to cracking temperature. No additional heat input is required.
- Bottom Outlet: The bottom outlet is used for withdrawing or discharging the slurry which is a mixture of char, liquid, metals and other inorganics. The size of this stream depends on the waste plastic composition and contamination.

The gas leaving the top of the reactor is cooled to cut off the heavy tail. The non-condensable gas enters into a distillation column where the boiling point of the product can be adjusted.

4. Upgrader Process

The pyrolysis oil produced from waste plastic contains contaminants that could hamper the steam cracker safety, performance, reliability and/or operability (e.g. increased corrosion, decrease of catalyst life). Therefore, Shell is investigating and working in collaboration with different partners to develop possible solutions (e.g. adsorption, hydroprocessing, extraction). In this presentation we will focus on the upgrader technology that will be used in the first demonstration plants and currently the most promising alternative: the solvent extraction process. A simplified block flow scheme of the 50,000 metric tons per annum capacity Upgrader is presented in Figure 3 and a 3D impression of the Shell upgrader in Bukom (Singapore) in Figure 4.

The various portions of the clean-up section of the process are described below, taking the following simplified Block Flow Diagram as a reference

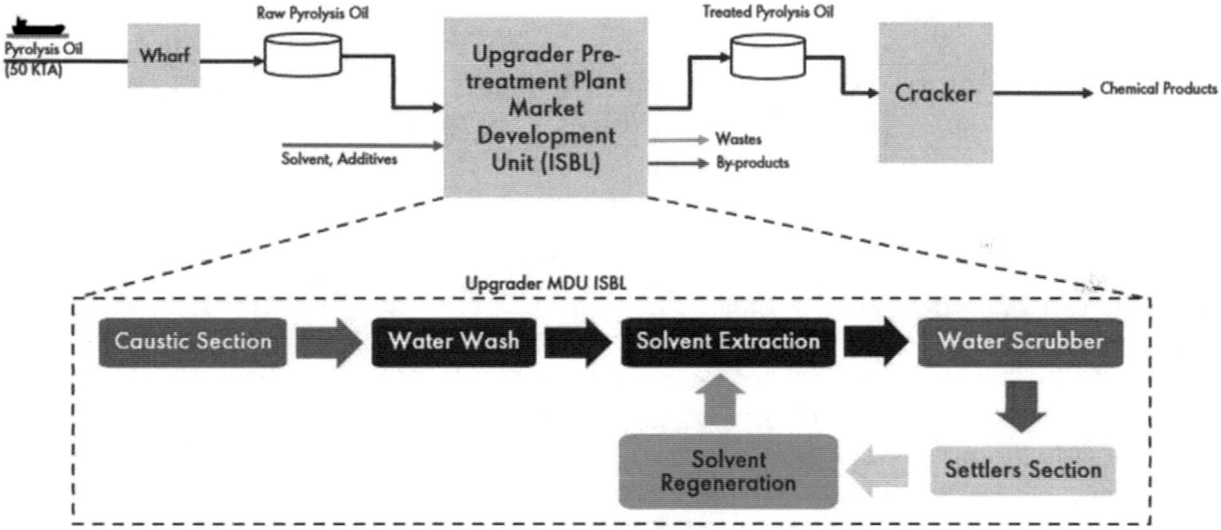

Figure 3: Simplified Block flow scheme Upgrade 50KTA capacity

Feedstock storage:

The pyrolysis oil feed coming by barge is discharged and stored in a heated tank at a temperature in the range 20-45°C depending on the pyrolysis oil pour point and on the required process temperature.

Caustic/water wash section:

The pyrolysis oil feed is first contacted with a concentrated solution of caustic soda in water (~10-20%wt NaOH). The objective is to remove phenols (e.g. p-isopropylphenol, p-tert-butylphenol) and other contaminants (e.g. carboxylic acids, caprolactam) that would otherwise be absorbed by the solvent in the solvent Extraction step and would accumulate in the solvent recycle loop.

The majority of the caustic solution in the caustic vessel is pumped back and will be reused to contact the pyrolysis oil feed again (recycle operation), The sodium salts of phenols and other contaminants removed by the caustic wash may build up to appreciable levels and create a third phase mainly comprising of sodium salts (ie, sodium phenoxide) and some sodium hydroxide dissolved in water.

The pyrolysis oil from the caustic wash step is routed to a water wash step for removal of any entrained caustic solution and other remaining water-soluble contaminants.

In this water wash step, the pyrolysis oil is fed from the bottom and contacted with water in countercurrent. Caustic entrained to the solvent extraction column would accumulate in the solvent recycle loop and cause severe corrosion, especially in the higher temperature solvent regeneration column bottom.

With a pyrolysis oil density lighter than water, the water will sink to the bottom of the tower. The oil phase is decanted from the water phase in the column top section. The interface level is controlled by adjusting the water flow rate. The washed oil from the column top is sent to the solvent extraction step and the used water from the bottom is sent to battery limit as wastewater for treatment or disposal.

Solvent extraction section:

The washed oil from the pre-wash section is sent to the solvent extraction column where the main cleaning process takes place. The solvent preferentially absorbs the contaminants present in the oil. The solvent is recycled from the regeneration section. The solvent to oil ratio, is a key parameter that determines the oil recovery in the extraction column. This ratio needs to be adjusted depending on the contaminant levels in the pyrolysis oil and the desired product purity. The solvent feed temperature is controlled to a desired level. As the solvent extracts some of the aromatic material from the pyrolysis oil, the extraction column temperature needs to be kept high enough to prevent clogging by solid waxy raffinate.

The column bottom stream containing the solvent and the extracted oil/contaminants will be sent to the settlers/coalescer drums.

Final water wash section (Scrubber):

The raffinate from the extraction column is fed to the bottom of the final wash column to remove the excess solvent. The final contaminant level achieved in the product oil depends heavily on the waste plastic quality (level of contaminants), the pyrolysis technology selected, and the solvent extraction operating parameters. Depending on the level of contaminants achieved, the oil can be fed directly to the cracker or mixed with other cracker feeds for further flexibility.

In the final extraction column, the oil/water contact will be enhanced by structured packing. While the water to oil ratio is controlled, the water will sink to the bottom of the tower.

The final wash operating temperature is around 30 to 45°C to avoid wax formation (same as the extraction column). The higher the operating temperature, the more water is required due to higher solvent solubility in the oil.

Settlers section:

The purpose of the section with settlers is to recover the oil and contaminants extracted with the solvent. This is the first step at recovering the solvent before the final solvent regeneration step. The separation is achieved by adding water to the solvent/extract to act as an antisolvent.

The first settler recovers hydrocarbon components that can be recycled back to the extraction column to increase raffinate yield. Such a solution is frequently applied in extraction designs.

The water required in the first settler is small. The subsequent settlers will separate various hydrocarbon streams, containing various contaminant levels. These streams are combined and recovered to the most economical and convenient location.

From the settlers, all the hydrocarbon-containing phases can be decanted and only the water with solvent and small amounts of hydrocarbons and contaminants are sent to the solvent regeneration section.

Solvent regeneration section:

The water and solvent mixture from the settlers are sent to the regeneration column to separate them.

The solvent is recycled back to the extraction column. A small amount of solvent is bled from the recycle loop to avoid accumulation of degraded solvent, hydrocarbons and contaminants. The solvent loop is provided with a fresh solvent make-up connection to compensate for the solvent bleed and other losses.

Figure 4: 3D impression of the Shell upgrader in Bukom (Singapore)

5. Integration with a steam cracker

Impact on the cracking furnaces

Defining acceptability criteria for the feed quality to the steam cracker is instrumental to a sustainable and reliable operation of the plant and depends greatly on the furnaces available in a given site. Like for other conventional feedstocks, the specific furnace design will determine the range of feedstock that can be safely processed together with conventional feeds. Some of the key elements to consider in that respect are:

Convection section fouling

After assessing convection section fouling experimentally in the Shell lab on various feed candidates at various degrees of dilution in conventional feeds, a full vaporization desktop study has been performed using the Shell cracker furnace simulation software suite and Aspen Plus®. The simulations are done for a specific coil configuration and specific conventional feed targeted to receive a blend of pyrolysis oil. As some pyrolysis oils have final boiling points (FBP) much higher than that of e.g. naphtha and boiling profiles close to that of gasoil, issues with vaporization would be most expected when pyrolysis oils are co-cracked with naphtha.

The graph below (Figure 5) illustrates the fact that al pyrolysis oils are not showing the same behaviors. With an operating window of the cracking furnaces between the green curve (naphtha) and the red curve (heavy feed), it can be seen that PO3 (blue curve) is out of it and likely will vaporize later than the heavy feed, potentially causing convection section deposits and plugging. PO1 (black curve) shows a vaporization behavior close to naphtha but with a heavy tail, likely triggering deposits and possible plugging in the convection section. PO2 (yellow curve) shows an in-between behavior, too heavy for co-cracking with naphtha, possibly too-light for co-cracking with a heavy feed. Various degrees of blending will move those vaporization curves to the left, making the operation feasible.

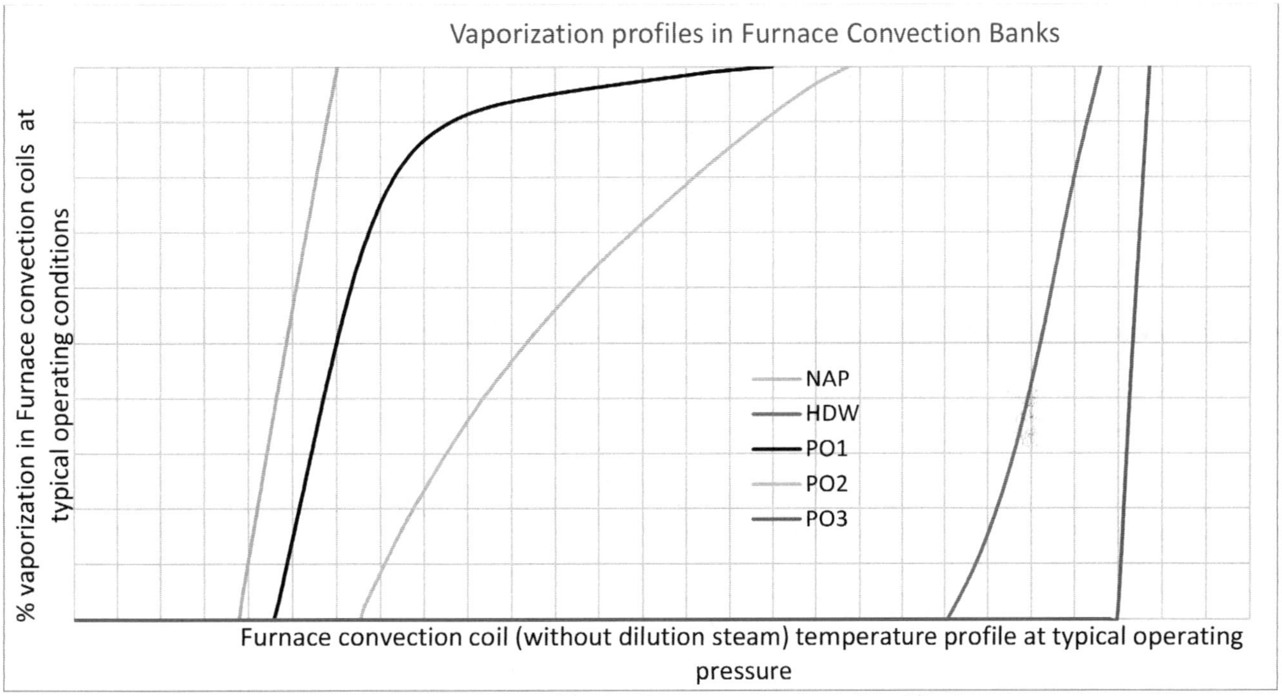

Figure 5: Vaporization Profile in Furnace Convection banks

Contaminant acceptability

The impact of pyrolysis oil contaminants have been abundantly discussed in the EPC[6] and other forums previously. We will not come back on a detailed review of what contaminants can be expected in pyrolysis oils and where they will end up, but we will emphasize two aspects:

- Understanding all contaminants coming with this new type of feedstock is essential. Some of them will be new exotic species not necessarily covered in the conventional feed specifications or with a behavior in the cracking process not well understood. For illustration purpose (see Figure 6), the typical chromatogram of a pyrolysis oil sample could look like a mountain range. Having a comprehensive experimental testing facility is of utmost importance to create new knowledge when not available. This work should not be underestimated. Experience tells that unlike e.g. naphtha, two sources of pyrolysis oil may

[6] EPC Proceedings, see references at the end

yield very different feedstocks, depending upon the waste plastic feedstock used and the process to generate that pyrolysis oil.

Figure 6: a "typical" chromatogram of a pyrolysis oil sample

- The contaminant levels are greatly reduced by the operation of the upgrader but not 100% for all contaminants equally. In any case, a proper in-depth dialogue with the cracker operators is required to thoroughly understand and assess the possible longer-term impact of remaining trace contaminants on the reliability of the cracker. This conversation is not different to those required when a new feedstock source is considered and will lead to a feed acceptance framework through relevant management of change.

Ultimately, the effect of contaminants which have a cumulative effect, e.g. silicon containing components, will also depend greatly on the configuration and availability of adequate guard beds.

Feed acceptance

The feed acceptance specifications set out in a certificate of acceptance (CoA) or "specification" consider not only the remaining contaminant levels encountered in the pyrolysis oil feed, but also the existing feed limits for the steam cracker considered. As far as possible, the CoA specifications have been set out to be the same as the cracker feed specifications, except in specific cases such as when the existing specifications do not have an existing limit on a given contaminant specific to pyrolysis oil processing, or if a higher limit than the specifications has been assessed to be acceptable with respect to the quantities to be processed, with the corresponding mitigations and monitoring measures in place.

Integration with an existing steam cracker

Adding the upgrader process plant next to the steam cracker comes with specific challenges, such as utilities integration, discharge of the reject streams to the existing facilities and the assessment of the impact these streams may have on the on existing assets such as e. g the wastewater treatment facility. Different solutions may apply, depending upon the proximity of other petrochemical assets or refinery.

6. Quality control issues in the longer term

In the longer term, this adventure will be a success when the following key aspects are well understood and under control:

Waste plastic quality: working with the pyrolysis oil supplier(s) to understand the waste plastic composition feeding the pyrolysis oil plant, how the quality of the feedstock affects the pyrolysis plant performance and the final quality of the steam cracker feed. Building up knowledge on the waste plastic source(s) and the sorting technology applied, including limitations, risks and future developments.

Pyrolysis plant design and operation: understanding how the waste plastic quality affects the pyrolysis plant operation and cracker products. Understanding the handles to deal with the contaminants and different qualities.

Upgrader performance: understanding how the waste plastic quality at the inlet affects the upgrader ability to produce on spec steam cracker feeds.

Maximizing the value of the pyrolysis oil: understanding how the waste plastic quality, pyrolysis plant operation and upgrader operation affects the quality and quantity of the cracker by-products and how the existing facilities can cope with them, maximize their value or minimize their cost.

Optimization of the pyrolysis oil boiling range vs cracking furnaces design and operation

Mitigation of the effects of pyrolysis oil contaminants with the steam cracker handles

7. Conclusion: Collaboration is critical to success

Shell is making good progress on its ambition. However, to achieve our goal and support the long-term growth of this promising industry, it is vital to foster industry collaboration to develop the solutions necessary to produce the large-scale volumes of pyrolysis oil customers' demand. This includes developing the waste management infrastructure needed to collect, sort and separate plastic waste.

The effort, technology/expertise and resources required to be able to process this new material (pyrolysis oil) safely, reliably and profitable in our crackers, can be achieved by companies like Shell with the resources, the heavy focus on technology development, the capability to develop collaborations with others, and the moral drive to materialize it.

8. References

Virtual AICHE 2021 Spring National Meeting, 33rd Ethylene Producers' Conference

- Challenges and opportunities for plastic waste pyrolysis oils as steam cracking feedstocks – paper Number 41a - Marvin Kusenberg & Kevin M. van Geem, Laboratory for Chemical technology (LCT), Ghent Univeristy, Ghent, Belgium
- How to produce oil from waste plastics and where to use it – paper Number 24a - Johnny Combs –Chief Executive Officer, New Hope Energy & Ilya Aranovich – Director Business Strategy, Lummus Technology

Ethylene Plant Technology Fundamentals, at the 2020 Spring National meeting

- Mixed plastic waste to olefins: from sorting, over Impurities to product yields and coke formation – paper 124a - Marvin Kusenberg ir.Ghent University, Ghent, Belgium et al.

Alliance to End Plastic Waste (AEPW): https://endplasticwaste.org/en

Blue Alp website: https://www.bluealp.nl/news

Shell's announcement:
https://www.shell.com.sg/media/2021-media-releases/investment-in-shell-energy-and chemicals-park-singapore-to-bring-circular-chemicals-to-asia-pacific-customers.html

BlueAlp Patents :

- WO2021/053139A1 Cracking Long Chained Hydrocarbons from plastic-containing waste and organic liquids
- US2018/0010050A1 Method and System for transferring plastic waste into a fuel having properties of diesel/heating oil

Shell Patents :

- WO2020212315 Method of treating plastics pyrolysis oil by liquid-liquid extraction
- WO2021115982 Method of treating plastics pyrolysis oil by solvent dewaxing
- WO2021213968 Separating unsaturated hydrocarbons by using dihydrolevoglucosenone (Cyrene) or derivative as extraction solvent

639927 Sustainability Certification for Ethylene Supply Chains Using Iscc PLUS

Abstract:

Sustainability certification for ethylene supply chains using ISCC PLUS

The circular economy and the bioeconomy continue to grow. Both have the goal of reducing the dependence on virgin, fossil resources and of reducing GHG emissions. Legal framework requirements like climate neutrality targets, recycling quotas, plastic taxes as well as consumer preferences, company commitments and stakeholder pressure are further driving the circular economy and bioeconomy, in North America, Europe as well as in other regions of the world. Along the complex supply chains, chain of custody concepts like mass balancing ensure the credible connection of sustainability information and the respective materials and products. Compliance with these concepts and its verification is crucial to ensure credibility for companies being involved. The International Sustainability and Carbon Certification (ISCC) scheme is increasingly being used by more than 5,000 system users worldwide and across entire supply chains from point of origin of the initial raw material via the petrochemical and chemical industry to converters and brandowners.

ISCC operates different schemes. ISCC EU is recognized by the European Commission as a voluntary certification scheme for the European biofuels market. ISCC CORSIA is recognized by ICAO (International Civil Aviation Organization) being a specialized agency by the United Nations for the aviation sector for the certification of sustainable aviation fuels. ISCC PLUS is a voluntary certification scheme for the circular economy and the bioeconomy being used by many companies in the chemical and downstream industry all the way to final producers.

Ethylene producers worldwide are already using ISCC PLUS certification. ISCC builds up credibility and acceptance for both, B2B partners and consumers and can potentially be used for reporting under regulatory frameworks or voluntary initiatives. The presentation will provide an overview on ISCC and ISCC PLUS, its sustainability and traceability requirements, ways to get certified under ISCC and benefits for ISCC certified companies.

640303 ESG Economics Lead to Third Party Verification of Greenhouse Gas Emissions

Abstract:

ESG – Environmental Social Governance – continues its increasing influence in the business world, especially in manufacturing segments where sustainability issues have a direct correlation to the financial health, risk, and value of the business. The environmental component of ESG is of particular interest to ethylene producers because the traditional method of ethylene production – thermal cracking of hydrocarbons – is an energy intense endothermic reaction whose needs have historically been met by the combustion of petroleum-based fuel that result in greenhouse gas (GHG) emissions to the environment. Mitigating this environmental impact is of utmost concern to the ethylene producer to satisfy the demands of various stakeholders – investors, insurers, customers, regulators and others. Mitigation can be in many forms but to be successful and have merit the accuracy of emission estimations and reductions is critical. Following the principles of financial accounting and reporting, standards for GHG quantification, verification and certification are quickly evolving. This paper will focus on these developing standards and how they can be applied to the ethylene producer that requires trusted data for important financial decisions.

Process Digital Twin as Enabler for Improving Operational Excellence in Ethylene Production Unit

S Mahesh Kumar[a], Tushar M. Kamerkar[b]
a - Global Cracker Technology, Olefins & Aromatics, MEA, SABIC
b - KBC Process Technology Ltd, United Arab Emirates

Abstract

Global consumption of olefins and its derivative is anticipated to be surging in near future. Along with the new capital developments, chemical production giants are focusing on maximizing existing asset utilization and feedstock utilization with the help of latest technologies. In volatile market environment, key business data from various sources is spread across the organization in random fashion, which makes decision-making cycles even more challenging. This is where Process Digital Twins (PDT) with advanced performance computing facilitates enables in-process performance analysis for non-linear, multi-dimensional trade-offs to maximize the profitability.

Conventionally process models and simulation tools have been utilized for design, engineering and offline analysis. Multiple models of the same facility create model maintenance and quality assurance issues making them futile when needed the most for sensitivity analysis, optimization, or revamp studies. On the other hand, PDT, a result of streamlined collaboration provided by bridging virtual and physical environments coupled with cognitive data enables engineers and process designers to build an immersive virtual replica for first-hand information on what is happening across the entire plant. Using PDTs, engineers can solve problems by reasoning alongside process models while they are in-operation, whether they are working on the field or in a remote location. Thus, PDT not only serve as the single source for design, optimization but also provide vital information for process control, virtual testing, predictive maintenance, and on-stream estimation of a physical process plant.

This paper discusses the development of a PDT tailored for an Ethylene Production Unit of SABIC that dynamically updates and adjusts in accordance with the real time operating data. Petro-SIM® process modelling and simulation tool has unique functionality to model associated utility systems on a unified platform that makes model development, validation, and updating simple. Integrated process and utility model provided a holistic view of the entire plant in real time. Purpose built utilities enabled plant engineers and operators to configure desired process (yield, losses etc.,) as well as sustainability performance Indicators within the Digital Twin architecture to monitor plant performance against business targets. Key aspect of PDT is its flexibility to convert it for conducting various objectives functions like equipment health check, bottleneck identification and overall optimization considering variable feedstock quantity / quality, operating and specific constraints in individual unit operations.

Keywords: Olefin, Ethylene, Process Digital Twin (PDT), Energy Efficiency, Petro-SIM

1.0 Introduction

Global consumption of olefins and its derivative is anticipated to be surging in near future. Along with the new capital developments, chemical production giants are focusing on maximizing the utilization of their existing assets with the help of latest technologies. In this volatile market environment, vital plant information and key business data from various sources are spread across the organization in discrete fashion. This makes decision-making cycles even more challenging in energy and chemical industries. Yet, at every instance, making faster & better decisions, for quick execution is essential for achieving superior results. A more sophisticated approach is required to maintain and enhance the profitability of a complex production unit. This is where Process Digital Twins (PDT) with advanced analytical tools comes to the fore to account non-linear, multi-dimensional trade-offs that make effective decision-making a challenge.

Conventionally, process simulation platforms have been utilized for design, engineering and offline analysis. As simulation models belong to certain individuals, full potential of process simulation may not be realized throughout the organization. Multiple simulation models of the same facility create model maintenance and quality assurance issues making them futile when needed the most for sensitivity analysis, optimization, or revamp studies. Digital replica of the processing facility in real time can provide proactive and complementary support to plant operators in their day-to-day activities. Up to date PDT with appropriate administration can promote knowledge transfer and collaborative workflow with greater ease. Utilization of such technological advancements to meet individual KPIs and overall organizational goals is key to SABIC's vision to be the preferred world leader in chemicals.

1.1 Concept & Scope of Process Digital Twin (PDT)

Even though PDT underpins the philosophy of 'One Model Many Uses', every twin should be developed with an intent to achieve specific organizational goals and should not be 'just another technological tool' available in the organization. It should bring added value to plant operators, process engineers and facility managers. Process Digital Twins integrated within the organizational workflows can assure maximum return on investment and enhanced profitability by increasing operational margins and reducing operational expenses.

PDT provides a single source of truth with improved visibility of information for better facility management, reduced downtime, and sustained plant safety. Effective utilization of Digital Twin, both online as well as offline can prove to be a quick and reliable tool for easy facility assessment, automated bottleneck identification and further optimization. It can significantly reduce the mechanical-iterative work, giving opportunity to engage valuable skills and time of plant engineers for most productive work.

PDT enhances unit monitoring by realizing & visualizing plant performance parameters that cannot be measured/ monitored directly. It acts as an advisory system that not only provides early warning of the performance deviations by continuous automated monitoring but is also useful for 'What-if' &'What's best' analysis during normal operation as well as during upset scenarios for speedy recovery. Additional profit improvement opportunities identified can be easily validated and justified using PDT for faster implementation and quick realization of sustained benefits.

Table-1: Scope and features of Process Digital model

Performance Monitoring	Process & Energy Optimization	Process Improvements
- Yield & Energy	- Optimized yield at varying feed rates	- Constraints Identification
- Mass balance	- Distillation optimization	- Debottleneck
- Instrument malfunction	- Performance gap identification	- Scenario testing
- Deterioration (fouling)	- Steam header optimization	- Sensitivity studies
- Regulatory targets	- Loss Minimization	- Equipment Adequacy check

KBC (A Yokogawa Company) is one of the leading simulation technologies providers in the world. Petro-SIM is a first principles-based process simulation platform developed by KBC for rigours modelling of variety of process units across the entire oil & gas supply chain including production, refining, petrochemicals etc. KBC is constantly delivering benefits and value to the clients for more than 42 years through its unique technology and consulting services. Over a period, significant developments in KBC simulation technologies have brought Petro-SIM beyond the conventional boundaries of process simulation platform. Because of these enhancements, Petro-SIM was recognised and awarded with Best Digital Twin Technology for year 2020 by Hydrocarbon Processing. Company leveraged on this advanced simulation technology to develop Digital Twin for one of their facilities in olefin complex.

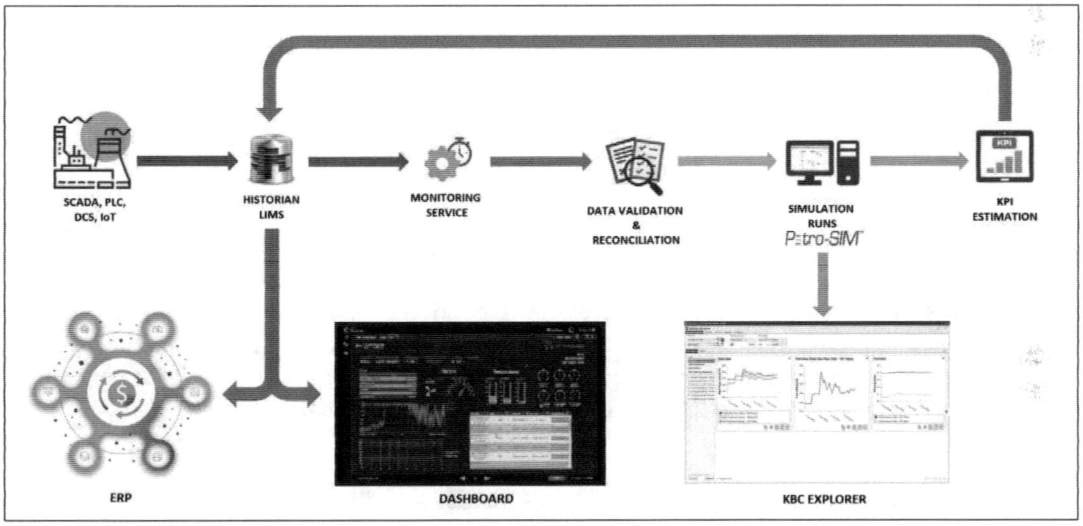

Figure-1: KBC Process Digital Twin Architecture

2.0 Process Digital Twin Development

A Single feed SABIC cracker was selected for development of PDT. A detailed process model consisting of all unit operations of an olefin steam cracker including ethylene, propylene refrigeration units and steam network was modelled. These units were configured on the same platform and connected to process users in real time to study the impact of these (steam and refrigeration suppliers) on process variabilities.

The furnace section was modelled by embedding already available Artificial Neural Network (ANN) model. The ANN model used in the current work was validated with plant data for multiple feed rates and cracking conditions. For a given conversion and dilution steam factor, the model would predict the yield, which was used as a Furnace output in the model. For major part of unit operations, Peng Robinson (PR) thermodynamic package was used. Quench tower was modelled using KBC's proprietary Multiflash™ CPA thermodynamic package. CPA (Cubic-Plus-Association) is an equation of state that is based on a combination of the Soave–Redlich–Kwong (SRK) equation with the association term of the Wertheim theory that offers accurate and robust CPA approach for predicting variety of complex phase equilibria, including mixtures containing alcohols, glycols, organic acids, water, and hydrocarbons.

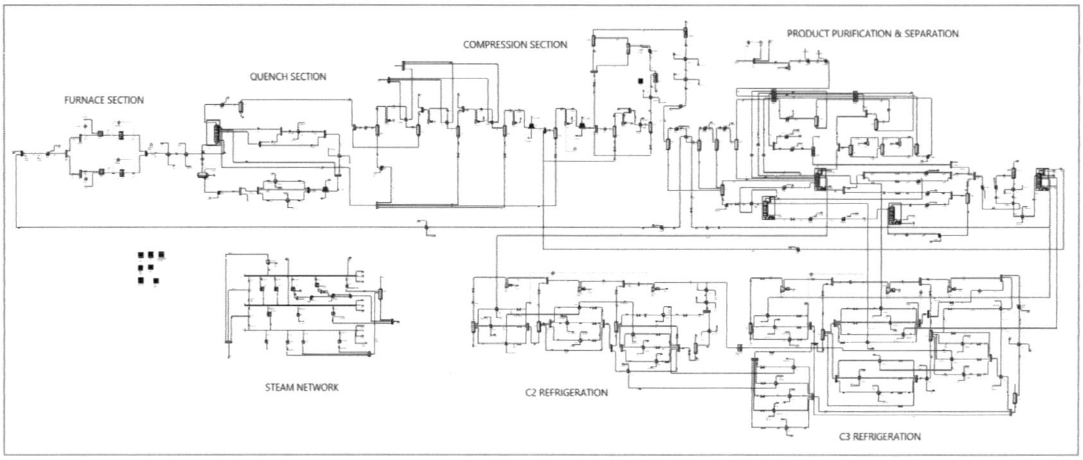

Figure-2: Petro-SIM Simulation Model

Petro-SIM Digital Twin has unique functionality to model associated utility systems on a unified platform. This integrated modelling approach ensures for every operating data; process utility demands are met when the model is converged. This further helps in comprehensive analysis of plant performance and supports in making operations or business planning decisions considering productivity vs. Yield vs. Energy trade-off.

The digital twin model was configured to calculate the KBC Best Technology (BT) Index value for cracker operations. The BT index basically compares current energy use against a thermodynamically and economically achievable (best available technology) minimum, with best technology designs and calculates the benefits that can be achieved from closing the gaps. Cracker target BT index is calculated based on an optimized process configuration, with the optimum target energy benchmark defined as 100. The actual BT index is calculated as the ratio of actual energy use divided by the target, in percentage. For example, if the subject cracker under interest is using twice as much energy as the benchmark, then the BT index is 200.

Utility section was modelled using ASME steam package. Liquid and gas driers were modelled as separation block and all Heat Exchangers were modelled in steady state rating mode. Compressor curves were incorporated to calculate the efficiency and the head. Following the model development, it was validated at first for the design case followed by plant operating case for multiple feed rates using plant data. Subsequently, efficiency factors for Turbines, Trays and Compressors were calibrated to reflect the observed plant conditions.

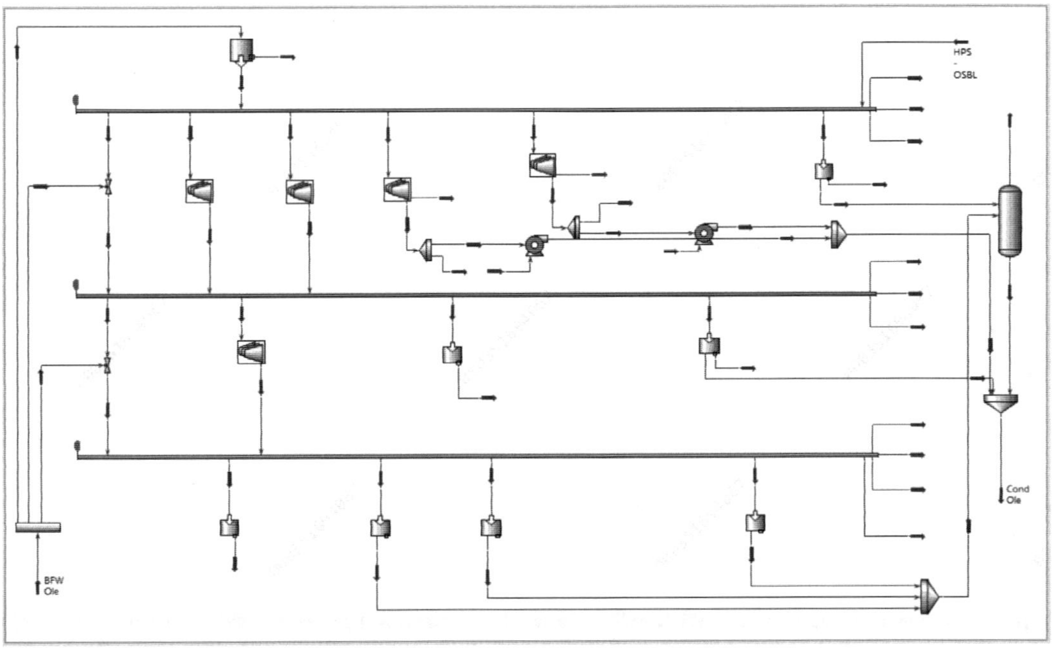

Figure-3: Integrated Utility Modelling

2.1 PI Historian Connection for Data Retrieval

First step for Digital Twin is to acquire necessary plant data in real time. Petro-SIM has native functionality to connect with process historian of different types as well as Lab Information Management Systems (LIMS). It is a common practice for plant engineers to fetch historian data in excel spreadsheet for offline utilization. Petro-SIM's Historian connectivity works seamlessly in either cases without requirement of additional plug-ins or third-party tools. Necessary process/ lab tag information can be easily configured and mapped to respective streams within the same simulation environment. Plant data for required time span can then be fetched manually or through automated scheduled runs.

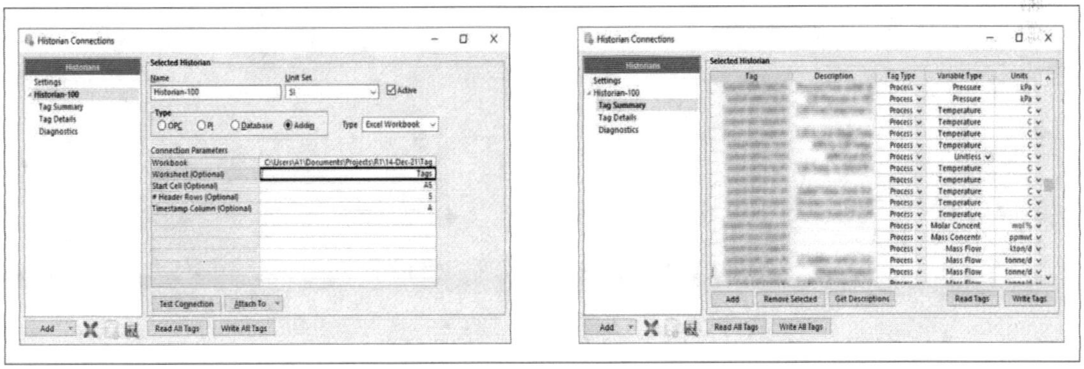

Figure-4: Native Connectivity with Plant Historian

2.2 Data Validation

Simply bringing in the plant data is not sufficient since raw data usually contains noise, disturbances. Data conditioning capabilities of Petro-SIM Digital Twin captures and highlights out-of-range plant measurements. In case of erroneous or no measurements from field instruments. Twin considers appropriate value for particular parameter ensuring sensible and uninterrupted operation of Digital

Twin. Configurable logic-based workflows available within Petro-SIM perform these Data Quality Parameter (DQPs) checks during each run and user warnings are issued for data quality deviations for suitable action.

The Model was thus configured to retrieve data from process historian, to feed process model and to compare the output data against the plant actual performance data. At first, the model fetches the raw data from the historian for designated time intervals. It is then cleansed and conditioned to remove unnecessary data like outliers (measurement from defective instrument), values and rate of changes that are physically not possible, etc.

2.3 Data Reconciliation and Simulation Run

Factors like equipment hold-up, controller action delays etc. introduce imbalances in raw plant data. Thus, plant data even though sanitized, cannot be directly used for the simulation without closing the material balance. Usual practices of data reconciliation call for operator intervention. Overall imbalance error is assigned to a single stream and approximate mass balance reports are generated manually on shift or hourly basis. The built-in data reconciliation tool performs overall material balance in automated fashion. It also provides flexibility to configure and perform individual element balance wherever required.

This step in the data conditioning involves an automated iterative process where the measurements, which are subject to biases in the model, are solved for minimum variance and a reconciled dataset is generated. With this approach a generalized reconciled dataset of plant-wide material balance, and energy balance is created. The goal of this step is to ensure the dataset is reconcilable and maximizes the information that can be extracted from the cleansed data using engineering principles, empirical rules, and elements of operational knowhow. Well-conditioned and reconciled plant data is then passed on to the simulation model.

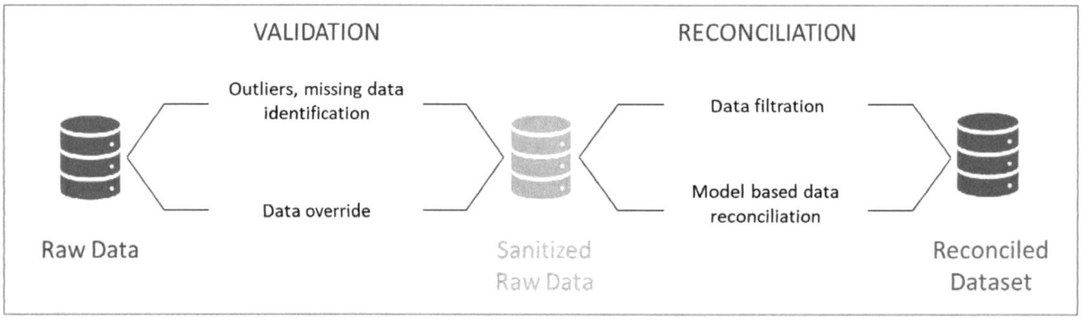

Figure-5: Data Conditioning Steps

2.4 Process KPI Monitoring

Petro-SIM offers integrated platform for modelling process and utilities for optimising design and operation. Integrated process and utility model provides a holistic view of the entire plant in real time. Purpose built utilities enable users to configure desired process (yield, product losses in recycles) as well as energy KPIs (Key Performance Indicators) within the Digital Twin architecture to monitor the plant performance against business planning targets and regulatory compliance targets like Energy Intensity targets and Feedstock utilization ratio as required by regional regulatory bodies (Saudi Energy Efficiency Program).

As actual plant performance varies over operational period, along with data quality, it is also important to keep a check on accuracy of the Digital Twin being used. Customizable Model Performance Indicators (MPIs) support Model Quality Assurance (MQA) to ensure that, results provided by the Twin are within tolerance limits and provide timely indication when model recalibration is required.

Simulation results along with the plant raw data, reconciled values and desired KPIs are stored in a central data repository called KBC Explorer. Process Engineers can perform deeper analysis using KBC Explorer as data is available in both tabular as well as graphical form. Area operators and engineers can directly interact with benchmarked calibrated model through KBC Explorer to understand the deviation and develop corrective actions to improve KPI's. Result values can be then easily populated on desired dash boarding tool.

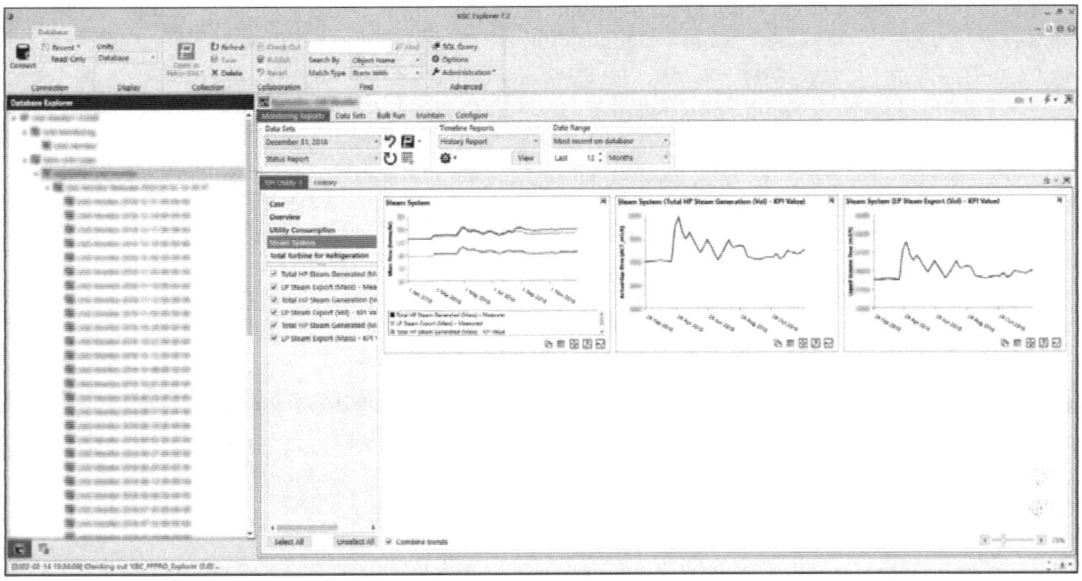

Figure-6: Simulation Result, KPI Visualization on KBC Explorer

2.5 Benefits and value addition

Continuously running automated digital twin model provides detailed insights for the evaluation of cracker performance in terms of Energy Intensity (EI), Capacity Utilization and Yield. These performance parameters are constantly monitored against industrial BT index and to its best-demonstrated performance. These comparative analyses help in identifying performance gaps for closure and overall improvements. Considering actual operating conditions; digital twin also keeps track of the individual equipment performance with respect to its design constraints to ensure, its operation within the desired operating envelop for varying feed rates or cracking conditions. Furthermore, digital twin continues to help in identification of opportunities for optimizing cracking conditions based on feedstock variability, seasonal variability (summer / winter; day /night), and constraints in specific unit operations in cracker and so on. Further, it also facilitates performance monitoring of major compressors like CGC and critical heat exchangers in the quench loop.

3.0 Case Example of Benefits at Ethylene Facilities

3.1 Case Study – 1: Optimization of Furnace Cracking Operations based on C2 Splitter constraints

In a reference, gas feed-based cracker under consideration, due to feedstock variability, corresponding changes are made in the operation of the plant especially in furnace cracking conditions and separation sections. When feed rate changes, furnace section operators modify the cracking conditions to gain yield (lower conversion) in case of low feedstock or capacity (higher conversion) at higher feed rates. Often these changes were made considering only furnace operability / capacity. Because of these changes, C2 fractionator experienced unstable operations and often the product quality goes off spec resulting in financial losses.

Effect of C2-Splitter reflux on profitability - Initially, decreasing reflux saves propylene refrigeration in the condenser, increasing profitability. However, decreasing reflux by even 1 - 2% below the optimum, results in costly C2= losses in the recycle

Hence, to avoid unstable conditions in C2 fractionator in the absence of good understanding between the feed rates, conversion and C2 splitter capacity, furnace operators were operating with comfortable margin and as a result underutilizing the opportunity to increase yield or capacity when feed rate changes. With the help of PDT model, detailed sensitivity studies were conducted based on actual plant operating conditions at varying feed rates and an operating window was developed to guide the operations to optimize the process at varying feed conditions. Further, process engineers validated the operating parameters based on prevailing conditions. Because of these collaborative efforts between the process engineering team and operators and with the help PDT model, cracker operations were optimized utilizing full capacity of the furnaces and separation section at varying feed rates.

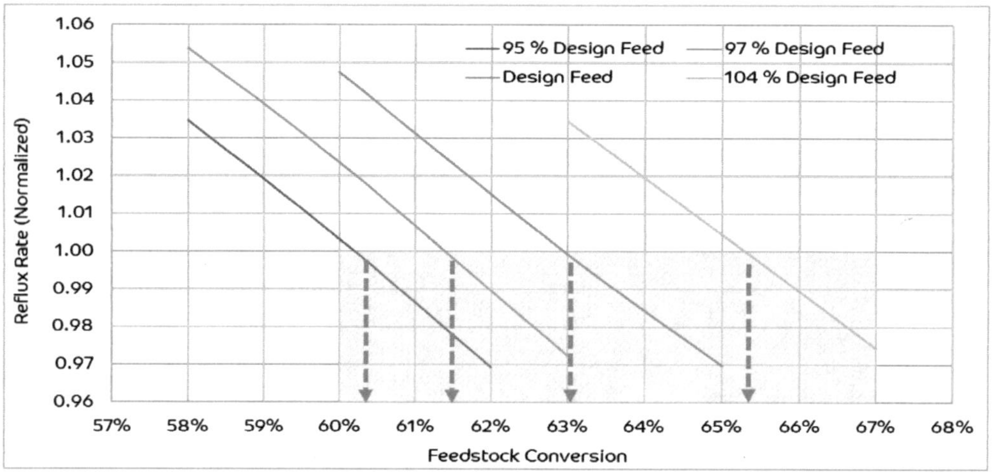

Figure-7: Sensitivity studies for identification of feasible region

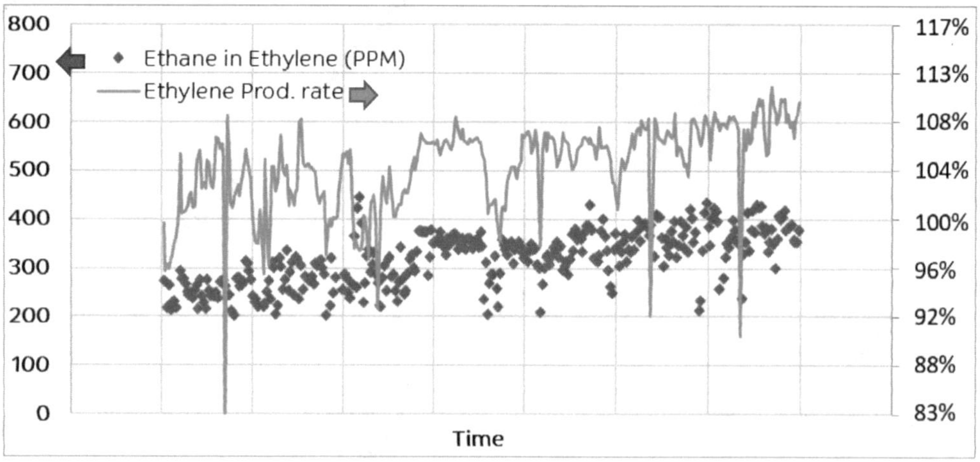

Figure-8: Trend of ethylene product quality improvement

3.2 Case Study – 2: Adequacy checks and Traffic comparison

Conventionally, adequacy checks and traffic comparisons were carried out using an offline model with the level of complexity needed to achieve a certain objective. Once the models are developed, they are validated for design and subsequently for operating conditions by updating the constraints. With the help of digital process twin, these steps are much simplified and executed at a faster time frame as compared to conventional process thus saving significant manpower / consultancy cost for the organization.

Thus, using PDT, equipment rating or adequacy check of existing equipment/unit/plant for changed operating conditions (e.g. pressure, temperature, flow, summer / winter etc.) was conducted to identify the plant constraints. Further relative plant traffic comparisons were made to identify the changes in hydrocarbon load to all parts of the cracker. The equipment current margins in comparison to original plant design margins were used to check for adequacy for new processing conditions and provide recommendations to operate the plant in stable manner while producing the target amount of ethylene product.

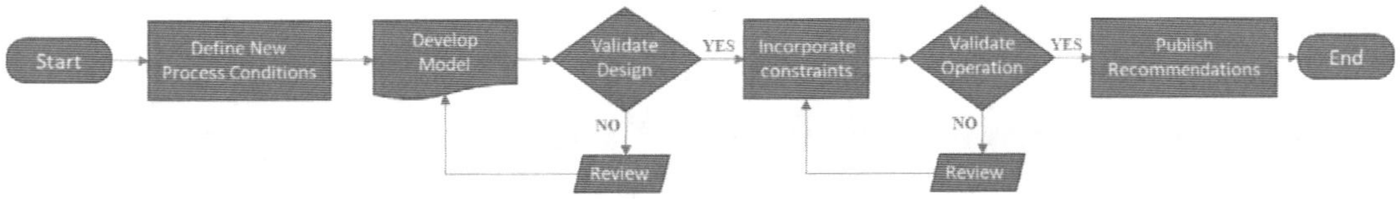

Figure-9: Conventional workflow process for Adequacy checks and Traffic comparison

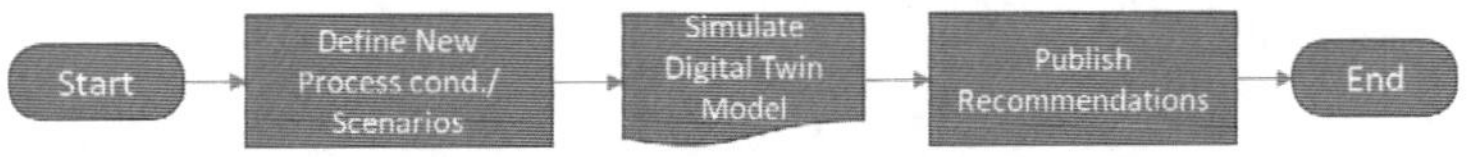

Figure-10: New workflow for Adequacy checks and Traffic comparison with process digital twin.

The advantage of the digital process twin is that the model readily compares the simulated operating conditions (what if) based on existing operating conditions (design margins) and visually highlights the constraints if any. These adequacy checks form the starting point for the study to identify new equipment requirements to overcome the equipment constraints in the processing areas to meet the study capacity targets.

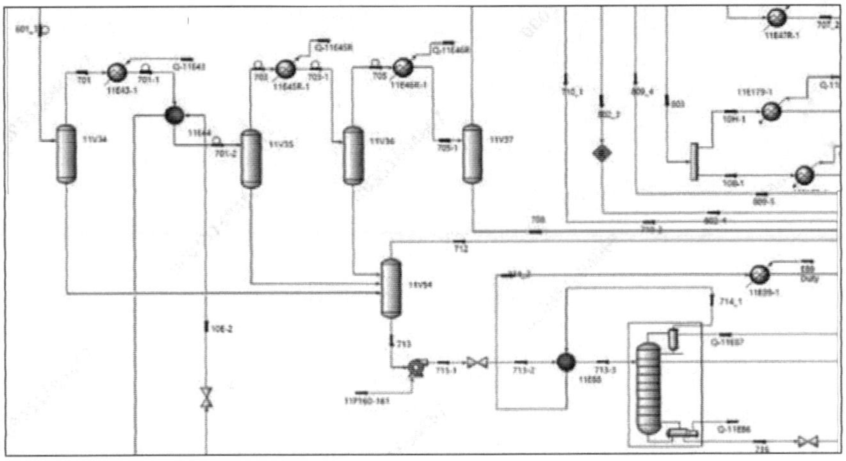

Figure-11: PetroSIM® process model snapshot indicating equipment bottleneck

Regular monitoring and comparison of desired process as well as energy KPIs help maintaining plant operation/ production as planned reducing unwanted downtime and production loss.

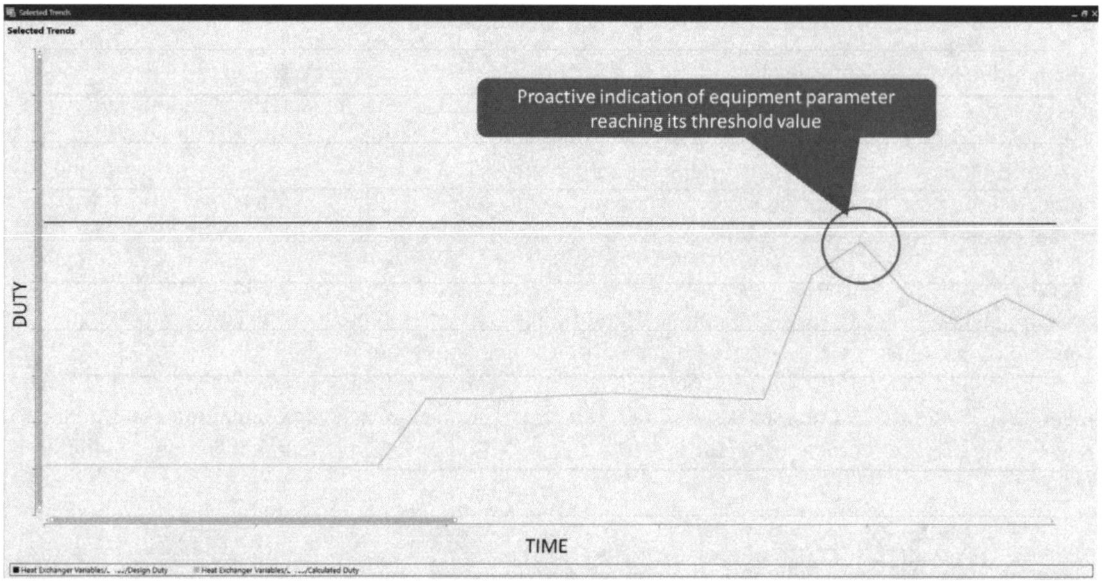

Figure-12: Equipment health check w.r.t design constraints

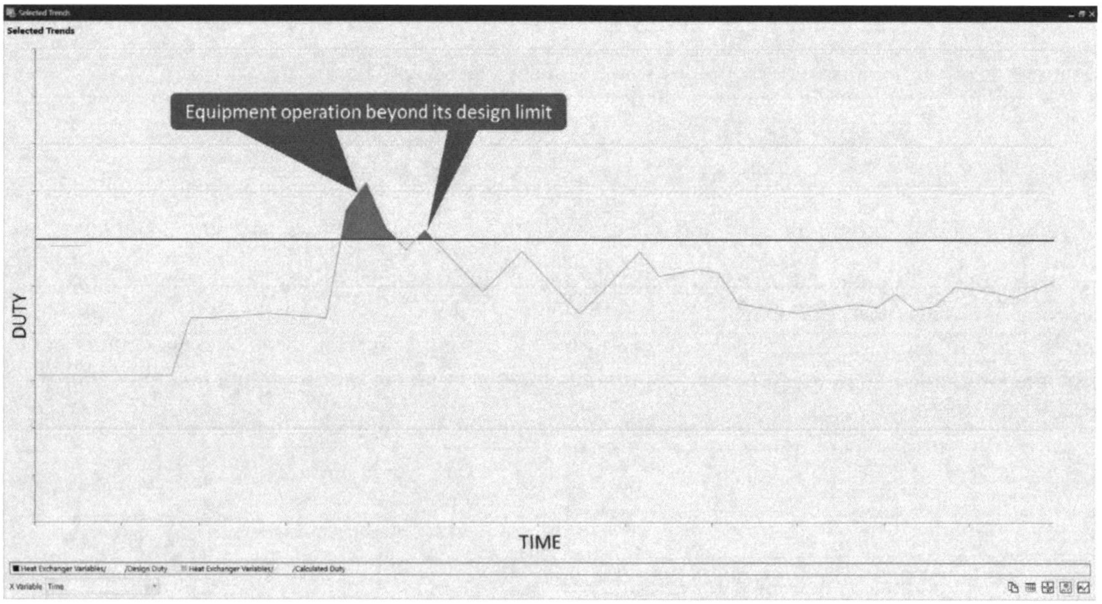

Figure-13: User alert for operation beyond threshold

3.3 Case Study – 3: Online Process & Energy Performance Monitoring

PDT enables engineers to benchmark actual plant operations against the optimum in real time environment. Digital / virtual twin will monitor the plant's performance and provide guidance for improving operational efficiency in real time. As a result, PDT steam cracker plant will help in increasing capacity and product quality while reducing energy consumption and emissions. As shown in the Table below, in PetroSIM ® we can continuously monitor and track all necessary KPI's to identify opportunity for improvement and/or to sustain optimum performance.

Table 2: KPI monitoring utility in PetroSIM

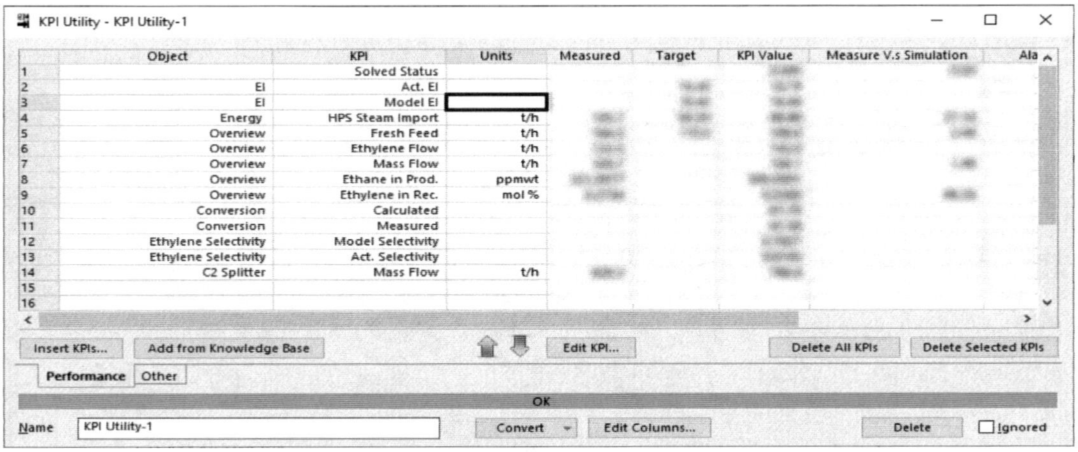

Unit's Energy Intensity (EI) was estimated and tracked on regular basis using PDT. Power consumed by unit for various operating scenarios was compared using BT Index to identify gaps and further enhance unit's energy efficiency.

Table 3: Online BT Index tracking

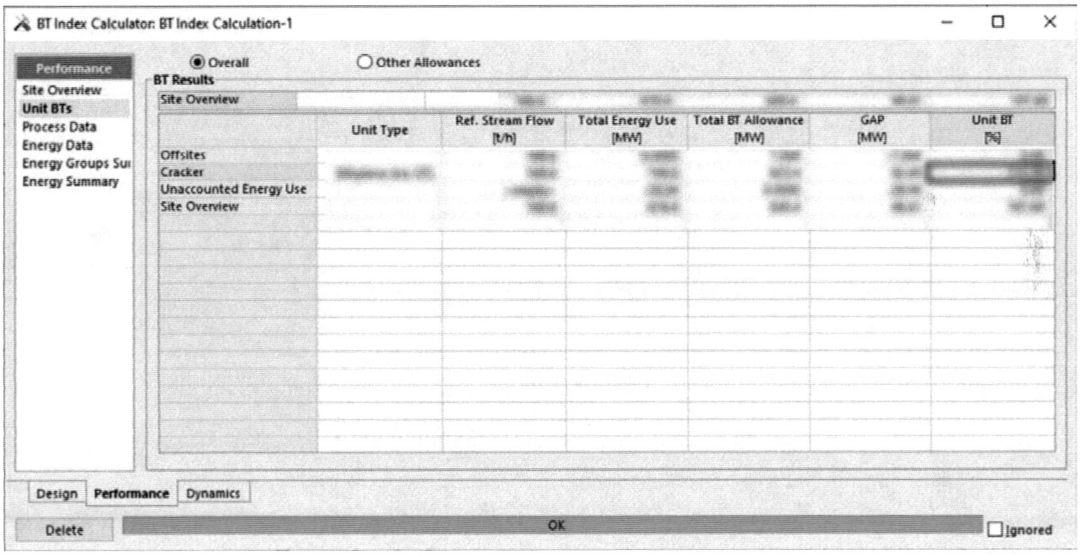

3.4 Benefits summary

Other key benefits from PDT is summarized in Table 4. As you may see, apart from regular plant monitoring and operations significant benefits are realized in process reporting, and engineering studies as well.

Table 4: Overall benefits achieved by deploying process digital twin at reference ethylene plant

Parameter	Relative benefits	Comments
Energy Reduction	Upto 1.5 %	steam header optimization
Capacity Increase	Upto 1.0 %	better assess utilization, C2 splitter
Man hours saved reporting	70%	automated KPI reporting to regulatory bodies
Case studies conducted	> 5	Route cause analysis and business planning
Consultancy man hour saved	40 – 60%	In house assessment and equipment rating

References

1. Pratap Nair, Big Data Analytics for Improved Overall Asset Effectiveness of Ethylene Plants, EPC Proceeding, Volume 30 – 2018.
2. Pratap Nair, Tailoring Big Data Analytics for Optimizing Ethylene Plant Asset Performance, EPC Proceeding, Volume 29 – 2017.
3. Ryan Mc Clarren John Adams, Artificial Intelligence and Machine Learning for Ethylene Plant Operation, EPC Proceeding, Volume 29 – 2017.
4. Daniel Renard, Re-Rating Ethylene Plant Equipment for Optimum Results, EPC Proceeding, Volume 24 – 2012.

639637 Benefits of Combining Digital Twins with AI/ML

Abstract:

Anyone who has watched the movie Apollo 13 witnessed one of the most successful execution of the term "Digital twin" we hear so often these days. That happened in April of 1970. What was unique about this event was the manner in which NASA mission controllers were able to adapt and modify the complex simulations that the astronauts trained on the ground to match the conditions of the crippled spacecraft and bring the astronauts back home safely.

With the dramatic advancement in modern communications, compared to those that existed over 50 years ago, 'Digital Twin' paired with advancements in AI and ML (Deep Learning) is an increasingly valuable tool for transforming a wide range of industrial operations to a 'high fidelity' profile. The digital twin, provides 'real-time' profile of the asset or process of an entire plant or part of the critical operations of a plant. It provides insights on performance across all aspects of the process – operations, reliability, environmental performance, and maintenance, to name a few life-cycle performance items.

By pairing a digital twin with AI/ML, insights through predictions and rule-based prescriptive advice are provided. Performance optimization potential is determined and the changes needed to reach the optimized state can be defined. And, in all cases, the science (i.e. the digital twin) explains the data. However instead of having a descriptive tool that must be physically manipulated for diagnostic and predictive purposes, real time prescriptive and predictive advice is being provided at the operations level.

The challenges of combining a 'high-fidelity' Digital Twin of an asset with AI/ML and successful use-cases will be presented.

638959 Olefin Plant Digitalization: Model-Based Coke Monitoring, Sensing of KPIs and Optimization for Steam Cracking Furnaces

Abstract:

The steam cracking furnace is at the heart of the ethylene production process, and its optimal operation is key to maximizing profitability. Accurate information on key process indicators, such as conversion, product yields, tube metal temperatures, and coking rates are essential for the optimal operation and control of furnaces. However, measuring these reliably in real time is often either impossible or problematic, requiring model based digital applications for reliable and timely information. As the furnace KPIs are affected by coking, it is also essential to have an accurate estimate of state of coking for accurate KPI prediction.

In addition, because feedstock availability changes with time, in order to maximize profitability it is necessary to periodically identify optimal operating conditions for the furnaces that nevertheless ensure operation within the plant constraints. Such optimization calculations should take into account the current state of coke build up in the furnaces to get correct results.

In this paper, we present three state-of-the-art Digital Applications for steam cracking furnaces to provide model-based coke build-up monitoring, real-time sensing of KPIs and optimization of operating conditions. Built on Siemens PSE's gPROMS Digital Applications Platform, they use sophisticated numerical manipulations to combine rigorous first principles models of steam cracking furnaces with plant operating data. Experience with the deployment of these applications on full-scale industrial plants will be presented. In addition, concepts of more advanced optimization applications such as multiperiod mixed-integer optimization for production and decoke scheduling over a near-term time horizon, and integrated optimization of multiple olefin crackers, will be briefly discussed.

SPRING22
+18TH GCPS
A Joint AIChE and CCPS Meeting

Using Thermal Imaging and A Cloud Platform To Optimize Cracking Furnace Efficiency In Real Time

Tareg Alsoudani, PhD
Senior Expert Advanced Automation
Advanced Operations Services, Linde plc
1585 Sawdust Rd Suite 300, The Woodlands, TX 77380
Tareg.Alsoudani@linde.com

Thomas Heinzerling
SVP Advanced Operations Services, Linde plc
Dr-Carl-v-Linde-Str 6-14, 82049 Pullach, Germany
Thomas.Heinzerling@linde.com

Prepared for Presentation at
American Institute of Chemical Engineers
2022 Spring Meeting and 18th Global Congress on Process Safety
San Antonio, TX
April 10 – April 14, 2022

Keywords: digital twin, cloud, kinetic model, camera, thermal imaging, efficiency, optimization, steam cracker, virtual furnace, TMT measurement

Abstract

Using kinetic models to predict the behavior and performance of cracking furnaces is state of the art. However, in many cases these predictions do not match reality in the plant resulting from problems with the measurements (e.g. coil outlet temperature) or the tuning of the model. Neither the predicted values from the model nor the manual tube metal temperature (TMT) measurements are really precise. Using infrared cameras with AI based image recognition software we can measure the actual temperature profile along the coils in real-time 24/7 during cracking and decoking. By continuously comparing the TMT profile predicted by the kinetic model with the TMT measured by the camera we can:

- Prevent coil overheating resulting from flame impingement or hot spots resulting from flow restrictions
- Extend furnace run-length closer to material limits
- Shorten decoking times through temperature controlled decoking
- Detect measurement and model errors and correct them so that the kinetic model output actually matches reality.

For a long time the ability of furnace experts from licensors to support furnace operators in real-time was limited by the possibility to share real-time process data. Modern cloud platforms like Microsoft Azure are providing the infrastructure for a collaboration of licensors and operators that meets the stringent IT security requirements for operators of critical infrastructure.
This paper will present Linde's solution combining their real-time kinetic model with continuous thermal imaging of the coils accessible through a cloud platform, and the first experiences with their implementation for operators of steam crackers in Europe, Middle East, Russia and Asia.

1 Linde Advanced Operations Services

Linde Advanced Operations Services provide solutions and services to process plant operators in order to improve efficiency

2 Using real-time digital twins in the cloud to optimize steam cracker efficiency

Real-time digital twins like the Linde Virtual Furnace (a kinetic model of a steam cracking furnace) run continuously in parallel to the real physical steam cracking furnace with the same input parameters (feed stream, fuel stream, steam).

The comparison between the measures process values and the process values calculated by the digital twin allows to optimize steam cracking furnace performance.

These models allow a continuous forecast of the key parameters (cracked gas composition, TMT, etc.) for the complete furnace run.

Cloud applications that make it possible for the plant engineer on site and the supporting engineer e.g. from Linde to simultaneously see the same values and forecasts in rel-time allow a seamless collaboration and support.

3 Using thermal imaging to optimize steam cracker efficiency

Measuring Tube Metal Temperatures (TMT) of coils manually with a pyrometer is a method that is not very accurate and reproducible.

Replacing this error prone manual measurements by a fixed installed infrared camera with a thermal imaging system generates accurate, reproductible continuous TMT measurements.

This allows to run furnaces with TMT closer to material limits. As a result the furnace runlength can be extended.

Comparing the TMT based on the thermal imaging with the TMT calculated by the digital twin allows to detect problems early and avoid coil damage or rupture.

4 Case study for the optimization of steam cracker efficiency

Case1: Optimizing the selection and procurement of liquid feed for a liquid cracker

Case 2: Optimising feed distribution to several furnaces in several steam crackers and the control of severity using Advanced Process Control for each plant and a Site Production Optimiser for all steam crackers.

SPRING22
+18TH GCPS
A Joint AIChE and CCPS Meeting

Augmented Intelligence for Ethylene Process

Presenter/Principal Author: Sameer Thorat
Director of Process Analytics and Engineering, North America. Ingenero, Inc.
Houston, Texas
sthorat@ingenero.com

Nishant Jaiswal
Manager: Process Technology, Ingenero Technology (I).
Thane, India
njaiswal@ingenero.com

Prepared for Presentation at
American Institute of Chemical Engineers
2022 Spring Meeting and 18th Global Congress on Process Safety
San Antonio, TX
April 10 – April 14, 2022

Augmented Intelligence for Ethylene Process

Presenter/Principal Author: Sameer Thorat
Director of Process Analytics and Engineering, North America. Ingenero, Inc.
Houston, Texas
sthorat@ingenero.com

Nishant Jaiswal
Manager: Process Technology, Ingenero Technology (I).

Keywords: Dynamic Benchmarking, Live Analytics, Furnace Run length predictions, Column Hydraulics, Soft Sensors.

Abstract

Augmented Intelligence provides the right information at right time to drive the right decisions without overwhelming operations with “data overload”. Driven by technology, Industry 4.0 is transforming the ethylene process and reshaping operation paradigms to create greater efficiency and synergies in operating strategy and solve challenging problems. An effective implementation of Machine Learning Operations (ML Ops) provides Augmented Intelligence to managers, engineers, and operators in the Oil & Gas Industry (Upstream, Midstream & Downstream), that would otherwise have taken years of operational experience.

This paper discusses the intelligent tools developed for an ethylene facility to gain insights for operational improvement through augmented intelligence. This also includes smart techniques and solutions to improve operational efficiency, addressing furnace run length and other reliability objectives, live insights to identify and close operational GAP’s. Real-time solution, developed with a state-of-art technology and advanced analytics application for ethylene process industry results in several benefits.

1 Introduction

Digital Transformation is driving major improvements in the process manufacturing industry, particularly seen in ethylene plants. The inherent complexities in the ethylene process makes application of data analytics a significantly challenging process, but the insights that create the returns are several. With the right domain expertise, in-depth process knowledge and appropriate ML techniques combined with first principal models, a customized data analytics solution can be applied in the ethylene process to meet individual plant needs and effectively drive improvement objective. ML/AI has made it possible to implement innovative solutions that efficiently provide Augmented Intelligence on real time basis for operations.

2 Role of AI in Ethylene Industry

Enhanced computation power provides the opportunity to process 1000's of data points in a matter of seconds and convert data to information. With real-time data analytics and the apt use of AI and Machine Learning techniques, this data can be converted into Augmented intelligence which can be applied on a live basis for process improvement. Augmented intelligence effectively utilizes artificial intelligence through human centered partnership to achieve operational excellence.

"Role of Augmented Intelligence is to enhance human intelligence by providing timely insights to aid learning, decision making and quick evaluation on process optimal."

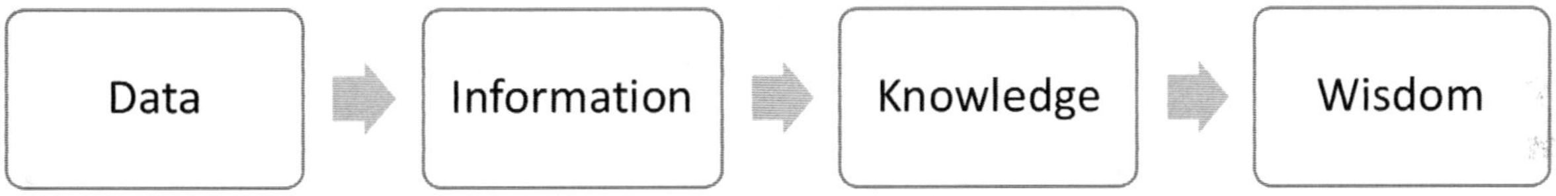

To improve any process, operations personnel need to take timely decisions. And to do so they need the right information and insights, so that they can drive towards optimal operational changes. For example, a distillation column. Ethylene process is dynamic and therefore its scenarios are ever changing. Ambient temperature, equipment condition and degradation, column feed and composition variation, etc. all these factors alter the column operating condition and ideal operating points.

An operator needs to know the optimal set point and conditions for a specific given scenario. How close is column to flooding point? Is any section limiting? What are the exact hydraulic issues and the right handles to address it? Any scope for capacity improvement? He needs details on reboiler fouling rate, and will it limit capacity in near term future. Are the indications reliable to ensure stable operation?

If this information is not available, then usually operators will take a conservative approach. Which will lead to suboptimal operation in terms of lower capacity utilization or lesser product recovery. Un-informed decisions may also lead to process disturbances. If indications are missed, it could lead to fouling or mechanical issues causing capacity constraints and un-timed shutdowns. A digitalized environment helps alleviates all this.

3 Timely insights to drive process excellence

Augmented Intelligence provides the right information at the right time to drive right decisions without overwhelming operations with data i.e., “data overload”.

Insightful information at right time	Overview of unit status, with capability to drill down details
	Prescriptive advice to make right decisions
	Automated alerts – with degree of significance
	Limitations, GAPS and options to meet optimum targets
	Optimal set points to maximize asset effectiveness and reliability
	Improving stability of process
	Deviations with indications for faulty instruments/readings
	Identifying next problem area
	Expected furnace run-length and residual life
	Optimal settings to deliver the highest profitability

Intelligent algorithms provides live GAP analysis providing changes required to achieve best achievable operation for the given scenario. Models find hidden data patterns and predict possible problems earlier. ML models predicts residual equipment life, provide prescriptions so as to meet desired run times. Outlier detection models track instrument accuracy, help identify instrument drifts and improve process control. Soft sensors with predictive models, drive decision-making and improve the reliability of current systems. Live tracking system can identify stability issues and eliminate constraints. Analytical models help optimize process control and improve APC performance. AI helps drive intelligent decisions to drive operational excellence.

- Benefits for Operators:
  - Provides insightful information and prescriptive advice to make right decisions in a timely manner, avoid suboptimal operation and achieve significant profit improvement.
  - A digitalized environment to address dynamic challenges and support safe and efficient operation
- Benefits for Engineers:
  - Data analytics provide a higher level of insights and understanding of process and equipment conditions
  - Helps engineers reduce data handling and model maintenance efforts and focus on plant improvement and troubleshooting

In the ethylene industry, Augmented Intelligence empowers operators and engineers to work *faster, better, and smarter.*

4 Dynamic process – A need for hybrid models

Applying analytics to gain right insights is a challenging process. Inherent intricacies of ethylene process makes it more complicated. For example, while improving furnace yields and capacity is economical, it needs to be ensured that it does not affect furnace availability i.e., optimal run length and coil life. Yield benchmarks needs to be dynamic to be effective, as plants may not meet same benchmarks at different conditions (summer vs. winter or clean vs. fouled). Controller effectiveness needs to be calculated and tracked. Instrument drifts needs to be captured to ascertain stable operation. Equipment problems needs to be identified as early as possible. Failures needs to be predicted and addressed to ensure reliability and continuous optimal operation. Considering economics and profitability is of prime importance. Process optimization is a dynamic process where opportunities need to be identified within given boundaries with respect to current economics, process capability, equipment health and safety.

Complex problems can be effectively managed by having the right mix of the approach, models, and techniques. Big Data Analytical techniques can be utilized to effectively analyze hundreds of minute wise tag data for a process equipment, in real time. Machine Learning Techniques in combination with first principle models have been applied to provide useful, non-intuitive insights that are usually hidden in the large volumes of data collected. Digital Twin model deployed with real-time analytics and applied machine learning techniques helps troubleshooting, addressing complex operational problems, and providing augmented intelligence to maintain operations at optimal levels.

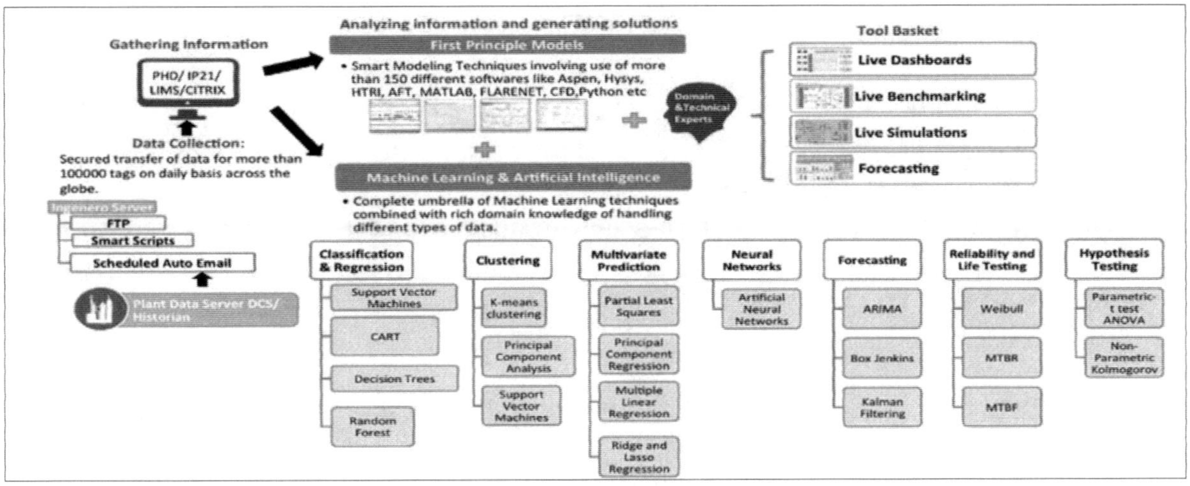

To ensure success, a right approach and utilization of the right technology is critical. An intelligent analytical solution offers real time decision insights to operations, driving actionable outcomes and creating new levels of operational excellence. The key is combining the right analytic tools and customizing a solution to your specific needs that delivers value.

5 Ethylene Challenges and Opportunities

Ethylene process has its own challenges, but due to complexities in process it offers tremendous opportunities. AI can help address those challenges efficiently and in a simple and direct way possible.

AI can provide meaningful insights and aid to ethylene process improvements through multiple means, such as:

1. Furnace Predictive Analytics – Run length predictions
    a. Improved yields, capacity and run length, coil life, etc.
2. Dynamic GAP Analysis
    a. Help improve Equipment performance to theoretical best
3. Equipment Health & Reliability
    a. Extension of plant T/A, Avoid unwanted load reductions
4. Process Stability and APC/DMC Performance
    a. Stable operations with improved asset performance
5. Instrument Drift & Soft Sensors
    a. Improves process control and avoids sub optimal operation

"Augmented Intelligence creates opportunities for greater efficiency and synergies in operating strategy."

6 Furnace Challenges

Furnaces are a core part of an ethylene plant operation and any improvement in its performance and reliability significantly helps in increasing production and reducing maintenance and operating costs. However, ethylene furnaces can have multiple issues:

- Furnace run lengths below expectation
- Rapid coking
- Inefficient furnace de-cokes
- Unexpected tube failures and lower than expected tube life
- Equipment reliability and lower than expected uptime
- Below capacity operation

- Recurring fouling in TLEs
- Utility limitations (dilution steam, fuel gas firing)
- Lower than expected conversion
- Higher than expected specific energy consumption

The severity of challenges can be different at each site or for a particular furnace, but each of the concerns can be effectively addressed through intelligent modeling solutions powered by "AI" and "ML". For example, using predictive modeling techniques to accurately forecast the furnace run length by estimating the coking rate based on key operating parameters and additionally providing key insights for better process optimization.

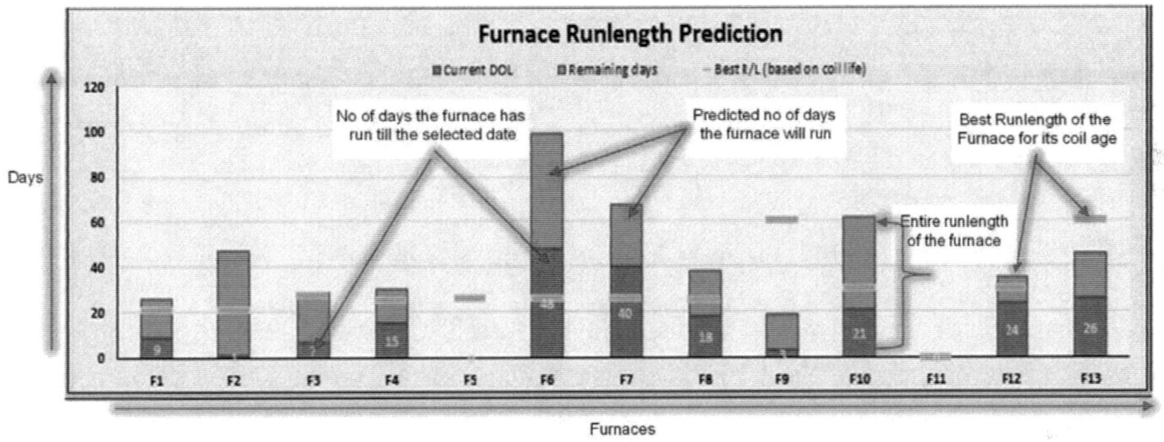

7 Smart Analytical Insights: Run Length Predictions

Hybrid furnace models (Kinetics and AI/ML) serves as an important tool to guide operations on various furnace aspects and provide timely insights to correct and improve furnace operations.

- Predicts end of the run based on CPR, TMT and TLE health
  - Based on kinetics based coking rates, past historical data and current age of all coils/passes
  - An algorithm coupled with Furnace model fine-tuned for actual performance is used to predict coking rate based on feed composition and rates, cracking severity, steam ratio etc.
  - Model estimates key furnace operating parameters Coil inlet and outlet pressures and TMT's which primarily determines furnace EOR criteria
- Identifies limiting coils/passes, TLE issues
  - Furnace Models help identify differential cracking among coils

        - Predictions provide insights on impact of feed, cracking severity and other production parameters and thereby allow appropriate adjustments to be made to optimize performance and production
    - Early detection and combination of other models helps identify and correct the problems much earlier than traditional monitoring
- Identifies opportunities for improvement
    - Ex. Feed biasing, Heat flux adjustments, conversion & yields etc.
- Helps in furnace scheduling
    - More scientific way provides opportunity to maximize the furnace performance without causing any furnace limiting scenarios
    - Model prediction results helps in detecting early anomalies, consider process changes to optimize performance, planning and scheduling
- Decoke efficiency
    - Decoke and startup improvement – to reduce CPR at SOR and improve coil life

7.1 Smart Analytical Insights: Furnace improvements and benefits

Apt application and usage of hybrid furnace models ensures that furnaces of that operating facility are consistently operated above their design capacities, making continuous improvements to operate closer to theoretical best.

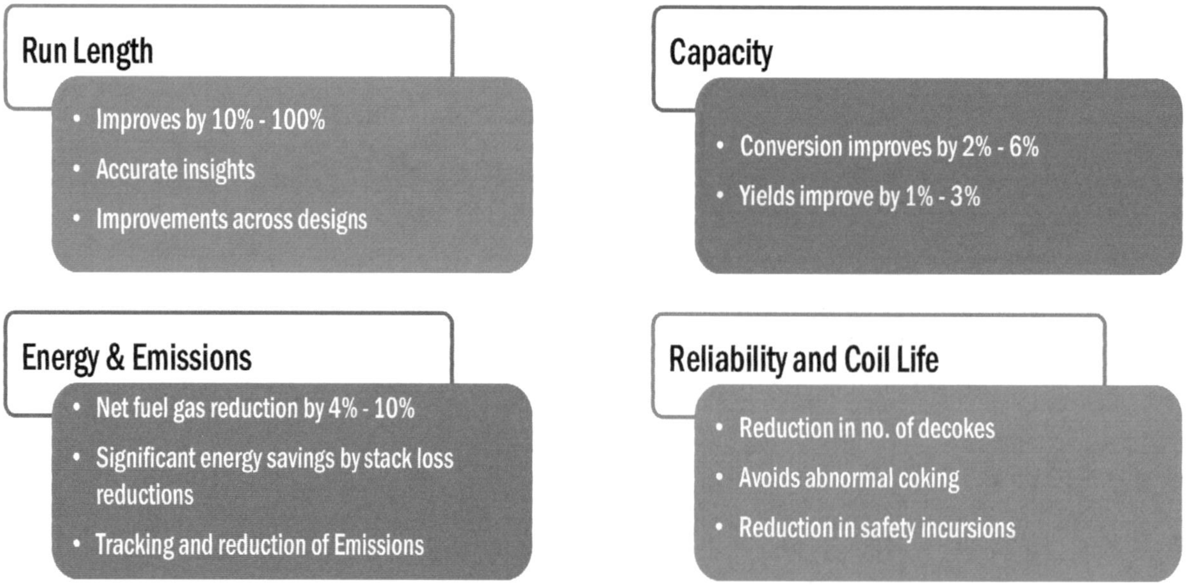

8 Successful Applications

Digital Transformation has the potential to create significant profit opportunities for manufacturers. Ingenero helps manufacturing units develop process software solutions to maximize asset performance through Augmented Intelligence. By optimizing data analytics/machine learning with simulation, Ingenero provides ethylene plants with solution insights to increase throughput, yield and availability while staying at the forefront of the industry 4.0+ Digitalization Age.

The section below demonstrates a few illustrations on how this process has helped in creating greater efficiency and synergies in operating strategy and solve problems including value potential and opportunities that can be achieved from real-time analytics.

8.1 Dynamic Benchmarking

Dynamic GAP analysis helps operations gain live insights and bridge the operational gaps per production plan and achieve equipment theoretical best capability. The first step is to get the solution objective aligned with the plant needs. Main user can define the desired operating target range, based on the overall planning inputs. Followed with the specific performance selection criteria for optimization which could be to maximize capacity, minimize energy, improve recovery, reduce stack temperature, reduce turbine/compressor load or even maximize yield depending on specific needs at a given time.

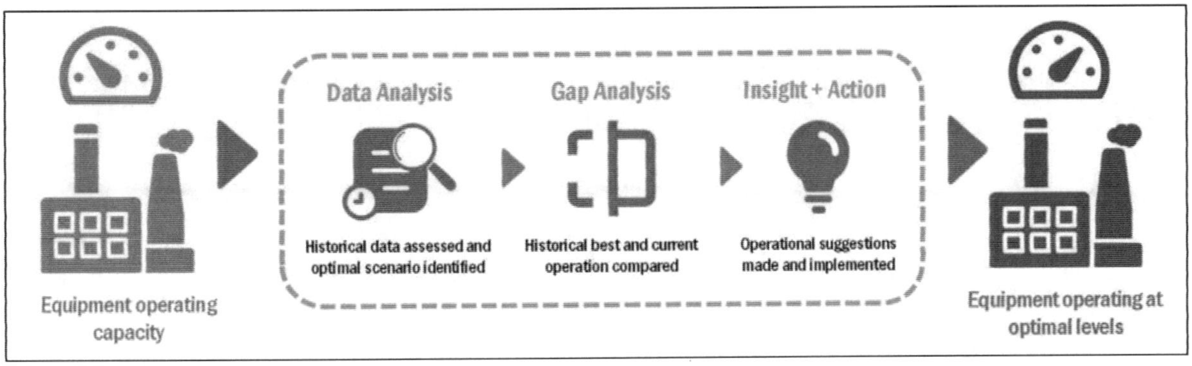

Intelligent algorithm powered with analytics then processes years of historical data to find out the best operating targets for the given operating scenario. The solutions also provide insights on key deviations and GAPS in operations, including for operational constraints and for specific performance criteria. It even highlights the improvement potential and benefits on a live basis. Smart solution is equipped with capability to highlight alerts providing operations to look at it only when it is critical thereby allowing simultaneous monitoring of several equipment effortlessly.

The solution does not stop here, but also provides additional insights to operators. It provides relevant details required to implement the right changes to bridge those operational gaps. While also highlighting parameters that needs to be optimized and provides features to access relevant specific details quickly. All this, to help make the right decisions and make changes to get to as optimized a process as possible.

Dynamic benchmarking helps to optimize operational parameters based on production goals, while gaining insights to address current bottlenecks.

8.2 Instrument Drift/Soft Sensors

Instrumentation plays and important role in all ethylene plants. Analytics and AI/ML models can be effectively utilized to track the health of the instruments, catch instrument drifts as well as estimate or predict an unknow parameters accurately on live basis.

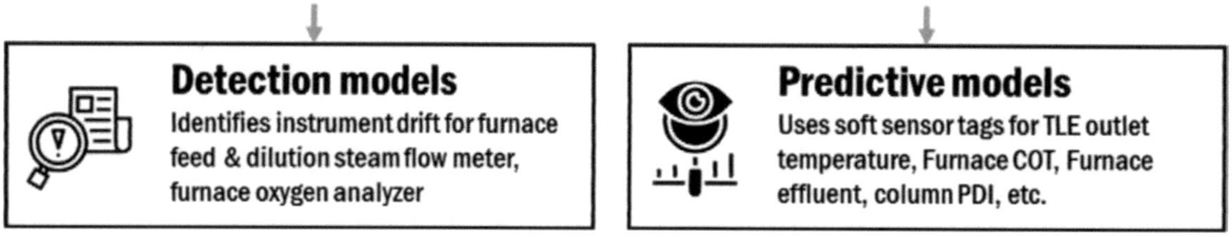

These soft sensors are customized to specific operations requirement. Few examples in ethylene industry where they have been effectively applied are as follows:

I. Furnaces
   a. Conversion and effluent analyzer
   b. Feed flow meter and Dilution steam flow meter
   c. Coil outlet temperature
   d. TLE Outlet Temperature
II. Surface Condenser pressure for Charge Gas Compressors, C2R and C3R compressors
III. Differential Pressure Indicator for columns like Ethylene Splitter, Deethanizer, Depropanizer and Debutanizer
IV. Product purity like Mix C4, Gasoline, Propylene product

8.3 Live Column Hydraulics

A real time analytics solution combines a first principle model and regression analytics with ML to provide live hydraulics and predictions to maintain and optimize column hydraulic operations. Hydraulic Model predicts operating points which helps monitor tray loadings and take pro-active actions to prevent flooding. Suitable ML techniques are used to utilize first principle results and find operating point on real time operation.

- ✓ Intuitive insights to address capacity constraints by addressing flooding areas with smart changes in operating parameters
- ✓ Ensuring reliable operation, tracking PDI and other critical instrumentation

8.3.1 Tray loadings at different sections of C2 Splitter Column

Live insightful details for operations to identify anomalies and take corrective actions

- ✓ Tray loading charts provides operational boundary limits
- ✓ Live section wise tray loadings and operating points helps identify current hydraulic loads

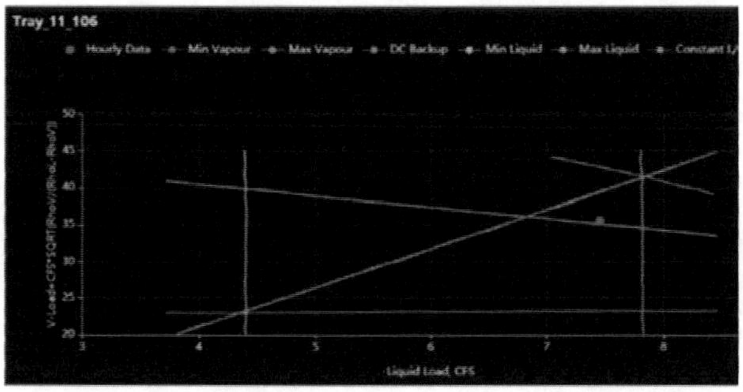

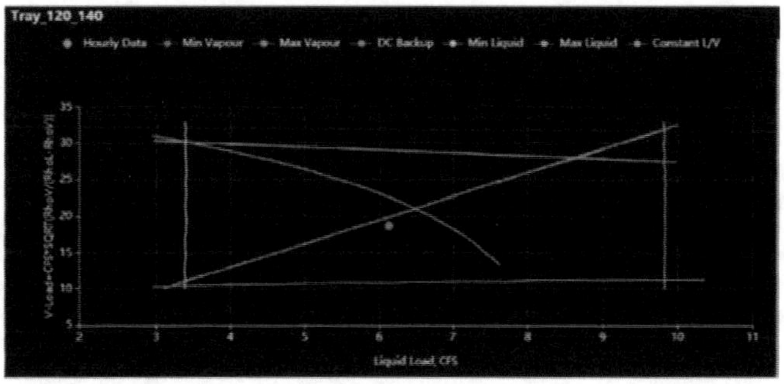

With alerts and insights to maximize column capacity while ensuring stable operation:

- ✓ Clear alerts and warnings on deviations in operating points
  - Flooding limits and predictive advice
  - Section closer to flooding
  - Type of flooding
  - Factors affecting flooding
- ✓ Benefits
  - Close monitoring and prevent flooding
  - Insights to improve column capacity
  - Guide proactive measures to ensure stable operations while maximizing capacity

Predictive furnace analytics, live benchmarking solutions, reliability predictions, soft sensors, column hydraulics etc., are few of the several examples which have demonstrated successful utilization and execution of digital transformation strategy meeting all the desired end objectives and goals of exponential gains in productivity, safety and reliability.

The aim of the digital transformation process should be to support traditional technical knowledge and manufacturing skills sets with new advanced techniques. Operations are largely well aware of unit complexities, even possible problems, and the impact of process changes. Powerful data analytical tools, supports their process improvement objectives by aiding their decision-making process.

Process analytics combines data, technology, and mathematical models to help operations make better and timely decisions, identify new opportunities, and become more competitive. Every aspect of process operation can benefit from using advance analytics, from understanding optimal targets faster, capturing non-linear relationships, providing predictive results on equipment all helps in improving operation, efficiencies and performance.

9 Accomplishments (Combining right analytics tools to customize a solution that delivers value)

Case Study 1:

Analytics and digitalization solutions helped address furnace issues and optimize overall plant performance resulting in significant production gains

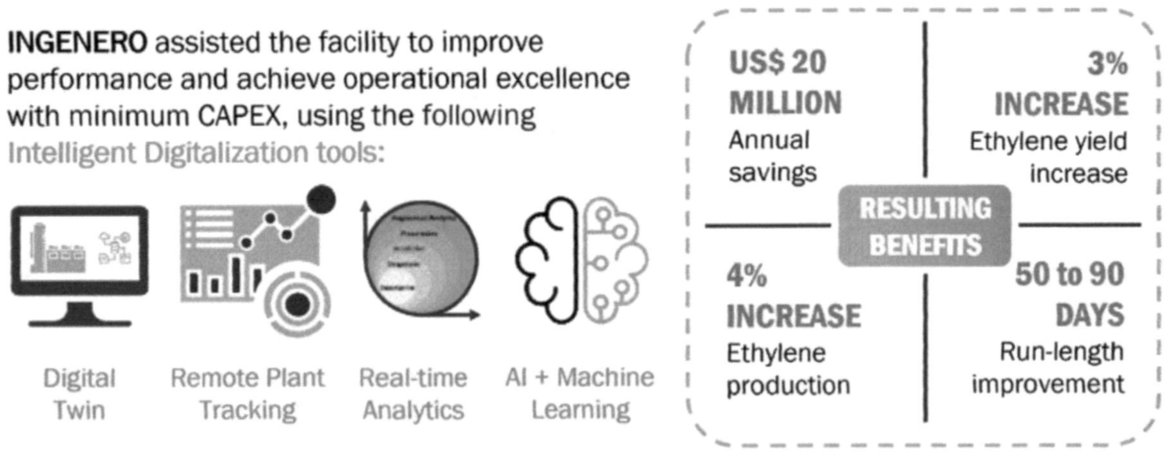

Case Study 2:

Improved furnace health which resulted in longer run length and improved capacity utilizations

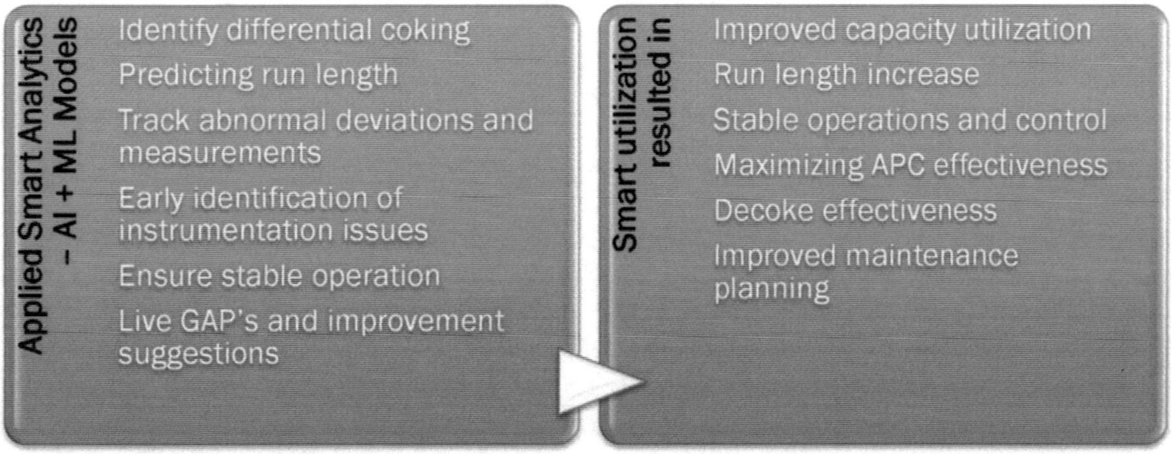

Impressive Results (~ 6% Capacity Increase): In a short period, the plant scenario changed from furnace capacity constrained to furnace capacity surplus

10 Conclusion: AI - Key to Success

People, methodology and technology remain the keys to long-term AI success.

People:
- Solutions should be based around meeting the needs of the people running the plant
- Experience is the hardest teacher, involving the right expertise is a key to success

Methodology:
- There are lots moving parts
- Identify the value proposition
- Chase the knowledge to be able to capture the value

Technology:
- Understand the analytics that need to be performed
- First principles simulation supports, validates, and augments data analytics
- Good visualization is important

Software solution powered by advanced analytics and AI/ML techniques provides real time insights to bridge operational gaps & maximize profits, improve reliability, process stability, performance benchmarks with descriptive and predictive insights. Power of analytics can be used to offer real time decision insight to operations, driving actionable outcomes creating new levels of operational excellence.

639562 Aligning Ethylene Plant Production Planning with Online Optimization

Abstract:

Ethylene producers are continuously faced with the challenge of reducing margin leakage that occurs between various levels of production execution — from production planning to real-time operations. This challenge arises from misalignment in models, assumptions or actual constraints that may occur over time across the layers of planning and operations. The units across the entire ethylene production process may not be running at their collective optimal to meet ambitious plans, even though individual process units may be pushed to their specific constraints by their respective APC applications. As the planning layer is optimized it is key to align the business process down below all the way to operations. Gaps may arise from slow or inaccurate information flow which leads to margin leakage. Also information about changing asset conditions from the operational layer needs to flow back up to planning. Lack of such information flow on a timely basis hinders the decision agility needed to close the gap between these layers.

AspenTech's General Dynamic Optimization Technology, Aspen GDOT™ addresses this long-standing issue by using an innovative modeling and optimization approach that combines fundamental planning models with dynamic APC models. This results in the ability to have consistent models, economics and objectives between offline planning and online optimization that ensures best site-wide economic results consistently and on a minute-by-minute basis. Plants can also boost production agility and responsiveness by giving planners a more accurate view of current process conditions and constraints via feedback on updated model parameters from GDOT. GDOT directly inherits APC models & can immediately adjust to any changes in the APC layer.

While real-time optimization applications for ethylene have existed for a long time, sustaining benefits from these applications has been a challenge and required significant process simulation experience. Aspen GDOT™ is designed to be run and maintained by APC engineers, making it significantly easier to adopt and sustain. Join us in this session for an overview of the technology and how it can benefit your ethylene production sites.

SPRING22
+18TH GCPS
A Joint AIChE and CCPS Meeting

LPG Dryer System Automation

Daryl Bitting
Eastman Chemical Company
Longview, TX
dbitting@eastman.com

Jeff Sutton
Eastman Chemical Company
Jerry L. Harris
Eastman Chemical Company

Prepared for Presentation at
American Institute of Chemical Engineers
2022 Spring Meeting and 18th Global Congress on Process Safety
San Antonio, TX
April 10 – April 14, 2022

LPG Dryer System Automation

Author

Daryl Bitting, Eastman Chemical Company

Abstract

LPG dryer systems can be prone to incident due to valve misalignment, partial movement of a valve, and partial completion of a step, to name a few. A well-designed Sequential Function Chart within the Emerson Distributed Control System can eliminate some of these hazards with properly selected instruments and switching valves.

This paper provides an example of how this concept was applied to a propylene dryer system capable of swapping beds, regenerating the offline bed, and returning the bed to standby with a push of a button. Independent verification of step completions is a key element to risk reduction.

Introduction

LPG dryer systems are typically comprised of at least one fixed bed operating in absorption mode, and at least one bed being regenerated or in standby mode. Members of the project team were very familiar with liquid propylene dryers, and some ground rules were established early in the design.

- Switching valves would fail "last"
- Operators would have minimal involvement during normal operation
- Controls would reside within the distributed control system
- Downstream units would not be impacted by dryer regeneration

Hazards associated with LPG systems can include low temperature, high pressure, flow in the wrong direction, and liquid carryover. Each of these hazards can be managed with a combination of procedures, trained operators, administrative controls, interlocks, process controls, and automation.

Manual regeneration systems are prone to error due to the number of steps involved, the time delay between steps, the communication between inside and outside operators, and communication between shifts for processes that extend more than 8-12 hours. A fully manual process requires acute attention to every detail.

Process interlocks can add a layer of protection to the manual system by forcing the position of a valve based on one or more process inputs. Interlocks are typically reserved for the most serious consequences. Unfortunately, interlocks can also add the unintended consequence of making it difficult to move a valve during non-routine operation without bypassing the interlock.

A semi-automated system can assist the operator with moving valves in the correct order, but still requires input from operations to make decisions based on observations or process knowledge. A fully automated system combines the valve movements with programmable logic to manage better manage the timing and the safety of the regeneration process.

The table below lists some of the common hazards associated with an LPG dryer regeneration system and the conditions that must be satisfied before proceeding to the next step.

Step	Hazard	Verification
Draining	• Liquids in fuel gas • Potential overpressure regeneration system	• Totalized minimum flow • Liquid level in bed • Bed pressure
Heating	• Incomplete regeneration • Reduced run length	• Regen gas flow rate • Bed temperature • Outlet temperature
Cooling	• Rapid vaporization in a hot bed • Potential overpressure absorber bed	• Regen gas flow rate • Bed temperature • Outlet temperature
Filling	• Liquids in fuel gas system • Bed Damage	• Bed level restored • Pressure equalized

Automating the Dryers

With the ground rules in place, a step-by-step animation of the regeneration process was constructed so that the actions of the proposed automation could be visualized and scrutinized during the design process and the Process Hazard Analysis (PHA). A key element to the safety review was understanding what needed to be verified before proceeding to the next step. Once the safety review was complete, the details of the animation were converted to Distributed Control System (DCS) logic.

Fundamental to the design of the automated system is the switching valves. Ground rules for the project required a "fail last" valve, keeping the valve from moving due to loss of power or instrument air. An Orbit rising stem ball valve with double acting pneumatic actuators was selected to meet this criterion. The valve only moves when instructed, and an energized solenoid valve maintains air pressure in the direction of the intended position. In the event of an air or solenoid valve failure, the valve does not move. Any valve that satisfies this fail last behavior would be a candidate for automation.

Emerson Sequential Function Chart (SFC) was selected as the logic controller for its ability to manage sequential events through "Steps" and "Transitions". Steps are the actions taken, and transitions are the confirmation the step is complete. A step can be a signal to move a valve, and the transition occurs when the limit switch confirms the position of the valve. An additional feature the SFC is the ability to retreat to the previous step if the operator needs to intervene, or if the step does not complete on its own.

Productivity was the initial motivation for automation. As the automation design evolved, the value of risk reduction became very apparent. With more than 50 discrete steps and transitions required to compete a dryer regeneration, multiple inputs were required for some of the more critical steps to verify the activity was fully complete before advancing to the next step. The following example will illustrate how multiple inputs are used to verify free liquids have been removed during the drain step.

Example – Drain Step Transition Verification

Liquids must be fully removed before an LPG dryer can be regenerated. The sequential function chart verifies the liquids are removed by combining flow, pressure and level measurements. Independent verification of all three variables is required to transition to the next step.

A flow totalizer would not be accurate to predict or measure the exact volume of material that needs to be drained, but it is more than adequate to ensure the bulk of the liquids has been removed. This is accomplished by setting the flow totalizer target so any remaining liquid would not exceed the free capacity of the knockout tank as shown in Figure 1.

Relying solely on the flow totalizer would require intervention by the operator when the total flow doesn't reach setpoint due to variations in density or composition. Operator intervention violates a ground rule for the project; therefore, the totalizer cannot be used independently to verify the dryer has been fully drained. Once the bulk of the liquid is removed, the SFC is allowed to consider the next two criteria.

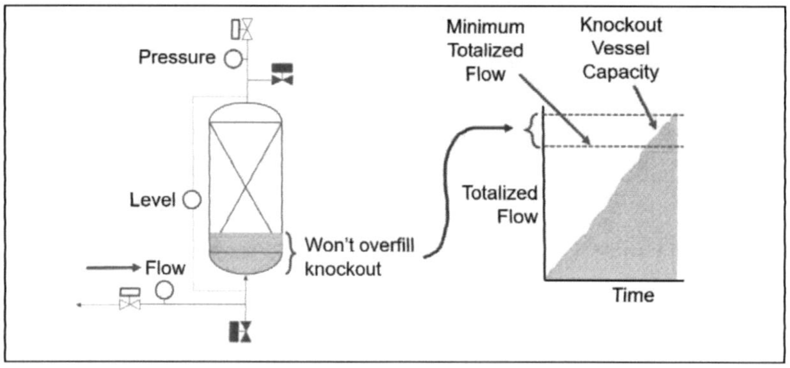

Figure 1 – Flow Totalizer

A liquid level transmitter across the bed provides the next layer of protection. Positioning the taps for the level transmitter in the piping ensures the level across the entire vessel is measured. As the level drops into the piping, the indicated level changes rapidly due to the difference in cross sectional area of the pipe and compared to the vessel as shown in Figure 2. The target level to indicate the dryer is empty is actually set below the bottom of the tank with this arrangement.

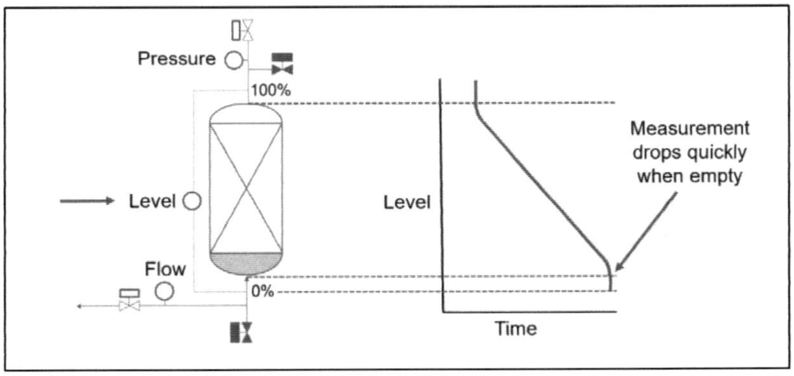

Figure 2 – Bed Level

The final confirmation occurs when the pressure in the bed decreases after the liquid is fully drained. Until the liquid is gone, the liquid will maintain the vapor pressure on the vessel. The third and final confirmation allows transition to the next step in the sequence.

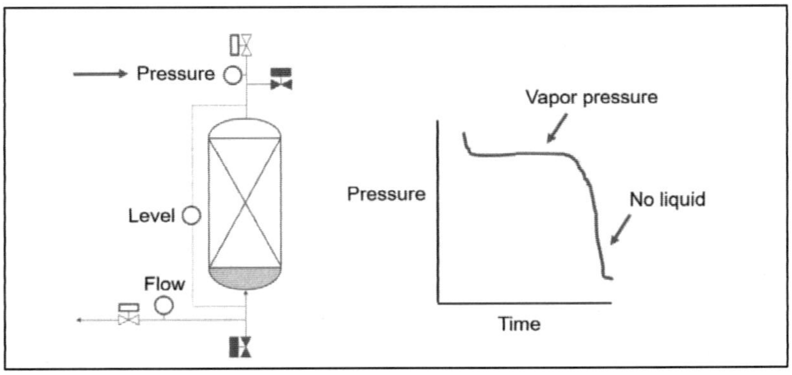

Figure 3 – Bed Pressure

Process Hazard Analysis

A tabletop model of the automation sequence was developed to visualize the steps required to complete the dryer regeneration from beginning to end. The model consisted of a valve position table for each step, additional actions beyond valve alignment, a description of when a step is complete, and a graphical display of the system. Once the preliminary design was completed, the model was reviewed as a control for the PHA process. If the controls were not adequate, changes were made to the automation to reduce the potential for an unsafe scenario.

Additional safeguards beyond dryer automation were added per a Layer of Protection Analysis, including a knock-out tank and interlocks to prevent liquids from entering the cracking plant regeneration gas system. The model improved the PHA review, and the PHA review improved the design. Once the design was finalized, the model was translated into a Sequential Function Chart and tested off-line before starting up the new process.

Conclusion

Dryer systems will likely continue to become more automated as the industry seeks to improve safety and productivity. The purpose of this paper is to share some of the fundamental concepts employed in a recent automation project.

Managing the transition from one step to the next is critical to enabling hands-free regeneration from start to finish. The success of the project is attributed to verifying each step is complete with the necessary redundancy before moving to the next step.

Reviewing the regeneration automation plan during the PHA review enhanced the value added by the automation.

Related EPC Articles:

Rob D. Fox, "To Err is Human: Mitigating the Hazards of Dryer Switching Operations", 2017 Spring Meeting (81e)

Bill O'Brien, "Cracked Gas Drier Survey", 2010 Spring Meeting (89e)

R.E. Sherman and A.W. Woltman, "Safeguarding Olefin Plant Drier Regen Systems", 1997 Spring Meeting (35b)

SPRING22
+18TH GCPS
A Joint AIChE and CCPS Meeting

Online Repacking of Valves in Liquid Hydrocarbon Service Risk Awareness & Mitigation

Andrea Dillon
Dow, Inc
2301 N Brazosport Blvd. Freeport, TX 77541
andillon@dow.com

Prepared for Presentation at
American Institute of Chemical Engineers
2022 Spring Meeting and 18th Global Congress on Process Safety
San Antonio, TX
April 10 – April 14, 2022

Online Repacking of Valves in Liquid Hydrocarbon Service Risk Awareness & Mitigation

Andrea Dillon
Dow, Inc.
2301 N Brazosport Blvd. Freeport, TX 77541
andillon@dow.com

Keywords: Process Safety, Release, Valve Packing

Abstract

Online repacking of valves in liquid hydrocarbon service is a practice that, while not common, is used on occasion to avoid the risks associated with shutting down an entire cracker facility. The repacking operation constitutes a period of heightened risk and it must be carefully planned, monitored, and managed. This paper aims to raise awareness of some less-obvious errors and failures which are possible by sharing a case where a significant release occurred from a manual valve on the recycle line to a LPG splitter column due to catastrophic failure of the valve stem packing, following attempted repacking of the valve. The paper will also share some mitigations which can be pursued to reduce the potential for a similar event and some measures to limit the risk to personnel and the plant should an event like this occur.

Introduction

To maintain a license to operate, all hydrocarbons facilities must have a robust leak detection and repair program to control and report fugitive emissions (FE) within air permit guidelines. Due to the significant safety risks and environmental and cost impacts associated with shutting down an ethylene cracker, when an FE leak is identified, a plan is developed, and a repair is attempted with the equipment remaining in service. If the repair cannot be made, that leak point is placed on a list to be repaired during the next turnaround or planned equipment outage.

A fugitive leak was identified from the packing of a six-inch manual gate valve in liquid ethane service. The leaking valve is on the discharge of one of the redundant reflux pumps on the LPG Splitter (Fig. 1). This valve was attempted to be repaired and after the online valve packing repair attempt, a significant ethane release occurred. Due to the release, the cracker was shutdown to protect plant personnel, stop the leak the leak and avoid a more catastrophic event should the ethane reach an ignition source.

During the root cause investigation, it was determined that the leak was due to damage on the stem and bushing of the valve caused by galling over time from uneven torquing during multiple FE leak repair attempts. Catastrophic events can be avoided by reviewing repetitive FE leakers and having a plan and procedures outlining effective in-service repacking and repair techniques.

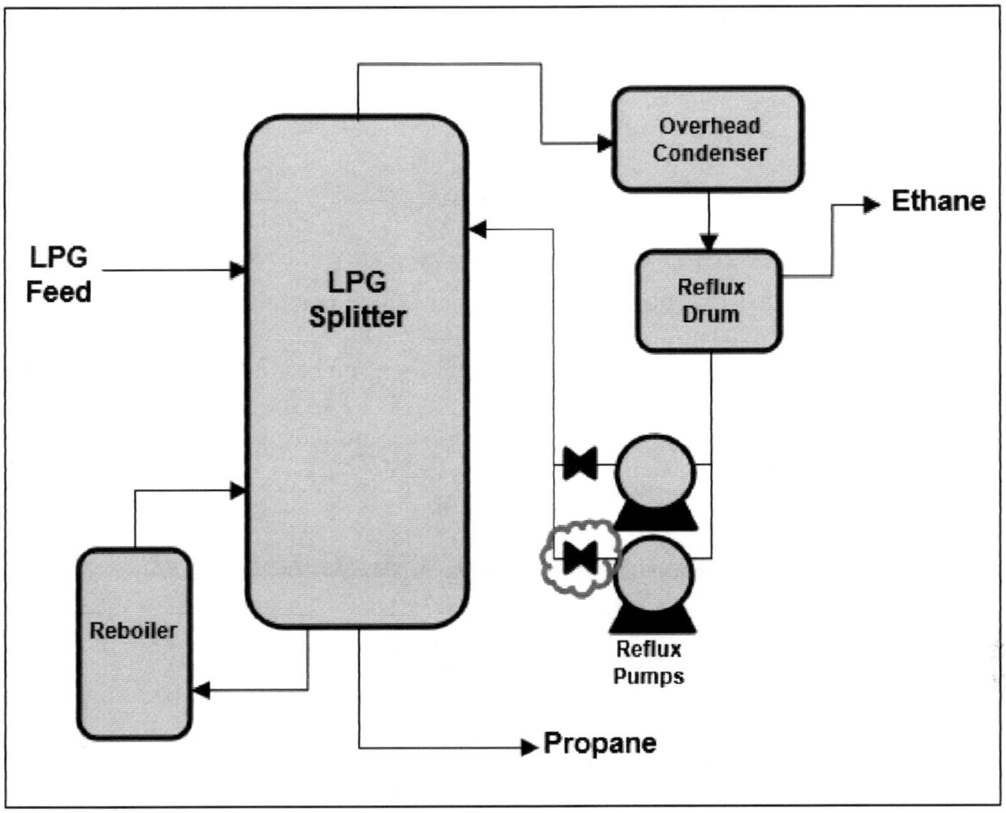

Figure 1. PFD of LPG Splitter system and location of leak point

Valve Repair History

This valve required eight FE repairs over nine years leading up to the catastrophic packing failure event. During each of the repairs the packing gland bolts are tightened. This is the method described as "tightening packing". Torquing the gland bolts pushes the gland flange down and applies more pressure to the packing.

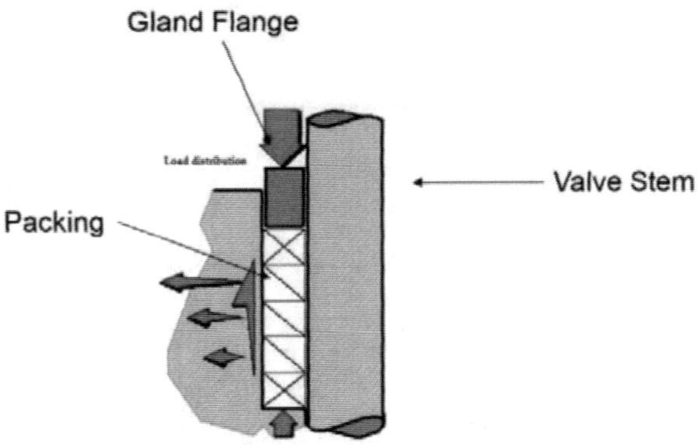

Figure 2. Diagram showing relationship between gland flange and packing

Each time that the valve packing was tightened due to an identified fugitive leak, the leak was fully resolved and removed from the LDAR list. The leak detection and repair program did not have a step to review fugitive emissions leaks from year to year to determine if there are patterns indicating a valve, fitting or piece of equipment needs replacement or re-design.

Valve Damage

Upon disassembly of the valve, the full length of the valve stem had gouge marks. The backseat bushing was worn and damaged from galling caused by contact with the stem. The backset bushing is supposed to create a tight seal between the packing and the process flow when in contact with the stem head. This is when the valve is fully open and *backseated*. The valve manufacture considers work on the packing to be non-invasive to the process because the backseat bushing provides isolation between the valve packing and process flow.

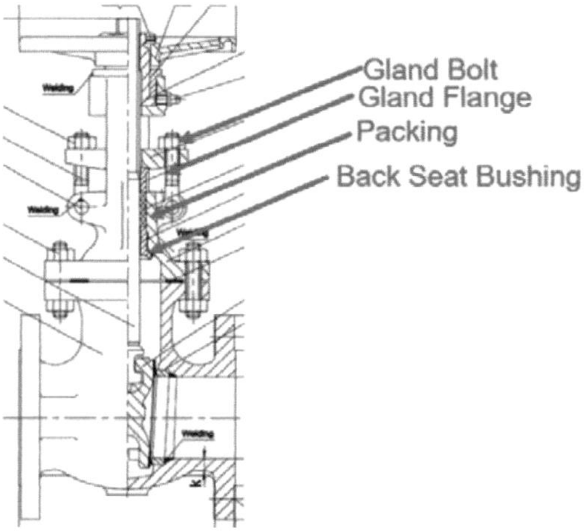

Figure 3. Diagram showing packing in relation to the back seat bushing and process flow

The galling of the stem and backseat bushing was likely due to uneven torquing of the gland bolts during one or multiple of the valve packing repairs. This causes the stem to become slightly misaligned and to contact the bushing. Over time the stem and bushing galled leading to the damage to the backseat bushing and inability to isolate the packing from the process fluid. There were gouge marks the full length of the valve stem from extreme galling over years of repairs and valve operation.

Figure 4. Damage in the ID of backseat bushing due to surface contact with valve stem.

The galling and damage to the backseat bushing continued to get worse with each repair attempt and each time the valve was operated. This is one of the reasons that this valve continued to leak from year to year.

Another finding during inspection of the valve was only one packing ring was present inside the stuffing box when it was inspected. This was a defect on valve installation eleven years prior to the catastrophic failure. A valve of this style and size is correctly fabricated with five packing rings installed.

Figure 5. Valve stuffing box missing five packing rings & single installed packing ring severely damaged

With an almost empty stuffing box, each time that the packing was tightened, the packing gland flange bottomed out against the valve body. The void in the stuffing box meant there was not enough packing to tighten. This mechanical defect was likely the primary reason that the valve originally leaked and multiple attempts going forward only worsened the condition of the valve and led to ongoing galling of stem and other components.

Opportunities to Avoid Similar Event

The fact that the valve was worked on eight times over a nine-year period could have been caught and investigated. Each time the leak was resolved and tested good, but each time the same repair method was used and the valve ultimately developed a leak again. These repair methods were likely making the damage worse from year to year. The LDAR program was not reviewed or critiqued at a multi-year level to determine and resolve repeat offenders.

The galling of the stem and backseat bushing had two primary causes. The first was the defective single packing ring installation by the original valve manufacturer. This caused inadequate packing for the gland to come in contact with during gland bolt tightening, which caused the gland to bottom out against the stuffing box.

The second cause was due to uneven torquing of the gland bolts which lead to misalignment of the valve stem. The stem and back seat bushing have very small clearances, so this misalignment resulted in significant galling over the years. This can be avoided by ensuring that prior to attempting to tighten packing, a procedure exists and is utilized. This procedure needs to include and emphasis on torquing evenly and what torque values to reach.

In addition to torque values, an in-service packing tightening procedure should include *acceptable criteria conditions.* These conditions should be clearly defined and if the valve is not deemed acceptable, no in-service repair attempt should be made due to the risk of a catastrophic packing failure and hydrocarbons release. The valve in question had stem damage, heavy rusting and corrosion of gland bolts and warping on gland bolt nuts.

Figure 6. Valve gland close-up showing the unacceptable condition for repair attempt

Examples of acceptable condition criteria include- light to moderate surface corrosion of valve gland, bonnet, bolts or stem; no visible stem damage or galling; nuts that make full contact with torque wrench (no blossoming or sever wear/corrosion); no visible signs of overpressure.

Finally, the repair history of a valve should be reviewed prior to making a repair attempt. This review should include both the frequency and type of repairs made in the past.

Outlining these critical steps in a packing tightening procedure to be utilized by employees and contractors when working in the facility can prevent catastrophic packing failures when repairing valves that are in service.

Conclusion

Making repairs to in service valves can avoid potential process and personal safety risks and the major environmental impact that can occur while shutting down a hydrocarbons plant. The first thing to When an in-service repair is going to be made it is important to ensure that a robust process or procedure is in place that includes steps to review valve repair history, inspect valve for acceptable condition and how much to evenly torque the gland bolts to effectively tighten the packing. Additionally, setting up an LDAR High Frequency Repair Review with key operations and maintenance staff at least every 3-5 years to identify and determine permanent solutions to resolve repeat offenders identified through the fugitive emissions program.

References

Newmans Valve. (2009). *NEWCO - Cast Steel Bolted Bonnet Valves Product Line Technical Data.* Stafford, TX: Newmans Valve.

Dow. (2017). *Root Cause Failure Analysis, LHC-7 6 inch Gate Valve Packing Failure.* Freeport, TX: Dow.

Dow. (2019). *Electronic Most Effecitve Technology Library (EMETL), Valve Packing Chamber Guidance.* Freeport, TX: Dow.

639761 Cold Safety Management of Specific Areas of a Steam Cracker

Abstract:

Cold safety considerations are a critical issue for all ethylene producers and can have a significant impact on many parts of an Olefins unit. Previous presentations have highlighted different aspects of the hazard including evaluation methods, mitigation approaches and safety training protocols. While this information has provided an overview of cold safety mitigation, close inspection indicates that practitioners can also benefit from deeper dive into other common questions and scenarios encountered such as piping assessment and mitigations and cold flare system monitoring and sustainment.

This presentation intends to provide a summary of the practices gathered from industry feedback via the EPC Low Temperature Safety Subcommittee. It highlights the cold safety concerns in these specific areas, offer considerations on how to address those identified concerns and, finally, provide an overview of how possible mitigations may fit into an overall cold safety management program.

AIChE Paper Number: 186d

Pyrolysis Furnace Flooding
Awareness, Recognition, and Response

Bryan McVicker

Site Pyrolysis Leader

Baton Rouge Chemical Plant

ExxonMobil Chemical

Prepared for Presentation at the 2022 Spring National Meeting

San Antonio, TX. , April 10 - 14, 2022

Pyrolysis Furnace Flooding
Awareness, Recognition, and Response

Bryan McVicker
Site Pyrolysis Leader
Baton Rouge Chemical Plant
ExxonMobil Chemical

Abstract:

Furnace flooding is a process safety concern characterized by an accumulation of incompletely combusted fuel in the firebox. Furnace flooding can lead to high consequence events, including but not limited to firebox explosions. Furnace flooding can occur at any point of furnace operation, from initial start-up to maximum fuel firing.

This presentation will summarize the basic principles of furnace flooding, early recognition tools, and proposed responses to these incidents. Proper recognition and response are key to avoid event escalation and potential firebox explosion.

Further, the author will review relevant ethylene industry examples of furnace flooding events, and will conclude discussing some potential misconceptions surrounding the subject of furnace flooding.

Furnace in the context of this paper is referencing Pyrolysis Furnaces, also known as Steam Crackers.

Pyrolysis Furnace Flooding Awareness, Recognition, and Response

Introduction

This presentation aims to explain the basics of pyrolysis furnace flooding and provide a relevant and informative safety topic to ethylene industry. Additionally, a DCS flooding recognition tool has been successfully demonstrated to facilitate swift and correct operator response to various flooding scenarios. A template version of the DCS flooding tool is shared to help enhance steam cracker process safety. The benefit of an early flooding recognition tool is illustrated through two common industry examples. Lastly, a few common flooding "misconceptions" are discussed that may assist scenario training.

Flooding Basics

Flooding is a hazardous accumulation of unburned fuel gas in a firebox. Hazards include burner flame-out and explosive re-ignition of unburned fuel, and the risk of furnace over-pressure due to an incorrect response (i.e. sudden addition of air).

Two primary event categories result in flooding events and near misses: not enough combustion air & too much fuel. Combustion air and fuel can be negatively impacted by manual and automatic control errors. For example, burner secondary air doors can be unbalanced across a furnace, or incorrect setpoints can be entered in to fuel or draft controls. Instrumentation and analyzers, such as O2 analyzer drifting, can fail and drive controls or operator moves into unsafe conditions. Olefins furnaces largely depend on quality of the tail gas sent by the Cold Ends Recovery units. Upsets can create significant heating value variation, which can be challenging for firing controls. The nature of steam cracking furnace requiring transient operations for decokes requires careful management of combustion air. If O2 is not in control, or left in manual control during firing ramp up, the fuel-to-air ratio could become unbalanced. Loss of primary containment inside the furnace casing can affect both of these categories. Last but not least, an uncontrolled hydrocarbon release may consume combustion air or inhibit air from entering the furnace, in addition to introducing excess fuel.

A generic furnace is shown in Figure 1 with a list of common and less common signs of flooding. The key process variables in bold text, O2 / Combustibles may be used for early detection and response to flooding. Usually a drop in O2 and rise in Combustibles will go hand-in-hand. A loss of draft can reduce the driving force for combustion air to the burners. In the field, smoke in the stack, hazy firebox, flame-out, and woofing sounds and vibrations are all later stage signs of flooding. For steam cracking furnaces, a drop in temperatures could be a sign of severe flooding and loss of combustion efficiency. Process variable correlations, also in bold text, can be very useful to detect flooding early, and these will be discussed later in more detail.

In general, pyrolysis furnaces run much hotter than the typical fired heater, well above fuel auto-ignition, and being much more instrumented, severe flooding is relatively rare as compared to non-pyrolysis industry heaters. Automatic SIS protections, such as trips and alarms (low draft protection, for example), are excellent front-line barriers against many conditions that can lead to severe flood and burner flame-out.

Signs of Flooding

1. **Low O2**
2. **High combustibles**
3. **Low draft / High bridgewall pressure**
4. Smoke observed from stack
5. Drop in temperatures
6. Temperatures not changing with O2
7. Erratic flame detected, or flame-out
8. Hazy firebox
9. Woofing sound, pulsating draft
10. **Process variable correlation**

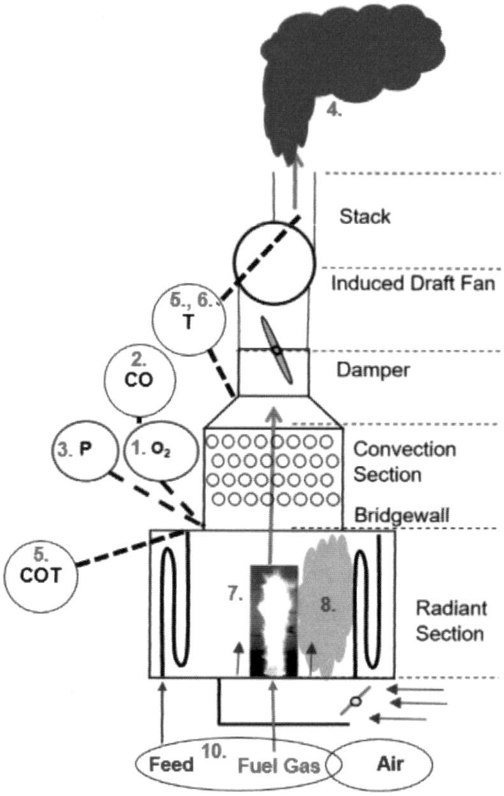

Figure 1: Furnace Signs of Flooding

Response to Flooding

Stay away - If furnace flooding has been initiated by draft control issues, there is a flame or hot flue gas exposure risk to personnel in the vicinity. Intuitive response to add air to an air-starved box is the WRONG action in a real flooding scenario.

Do not add air - Draft controls could be the cause of the flooding, or dampers could possibly stick and be trying to open. Placing the draft control damper in manual and holding last good position is recommended.

Reduce fuel - Cut fuel to recover from or prevent going into flood. 25% initial cut is a good rule of thumb. Cut fuel as-needed thereafter. If truly in flood, temperatures may begin to rise or recover as fuel is decreased.

In normal operation, cold tripping is undesirable for a pyrolysis furnace. Steam cracking furnace tubes may be carburized and susceptible to brittle fracture if cooled too fast without decoking first. In normal draft operation a 100% fuel cut can result in high draft and a large amount of combustion air. Therefore, full furnace trip is a last resort in a flooded scenario.

In cold startup however, the risk of a complete flame-out is higher due to the relatively cold firebox. A 100% trip is desirable if the firebox is flooded in a cold startup scenario. If the burners are flamed out, simply cutting fuel will be ineffective. The tubes are generally clean after decoke prior to shut down, so coil fracture risk is minimized and full trip is the appropriate action.

Early Recognition Tool in DCS

A flooding DCS tool has been successfully applied at several sites. Figure 2 illustrates how process variables are monitored. A voting hierarchy decides if a furnace is in "suspect flood" or "flooded – stay away". Note that the voting weight of any variable may be site specific (hierarchy may vary), and conditional values will be specific to each furnace/burner type.

Not only does the tool provide early troubleshooting signals to the console operators, it takes actions to generate Operator & Technical attention and drive further Operator action. Note the lower-right box. "Suspect Flood" generates a DCS alert and sends an email to Technical. "Flooded" generates a DCS emergency alarm and switches DMC off and holds damper in manual last position. The Operator pulls up the flooding screen and can quickly troubleshoot which conditions are voting for "Flooded", and helpful trends are available on the same screen. Recurring flooding training drives the desired operator response to the potential flooding event signaled by the early detection tool.

Process correlations are a great feature of the tool for older sites that may not have the built-in redundancy or desired availability of bridgewall or stack O2/Combustibles analyzers. To develop the correlations, site Technical trends the process variables, such as firing-to-feed ratio, and sets the bounds that could trigger a vote for flooding. Note that process correlations are feed type dependent and may change over time due to furnace configuration or process condition changes. To prevent nuisance alerts or alarms, the Operator has the option to bypass any variable that is a known bad actor or needs to be updated.

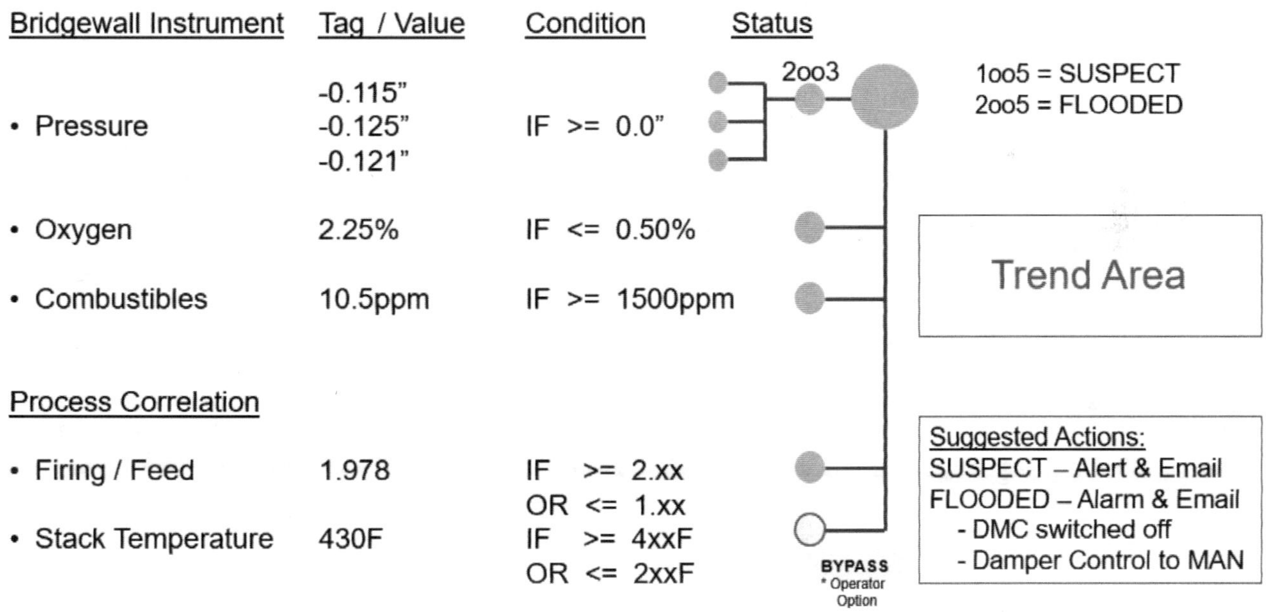

Figure 2: Early Flooding Recognition Tool in DCS

Examples

Consider a failed O2 analyzer example in Figure 3. The furnace is equipped with only one flue gas analyzer and it is a combination O2/Combustible analyzer with a common sample line. Operators attempt to optimize stoichiometric excess O2 by adjusting/closing burner air doors, however O2 remains elevated. Subsequent Operations shifts attempt the same optimizations and drive the furnace into a hazardous condition over time.

Early flooding recognition can be achieved with process variable correlations (adjusted for feed type in multi-feed capable furnaces). Note how the DCS tool alerted "Suspect" flood in Figure 3 when the firing-to-feed correlation drifted out of bounds. A trained operator would investigate the indication and possibly consider how the O2 measurement is trending with overall firing moves. If the furnace is not believed to be flooded, the O2 analyzer could be checked against portable O2 analyzer readings. The proper response would be to recognize the potential analyzer issue, proceed with caution, and trim fuel back to a more typical Firing-to-Feed ratio.

Late recognition could be more reactive and the resulting flooding more severe – e.g. unhealthy burner flames, smoke in the stack, falling control temperatures. If falling temperatures or visual signs in the field confirm flooding, then emergency response actions discussed previously are required.

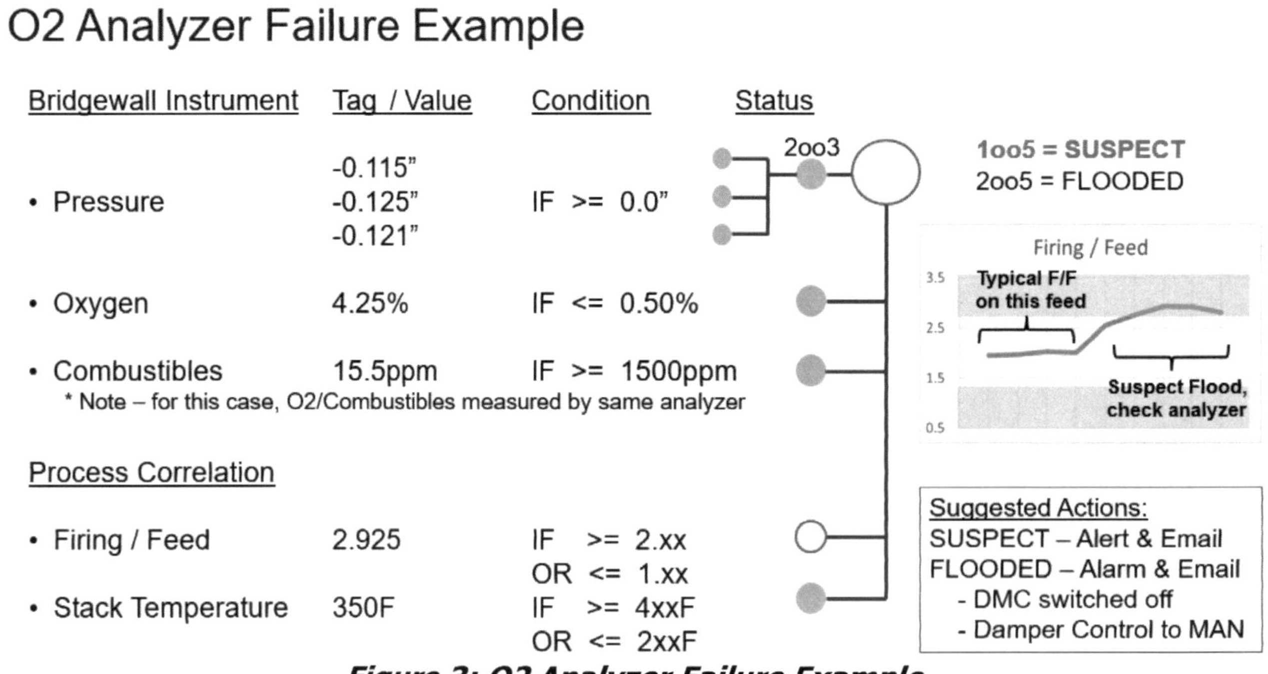

Figure 3: O2 Analyzer Failure Example

Consider the loss of tail gas hydrogen example in Figure 4 that results in a sudden swing in fuel heating value. The Cold Ends Recovery unit experiences an upset and tail gas is removed from furnace fuel. Natural gas or other backup fuel source automatically opens to maintain header pressure. Excess O2 is quickly depleted and Combustibles begin to spike.

Firing-to-Feed ratio will change when the fuel analyzer registers the change. At least two votes for flooding occur in this scenario.

Recognition and response speed depends on the type of analyzers employed. For instance, a fuel densitometer could provide fuel molecular weight (MW) results in seconds. A fuel GC analyzer would be slower to respond, say if a densitometer is unavailable, to correct the fuel heating value. The issue may be compounded by the furnace fuel control valve opening in response to a low fuel pressure dip, which is generally the opposite valve action needed for a large fuel mass flow increase due to higher fuel MW/density. For high H2 fuel (where sustainable growth is heading), loss of tail gas could be at least double the BTU value and half the required stoichiometric air (i.e. 80mol% H2 cut to 10mol%H2 in fuel).

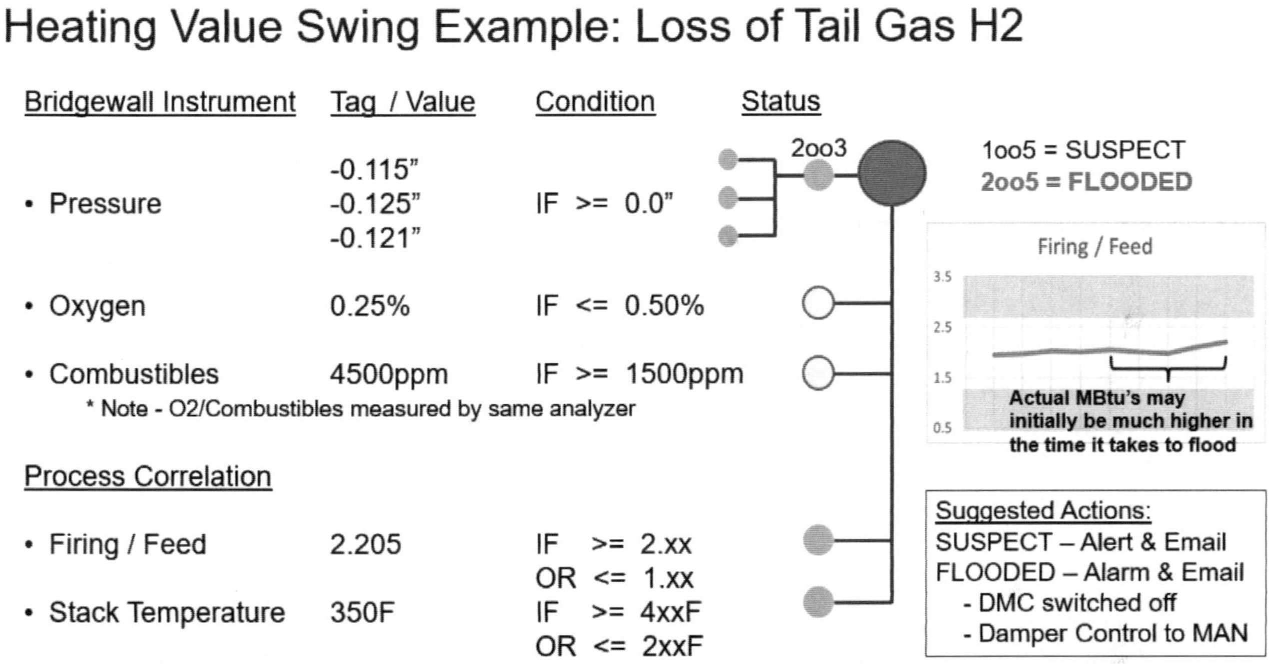

Figure 4: Heating Value Swing due to Loss of Tail Gas H2 Example

The DCS tool shown in Figure 4 may catch the flood indicators quickly, depending on the types of analyzers a site is using. To avoid severe flood however, feed forward controls may be needed that take action within seconds, not minutes. Fuel densitometers can report a fuel composition change very quickly. When employing densitometers in a control scheme, consider failure mode clamps (PV diagnostics and high/low clamps) as well as an obsolescence plan (redundancy or spare parts available). Feed forward control schemes will be site dependent. Fuel analyzer types, backup fuel supply and controllability, flue gas analyzer response time, furnace fuel CV tuning are all important factors.

A couple final comments on the loss of tail gas scenario, and flooding detection and response in general: High Availability Flooding alarms and DCS voting tools greatly improve flooding response. Train personnel often enough that the human response to cut fuel before adding air is essentially "automatic". Remember the three key actions to take if flooded.

Common Misconceptions

A low draft alarm is not necessarily a "flooding" alarm. Bridgewall pressure can be slightly positive without the furnace being flooded. Sites with wall burner furnaces may be justifiably more sensitive to high BW pressure due to potential for personnel exposure on upper level furnace decks. If merely a one-vote "suspect" flood, it may still be acceptable to increase draft and/or add air to remedy the alarm. An overly aggressive flooding alarm and fuel cut response can result in a needless furnace, and possibly entire unit, swing.

As noted previously, high firebox pressure / low draft is a flooding scenario in which a reliable Safety Instrumented Control System (SIS) may provide a significant barrier to severe flooding situations.

High draft does not necessarily mean sufficient O2. At low firing, say startup mode or turndown operation, pinching burner air doors to optimize O2 can lead to flooding potential. If a furnace is equipped with a forced draft fan, pinching combustion air doors to increase draft could further starve O2 and have an undesired effect.

Combined O2/Combustible analyzers have a significant overlap of failure modes due to the common sample line. If the O2 measurement is suspect, CO/Combustibles measurement may be considered suspect as well. Redundant bridgewall analyzers or independent stack analyzers can be used to validate flooding alarms.

Conclusion

Periodic flooding hazard refresher training is important. This paper covered flooding basics with a special focus on detection and response. Remember the three key responses, in this order: Stay Away, <u>Do Not</u> Add Air, and Cut Fuel. Consider common flooding misconceptions in flooding recognition and scenario response training.

A recommended DCS-based early flooding recognition tool has successfully facilitated swift operator response to potential flooding scenarios. Failed O2 Analyzer and Loss of Tail Gas examples were used to demonstrate the tool.

Plants using modern fuels, as well as existing plants with increased gas cracking, may benefit from flooding recognition tools and training. A swift controls response to loss of tail gas hydrogen should be considered.

AIChE Paper Number: 186e

FATIGUE CRACKING OF SMALL DIAMETER BRANCH CONNECTION AT PRESSURE RELIEF VALVE

Stephen Hagberg
Lead Mechanical Engineer
ExxonMobil Chemical Company Gulf Coast Growth Ventures
stephen.l.hagberg@exxonmobil.com

David Dutschmann
Recovery Sr. Engineering Associate
ExxonMobil Chemical Company Olefins Technology
david.c.dutschmann@exxonmobil.com

Prepared for Presentation at the American Institute of Chemical Engineers
2022 Spring Meeting and 18th Global Congress on Process Safety
San Antonio, TX. April 10-14, 2022

FATIGUE CRACKING OF SMALL DIAMETER BRANCH CONNECTION AT PRESSURE RELIEF VALVE

Stephen Hagberg
Lead Mechanical Engineer
ExxonMobil Chemical Company Gulf Coast Growth Ventures
stephen.l.hagberg@exxonmobil.com

David Dutschmann
Recovery Sr. Engineering Associate
ExxonMobil Chemical Company Olefins Technology
david.c.dutschmann@exxonmobil.com

Abstract

Small diameter branch connections for instrumentation and bleeders are a necessity in piping systems but also a vulnerability, in particular around sources of vibration. Operating plants have multiple sources of vibration, for example, rotating equipment often provides sufficient input frequency to excite natural harmonics of small branch connections. A discharging Pressure Relief Valve (PRV) can also lead to similar issues.

These vibrations can lead to fatigue cracking and separation resulting in significant downtime and Process Safety Risk. This paper will present one such case, where a small diameter connection adjacent to a PRV cracked at the pipe wall following a series of relieving events during unstable operations.

This paper aims to raise awareness by sharing a case where vibrations are generally not observed (during the act of PRV relieving) and can result in a small diameter piping failure in a relatively short duration. The paper will also share some mitigations pursued following this event to reduce the potential for vibrations and some measures to limit the risk should they occur.

Introduction

Small diameter branch connections are a necessity in piping systems but can also be a weak link when subjected to unexpected vibrations. Typical sizes most vulnerable are in the 1/2" to 1-1/2" pipe size range which are commonly referred to as small diameter branch connections. These often take the form of valve connections such as bleeders used for zero energy verification or taps for instrumentation.

When subjected to vibrations such connections are particularly prone to bending stress at their attachment points due to their cantilevered geometry for supported weight. Often times the weight of such valve connections can be in 6 lbs. to 10 lbs. range for commonly used ¾" and 1" sizes in 800# class ratings, up to 12 lbs. to 20 lbs. for higher 1500# pressure class ratings [1] making them significant but not enough to warrant special supports or bracing for most non-vibrating applications.

Sources of plant vibrations often occur at a frequencies of many cycles per second which can quickly build to thousands of cycles in a short period of time. When the stress of each cycle is sufficiently high an accumulation of strain in the base metal occurs which ultimately results in a fatigue failure if cycles continue. The location of failure occurs at the point of highest stress often originating from the toe of a weld at the base of the branch connection [Figure 1].

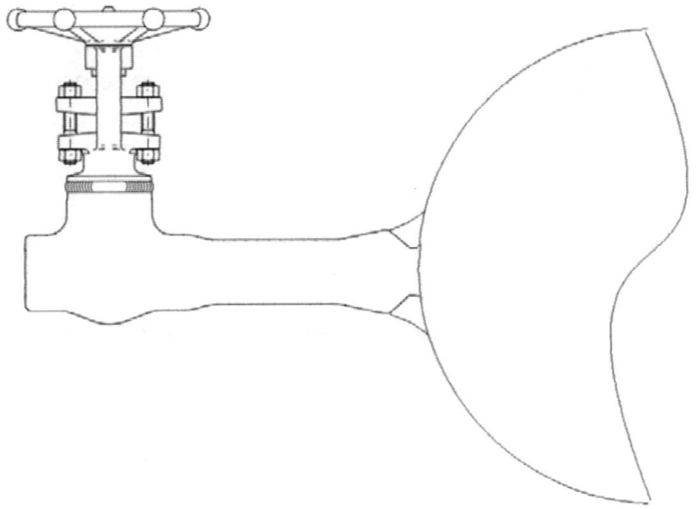

Figure 1. Typical Small Branch Connection Diagram

Industry guidance for reinforcing such connections does exist and can be applied when the process conditions and piping layout are defined well enough to anticipate vibrations. Such guidance can be found in documents published by the Energy Institute [2] and Gas Machinery Research Council [3] where common applications often involve cases of flow induced vibration from high velocity fluid flow, acoustic energy vibrations from a combination of high flow rates together with pressure drop, and mechanically induced vibrations occurring near reciprocating machinery. Typical actions from these methods often result in the addition of bracing, or gussets, to reinforce the branch connection to help support the overhung weight.

Pressure Relief Valves (PRV) introduce another unique potential source for vibration. Common examples are, Chatter which is a condition where the disk opens and closes rapidly making contact with the seat, and Cycling which occurs when multiple PRV's are set at a common pressure allowing the valves to randomly take turns opening and closing rather than in a predictable fashion. These behaviors have the potential to generate mechanical vibrations in piping which can lead to fatigue failures. Application of Industry Guidance such as that found in API 520 aims to avoid these conditions in sizing and selection of PRVs for given application [4].

Another consideration with PRV's is that they are seldom witnessed in the act of relieving which makes vibration issues particularly hard to identify. As a result cases often go unreported should un-anticipated vibrations begin to occur. PRV applications also see a wide range of possible real world scenarios from maximum relief capacity cases, to turn down, and even alternative fluid, or two phase compositions. While such considerations are taken into account to select PRV's which are less prone to instability, un-expected cases can occur during abnormal operations and additional methods to make piping connections more robust are worthy of consideration.

Incident Description

In March of 2020 at the Baytown Olefins Plant an Ethylene Refrigeration Machine tripped due to motor phase ground fault. Throughput on the affected unit was reduced with Major Machines left running while troubleshooting commenced. Following up on reports of an unexpected PRV lifting on the Process Gas Compressor PRV deck an Operator discovered a leak in the area via their personal gas monitor. Closer investigation revealed that a ¾" bleeder on bypass piping for the process gas compressor PRV's had visual linear indications along its sides where the bleeder valve attaches to the piping at the base of its connection [Figure 2].

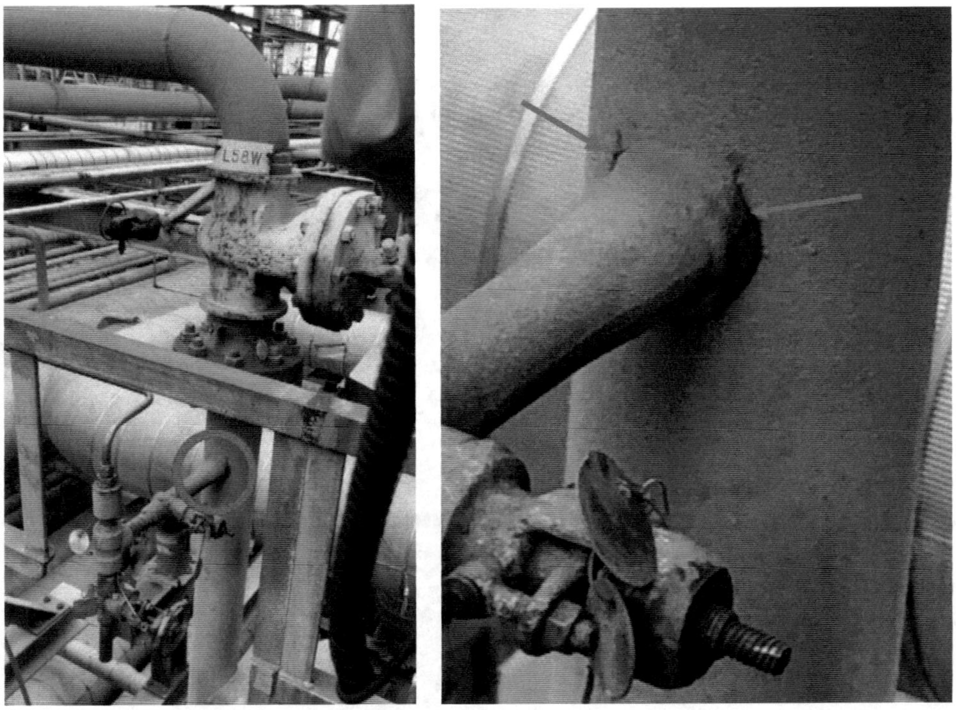

Figure 2. Failed Small Branch Connection

The section of piping had been in service approximately 40 years and did not show any signs that it was compromised from corrosion or otherwise unfit for service. Other additional noteworthy damage found were broken vent piping for the PRV bonnet bellows and a pipe support shoe which shifted off its location, both of which are suspected to have occurred as a result of this incident and not causal factors for the event.

Pressure Relief Valves

Five pressure relief valves protect discharge of the process gas compressor, three of which are co-located on a PRV deck where the leak occurred. All five valves are balanced bellows type with a common set point. Post incident the middle of the three on the PRV deck was found to have been severely damaged internally. Shop inspection found disc guides and other parts in the PRV severely galled and the bellows broken with multiple cracks and missing pieces from its convolutions [Figure 3].

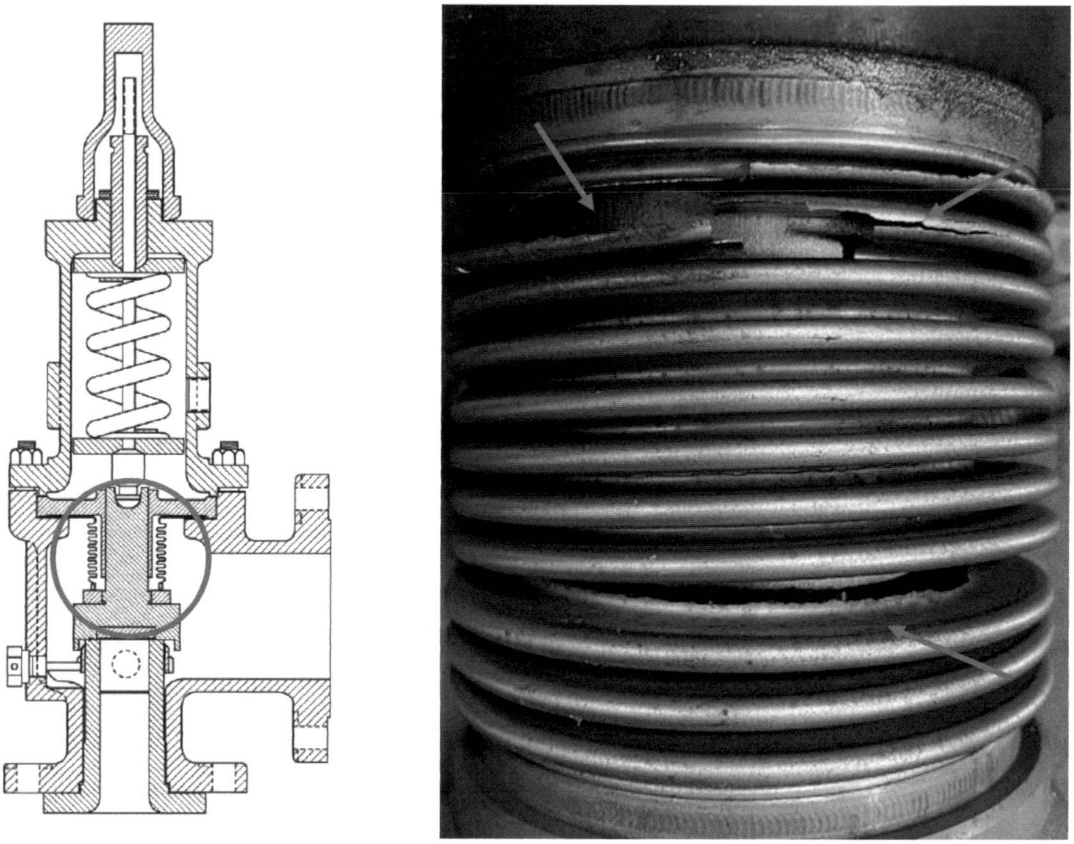

Figure 3. Failed PRV Bellows Evidence of Extreme Cyclic Behavior

Such was the condition of the valve that it could not be fully torn down due to parts being galled together and the standard receiving pop-test could not be performed. These were all indicators of rapid cycling occurring within this PRV. The other PRV's in the system did not experience damage despite indications by upstream rupture discs that they also relived.

Metallurgical Lab Analysis

The failed pipe section was sent to laboratory where closer examination confirmed fatigue failure. The cracks in base pipe along both sides of the branch connection revealed fracture surface patterns initiating on the inside and outside surface of the pipe wall. This is typical of reverse bending fatigue and an indication that the small branch connection was whipping back and forth in a side to side motion [Figure 4]. Additionally the distance between each fatigue striation was small which is an indication of many cycles propagating until thru wall failure occurred [Figure 5].

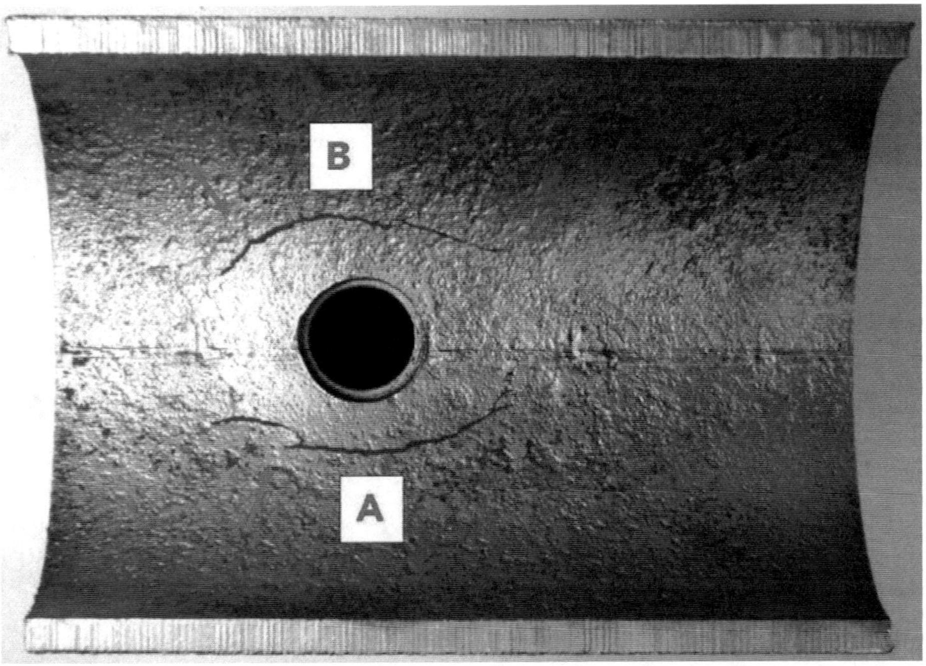

Figure 4. Internal view of pipe cross section showing thru wall cracks

Figure 5. Cross section of facture surface showing fatigue striations

Rapid cycling of the pressure relief valve was determined to be the source of mechanical vibrations in piping system causing the failure of the branch connection. Momentum changes as the PRV opened and closed repeatedly are suspected to have contributed to a side to side motion that was transmitted to the branch connection.

Process Timeline

Following an ethylene refrigeration compressor trip unit throughput was reduced and major machines left running while troubleshooting occurred. During the ensuing time period the PRV's protecting the outlet of the process gas compressor discharged. Relief initially occurred at set point during upset conditions and then again but below set point while the unit was slowly ramping up pressure to try and restart which was immediately followed by significant pressure fluctuations and rapid cycling [Table 1 & Figure 6].

Table 1. Key Initiating Events

3/29 - 9 pm: Ethylene Refrigeration Machine tripped due to motor ground fault
3/29 - 12 pm: Unit throughput reduced Major Machines left running
3/30 - 1 pm: PRVs lifted on Process Gas Compressor Discharge below set point & rapid cycling
3/30 - 2 pm: Unit shut down following discovery of cracked branch by Operator attending to PRV lift

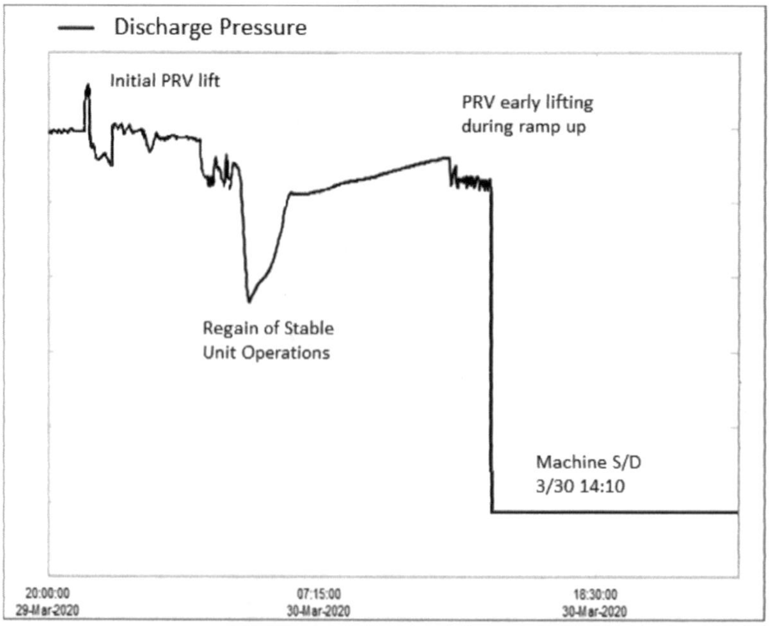

Figure 6. Trace Process Gas Compressor Discharge Pressure

Additional high speed data was recorded which shows cyclic pressure fluctuations with indications of higher frequency fluctuations imposed onto more visible lower frequency pressure cycles [Figure 7].

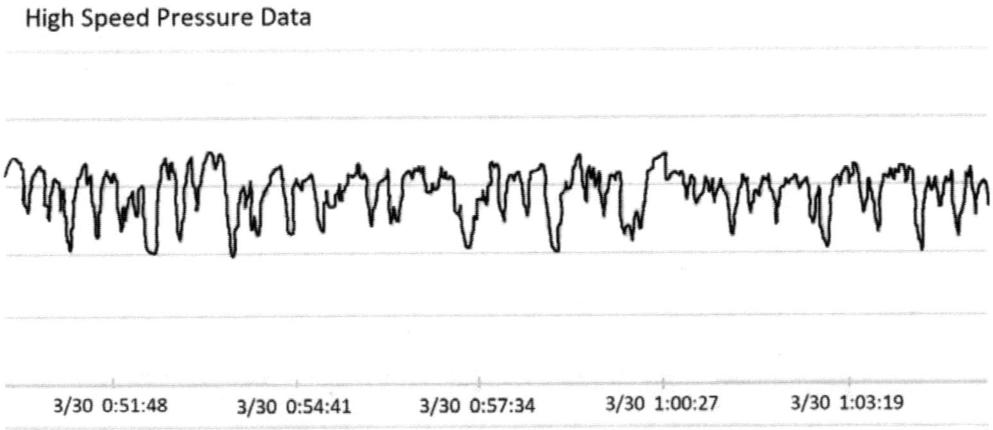

Figure 7. High Speed PRV Inlet Pressure Data

While it is not clear why the PRV relieved at a lower pressure than expected it is suspected to have been damaged by un-noticed earlier pressure cycling during initial hours of the unit upset while operator attention was elsewhere in the field. Once the condition was noticed and the leak discovered the unit was safely taken down for repairs. Post incident review of the pressure data was indicative of PRV instability creating the vibrations in the piping system which was concluded to have led to the branch connection failure.

Mitigations

Several actions were taken in the immediate repair including reinforcing of the new bleeder valve and reducing the overhung weight on the connection [Figure 8]. Other recommendations for further study specifically related to the PRV's were noted. Of particular interest were causes of instability and methods of avoidance. Staggering set points and effects of inlet pressure drop were two areas flagged for review. Inspection of small diameter branch connections for the other PRV's protecting the process gas compressor discharge was performed and while no further damage was found some connections were reinforced either temporarily by bolt on bracing, or permanently by additional welded on reinforcement for long term robustness.

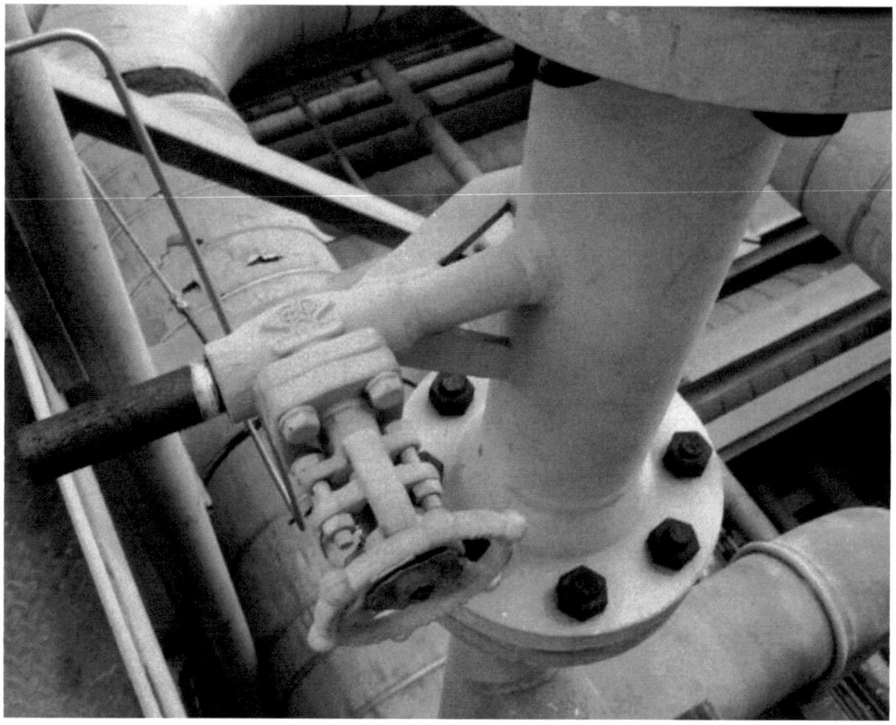

Figure 8. Repaired Small Diameter Branch Connection

Additionally another event in Industry at Chempark Zaluzi Complex in Czech Republic recently published in 2021 concluded that PRV instability can lead to standing pressure waves that have the ability to generate traverse motions in piping [5]. In this incident it was a flange connection that loosened as a result of the vibrations after which among other mitigations they applied the use of wedge lock washers on the PRV and associated isolation valve flanges.

The paper further mentions acoustic interaction as a phenomenon that can also lead to PRV instability. Depending on the piping layout geometry the sound waves generated when a PRV lifts can be reflected backwards to the PRV leading to rapid opening and closing cycles similar to Chattering. Improvements to evaluation methods for acoustic length criteria have been included in recent updates to API 520 and have also been shared in industry forums which focus on plant engineer audiences to raise awareness of the potential role this may play in PRV instability [6].

Given that PRV stability is a not guarantee for all possible abnormal and upset scenarios, methods to improve robustness of small diameter branches connections, or connections in general as shown by the flange example in Czech Republic, are additional considerations to take into account in piping design adjacent to PRV's where factors for potential instability exist.

Conclusions

Unstable and abnormal operations will test plant systems. During such scenarios assurance that Pressure Relief Valves and associated piping will function without failure is essential. As these events are unplanned they often occur without direct observation. As such it can be hard to recognize if an abnormal PRV operation or excessive vibration may have occurred during the act of relieving.

PRV cycling and chattering can generate high frequency vibrations which can subject components to many thousands of stress strain cycles in only a few hours. As such immediate follow up action is warranted when an abnormal behavior is suspected. Some suggestions for field observations after relief event are below.

- Operator rounds to look for clues (ie. chipped paint, lose or missing handwheels)
- Use of personnel monitors to aid in leak detection & safety of responding operators
- Review Process data for signs of PRV instability (ie. pressure cycling)
- Pipe supports, shifted, no longer resting, or rub marks between steel & support
- Previously attached tubing of vent piping bent or broken, loose parts on deck

While design guides and selection methods exist with the goal to avoid vibrations in relief systems, measures to make piping and in particular small diameter branch connections more robust near relief valves prone to instability is prudent. Some example cases prone to instability could be but not limited to; multiple non-staged PRV's, installations with higher inlet piping pressure drops, two phase and liquid flow. In such cases some actions to consider are below.

- Reducing overhung weight used for instrumentation from branch connections
- Addition of bracing to bleeder valves and minimizing cantilevered length
- Use of wedge lock washers for PRV flanges and their associated isolation valves

Finally it should be said that learning comes from observation. It is important for plant Operators and Engineers to talk to each other and go in the field after events, along with reviewing plant data when available together, to continue learning and building their understanding of the systems under their responsibility.

References

[1] Bonney Forge Product Catalogue; "Forged Steel Valves", pg.63 https://bonneyforge.com/resources/FSV.pdf

[2] Energy Institute, "Guidelines for the Avoidance of Vibration Induced Fatigue Failure in Process Pipework", *Energy Institute, 2nd Edition,* January 2008

[3] Wilcox, Melissa, et al. (2011) "Design Guideline for Small Diameter Branch Connections", Releases 1.0, *Gas Machinery Research Council*, March 2011

[4] API 520, "Sizing, Selection and Installation of Pressure-Relieving Devices, American Petroleum Institute", *API STD 520: Parts 1 & 2*

[5] Herink, Tomas, et al. (2021), "Propylene column pressure relief valves chattering resulting in explosion and fire of the Steam Cracker unit", *Journal of Loss Prevention in the Process Industries*. October, 2021

[6] Henry, Phil (2022), "Pressure Relief for Non-Process Engineers – 10 Things All Plant Personnel Should Know", *Inspectioneering & The Equity Engineering Group, Inc.* February, 2022

640393 Distillation Column Fundamentals in Olefins Plants

Abstract:

Ethylene and Propylene are generally produced by cracking various hydrocarbons. The cracked gas is processed in the hot and cold sections of an olefins plant where it goes through a series of process equipment before getting converted into polymer grade ethylene/propylene. The distillation and absorption columns play a very important role to recover and purify the key components from the cracked gas. Each column in this process has a specific purpose and some of them must tackle severe fouling/foaming conditions. The columns operate from atmospheric to high pressure depending on the process requirements. The design of the column internals is an interesting challenge. The main purpose of this presentation is to discuss the design requirements for the key columns in an olefins plant. In addition, there will also be discussions about the various types of column internals and the methodology used to select the most optimum one. The discussion will include the revamp potential when the plant needs to increase capacity.

640401 Gamma Scanning in Ethylene Plants

Abstract:

Modern day gamma scanning is a powerful diagnostic tool which is used routinely to troubleshoot trayed and packed towers throughout the petrochemical, refining and gas processing industries. Nowadays most towers can be scanned quickly with several towers being scanned in a single day oftentimes with multiple scans per tower. Scan data can then be plotted and interpreted within minutes giving the operator a quick diagnosis, which can be used to make critical process adjustments. In this presentation, ways of interpreting the scans and various examples for applications of scanning including flooding, fouling, tray damage and location will be discussed.

640395 Expanding the Capacity of Ethylene Plant Distillation Towers

Abstract:

Incremental expansions of existing olefins plants often show very attractive net present value (NPV) for invested capital. Improved furnace designs, or the addition of a furnace, can provide the incremental capacity at the front end of the plant, but then the downstream equipment also needs to be capable of receiving the increased product flow. The fractionation towers are a very important part of the downstream equipment that need to be carefully reviewed when considering such a plant expansion. Another area where the fractionation equipment is likely to be implicated is in changes to the plant feed, for example switching from cracking naphtha to ethane, which can significantly alter the hydraulic loading of the fractionation equipment. This presentation will review the current state of the art in revamps of fraction internals, focussing on the high pressure, cold section of the fractionation train, including deethanizers and ethylene splitters with some examples to increase capacity

Such tower revamps typically take place during scheduled plant outages and are therefore often time sensitive, which adds another constraint. We will also describe some of the special mechanical design techniques that have been developed to simplify the installation of the improved fractionation internals, allowing complex revamps to be completed safely and on schedule.

640470 Liquid Cracker Quench Water Tower Choice of Internals and Operation I

Abstract:

Quench tower in a liquid based olefins plant refers to both Quench Oil tower and Quench Water tower. Gas cracker may have only Quench Water towers. Generally, the Quench Oil tower is subjected to heavy fouling tendency depending on the nature of feed used for cracking. However, the Quench Water tower has much lower potential for fouling in liquid cracker. Large number of Quench Water towers in liquid crackers that are in operation worldwide today are equipped with random packing. While generally agreed that random packing is suitable to handle the fouling tendency, another option is available when studying the problem in detail. In this paper, a different design concept is introduced, which uses structured packings to achieve a great success for the Quench Water tower in liquid crackers for more than 10 years in operation. Based on the recent site inspection, the original structured packings are still in good working conditions to achieve its heat transfer requirement and operable without any fluctuation in the pressure drop. No replacement of either the structured packings or the liquid distributors and other column internals were needed to-date. Besides using the right choice of tower internals, operational excellence which includes running and maintaining the tower by the end-user with the highest standards and broad experience in the Olefins plant operation determines the success of the Quench tower.

640397 Distillation in Ethylene Plants – An Operators' Overview I

Abstract:

Distillation towers are major components of an ethylene plant. Located downstream the pyrolysis furnaces, there purposes can be divided in 3 categories: heat transfer, absorption, and fractionation. The operation functionality and design characteristics are driven by the service in which they operate and the defined process objectives. At a high level, the ultimate objective is to remove contaminants/ impurities and recover value chemicals to purities meeting saleable products. Problems associated with distillation tower operation typically involved fouling, controls and startup difficulties. This presentation discusses functional characteristics, internals, and typical problems by service. The sensitivity to feedstock variations is also discussed.

640399 Systematic Approach to Tower Internals in Olefins Plants

Abstract:

Utilizing high performance mass transfer equipment in the cold section of an ethylene plant provides several avenues to optimize process operations. Many factors are considered to ensure the correct technology is selected to meet the operating and capital cost objectives of a given project. This presentation will give a view of the decision-making process from the end user's perspective that drove the final decision on the type of tower internals used in light hydrocarbon applications.

AUTHOR INDEX